Green Materials for Energy, Products and Depollution

Environmental Chemistry for a Sustainable World

Volume 3

Series Editors

Eric Lichtfouse
UMR, Agroécologie, Dijon, France

Jan Schwarzbauer
Institute of Geology and Geochemistry of Petroleum and Coal, RWTH Aachen University, Aachen, Germany

Didier Robert
Université de Lorraine, ICPEES, Saint-Avold, France

For further volumes:
http://www.springer.com/series/11480

Eric Lichtfouse • Jan Schwarzbauer • Didier Robert
Editors

Green Materials for Energy, Products and Depollution

Springer

Editors
Eric Lichtfouse
UMR, Agroécologie
Dijon, France

Didier Robert
Université de Lorraine
ICPEES
Saint-Avold, France

Jan Schwarzbauer
Institute of Geology and Geochemistry of Petroleum and Coal
RWTH Aachen University
Aachen, Germany

ISSN 2213-7114 ISSN 2213-7122 (electronic)
ISBN 978-94-007-6835-2 ISBN 978-94-007-6836-9 (eBook)
DOI 10.1007/978-94-007-6836-9
Springer Dordrecht Heidelberg New York London

Library of Congress Control Number: 2013945557

Printed on acid-free paper

Springer is part of Springer Science+Business Media (www.springer.com)

Preface

There are always flowers for those who want to see them.

Henri Matisse

Nature is the best green and environmental chemist. Most biological reactions occur at room temperature and thus save energy, whereas many synthetic reactions need high temperature. Most biological reactions are quantitative and thus do not produce undesired by-products, whereas synthetic reactions are rarely quantitative and thus generate substantial amounts of undesired, sometimes toxic, by-products. Most biological reactions occur in water, a safe solvent, whereas many synthetic reactions are still often performed in hazardous solvents. Most biological products are useful, recyclable and recycled, whereas many synthetic products are not degradable and end up polluting the environment for centuries. Living organisms possess the best catalysts called enzymes, which should be awarded the Nobel Prize in chemistry if Nobel Prizes were given to natural substances, whereas synthetic catalysts almost never reach the natural efficiency and often contain toxic metals. Biological reactions use renewable products that are usually available in the very near surroundings of the living organism, whereas human chemistry often use non-renewable substances extracted from deep earth ores then shipped thousands of miles, thereby wasting time, money and energy. In short, despite several centuries of advanced scientific progress, human chemistry has still a lot to learn from nature chemistry. Therefore the positive side of actual society issues – or seeing flowers as suggested by Henri Matisse – is that there is a huge progress margin for the imagination of green chemists.

This book presents the following key examples of green materials and safe remediation:

- Biofuel from microalgae is a very promising fuel because microalgae is rapidly renewable, does not compete with food production and does not need huge surface area for production (Fig. 1, Chap. 1).

Fig. 1 Biofuel production from microalgae is fast and does not compete with food production from agricultural crops. Source: Gonçalves et al. Biodiesel from microalgal oil extraction, Chap. 1

- Water and soil cleaning can be performed with unprecedented speed and efficiency by novel techniques of electrochemistry and photocatalysis that remove toxic metals and gas, pharmaceuticals, pathogens, chlorinated polycyclic aromatic hydrocarbons (PAH) and other pollutants (Chaps. 2–4, 10). Chapter 8 presents cheap natural materials such as rice husk, wheat straw and fly ash to remove toxic metals. Chapter 7 shows how surfactants can be used to decrease pesticide toxicity.
- Food security can be improved by a wide number of bioindicators including plants, animals and microbes, which betray the presence of pollutants in air, water and soil (Chap. 5).
- Sustainable and safe clothes can be designed using natural dyes and antimicrobials, an old practice that is regaining interest in a fossil-free society (Chap. 6).

Thanks for reading

Eric Lichtfouse, Jan Schwarzbauer and Didier Robert
Founders of the journal Environmental Chemistry Letters and of the European Association of Environmental Chemistry
E-mail: Eric.Lichtfouse@dijon.inra.fr, Jan.Schwarzbauer@emr.rwth-aachen.de, Didier.Robert@univ-lorraine.fr

Other Publications by the Editors

Books

Scientific Writing for Impact Factor Journals
http://fr.slideshare.net/lichtfouse/scientific-writing-for-impact-factor-journals
https://www.novapublishers.com/catalog/product_info.php?products_id=42211

Environmental Chemistry
http://www.springer.com/book/978-3-540-22860-8

Organic Contaminants in Riverine and Groundwater Systems. Aspects of the Anthropogenic Contribution
http://www.springer.com/book/978-3-540-31169-0

Sustainable Agriculture
Vol. 1: http://www.springer.com/book/978-90-481-2665-1
Vol. 2: http://www.springer.com/book/978-94-007-0393-3

Rédiger pour être publié!
http://www.springer.com/book/978-2-8178-0288-6

Journals

Environmental Chemistry Letters
http://www.springer.com/10311

Agronomy for Sustainable Development
http://www.springer.com/13593

Book Series

Environmental Chemistry for a Sustainable World
ISSN: 2213-7114
http://www.springer.com/series/11480

Sustainable Agriculture Reviews
http://www.springer.com/series/8380

Contents

Contributors

Mariano J.L. Castro Facultad de Ciencias Veterinarias, Centro de Estudios Transdisciplinarios del Agua (CETA-INBA-CONICET), Universidad de Buenos Aires, Ciudad de Buenos Aires, Argentina

Alicia Fernández Cirelli Facultad de Ciencias Veterinarias, Centro de Estudios Transdisciplinarios del Agua (CETA-INBA-CONICET), Universidad de Buenos Aires, Ciudad de Buenos Aires, Argentina

Dionysios D. Dionysiou Environmental Engineering and Science Program, 705 Engineering Research Center, University of Cincinnati, Cincinnati, OH, USA

Ruey-an Doong Department of Biomedical Engineering and Environmental Sciences, National Tsing Hua University, Hsinchu, Taiwan

Gana Gecheva Faculty of Biology, University of Plovdiv, Plovdiv, Bulgaria

Beverley D. Glass School of Pharmacy and Molecular Sciences, James Cook University, Townsville, QLD, Australia

Ana L. Gonçalves Faculdade de Engenharia, LEPAE, Departamento de Engenharia Química, Universidade do Porto, Porto, Portugal

Sie-Tiong Ha Faculty of Science, Universiti Tunku Abdul Rahman, Kampar, Perak, Malaysia

Chun He School of Environmental Science and Engineering, Sun Yat-sen University, Guangzhou, China

Kazuo Higashino The Tokyo Metropolitan Research Institute for Environmental Protection, Koto-ku Tokyo, Japan

Yung-Tse Hung Department of Civil and Environmental Engineering, Cleveland State University, Cleveland, OH, USA

Devagi Kanakaraju School of Pharmacy and Molecular Sciences, James Cook University, Townsville, QLD, Australia

Masoud B. Kasiri Faculty of Applied Arts, Tabriz Islamic Art University, Tabriz, Iran

Pei-Sin Keng Department of Pharmaceutical Chemistry, International Medical University, Kuala Lumpur, Malaysia

Siew-Ling Lee Ibnu Sina Institute for Fundamental Science Studies, Universiti Teknologi Malaysia, Skudai, Johor, Malaysia

Michael Oelgemöller School of Pharmacy and Molecular Sciences, James Cook University, Townsville, QLD, Australia

Takeshi Ohura Faculty of Agriculture, Department of Environmental Bioscience, Meijo University, Nagoya, Japan

Carlos Ojeda Facultad de Ciencias Veterinarias, Centro de Estudios Transdisciplinarios del Agua (CETA-INBA-CONICET), Universidad de Buenos Aires, Ciudad de Buenos Aires, Argentina

Siew-Teng Ong Faculty of Science, Universiti Tunku Abdul Rahman, Kampar, Perak, Malaysia

Mehmet A. Oturan Laboratoire Géomatériaux et Environnement (LGE), Université Paris-Est, EA 4508, UPEMLV, 77454 Marne-la-Vallée, France

José C.M. Pires Faculdade de Engenharia, LEPAE, Departamento de Engenharia Química, Universidade do Porto, Porto, Portugal

Siyamak Safapour Faculty of Carpet, Tabriz Islamic Art University, Tabriz, Iran

Yuko Sasaki The Tokyo Metropolitan Research Institute for Environmental Protection, Koto-ku Tokyo, Japan

Virender K. Sharma Center of Ferrate Excellence and Chemistry Department, Florida Institute of Technology, Melbourne, FL, USA

Manuel Simões Faculdade de Engenharia, LEPAE, Departamento de Engenharia Química, Universidade do Porto, Porto, Portugal

Slavka Stankovic Faculty of Technology and Metallurgy, University of Belgrade, Belgrade, Serbia

Ana R. Stankovic Faculty of Technology and Metallurgy, University of Belgrade, Belgrade, Serbia

Subramanyan Vasudevan Electroinorganic Chemicals Division, CSIR- Central Electrochemical Research Institute, Karaikudi, India

Teru Yamamoto The Tokyo Metropolitan Research Institute for Environmental Protection, Koto-ku Tokyo, Japan

Lilyana Yurukova Institute of Biodiversity and Ecosystem Research, Bulgarian Academy of Sciences, Sofia, Bulgaria

Chapter 1
Biodiesel from Microalgal Oil Extraction

Ana L. Gonçalves, José C.M. Pires, and Manuel Simões

Abstract The rapid development of the modern society has resulted in an increased demand for energy, and consequently an increased use of fossil fuel reserves. Burning fossil fuels is nowadays one of the main threats to the environment, especially due to the accumulation of greenhouse gases in the atmosphere, which are responsible for global warming. Furthermore, the continuous use of this non-renewable source of energy will lead to an energy crisis because fossil fuels are of limited availability. In response to this energy and environmental crisis, it is of extreme importance to search for different energy supplies that are renewable and more environmentally friendly. Microalgae are a promising sustainable resource that can reduce the dependence on fossil fuel. Biodiesel production through microalgae is actually highly studied. It includes several steps, such as cell cultivation and harvesting, oil extraction and biodiesel synthesis. Although several attempts have been made to improve biodiesel yields from microalgae, further studies are required to optimize production conditions and to reduce production costs.

This chapter reviews recent developments on oil extraction for biodiesel production. Two different processes are distinguished: (i) an indirect route, in which microalgal oil is recovered in an appropriate solvent and then converted into biodiesel through transesterification; and (ii) a direct route, in which the production of biodiesel is performed directly from the harvested biomass. Both routes, direct and indirect, should be preceded by cell wall disruption because this step facilitates the access of solvents to microalgal oil. The most advantageous disruption methods for lipid extraction are enzymatic and pulsed electric field disruption because enzymes present higher selectivity towards cell walls. In addition pulsed electric field requires less energy than other disruption methods.

A.L. Gonçalves • J.C.M. Pires (✉) • M. Simões
LEPAE, Departamento de Engenharia Química, Universidade do Porto,
Rua Dr. Roberto Frias, Porto 4200-465, Portugal
e-mail: jcpires@fe.up.pt

E. Lichtfouse et al. (eds.), *Green Materials for Energy, Products and Depollution*,
Environmental Chemistry for a Sustainable World 3, DOI 10.1007/978-94-007-6836-9_1,

For the indirect route, it is possible to use three different types of solvents to recover microalgal oil. Although extraction with supercritical fluids has higher extraction efficiencies and is safer for the environment, costs are very high. The use of ionic liquids is also safer for the environment, but their cost is also very high. An alternative is the use of organic solvents such as n-hexane because it is less harmful and has a higher selectivity for neutral oil fractions than other organic solvents. We conclude that the direct route, which involves production of biodiesel directly from the microalgal biomass, is more efficient. Indeed, the application of the direct route to the microalga *Schizochytrium limacinum* resulted in a biodiesel yield of 72.8 %, while the indirect route, in the same conditions, has resulted in a biodiesel yield of 63.7 %.

Keywords Biofuel • Algae • Microalgae • Oil extraction • Liquid • Transesterification • Chlorella vulgaris • Schizochytrium • Limacinum • Cell wall disruption • Pulse electric field • Enzymatic disruption

Contents

1.1 Introduction

The depletion of fossil fuels reserves and the effect of exhaust gas emissions on global climate change have stimulated the search for sustainable sources of energy that are carbon neutral or renewable. As an alternative energy source, much attention has been paid to biodiesel production from vegetable oil crops, such as palm, rapeseed and soybean, and animal fats (Demirbas 2011; Ranjan et al. 2010). However, biodiesel production yields from oil crops and animal fats do not achieve the current demand for transport fuels (Chisti 2007; Demirbas 2011). Furthermore, producing biodiesel from vegetable crops is time consuming and requires great areas of arable land that would compete with the one used in food crops, leading to starvation in developing countries (Costa and de Morais 2011; Demirbas 2011; Demirbas and Demirbas 2011).

Avoiding the competition between energy and food production, attentions are now focused on evaluating the potential of microalgae as oil source for biodiesel production. Microalgae are eukaryotic photosynthetic microorganisms that can be found in aquatic or terrestrial ecosystems (Fig. 1.1). They present several advantages

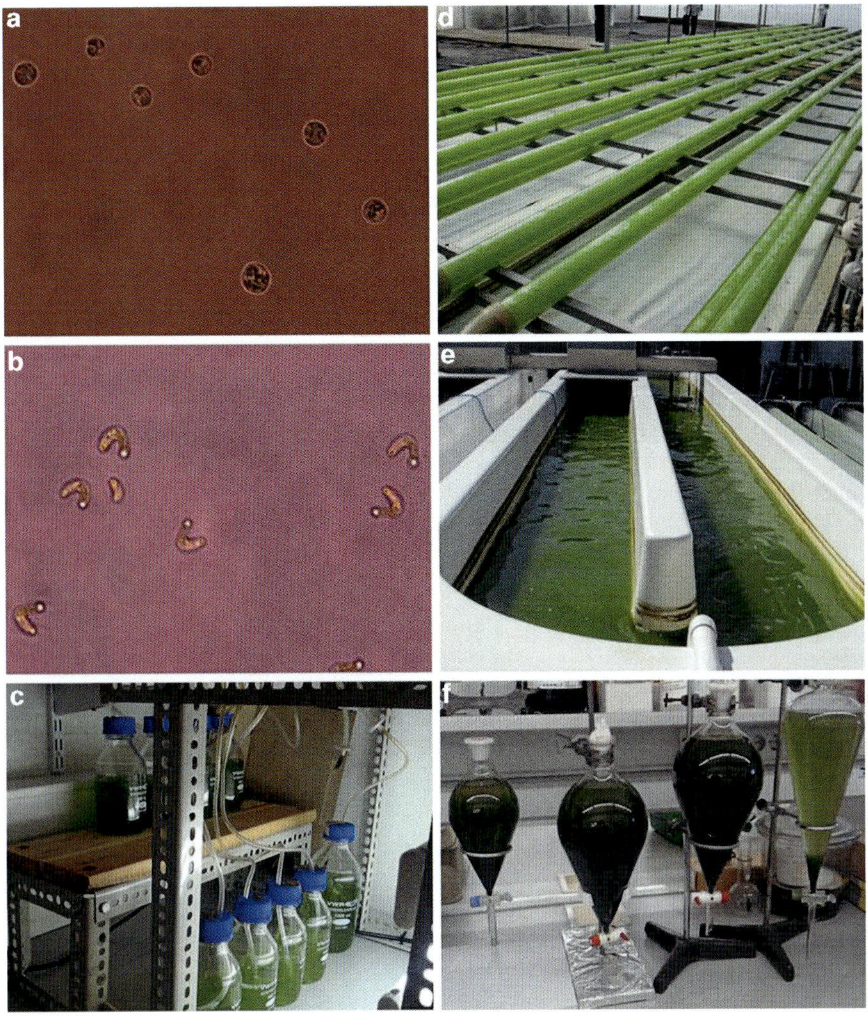

Fig. 1.1 Microalgae: (**a**) microscopic photograph of the microalga *Chlorella vulgaris*; (**b**) microscopic photograph of the microalga *Pseudokirchneriella subcapitata*; (**c**) microalgal culturing in small flasks; (**d**) microalgal culturing in horizontal tubular photobioreactors; (**e**) microalgal culturing in raceway ponds; and (**f**) harvesting of microalgae through sedimentation. (**a**), (**b**), (**c**), and (**f**) were obtained from our research group; (**d**) from http://badger.uvm.edu; and (**e**) from http://www.abc.net.au

over oil crops, including: (i) higher oil contents; (ii) higher growth and biomass production rates; (iii) shorter maturity rates; and (iv) require far less land due to higher oil productivities (Chisti 2007; Mercer and Armenta 2011). As well as microalgae, there are other photosynthetic microorganisms with potential interest. These prokaryotic microorganisms, known as cyanobacteria, behave similarly to microalgae and present the same advantages. Despite the referred advantages, the production of biodiesel from microalgae is not economically viable. Technological improvements should be performed to reduce the associated costs, including: (i) improvement of photosynthetic efficiency by the study of photobioreactor design; (ii) reduction of water and carbon dioxide losses in microalgal cultures; (iii) improvement of energy balance for water pumping, CO_2 transfer, biomass harvesting, oil extraction and biodiesel synthesis; and (iv) use of flue gas as CO_2 source. This review focuses on the oil extraction and biodiesel synthesis, presenting the research advances in the associated processes.

1.2 Lipid Recovery

After cell cultivation, the downstream process towards the production of biodiesel includes: (i) harvesting of microalgae; (ii) drying or dewatering; (iii) cell disruption and oil extraction; and (iv) transesterification reaction (Amaro et al. 2011; Brennan and Owende 2010), as it is possible to see in Fig. 1.2.

Oil extraction from biomass requires a specific solvent with great affinity to microalgal oil. Extraction procedures involving organic solvents, supercritical fluids and ionic liquids are the most common applied methods to recover oil from microalgae (Amaro et al. 2011; Mercer and Armenta 2011; Taher et al. 2011; Kim et al. 2012). Although these procedures can be applied directly to the dewatered biomass, it is reported that their efficiency is very low because microalgae present cell walls that block the access of solvents to cytosol (Cravotto et al. 2008; Lee et al. 2010), the cell compartment where the majority of microalgal lipids accumulate (Chen et al. 2009). To overcome the low efficiencies associated to the application of solvent extraction methods, some authors have reported the application of cell wall disruption methods, such as (i) enzymatic disruption; (ii) pulsed electric field; (iii) ultrasound and microwave; and (iv) expeller pressing (Amaro et al. 2011; Cravotto et al. 2008; Lee et al. 2010; Mercer and Armenta 2011; Taher et al. 2011). Table 1.1

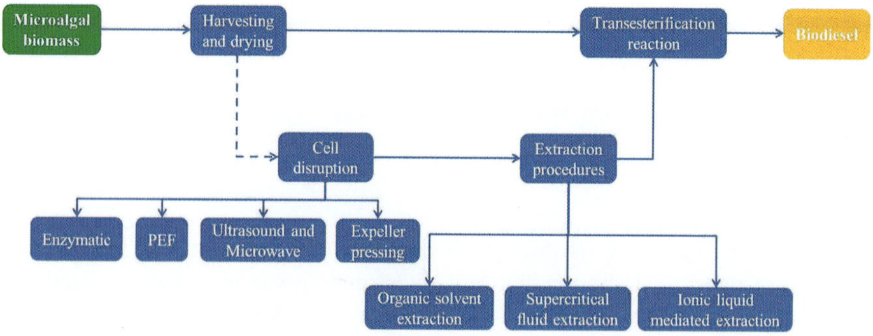

Fig. 1.2 Steps involved in the production of biodiesel from microalgal biomass. *PEF* pulsed electric field

Table 1.1 Effectiveness of some oil extraction methods applied in fatty acid recovery from different microalgae

Extraction method	Microorganism	Total fatty acids[a] % (w_{FA}/w_{DW})	Reference
Organic solvent (n-hexane)	*Crypthecodinium cohnii*	4.8	Cravotto et al. (2008)
Ultrasonic-assisted + Organic solvent (n-hexane)	*Crypthecodinium cohnii*	25.9	
Microwave-assisted + Organic solvent (n-hexane)	*Crypthecodinium cohnii*	17.8	
Organic solvent (ethanol:n-hexane 1:1 v/v)	*Scenedesmus dimorphus*	6.3	Shen et al. (2009)
Ultrasonic-assisted + Organic solvent (ethanol:n-hexane 1:1 v/v)	*Scenedesmus dimorphus*	21.0	
French press + Organic solvent (ethanol:n-hexane 1:1 v/v)	*Scenedesmus dimorphus*	21.2	
Organic solvent (ethanol:n-hexane 1:1 v/v)	*Chlorella protothecoides*	5.6	
Ultrasonic-assisted + Organic solvent (ethanol:n-hexane 1:1 v/v)	*Chlorella protothecoides*	10.7	
French press + Organic solvent (ethanol:n-hexane 1:1 v/v)	*Chlorella protothecoides*	14.9	
Organic solvent (chloroform:methanol, 3:1 v/v)	*Scenedesmus* sp.	2.0	Ranjan et al. (2010)
Ultrasonic-assisted + Bligh and Dyer (chloroform:methanol, 3:1 v/v)	*Scenedesmus* sp.	6.0	
Organic solvent (n-hexane)	*Scenedesmus* sp.	0.6	
Organic solvent (chloroform:methanol, 1:1 v/v)	*Botryococcus* sp.	7.9	Lee et al. (2010)
Microwave-assisted + Organic solvent (chloroform:methanol, 1:1 v/v)	*Botryococcus* sp.	28.6	
Organic solvent (chloroform:methanol, 1:1 v/v)	*Chlorella vulgaris*	4.9	
Microwave-assisted + Organic solvent (chloroform:methanol, 1:1 v/v)	*Chlorella vulgaris*	9.9	
Organic solvent (chloroform:methanol, 1:1 v/v)	*Scenedesmus* sp.	1.9	
Microwave-assisted + Organic solvent (chloroform:methanol, 1:1 v/v)	*Scenedesmus* sp.	10.4	
Organic solvent (n-hexane; 10 h)	*Scenedesmus obliquus*	46.9	Balasubramanian et al. (2011)
Microwave-assisted + Organic solvent (n-hexane, 0.5 h)	*Scenedesmus obliquus*	77.1	
Organic solvent (water:methanol:chloroform, 0.8:2:1 v/v/v; 18 h)	Marine microheterotroph	25.8	Lewis et al. (2000)
Organic solvent (chloroform:methanol:water, 1:2:0.8 v/v/v; 18 h)	Marine microheterotroph	35.0	
Organic solvent (n-hexane)	*Spirulina maxima*	4.1	Gouveia and Oliveira (2009)

(continued)

Table 1.1 (continued)

Extraction method	Microorganism	Total fatty acids[a] % (w_{FA}/w_{DW})	Reference
Organic solvent (n-hexane)	*Chlorella vulgaris*	5.1	
Organic solvent (n-hexane)	*Scenedesmus obliquus*	17.7	
Organic solvent (n-hexane)	*Dunaliella tertiolecta*	16.7	
Organic solvent (n-hexane)	*Nannochloropsis* sp.	28.7	
Organic solvent (n-hexane)	*Neochloris oleoabundans*	29.0	
Organic solvent (ethanol, 96 % v/v)	*Phaeodactylum tricornutum*	6.4	Fajardo et al. (2007)
$scCO_2$ (70.0 MPa; 55 °C; 10 kg.h^{-1}; 1.4 h)	*Nannochloropsis* sp.	23.0	Andrich et al. (2005)
Organic solvent (n-hexane; 8 h)	*Spirulina platensis*	7.8	Andrich et al. (2006)
$scCO_2$ (70.0 MPa; 55 °C; 10 kg.h^{-1}; 0.75 h)	*Spirulina platensis*	7.7	
Organic solvent (chloroform:methanol, 2:1 v/v)	*Crypthecodinium cohnii*	19.9	Couto et al. (2010)
$scCO_2$ (20.0 MPa; 40 °C; 0.6 kg.h^{-1}; 3 h)	*Crypthecodinium cohnii*	6.9	
$scCO_2$ (25.0 MPa; 40 °C; 0.6 kg.h^{-1}; 3 h)	*Crypthecodinium cohnii*	6.3	
$scCO_2$ (30.0 MPa; 40 °C; 0.6 kg.h^{-1}; 3 h)	*Crypthecodinium cohnii*	5.7	
$scCO_2$ (20.0 MPa; 50 °C; 0.6 kg.h^{-1}; 3 h)	*Crypthecodinium cohnii*	5.6	
$scCO_2$ (25.0 MPa; 50 °C; 0.6 kg.h^{-1}; 3 h)	*Crypthecodinium cohnii*	7.1	
$scCO_2$ (30.0 MPa; 50 °C; 0.6 kg.h^{-1}; 3 h)	*Crypthecodinium cohnii*	8.6	
Organic solvent (n-hexane; 5.5 h)	*Chlorococcum* sp.	3.2	Halim et al. (2011)
$scCO_2$ (30.0 MPa; 60 °C; 18.5 kg.h^{-1}; 1.3 h)	*Chlorococcum* sp.	5.8	
$scCO_2$ (30.0 MPa; 80 °C; 18.5 kg.h^{-1}; 1.3 h)	*Chlorococcum* sp.	4.8	
$scCO_2$ (35.0 MPa; 70 °C; 6 kg.h^{-1}; 4.5 h)	*Nannochloropsis granulata*	2.8	Bjornsson et al. (2012)
IL ([Emim]$MeSO_4$) and methanol	*Dunaliella* sp.	8.6	Young et al. (2010)
IL ([Emim]$MeSO_4$) and methanol	*Chlorella* sp.	38.0	
Organic solvent (chloroform:methanol, 2:1 v/v)	*Chlorella vulgaris*	11.1	Kim et al. (2012)
IL ([Bmim]CF_3SO_3) and methanol	*Chlorella vulgaris*	19.0	
IL ([Bmim]$MeSO_4$) and methanol	*Chlorella vulgaris*	17.4	

[a]Total fatty acids recovered are represented as a fraction (in %) of oil weight in the biomass dry weight

Table 1.2 Advantages and disadvantages of the cell wall disruption methods and oil extraction procedures applied to microalgae

	Procedure	Advantages	Disadvantages
Cell wall disruption methods	Enzymatic disruption	Higher degradation selectivity Requirement of less energy than mechanical methods	Enzymes are very expensive
	Pulsed electric field	Requirement of less time and energy than other applied methods	Difficulties in operating at large scale High operational and maintenance costs
	Ultrasound and Microwave	Higher efficiencies and reduced extraction times Increased yields	Moderate to high energetic costs
	Expeller pressing	Simple method Useful for large scale applications	High power consumption and maintenance costs
Lipid extraction methods	Organic solvent extraction	Simple and inexpensive method	The majority of organic solvents are toxic, harmful and non-reusable Time-consuming
	Supercritical fluid extraction	Supercritical fluids are non-toxic and present higher mass transfer rates Recovery of fatty acids is easier Requirement of less time than organic solvent extraction	High energetic and maintenance costs Difficulties in scale-up
	Ionic liquid mediated extraction	Ionic liquids are non-toxic and non-volatile and present higher solvatation capacities Ionic liquids can be produced specifically, according to their application	Ionic liquids are expensive

shows the most applied extraction methods and the achieved mass percentages of recovered oil and Table 1.2 summarizes the main advantages and disadvantages of the presented cell disruption methods and oil extraction procedures.

1.2.1 Cell Disruption Methods

Cell disruption methods aim to disintegrate cell walls to allow the release of intracellular components into an adequate solvent. The methods used in cell wall disruption can be classified into mechanical, where cell wall destruction is non-specific, and non-mechanical, where methods are more specific. Mechanical methods include bead mill, French press, ultrasound, microwaves and high pressure homogenizer, whereas non-mechanical methods comprise the physical methods thermolysis, decompression and osmotic shock, the chemical methods, where antibiotics, chelating agents, solvents and detergents are applied, and the enzymatic methods, which use lytic enzymes (Geciova et al. 2002; Chisti and Moo-Young 1986).

Enzymatic Disruption

Enzymes can be applied in oil extraction from microalgae, as they can mediate the hydrolysis of cell walls, enabling the release of their content into an appropriate solvent. Application of lytic enzymes with little volumes of organic solvent can improve oil recovery yields, as well as extraction times (Mercer and Armenta 2011). For cell wall degradation, cellulases are the most applied enzymes (Mercer and Armenta 2011; Sander and Murthy 2009). Although enzymatic extraction has not yet been applied to microalgae, it has been successfully used in oil extraction from *Jatropha curcas* L. seeds (Shah et al. 2004). Three phase partitioning (TPP) method and an enzyme-assisted TPP (EATPP) method were applied to recover oil from these seeds. The TPP consisted in the addition of three solvents to the seeds, in order to form a three phase system. The applied solvents were water, ammonium sulphate, and *t*-butanol. The three phases were separated by centrifugation and the phase containing the recovered oil was the one containing *t*-butanol, which was eliminated through evaporation. TPP and EATPP were performed at different pH conditions: 4.0, 7.0, and 9.0. Higher oil yields were obtained at pH 9.0: (i) 32.0 % (wt.) for TPP; and (ii) 36.8 % (wt.) for EATPP (Shah et al. 2004).

Despite being expensive, enzymes offer several advantages over other cell wall disruption methods. They present higher degradation selectivity than mechanical disruption methods. Furthermore, microalgal cell walls are more recalcitrant than the ones of other microorganisms, being very resistant to degradation. Thus, the use of mechanical disruption methods requires higher energy amounts (Sander and Murthy 2009).

Pulsed Electric Field

The pulsed electric field (PEF) technology seems to be a potential alternative for oil extraction from microalgae (Taher et al. 2011). This technique applies brief pulses of a strong electric field to cells, which induces non-thermal permeabilization of membranes (Guderjan et al. 2005; Taher et al. 2011). In determined conditions, PEF can also cause significant damage in microalgal cell walls (main barrier for oil extraction in most microalgae), membrane and it can led to complete disruption of cells into fragments. PEF is a relatively new method that has not yet been applied to extract microalgal oil. However, evidence of high extraction efficiencies in plant products, such as maize, olives (Guderjan et al. 2005) and *Brassica napus* cells

(Guderjan et al. 2007), suggests that this can be a suitable method to improve the permeabilization of microalgal membranes and efficiently extract their oil (Mercer and Armenta 2011). Guderjan et al. (2005) used PEF to induce stress in plant cells and thus recover functional food ingredients, such as phytosterols and polyphenols. The authors applied 120 pulses with field strength of 0.6 and 7.3 $kV.cm^{-1}$ to maize and they added a small amount of n-hexane to perform the oil extraction. On the other hand, olives were treated using the following conditions: 30 pulses with field strength of 0.7 $kV.cm^{-1}$ and 100 pulses with field strength of 1.3 $kV.cm^{-1}$. After application of PEF the membranes were completely disintegrated and the oil content was recovered by centrifugation. Oil recovery obtained for dried maize was 23.5 and 23.9 % (wt.) for pulses with 0.6 and 7.3 $kV.cm^{-1}$, respectively. These results were obtained for maize treated with PEF and n-hexane and further drying for 24 h, at 38 °C. Application of electric pulses with field strength of 0.6 $kV.cm^{-1}$, followed by n-hexane addition, an incubation period of 24 h and drying for 24 h at 38 °C, resulted in an oil yield of 43.7 %, which means that incubation of the mixture with the organic solvent allows higher oil recovery. Regarding olives, the application of PEF with strength of 0.7 and 1.3 $kV.cm^{-1}$ followed by centrifugation resulted in an oil yield of 6.5 and 7.4 g_{oil} per g_{mash}. Guderjan et al. (2007) applied 120 pulses with a field strength of 7.0 $kV.cm^{-1}$ and a duration of 30 μs to hulled rapeseed, followed by drying at 50 °C for 10 h and an extraction step with n-hexane. With these extraction procedures, the authors obtained an oil yield of 32 % (wt.), against 23 % obtained without PEF application.

PEF requires less time and energy than other applied methods (Guderjan et al. 2005) and its use as a pre-treatment for organic solvent extraction requires far less organic solvents (usually presenting high toxicity) than the conventional organic solvent extraction methods, which reduces the energy needs in the extraction process (Guderjan et al. 2007).

Ultrasound and Microwave

Another method that can be used to promote cell wall disruption of microalgal cells is the application of ultrasounds and microwaves. In ultrasonic-assisted method, microalgal oil can be recovered by cavitation. This phenomenon occurs when vapour bubbles of the liquid are formed when the pressure is lower than its vapour pressure. As these bubbles grow when pressure is negative and compress under positive pressure, a violent collapse of the bubbles is promoted. When bubbles collapse near cell walls, damage can occur, leading to the release of cell contents (Mercer and Armenta 2011; Taher et al. 2011). Application of this ultrasound-assisted method to microalgal biomass can improve extraction efficiencies by reducing extraction times and increasing oil recovery yields. The experiments performed by Cravotto et al. (2008) with the microalga *Crypthecodinium cohnii* showed that cell disruption using ultrasounds increased oil extraction yields from 4.8 %, when applying Soxhlet extraction with n-hexane, to 25.9 % (wt.). Shen et al. (2009) used the microalgae *Scenedesmus dimorphus* and *Chlorella protothecoides* to evaluate the effect of sonication before solvent extraction using a mixture of ethanol:hexane in a ratio of 1:1 (v/v). Application of ultrasound-assisted disruption increased the oil yields from 6.3 % to 21.0 % (wt.) for *S. dimorphus* and from 5.6 % to 10.7 % (wt.) for *C. protothecoides*. Additionally, Ranjan et al. (2010) compared oil extraction yields from *Scenedesmus* sp. using the following methods: (i) Bligh and Dyer's method (1959), organic solvent extraction using a solvent mixture of chloroform, methanol and water, where solvents were applied in a ratio of 3:1:0 (v/v); (ii) ultrasound-assisted extraction followed by the Bligh and Dyer's method, using the same mixture of chloroform and methanol. Results obtained with these experiments showed an increase in oil yields from 2.0 % to 6.0 % (wt.), when applying method (i) and (ii), respectively. One possible reason for this increase in oil recovery is that when both methods are applied, oil extraction is a result of the interaction between two phenomena: oil diffusion across the cell wall and disruption of the cell wall with release of its contents to the solvent (Ranjan et al. 2010).

Microwave-assisted method is supported by the principle that microwaves directly affect polar solvents and materials. Even when they are applied to dried cells, trace amounts of moisture are affected: temperature increases due to microwaves, moisture is evaporated, and pressure in the cells increases, leading to a damage or rupture of the cell wall followed by the release of its contents. Microwave theory and the extraction principle are described in detail by Mandal et al. (2007). The use of microwaves followed by organic solvent extraction using small amounts of solvent contributes to an efficient and inexpensive extraction procedure, which does not require previous biomass dehydration. Cravotto et al. (2008) applied organic solvent extraction with n-hexane and microwave-assisted solvent extraction (using the same solvent) to the microalga *C. cohnii*, achieving oil recovery yields of 4.8 % and 17.8 % (wt.), respectively. Furthermore, Lee et al. (2010) used the Bligh and Dyer's method (1959) with a mixture of chloroform:methanol in the ratio of 1:1 (v/v) preceded by the application of a microwaves treatment to the microalgae *Botryococcus* sp., *Chlorella vulgaris*, and *Scenedesmus* sp. With this microwave-assisted method, the oil extraction yields obtained for these organisms were 28.6, 9.9, and 10.4 % (wt.), respectively, against the 7.9, 4.9, and 1.9 % (wt.) obtained for the control method – Bligh and Dyer's method (1959; Lee et al. 2010). Recently, Balasubramanian et al. (2011) promoted cell wall disruption of *Scenedesmus obliquus* using the microwave-assisted method and compared the achieved results with organic solvent extraction with n-hexane. Disruption using microwaves was performed for 30 min, while organic solvent extraction was performed by 10 h. Oil recovery yields obtained with solvent extraction and the microwave-assisted method were 46.9 % and 77.1 % (wt.), respectively (Balasubramanian et al. 2011).

Both methods improve significantly oil extraction from microalgae, presenting higher efficiency, reduced extraction times and increased yields, as well as moderate costs and negligible added toxicity.

Expeller Pressing

Pressing techniques lie on the principle that when microalgal cells are submitted to high pressures, they start to crush, releasing their contents to an adequate solvent. As the methods described before, pressing techniques can be advantageous when using as a pre-treatment for organic solvent extraction. A pre-treatment using French press was applied by Shen et al. (2009) to the microalgae *S. dimorphus* and *C. protothecoides*. After pressing microalgae, the oil was recovered using a solvent system containing ethanol and n-hexane in a 1:1 (v/v) ratio. Comparing extracted oil yields with those obtained without pre-treatment, oil content achieved for *S. dimorphus* raised from 6.3 % to 21.2 % (wt.), while for *C. protothecoides* it raised from 5.6 % to 14.9 % (wt.).

Although this method is very simple and has reduced equipment costs, it presents some disadvantages when compared to other cell disruption methods, such as high power consumption and maintenance costs.

1.2.2 Lipid Extraction Methods

Extraction of microalgal oil can be performed directed to the harvested biomass or in addition to a cell wall disruption method. The second methodology generally presents higher lipid recoveries, as cell contents are released to the solvent applied. In the recovery of microalgal oil, it is very important to choose an appropriate solvent because this choice can improve lipid recovery yields and reduce process costs. Additionally, the majority of solvents applied are harmful and toxic, meaning that the selection of the solvents used should take into account their impact in the environment and in public health. The extraction procedures commonly applied to extract microalgal oil include the use of organic solvents, supercritical fluids and ionic liquids (Mercer and Armenta 2011; Kim et al. 2012).

Organic Solvent Extraction

The use of organic solvents to extract microalgal oil is the most applied extraction method. The main organic solvents applied include hexane, cyclohexane, benzene, ethanol, acetone, and chloroform (Brennan and Owende 2010; Mercer and Armenta 2011; Grima et al. 2003). These solvents have shown to be quite effective in oil extraction from microalgae. A good solvent for oil extraction may present the following characteristics: (i) to be insoluble in water; (ii) to have high affinity for oil, i.e. non-polar, to increase its permeability to cell membrane and also to solubilise the target compounds; (iii) to have a low boiling point to facilitate its removal after extraction; (iv) to have a considerably different density from that of water. Furthermore, the organic solvent applied should be inexpensive, non-toxic and reusable (Mercer and Armenta 2011).

Several studies have reported the use of a chloroform, methanol and water mixture, known as the Bligh and Dyer's method (1959), to extract microalgal oil (Mercer and Armenta 2011). Lewis et al. (2000) studied the effect of applying the solvents chloroform, methanol and water in different sequences and proportions on an oil-producing strain of a marine microheterotroph, Thraustochytrid ACEM 6063. The authors used the following sequences and ratios: (i) water:methanol:chloroform (0.8:2:1, v/v/v); (ii) chloroform:methanol:water (1:2:0.8, v/v/v); (iii) chloroform:methanol:water (1:4:0.8, v/v/v). Total fatty acids extracted with these three solvent systems were 258.5, 350.0, and 326.5 $mg.g^{-1}{}_{dry\ weight}$, respectively. With this study, the authors concluded that changing solvent sequence can have significant effects on extraction yields and that increasing the proportion of methanol did not significantly affect the extraction efficiency. Later, Lee et al. (2010) used a mixture of chloroform and methanol (1:1, v/v) to extract oil from the organisms *Botryococcus* sp., *C. vulgaris*, and *Scenedesmus* sp. Extraction yields obtained with this method were 12.0, 24.9, and 18.8 $mg.g^{-1}{}_{dry\ weight}$ for *Botryococcus* sp., *C. vulgaris*, and *Scenedesmus* sp., respectively.

Another common organic solvent applied to extract microalgal oil is n-hexane. Gouveia and Oliveira (2009) used n-hexane to determine oil contents of the microalgae *Spirulina maxima*, *C. vulgaris*, *S. obliquus*, *Dunaliella tertiolecta*, *Nannochloropsis* sp., and *Neochloris oleoabundans*. Oil yields obtained with this solvent ranged between 4.1 % (wt.) from *S. maxima* and 29.0 % (wt.) from *N. oleoabundans* (Gouveia and Oliveira 2009). Shen et al. (2009) applied a solvent system composed by a mixture of ethanol:n-hexane (1:1, v/v) to the microalgae *S. dimorphus* and *C. protothecoides*. Oil contents obtained with this method were 6.3 % and 5.6 % (wt.) for *S. dimorphus* and *C. protothecoides*, respectively (Shen et al. 2009).

Ranjan et al. (2010) compared oil extraction from the *Scenedesmus* sp. using two organic solvent methods: Soxhlet extraction with n-hexane, and the Bligh and Dyer's (1959), using a chloroform and methanol mixture in a ratio of 3:1 (v/v). The achieved oil contents were 0.6 % (wt.) for extraction with n-hexane and 2.0 % (wt.) for extraction with the Bligh and Dyer's method, showing that the last method is most efficient. This can be explained by the non-polar character of n-hexane, which results in a lower selectivity of microalgal oil, mainly composed by unsaturated fatty acids, toward n-hexane. On the other hand, chloroform has a polar nature, which allows a higher selectivity of microalgal oil toward this organic solvent (Ranjan et al. 2010).

Fajardo et al. (2007) used ethanol as an organic solvent for oil extraction from the microalga *Phaeodactylum tricornutum*. The authors applied an ethanol solution (96 % v/v) to freeze dried biomass, obtaining a oil yield of 6.4 % (wt.) (Fajardo et al. 2007).

Although n-hexane may be less efficient than chloroform, it is less toxic and it has an apparently higher selectivity for neutral oil fractions (Amaro et al. 2011). The application of this organic solvent coupled with an efficient cell wall disruption method could be a promising alternative to avoid the harmfulness of chloroform.

Supercritical Fluid Extraction

An alternative to the use of volatile and toxic organic solvents in microalgal oil extraction is the application of supercritical fluids as solvents (Amaro et al. 2011; Halim et al. 2011; Mercer and Armenta 2011). Supercritical fluids are compounds that behave both as a liquid or a gas when exposed to temperatures and pressures above their critical values. The most used supercritical fluid for oil extraction is CO_2 ($scCO_2$) because it presents low critical temperature (31.1 °C) and pressure (72.9 atm) (Mercer and Armenta 2011).

The $scCO_2$ extraction presents several advantages over the traditional organic solvent extraction procedures, such as: (i) tuneable solvating power; (ii) low toxicity and flammability; (iii) favourable mass transfer rates; and (iv) production of solvent free extracts because at room temperature CO_2 behaves as a gas (Amaro et al. 2011; Crampon et al. 2011; Halim et al. 2011; Macías-Sánchez et al. 2007). The main disadvantage is the associated cost, which is mainly due to the required infrastructure and operational conditions (Halim et al. 2011).

Efficiencies of oil extraction using $scCO_2$ depend on the following factors: (i) pressure; (ii) temperature; (iii) CO_2 flow rate; and (iv) extraction time. Furthermore, the use of modifiers or co-solvents, such as ethanol can be adjusted to optimize extractions. When ethanol is applied as a co-solvent, polarity of CO_2 increases and its viscosity is altered, increasing the fluid solvating power. In these conditions, lower temperature and pressure are required, improving the extraction efficiency. Another limiting factor of $scCO_2$ extraction is the level of moisture in the sample. High moisture content can reduce contact time between the solvent and biomass, making difficult the diffusion of CO_2 into the sample and the diffusion of oil out of the cell, because microalgal biomass tends to gain a thick consistency (Halim et al. 2011).

Studies performed by Mendes et al. (1995) showed that application of $scCO_2$ with a gas flow rate of 21.4 $kg.h^{-1}$ at 35.0 MPa and 55 °C to *C. vulgaris* cells resulted in an oil yield of 13.3 % (wt.). Application of organic solvents, such as acetone and n-hexane, resulted in an oil yield of 16.8 % and 18.5 % (wt.), respectively (Mendes et al. 1995). Andrich et al. (2005) applied different extraction conditions using $scCO_2$ and also organic solvent extraction with n-hexane to the microalga *Nannochloropsis* sp. Extraction procedures allowed the achievement of 250 $mg.g^{-1}{}_{\text{dry weight}}$ (23.0 %) using $scCO_2$ (gas flow rate of 10 $kg.h^{-1}$, 70.0 MPa, and 55 °C) and 120 $mg.g^{-1}{}_{\text{dry weight}}$ (12.0 %) using n-hexane at both 52 °C and room temperature (Andrich et al. 2005; Crampon et al. 2011). Later, Andrich et al. (2006) used *Spirulina platensis* to verify the extraction efficiency of $scCO_2$ technique using four different pressures (25.0, 40.0, 55, and 70.0 MPa) and two different temperatures (40 and 55 °C), with a gas flow rate of 10 $kg.h^{-1}$. In addition to this method, the authors also tested solvent extraction with n-hexane. Results showed that after 45 min, the amount of oil extracted reached its maximum (78.2 $mg.g^{-1}{}_{\text{dry weight}}$) for extraction performed at 55 °C and 70.0 MPa. The same amount of extracted oil was obtained after 2.5 h and after 3.5 h, for extraction performed at 55 °C and 40.0 MPa and at 40 °C and 40.0 MPa, respectively. For solvent extraction with n-hexane, the highest oil recovery (77.7 $mg.g^{-1}{}_{dry\ weight}$) was achieved after 8 h of reaction (Andrich et al. 2006). Couto et al. (2010) performed supercritical fluid extraction from the microalga *C. cohnii* at temperatures of 40 and 50 °C and pressures of 20.0, 25.0 and 30.0 MPa. Gas flow rate was 0.6 $kg.h^{-1}$ and extraction time was 3 h. Optimum extraction conditions were found to be 30.0 MPa and 50 °C (8.6 %), against the 19.9 % (wt.) achieved by application of Bligh and Dyer's method (1959). With this work, it was possible to state that at the highest pressures (25.0 and 30.0 MPa) the extraction yield increases with the temperature, while at the lowest pressure (20.0 MPa), an increase in temperature leads to a decreased yield. Temperature influence on the extraction efficiency results from the combination of the following antagonic thermodynamic effects: (i) at constant pressure, an increase in temperature leads to a decrease in the density of the supercritical fluid and thus its solvatation capacity; (ii) vapour pressure of solutes increase with the temperature, resulting in an high solubility in the supercritical fluid (Couto et al. 2010). Using $scCO_2$ to extract oil from the microalga *Chlorococcum* sp., Halim et al. (2011) achieved an oil recovery of 58 $mg.g^{-1}{}_{dry\ weight}$ (5.8 %) at a flow rate of 18.5 kg h^{-1}, a pressure of 30.0 MPa and a temperature of 60 °C, during 80 min. By increasing temperature to 80 °C, oil yield was 48 $mg.g^{-1}{}_{dry\ weight}$ (4.8 %). The authors also applied organic solvent extraction (using n-hexane) obtaining a oil yield of 32 $mg.g^{-1}{}_{dry\ weight}$ (3.2 %) after a reaction time of 5.5 h (Halim et al. 2011). Bjornsson et al. (2012) used the microalga *Nannochloropsis granulata* to study the effect of pressure, time and temperature in oil extraction yields. Firstly, the authors evaluated the effect of pressure (35, 45, and 55 MPa; 50 °C; 6 kg h^{-1}; 3 h), concluding that no significant differences in oil yields were obtained by varying this process variable. Later, different extraction times were applied (3, 4.5, and 6 h), maintaining pressure, temperature and gas flow rate constant (35 MPa; 50 °C; 6 $kg.h^{-1}$). The increase of extraction time resulted in differences in oil yield that ranged from 8.67 $mg.g^{-1}{}_{dry\ weight}$ (over 180 min) to 15.56 $mg.g^{-1}{}_{dry\ weight}$ (over 270 min) and to 16.91 $mg.g^{-1}{}_{dry\ weight}$ (over 360 min). However, the differences in yields were not statistically significant. Finally, $scCO_2$ extraction was performed by keeping pressure, gas flow rate, and time constant (35 MPa, 6 $kg.h^{-1}$, and 4.5 h), and by varying temperature (50, 70, and 90 °C). The increase of extraction temperature resulted in a statistically significant increase in oil yield from 15.56 $mg.g^{-1}{}_{dry\ weigh}$ at 50 °C to 28.45 $mg.g^{-1}{}_{dry\ weight}$ at 70 °C and to 25.75 $mg.g^{-1}{}_{dry\ weight}$ at 90 °C. Although oil yield decreased by increasing temperature from 70 to 90 °C, this decrease was not statistically significant (Bjornsson et al. 2012).

Application of $scCO_2$ to extract microalgal oil is very attractive, as it is a green technology and it allows a complete characterization of the extracted oil and the resulting biofuel. However, it still needs to be improved, because of its high cost-effectiveness and high energy-consuming drying step required before supercritical fluid extraction (Crampon et al. 2011).

Ionic Liquid Mediated Extraction

Ionic liquids (ILs) have been reported as an attractive alternative for volatile and toxic organic solvents because of their non-volatile character, thermal stability, and high solvatation capacity (Kim et al. 2012; Lateef et al. 2009). ILs are salts of relatively large asymmetric organic cations coupled with smaller organic or inorganic anions. These organic salts can be liquid at room temperatures or low melting point solids (<100 °C) (Lateef et al. 2009; Young et al. 2010; Khodadoust et al. 2006). Cations are generally composed of a ring structure containing nitrogen, such as imidazolium or pyrimidine. On the other hand, anions can vary from single ions, such as chloride, to larger complex ions, such as $[N(SO_2CF_3)_2]^-$ (Young et al. 2010). By altering the nature of both cation and anion of the ionic liquid, either hydrophilic or hydrophobic ionic liquids can be prepared, in order to make them suitable for different applications (Lateef et al. 2009).

The use of ionic liquids for oil extraction has been reported by Young et al. (2010). The authors applied mixtures of 1-ethyl-3-methylimidazolium methyl sulphate [Ethyl-mim]$MeSO_4$ and methanol to extract oil from microalgal biomass and different seeds. Ionic liquid extraction from the microalgae *Dunaliella* sp. and *Chlorella* sp. resulted in an oil yield of 8.6 % and 38.0 % (wt.), respectively (Young et al. 2010). A wide variety of ILs has also been used by Kim et al. (2012) for oil extraction from *C. vulgaris*. Oil was extracted from harvested biomass using systems of methanol and the following ILs: [Butyl-mim]CF_3SO_3, [Butyl-mim]$MeSO_4$, [Butyl-mim]CH_3SO_3, [Butyl-mim]BF_4, [Butyl-mim]PF_6, [Butyl-mim]$MeSO_4$, [Butyl-mim]Tf_2N, [Butyl-mim]Cl, [Ethyl-mim]$MeSO_4$, [Ethyl-mim]Cl, [Ethyl-mim]Br, and [Ethyl-mim]Ac. Bligh and Dyer's method (1959) was also used in terms of comparison. Application of these methods resulted in a total fatty acid content of 106.2 (11.1 %), 125.4 (19.0 %), and 118.4 (17.4 %) $mg.g^{-1}{}_{dry\ weight}$ for Bligh and Dyer's method, [Butyl-mim]CF_3SO_3, and [Butyl-mim]$MeSO_4$, respectively. It was also shown that fatty acids profiles were very similar for the three extraction methods. The use of an IL-methanol system reduces high prices and aquatic toxicity of imidazolium-based ILs, providing an efficient and environmentally-friendly system for oil extraction from microalgal biomass (Kim et al. 2012).

Although ILs are more expensive than the conventional organic solvents, application of these compounds as solvents for microalgal oil can be a promising alternative because of their higher affinities and non-toxic character.

1.3 Biodiesel Production

Biodiesel is a renewable fuel produced from vegetable oils, animal fats, microorganisms' oils, or waste cooking oil (Wahlen et al. 2011). It constitutes the best candidate to substitute diesel fuel, as it can be used directly as a fuel requiring some engine modifications, or blended with petroleum diesel and used in diesel engines with few or no modifications (Leung et al. 2010). Chemically, biodiesel is a mixture of esters with long-chain fatty acids, such as lauric, palmitic, stearic, oleic, etc. (Demirbas and Demirbas 2010). Recently, this biofuel has become more attractive due to its environmental benefits: it is biodegradable and it has lower sulphur and aromatic content than diesel fuel, meaning that it will emit less toxic gases. Furthermore, it presents several advantages over conventional petroleum diesel, such as higher combustion efficiency and cetane number. The main disadvantages of biodiesel include the high production costs, its higher viscosity, lower energy content and higher nitrogen oxide (NO_x) emissions (Demirbas and Demirbas 2010; Leung et al. 2010).

Biodiesel can be produced from extracted oil through four different methods: (i) direct use or blending of oils; (ii) microemulsification of oils; (iii) thermal cracking or pyrolysis; and (iv) transesterification, also known as alcoholysis (Balat and Balat 2010; Leung et al. 2010; Ma and Hanna 1999). Additional information of the referred methods can be obtained in Balat and Balat (2010) and Ma and Hanna (1999). Transesterification process, schematically represented in Fig. 1.3, constitutes the most applied method for biodiesel production, as it presents several advantages over the other methods. For example, blending and microemulsification may have some problems, such as carbon deposition and oil contamination, whereas pyrolysis is responsible for the production of low valuable products, as well as the production of gasoline instead of diesel (Sharma and Singh 2009). Therefore, transesterification, the chemical conversion of triglycerides in glycerol and esters in the presence of an alcohol, seems to be the most appropriate method for biodiesel production and it will be studied in detail in the next sections.

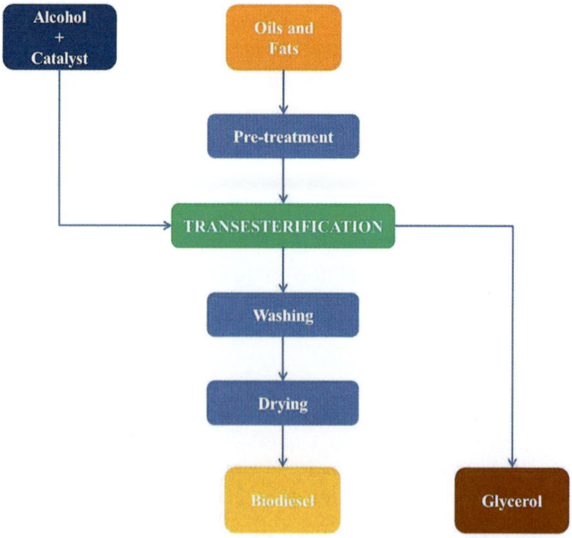

Fig. 1.3 Biodiesel production process through transesterification

1.3.1 Transesterification Reaction

As shown in Eqs. 1.1, 1.2, and 1.3, the transesterification is a multi-step reaction where triglycerides are converted into diglycerides, monoglycerides and finally into glycerol during three reaction steps. These reactions are reversible and each one results in the formation of 1 mol of fatty acid ester (Leung et al. 2010; Ma and Hanna 1999).

$$\text{Triglyceride (TG)} + \text{R}'\text{OH} \overset{\text{catalyst}}{\longleftrightarrow} \text{Diglyceride (DG)} + \text{R}'\text{COOR}_1 \tag{1.1}$$

$$\text{Diglyceride (DG)} + \text{R}'\text{OH} \overset{\text{catalyst}}{\longleftrightarrow} \text{Monoglyceride (MG)} + \text{R}'\text{COOR}_2 \tag{1.2}$$

$$\text{Monoglyceride (MG)} + \text{R}'\text{OH} \overset{\text{catalyst}}{\longleftrightarrow} \text{Glycerol (GL)} + \text{R}'\text{COOR}_3 \tag{1.3}$$

R represents a small hydrocarbon chain, whereas R_1, R_2, and R_3 represent long-chain hydrocarbons, also known as fatty acid chains.

Variables affecting biodiesel yields during transesterification include the alcohol and molar ratio employed, type of used catalyst, amount of free fatty acids (FFA), water content, and reaction temperature and time (Ehimen et al. 2010; Ma and Hanna 1999; Miao et al. 2009; Sharma and Singh 2009; Wahlen et al. 2011). Due to the reversibility of the above mentioned reactions, an excess of alcohol is used to shift the equilibrium towards fatty acid esters formation. In a study performed by Miao and Wu (2006), transesterification reaction was applied to oil recovered from the microalga *C. protothecoides* at 30 °C for 7 h, using 100 % (v/v) of catalyst (sulphuric acid) and the following molar ratios of methanol to oil: 25:1, 30:1, 45:1, 56:1, 70:1, and 84:1 (v/v). Results showed that the highest biodiesel yields (68.0 % and 63.0 %) were obtained using the molar ratios of alcohol to oil of 45:1 and 56:1, respectively (Miao and Wu 2006). In the transesterification process, short-chain alcohols such as methanol, ethanol, propanol, butanol, and amyl alcohol are used. However, the most used ones are methanol and ethanol. Methanol has been extensively applied in transesterification reactions because of its low cost (Gong and Jiang 2011; Leung et al. 2010; Ma and Hanna 1999) and physical and chemical properties, including its high polarity and small chain length (Ma and Hanna 1999). When methanol is used as alcohol the produced esters are known as fatty acid methyl esters (FAME). Wahlen et al. (2011) used five different alcohols (methanol, ethanol, butan-1-ol, 2-methylpropan-1-ol, and 3-methylbutan-1-ol) to produce biodiesel from the microalga *Chaetoceros gracilis* through transesterification. The assays were performed at 60 °C for 100 min, using sulphuric acid as a catalyst in a proportion of 1.8 % (v/v). The amount of fatty acid esters produced with the different alcohols was not significantly different, meaning that methanol, the cheapest one, is suitable for application in the transesterification process (Wahlen et al. 2011).

Three types of catalysts can be used in the transesterification reaction: alkalis, acids, and enzymes (Drapcho et al. 2008; Ma and Hanna 1999). Alkalis and acids are the most commonly used catalysts, both including homogeneous and heterogeneous catalysts (Drapcho et al. 2008; Leung et al. 2010). The alkalis used in this process include NaOH, KOH, carbonates and corresponding sodium and potassium alkoxides like sodium methoxide, sodium ethoxide, sodium propoxide, and sodium butoxide (Ma and Hanna 1999). The main acids used as catalysts include sulphuric acid, sulfonic acid, and hydrochloric acid. Normally, alkali catalysts are preferred over acid catalysts because reactions catalysed by alkali catalysts are faster than reactions catalysed by acids (Drapcho et al. 2008; Leung et al. 2010). However, transesterification reaction using an alkali catalyst results in the formation of small amounts of water (Sharma and Singh 2009). Water is undesirable in this process because it promotes the hydrolysis of the glycerides, forming FFA. FFA produce soap and water through a saponification reaction with the alkali catalyst (Leung et al. 2010). Soap formation must be avoided because it lowers the fatty acid esters yield and inhibits the separation of the esters from glycerol (Leung et al. 2010; Ma and Hanna 1999). Thus, when using an alkali catalyst, glycerides and the used alcohol must be substantially anhydrous (Ma and Hanna 1999) and FFA content of glycerides must be below 0.5 % (wt.) (Gong and Jiang 2011). When FFA contents in glycerides exceed 0.5 %, an acid catalyst should be employed (Drapcho et al. 2008; Leung et al. 2010; Ma and Hanna 1999). Acids catalyse the formation of water and fatty acid esters, i.e. biodiesel, from FFA and alcohol. The main disadvantage of acid catalyst application is the slow reaction rate and the high methanol to oil molar ratio that is required (Leung et al. 2010). Alternatively, triglycerides can be purified by saponification by alkali treatment before transesterification (Ma and Hanna 1999). The use of enzymes, e.g. lipases, as catalysts constitutes a promising alternative in biodiesel production. They avoid soap formation and facilitate the downstream process of purification, i.e. recovery of glycerol at the end of the reaction is easier when using an enzymatic catalyst. Furthermore, the reactions catalysed by lipases are not affected by FFA and water content (Gong and Jiang 2011; Leung et al. 2010). The main obstacles in using enzymes as catalysts are related with longer reaction times (Leung et al. 2010) and higher costs (Gong and Jiang 2011; Leung et al. 2010).

Reaction temperature and time also influence biodiesel yields through transesterification. Increasing reaction time normally increases conversion rate. Alternatively, different optimum temperatures of transesterification can be determined, depending on the used oil (Ma and Hanna 1999). It is very common to study these variables together because an increase in temperature reaction allows the production of higher amounts of biodiesel in a shorter period of time. This conclusion is supported by a study of Ehimen et al. (2010), who studied the effect of reaction temperature and time on the production of biodiesel from *Chlorella* sp. Transesterification reaction was performed at four different temperatures (23, 30, 60, and 90 °C), the used alcohol was methanol and the catalyst was sulphuric acid. Results showed faster conversion rates at higher temperatures (60 and 90 °C). FAME conversion rates reached similar asymptotic values after 2 and 4 h of reaction for 60 and 90 °C (Ehimen et al. 2010).

At the end of the transesterification process, a mixture containing essentially esters and glycerol is obtained. As the phase containing glycerol has higher density than the one containing esters, the separation of the two phases becomes easier because glycerol phase tends to settle at the bottom. However, both glycerol and esters phase may contain residues of the alcohol, catalyst and oil that did not react during the transesterification, and also soap that has been formed (Leung et al. 2010; Ma and Hanna 1999). The presence of these contaminants indicates that crude biodiesel should be purified before its use in diesel engines. It is also important to refine the crude glycerol because it has a wide variety of industrial applications, such as soaps, cosmetics, medicines, and others (Leung et al. 2010). The procedures used in the purification of crude biodiesel and glycerol were described in detail by Leung et al. (2010).

There are two types of transesterification: transesterification applied to the extracted oil and transesterification directly applied to the oil source, without previous oil extraction, also known as transesterification *in situ*. The following sections compare current attempts in producing biodiesel from microalgae using these two types of transesterification. Table 1.3 shows a resume of the research studies about transesterification reaction, presenting the achieved biodiesel yields.

Transesterification from the Recovered Oil

Several studies have reported the use of transesterification to convert fatty acids extracted from microalgae into fatty acid esters. The applied reaction follows the above mentioned principles and can be influenced by the referred variables. Miao and Wu (2006) have applied acid-catalysed transesterification to oil extracted from the microalga *C. protothecoides* with n-hexane. The authors adopted an acid catalyst, e.g. sulphuric acid, because of the high acid value of microalgal oil due to high FFA content. Transesterification was performed at different conditions, to evaluate the effect of catalyst quantity, methanol to oil molar ratio, and reaction time and temperature on the yield and properties of biodiesel product. The reaction was then carried out using: (i) four levels of catalyst quantity – 25, 50, 60, and 100 % sulphuric acid based on oil weight; (ii) six different molar ratios of methanol to oil – 25:1, 30:1, 45:1, 56:1, 70:1, and 84:1 (v/v); and (iii) three different temperatures – 30, 50, and 90 °C. In each experiment, 9.12 g of microalgal oil was used. Results after oil extraction showed that the *C. protothecoides* produced 55.2 % (wt.) of oil under heterotrophic conditions. Transesterification of this oil resulted in maximum biodiesel yields of 68.0 and 63.0 % for a molar ratio of methanol to oil of 45:1 and 56:1 (v/v), respectively (Miao and Wu 2006). Later, in a study performed by Johnson and Wen (2009), oil extracted from *Schizochytrium limacinum* was submitted to transesterification reaction using methanol and sulphuric acid. Firstly, the authors applied the Bligh and Dyer's method (1959) to extract oil from 1 g of microalgal biomass. Chloroform and methanol were added to biomass in the ratio of 1:2 (v/v). After the extraction step, oils were transesterified using a mixture of methanol, sulphuric acid and chloroform at 90 °C with the reaction time of 40 min.

Table 1.3 Biodiesel yields as a fraction of total oil obtained after transesterification of microalgal oils or after transesterification of microalgal biomass

Transesterification of the recovered oils		Transesterification *in situ*			
Conditions	Biodiesel yield (%)	Conditions	Biodiesel yield (%)	Microorganism	Reference
Methanol (45:1), H_2SO_4 (100 %), 30 °C, 7 h	68.0	–	–	*Chlorella protothecoides*	Miao and Wu (2006)
Methanol (3.4 mL), H_2SO_4 (0.6 mL), chloroform (4.0 mL) 90 °C, 40 min	63.7	Methanol (3.4 mL), H_2SO_4 (0.6 mL), chloroform (4.0 mL) 90 °C, 40 min	72.8	*Schizochytrium limacinum*	Johnson and Wen (2009)
–	–	Methanol (60 mL), H_2SO_4 (2.2 mL), 90 °C, 1 h	92.0	*Chlorella* sp.	Ehimen et al. (2010)
–	–	Methanol (2 mL), H_2SO_4 (1.8 %, v/v), 80 °C, 20 min	82.0	*Chaetoceros gracilis*	Wahlen et al. (2011)
–	–	Methanol (2 mL), H_2SO_4 (1.8 %, v/v), 80 °C, 20 min	78.0	*Tetraselmis suecica*	
–	–	Methanol (2 mL), H_2SO_4 (1.8 %, v/v), 80 °C, 20 min	77.0	*Chlorella sorokiniana*	
–	–	Methanol (2 mL), H_2SO_4 (1.8 %, v/v), 80 °C, 20 min	39.0	*Synechocystis* sp.	
–	–	Methanol (2 mL), H_2SO_4 (1.8 %, v/v), 80 °C, 20 min	40.0	*Synechococcus elongatus*	

From the total fatty acids present in *S. limacinum* representing 40–50 % of dry biomass, transesterification reaction resulted in biodiesel yields of 63.7 % (Johnson and Wen 2009).

Transesterification of microalgal oil seems to be a promising alternative in biodiesel production, as conversion rates obtained with this method are very high. However, further improvements in operation conditions are needed, to reduce production costs and increase biodiesel yields. One possible alternative is the transesterification *in situ*, as it overtakes the high-costly extraction step.

Transesterification *In Situ*

Transesterification *in situ* is very similar to the previously referred method, but instead of being applied to oils, it is applied directly to the biomass containing the oils. As this process can produce biodiesel without the extraction step, it is thought that direct transesterification could lower the production costs of biodiesel fuel (Ma and Hanna 1999; Patil et al. 2011, 2012). Several authors have reported the use of transesterification *in situ* to produce biodiesel from micro and macroalgae. For instance, Ehimen et al. (2010) applied transesterification *in situ* to the microalga *Chlorella* sp. Different reaction conditions were applied to identify the main variables that affect biodiesel yields. Transesterification of 15 g of dried biomass was performed using: (i) 2.2 mL of sulphuric acid as a catalyst; (ii) different volumes of methanol – 20, 40, 60, 80, and 100 mL; (iii) different reaction temperatures – 23, 30, 60, and 90 °C; and (iv) different reaction times – 0.25, 0.5, 1, 1.5, 2, 4, 8, and 12 h. Conversion of microalgal biomass into biodiesel reached 92.0 % at 90 °C after 1 h of reaction, using 2.2 mL of catalyst and 60 mL of methanol (Ehimen et al. 2010). Carvalho Júnior et al. (2011) applied *in situ* methanolysis to *Nannochloropsis oculata*. The transesterification reaction of 2 g of biomass was carried out using a mixture of methanol:chlorydric acid:chloroform in a 10:1:1 (v/v/v) ratio, at 80 °C for 2 h under continuous stirring. These conditions allowed the production of 23.2 % of biodiesel in $w_{esters}/w_{biomass}$. Considering that microalgae has an oil content ranging from 20 % to 50 % (Chisti 2007), the performance obtained is quite satisfactory (Carvalho Júnior et al. 2011). Wahlen et al. (2011) used transesterification *in situ* to produce biodiesel from the microalgae *C. gracilis*, *Tetraselmis suecica*, and *Chlorella sorokiniana*, and from the cyanobacteria *Synechocystis* sp., and *Synechococcus elongatus*. Transesterification reaction was applied to 100 mg of biomass using 2 mL of methanol and sulphuric acid in a volume fraction of 1.8 %. Reaction temperature and time were 80 °C and 20 min. Levels of biodiesel per extractable oil were 82.0, 78.0, 77.0, 39.0 and 40.0 % for the organisms *C. gracilis*, *T. suecica*, *C. sorokiniana*, *Synechocystis* sp. and *S. elongatus*, respectively (Wahlen et al. 2011). To compare the two types of transesterification, Johnson and Wen (2009) also applied direct transesterification to cells of *S. limacinum*. Using a mixture of methanol, sulphuric acid, and chloroform at 90 °C and during 40 min, the authors obtained an ester yield of 72.8 %.

These studies show that biodiesel conversion yields are similar for both types of transesterification, meaning that transesterification *in situ* should be adopted instead of the double-step procedure of extraction and transesterification. However, these results also suggest that application of direct transesterification require higher volumes of alcohol and catalyst.

1.4 Conclusion

The extraction methods described in this review constitute promising alternatives to recover microalgal oil. Experiments conducted in the last decade have showed that organic solvent extraction is the most efficient method for microalgal oil extraction. The main drawback of applying organic solvents for oil extraction is related to the harmfulness of these compounds. The amount of organic solvent required can be reduced by previous disruption of cell walls using enzymatic, PEF, ultrasound, microwave and expeller pressing methods. These procedures are also responsible for an increase in oil extraction efficiencies. Enzymatic and PEF extraction could be promising techniques because enzymes present higher selectivity towards cell walls and PEF has reduced energetic costs compared to the other disruption methods presented in this chapter. Application of $scCO_2$ extraction has shown to have high extraction efficiencies and to be safer to the environment, but the costs associated are extremely high. Another possibility to avoid the use of organic solvents is to use ionic liquids as solvents. Although extraction efficiencies are lower, these compounds are more environmentally friendly than organic solvents.

An important lack in the research of microalgal oil extraction is the process scale-up, including the analysis of the process cost and efficiency. The majority of studies concerning oil extraction and biodiesel production from microalgae have been performed for lab-scale. As a result, little is known about the feasibility of these processes in large scale. Further studies should be conducted using larger amounts of microalgal biomass to analyse the oil extraction efficiency and to compare with results already known for lab-scale.

In the transesterification reaction, the different variables affecting this reaction must be taken into account. As referred earlier in this review, the alcohol and its employed amount, the type of catalyst and its concentration, temperature and reaction time have a great influence in the alcoholysis reaction. As well as in the oil extraction procedures, the reaction scale is also a determining factor, as it has a great impact on the volumes of alcohol and catalyst employed and on reaction time, which reflects in the total costs of the reaction. Therefore, transesterification reaction should be considered using higher volumes of oil and also different combinations of operational conditions. Additionally, conditions that maximize oil extraction efficiencies should be applied together with those responsible for higher oil to biodiesel conversion rates, to produce biodiesel able to compete with current diesel fuel with reduced costs and higher productivities.

Finally, attention should also be paid to the transesterification *in situ* process, as it eliminates the oil extraction step. Future studies should focus on this method to verify if costs can be really reduced in the absence of oil extraction steps. Operational conditions at large scale should also be addressed in order to achieve higher productivities of fatty acids esters.

Acknowledgements Ana L. Gonçalves and José C.M. Pires are grateful to Foundation for Science and Technology (FCT), POPH-QREN and FSE for their fellowships SFRH/BD/88799/2012 and SFRH/BPD/66721/2009, respectively.

References

Amaro HM, Guedes AC, Malcata FX (2011) Advances and perspectives in using microalgae to produce biodiesel. Appl Energy 88(10):3402–3410. doi:10.1016/j.apenergy.2010.12.014

Andrich G, Nesti U, Venturi F, Zinnai A, Fiorentini R (2005) Supercritical fluid extraction of bioactive lipids from the microalga *Nannochloropsis* sp. Eur J Lipid Sci Technol 107(6):381–386. doi:10.1002/ejlt.200501130

Andrich G, Zinnai A, Nesti U, Venturi F (2006) Supercritical fluid extraction of oil from microalga *Spirulina (Arthrospira) platensis*. Acta Aliment 35(2):195–203

Balasubramanian S, Allen JD, Kanitkar A, Boldor D (2011) Oil extraction from *Scenedesmus obliquus* using a continuous microwave system – design, optimization, and quality characterization. Bioresour Technol 102(3):3396–3403. doi:10.1016/j.biortech.2010.09.119

Balat M, Balat H (2010) Progress in biodiesel processing. Appl Energy 87(6):1815–1835. doi:10.1016/j.apenergy.2010.01.012

Bjornsson WJ, MacDougall KM, Melanson JE, O'Leary SJB, McGinn PJ (2012) Pilot-scale supercritical carbon dioxide extractions for the recovery of triacylglycerols from microalgae: a practical tool for algal biofuels research. J Appl Phycol 24(3):547–555. doi:10.1007/s10811-011-9756-2

Bligh EG, Dyer WM (1959) A rapid method of lipid extraction and purification. Can J Biochem Physiol 37:911–917

Brennan L, Owende P (2010) Biofuels from microalgae – a review of technologies for production, processing, and extractions of biofuels and co-products. Renew Sust Energy Rev 14(2):557–577. doi:10.1016/j.rser.2009.10.009

Carvalho Júnior RM, Vargas JVC, Ramos LP, Marino CEB, Torres JCL (2011) Microalgae biodiesel via *in situ* methanolysis. J Chem Technol Biotechnol 86(11):1418–1427. doi:10.1002/jctb.2652

Chen W, Zhang C, Song L, Sommerfeld M, Hu Q (2009) A high throughput Nile red method for quantitative measurement of neutral lipids in microalgae. J Microbiol Methods 77(1):41–47. doi:10.1016/j.mimet.2009.01.001

Chisti Y (2007) Biodiesel from microalgae. Biotechnol Adv 25(3):294–306. doi:10.1016/j.biotechadv.2007.02.001

Chisti Y, Moo-Young M (1986) Disruption of microbial cells for intracellular products. Enzyme Microb Technol 8(4):194–204. doi:10.1016/0141-0229(86)90087-6

Costa JAV, de Morais MG (2011) The role of biochemical engineering in the production of biofuels from microalgae. Bioresour Technol 102(1):2–9. doi:10.1016/j.biortech.2010.06.014

Couto RM, Simões PC, Reis A, da Silva TL, Martins VH, Sánchez-Vicente Y (2010) Supercritical fluid extraction of lipids from the heterotrophic microalga *Crypthecodinium cohnii*. Eng Life Sci 10(2):158–164. doi:10.1002/elsc.200900074

Crampon C, Boutin O, Badens E (2011) Supercritical carbon dioxide extraction of molecules of interest from microalgae and seaweeds. Ind Eng Chem Res 50(15):8941–8953. doi:10.1021/ie102297d

Cravotto G, Boffa L, Mantegna S, Perego P, Avogadro M, Cintas P (2008) Improved extraction of vegetable oils under high-intensity ultrasound and/or microwaves. Ultrason Sonochem 15(5):898–902. doi:10.1016/j.ultsonch.2007.10.009

Demirbas A (2011) Biodiesel from oilgae, biofixation of carbon dioxide by microalgae: a solution to pollution problems. Appl Energy 88(10):3541–3547. doi:10.1016/j.apenergy.2010.12.050

Demirbas A, Demirbas MF (2010) Biofuels. In: Demirbas A, Demirbas MF (eds) Algae energy: algae as a new source of biodiesel. Springer, New York, pp 56–59

Demirbas A, Demirbas MF (2011) Importance of algae oil as a source of biodiesel. Energy Convers Manage 52(1):163–170. doi:10.1016/j.enconman.2010.06.055

Drapcho CM, Nhuan NP, Walker TH (2008) Biodiesel. In: Drapcho CM, Nhuan NP, Walker TH (eds) Biofuels engineering process technology. McGraw Hill, New York, pp 201–208

Ehimen EA, Sun ZF, Carrington CG (2010) Variables affecting the *in situ* transesterification of microalgae lipids. Fuel 89(3):677–684. doi:10.1016/j.fuel.2009.10.011

Fajardo AR, Cerdán LE, Medina AR, Fernández FGA, Moreno PAG, Grima EM (2007) Lipid extraction from the microalga *Phaeodactylum tricornutum*. Eur J Lipid Sci Technol 109(2):120–126. doi:10.1002/ejlt.200600216

Geciova J, Bury D, Jelen P (2002) Methods for disruption of microbial cells for potential use in the dairy industry – a review. Int Dairy J 12(6):541–553. doi:10.1016/s0958-6946(02)00038-9

Gong Y, Jiang M (2011) Biodiesel production with microalgae as feedstock: from strains to biodiesel. Biotechnol Lett 33(7):1269–1284. doi:10.1007/s10529-011-0574-z

Gouveia L, Oliveira AC (2009) Microalgae as a raw material for biofuels production. J Ind Microbiol Biotechnol 36:269–274

Grima EM, Belarbi EH, Fernández FGA, Medina AR, Chisti Y (2003) Recovery of microalgal biomass and metabolites: process options and economics. Biotechnol Adv 20(7–8):491–515. doi:10.1016/s0734-9750(02)00050-2

Guderjan M, Töpfl S, Angersbach A, Knorr D (2005) Impact of pulsed electric field treatment on the recovery and quality of plant oils. J Food Eng 67(3):281–287. doi:10.1016/j.jfoodeng.2004.04.029

Guderjan M, Elez-Martínez P, Knorr D (2007) Application of pulsed electric fields at oil yield and content of functional food ingredients at the production of rapeseed oil. Innov Food Sci Emerg Technol 8(1):55–62. doi:10.1016/j.ifset.2006.07.001

Halim R, Gladman B, Danquah MK, Webley PA (2011) Oil extraction from microalgae for biodiesel production. Bioresour Technol 102(1):178–185. doi:10.1016/j.biortech.2010.06.136

Johnson MB, Wen Z (2009) Production of biodiesel fuel from the microalga *Schizochytrium limacinum* by direct transesterification of algal biomass. Energy Fuel 23(10):5179–5183. doi:10.1021/ef900704h

Khodadoust AP, Chandrasekaran S, Dionysiou DD (2006) Preliminary assessment of imidazolium-based room-temperature ionic liquids for extraction of organic contaminants from soils. Environ Sci Technol 40(7):2339–2345. doi:10.1021/es051563j

Kim Y-H, Choi Y-K, Park J, Lee S, Yang Y-H, Kim HJ, Park T-J, Hwan Kim Y, Lee SH (2012) Ionic liquid-mediated extraction of lipids from algal biomass. Bioresour Technol 109:312–315. doi:10.1016/j.biortech.2011.04.064

Lateef H, Grimes S, Kewcharoenwong P, Bailey E (2009) Ionic liquids in the selective recovery of fat from composite foodstuffs. J Chem Technol Biotechnol 84(11):1681–1687. doi:10.1002/jctb.2230

Lee J-Y, Yoo C, Jun S-Y, Ahn C-Y, Oh H-M (2010) Comparison of several methods for effective lipid extraction from microalgae. Bioresour Technol 101(1 Suppl):S75–S77. doi:10.1016/j.biortech.2009.03.058

Leung DYC, Wu X, Leung MKH (2010) A review on biodiesel production using catalyzed transesterification. Appl Energy 87(4):1083–1095. doi:10.1016/j.apenergy.2009.10.006

Lewis T, Nichols PD, McMeekin TA (2000) Evaluation of extraction methods for recovery of fatty acids from lipid-producing microheterotrophs. J Microbiol Methods 43(2):107–116. doi:10.1016/s0167-7012(00)00217-7

Ma F, Hanna MA (1999) Biodiesel production: a review. Bioresour Technol 70(1):1–15. doi:10.1016/s0960-8524(99)00025-5

Macías-Sánchez MD, Mantell C, Rodríguez M, de la Ossa EM, Lubián LM, Montero O (2007) Supercritical fluid extraction of carotenoids and chlorophyll *a* from *Synechococcus* sp. J Supercrit Fluids 39(3):323–329. doi:10.1016/j.supflu.2006.03.008

Mandal V, Mohan Y, Hemalatha S (2007) Microwave assisted extraction – an innovative and promising extraction tool for medicinal plant research. Pharmacog Rev 1(1):7–18

Mendes RL, Fernandes HL, Coelho J, Reis EC, Cabral JMS, Novais JM, Palavra AF (1995) Supercritical CO_2 extraction of carotenoids and other lipids from *Chlorella vulgaris*. Food Chem 53(1):99–103. doi:10.1016/0308-8146(95)95794-7

Mercer P, Armenta RE (2011) Developments in oil extraction from microalgae. Eur J Lipid Sci Technol 113(5):539–547. doi:10.1002/ejlt.201000455

Miao X, Wu Q (2006) Biodiesel production from heterotrophic microalgal oil. Bioresour Technol 97(6):841–846. doi:10.1016/j.biortech.2005.04.008

Miao X, Li R, Yao H (2009) Effective acid-catalyzed transesterification for biodiesel production. Energy Convers Manage 50(10):2680–2684. doi:10.1016/j.enconman.2009.06.021

Patil PD, Gude VG, Mannarswamy A, Deng SG, Cooke P, Munson-McGee S, Rhodes I, Lammers P, Nirmalakhandan N (2011) Optimization of direct conversion of wet algae to biodiesel under supercritical methanol conditions. Bioresour Technol 102(1):118–122. doi:10.1016/j.biortech.2010.06.031

Patil PD, Reddy H, Muppaneni T, Mannarswamy A, Schuab T, Holguin FO, Lammers P, Nirmalakhandan N, Cooke P, Deng SG (2012) Power dissipation in microwave-enhanced *in situ* transesterification of algal biomass to biodiesel. Green Chem 14(3):809–818. doi:10.1039/C2gc16195h

Ranjan A, Patil C, Moholkar VS (2010) Mechanistic assessment of microalgal lipid extraction. Ind Eng Chem Res 49(6):2979–2985. doi:10.1021/Ie9016557

Sander KB, Murthy GS (2009) Enzymatic degradation of microalgal cell walls. Paper presented at the 2009 ASABE annual international meeting, Reno

Shah S, Sharma A, Gupta MN (2004) Extraction of oil from *Jatropha curcas* L. seed kernels by enzyme assisted three phase partitioning. Ind Crop Prod 20(3):275–279. doi: citeulike-article-id:4962402

Sharma YC, Singh B (2009) Development of biodiesel: current scenario. Renew Sust Energ Rev 13(6–7):1646–1651. doi:10.1016/j.rser.2008.08.009

Shen Y, Pei Z, Yuan W, Mao E (2009) Effect of nitrogen and extraction method on algae lipid yield. Int J Agric Biol Eng 2(1):51–57. doi:10.3965/j.issn.1934-6344.2009.01.051-057

Taher H, Al-Zuhair S, Al-Marzouqui AH, Haik Y, Farid MM (2011) A review of enzymatic transesterification of microalgal oil-based biodiesel using supercritical technology. Enzyme Res. doi:10.4061/2011/468292

Wahlen BD, Willis RM, Seefeldt LC (2011) Biodiesel production by simultaneous extraction and conversion of total lipids from microalgae, cyanobacteria, and wild mixed-cultures. Bioresour Technol 102(3):2724–2730. doi:10.1016/j.biortech.2010.11.026

Young G, Nippgen F, Titterbrandt S, Cooney MJ (2010) Lipid extraction from biomass using co-solvent mixtures of ionic liquids and polar covalent molecules. Sep Purif Technol 72(1):118–121. doi:10.1016/j.seppur.2010.01.009

Chapter 2
Electrochemistry and Water Pollution

Subramanyan Vasudevan and Mehmet A. Oturan

Abstract This article reviews both the pollution by the electrochemical industry and the use of electrochemistry to clean water, air and soils. Main pollutants include Pd, Cd, Ni, Hg and other metals, SO_2, CO_2 and cyanide. The cause for water pollution by electrochemistry is due to the effluents from different electrochemical industries such as mercury from chlor-alkali industry; lead, cadmium and mercury from battery industry; heavy metals and organic contaminants from electroplating wastes; and contaminants from corrosion processes. Most pollutants can be successfully eliminated or converted to non-toxic materials by methods based on electrochemical principles. Electrochemical depolluting methods are mainly electrodialysis, electrocoagulation, electroflotation, anodic processes, cathodic processes, electrokinetic remediation, and electrochemical advanced oxidation processes.

Keywords Water and soil pollution • Hg • Cd • Pd • Ni • Electrochemical remediation • Electrodialytic • Electrocoagulation • Electroflotation • Battery • Chlor-alkali

---

Dr. S. Vasudevan, has been awarded **Eminent Scientist Award – 2013** by Indian Society for Electro Analytical Chemist (ISEAC) for his contribution in the area of Electrochemical Science and Technology.

S. Vasudevan (✉)
Electroinorganic Chemicals Division, CSIR- Central Electrochemical Research Institute, Karaikudi 630006, India
e-mail: vasudevan65@gmail.com

M.A. Oturan
Laboratoire Géomatériaux et Environnement (LGE), Université Paris-Est, EA 4508, UPEMLV, 77454 Marne-la-Vallée, France
e-mail: mehmet.oturan@univ-paris-est.fr

E. Lichtfouse et al. (eds.), *Green Materials for Energy, Products and Depollution*, Environmental Chemistry for a Sustainable World 3, DOI 10.1007/978-94-007-6836-9_2,

Contents

2.1 Introduction

Water is one of the essential companion of life on earth and the nature has iterated the vitality of water for life. It is very appropriate to say that existence of life on Earth is largely owed to the presence of water, but it is fast becoming a scare commodity in most part of the world. The earliest form of life appeared in sea water about 3.5 billion years ago and was transferred to land only 380 million years ago. There is no aspect of our life that is not touched by water. Water is one of the clear signs of prosperity, health, serenity, beauty, artistry, purity and many other attributes. Leonardo Da Vinci had described water as "the vehicle of nature" ("vetturale di natura"). Water resources of a country constitute one of its vital assets. Groundwater is a more reliable and less expensive water supply than those obtained from surface water (lakes, rivers and streams), marine and coastal; therefore, it is withdrawn in large quantities for domestic and agricultural applications all over the world. It is also reported that more than 1.2 billion people lack access to safe drinking water, 2.6 billion have little or no sanitation; millions of people die annually (3,900 children a day) from diseases transmitted through unsafe water or human excreta (Mark Shannon et al. 2008; Ball 2008; Montgomery and Elimelech 2007; Kumar et al. 2005; Ayoob and Gupta 2006). As per the international norms, if per-capita water availability is less than 1,700 m^3 per year then the country is categorized as water stressed and if it is less than 1,000 m^3 per capita per year then the country is classified as water scarce Hence, there is a need for proper planning, development and management of the greatest assets of the country, viz. water and land resources for raising the standards of living of the millions of people, particularly in the rural areas (Chakraborti et al. 2009; Roy Chowdhury et al. 1999).

Though water is available in the universe in huge quantity in the order of 1,400 × 10^6 km^3, only 2.5 % of the global water resources are consisted of fresh water. Among the fresh waters, only about 5 % of them or 0.15 % of the total world waters are readily available for beneficial use, the rest being contained in the polar glaciers or glaciers in altitude. Global freshwater reserves are rapidly depleting and this is expected to significantly impact many densely populated areas of the world. Low to middle income developing regions as well as highly developed countries will face water stress in the future, unless existing water reserves are managed effectively. Although low and middle income developing countries currently have low per capita water consumption, rapid growth in population and inefficient use of water across sectors is expected to lead to a water shortage in the future. Developed countries traditionally have high per capita water consumption and need to focus on reducing their consumption through improved water management techniques and practices. By 2025, India, China and select countries in Europe and Africa will face water scarcity if adequate and sustainable water management initiatives are not implemented.

The present review article presents the causes for water pollution by electrochemistry which are due to the effluents from different electrochemical industries viz., (i) mercury from chlor-alkali industry, (ii) lead, cadmium and mercury from battery industry, (iii) heavy metals from electroplating wastes and (iv) organic pollutants from dye, pesticide, pharmaceuticals industry etc., and electrochemical treatment technologies viz., (i) electrodialysis, (ii) electrocoagulation/electroflotation, (iii) anodic processes, (iv) cathodic processes, (v) electrokinetic remediation, (vi) electrochemical advanced oxidation processes and the details of the same are described.

2.2 How Electrochemistry Is the Cause for Water Pollution?

The modern definition of electrochemistry, as per Bockris, is that it is the interdisciplinary activity which has grown up around phenomena connected with charge or electron transfer at interfaces (Bockris and Nagy 1974). Some of the industries like caustic soda-chlorine, aluminum, electroplating, electrowining of metals, batteries etc., are based on electrochemical science and form the typical electrochemical industries – some of them being multibillion dollar industries.

i. The battery industry can cause lead, cadmium and mercury poisoning of water.
ii. The disposal of electroplating solutions can lead to water pollution especially due to cadmium, zinc, lead, copper, chromium and cyanides.
iii. Mercury which enters water from chlor-alkali mercury cells has been found to exceed tolerable limits (this type of industry is operated still in some places).
iv. Chlorides, carbonates etc. found in water will be the cause for corrosion of metals which can get included in water and corrosion is naturally an electrochemical reaction.

These will be dealt with in a more detailed manner in the following.

2.2.1 Pollution from Battery Industry

There will be a possibility of one or more of the eight metals like mercury, cadmium, nickel, lead, zinc, manganese, silver are contaminating the water from battery industry. At present mercury pollution due to battery industry is very limited or otherwise almost stopped. Metal pollutants like lead, cadmium and nickel are major contaminant from battery industry rather than zinc, manganese etc., in ground water.

2.2.2 Lead

Lead acid batteries are considered to have one of the fastest global growth rates [16–20]. The emerging new concept of zero emission pollution control in the West, has paved way for development of electric vehicles (EVs) and Hybrid Electric Vehicles (HEVs). Usage of lead acid battery is expected to grow further with technological advancements in the electric vehicles market. Rapidly developing markets within Asia-Pacific, particularly China and India, hold ample opportunities for future growth (Wang et al. 1998; Chen et al. 2012).

But in lead acid batteries, lead poisoning is one of the most ubiquitous environmental health threats to children and a significant contributor to occupational disease (Chen et al. 2012). Lead causes a range of symptoms ranging from the loss of neurological function to death depending upon the extent and duration of exposure. In children, moderate lead exposure is responsible for a significant decrease in school performance, lower standardized test scores (including IQ test scores), and is linked with hyperactive and violent behavior. Both children and adults can suffer from a range of illnesses including effects on the central nervous system, kidneys, gastrointestinal tract, muscle and joint pains, sleep problems, blood pressure, anemia, and the reproductive system in both men and women. Fetuses, infants, and children are especially vulnerable to lead exposure compared with adults, since lead is more easily absorbed into growing bodies. Also, the tissues of small children are more sensitive to the damaging effects of lead (Srianujata 1998).

Despite the potentially harmful effects of lead, it continues to be used in a wide variety of industries in the manufacture of thousands of products. But the battery industry is by far the principle consumer of lead, using an estimated 76 % of annual primary and secondary lead (mined and recycled metallic lead) production. Lead-acid batteries are primarily used in automobiles for Starting, Lighting and Ignition purposes (SLI), but are also used as back-up power supplies. Battery composition by weight is composed as follows: 50 % lead salts and oxides, 24 % acid, 17 % metallic lead, 5 % plastics, and 4 % ebonite and separators (Dagogo 1994).

Lead-acid batteries contain concentrated sulphuric acid and large amounts of lead. The acid is extremely corrosive and also a good carrier for soluble lead and lead particulate. If the acid leaks onto the ground, it may contaminate the soil and then the soil will become a source of lead particulate as the solution dries out and the lead becomes incorporated into soil particles which may be blown by wind or raised by vehicle transit. From the soil, the lead will go into ground water resulting the water will be contaminated with lead (Haiao et al. 2001).

2.2.3 Cadmium and Nickel

Cadmium is considered to be highly toxic and therefore its permitted concentration in water is limited to 5–10 ppb. In recent years cadmium has become of major environmental concern due to its introduction to natural water reservoirs. The main sources of cadmium contamination is Ni/Cd battery production which may deliver both cadmium and nickel to groundwater in untreated aqueous wastes, or because of uncontrolled- led disposal of used batteries (Friberg 1974).

2.2.4 Pollution from Metal Finishing Industry

The electroplating or metal finishing industry uses a wide variety of materials and processes to clean, etch, and plate metallic and nonmetallic surfaces to provide desired surface properties (Weiner 1963). The materials include solvents and surfactants for cleaning, acids and bases for etching, and solutions of metal salts and other compounds to plate a finish onto a substrate.

The electroplating or metal finishing industry has been playing a momentous role in the development and growth of numerous metal manufacturing and other engineering industries since the early part of this century. While electroplating operations have, in the course of time, become an essential and integral part of many engineering industries throughout the world, there has also been a steady growth of independent and small- to medium-scale electroplating industries, especially in the developing countries including those of South East Asia after World War II. The growth of these independent small-scale electroplating industries in the developing countries may be attributable to the growth of light and medium engineering industries which found it more convenient and economical to have their products metal plated by independent electroplaters (Nagendran and Parthasaradhy 1966).

In comparison with other industries, however, the electroplating industry uses much less water; hence the volume of the waste waters produced by this industry is also comparatively much smaller. Nevertheless, the waste waters are highly toxic in nature due to the presence of metals such as Ag, As, Be, Cd, Pb, Cr(VI), Cu, Zn and Ni etc., and of acids and the highly dangerous cyanides. The plating operations are preceded by cleaning and stripping of metal surfaces by alkali and/or other solvents and acids respectively. After plating is completed, the plated objects are rinsed with water. The cleaning, stripping and rinsing are the main sources of waste waters in an electroplating industry. The volume and characteristics of electroplating waste waters vary widely from industry to industry, depending on the operating practices and water conservation measures adopted in the industry.

Among the cations, most heavy metal compounds such as Ag, As, Be, Cd, Pb, Cr(VI), Cu, Zn and Ni are toxic to aquatic life. Ammonia would be another cation toxic to fish even in low concentrations. As far as anions are concerned, very low concentrations of cyanides, chromic acid and soluble chromates have been found to be toxic to aquatic life. However, nitrates and phosphates are anions coming under the heading of nuisance effects rather than toxicity (Drogen and Passek 1965).

2.2.5 *Mercury Pollution from Chlor-Alkali Industry*

While about 23 % of total mercury consumption in USA is in chlor-alkali production, the consumption of mercury in United Kingdom for chlor-alkali is 49 %. Since purer and concentrated caustic soda is obtained from the mercury cells, these cells are preferred. Until 2006, in India, the production capacity of caustic soda is 2.33 million tons per year. In this total capacity, 1.90 and 0.43 million tons per year is from membrane and mercury cell technologies respectively. In general, in India the membrane and mercury cell technology share is 82 % and 18 % respectively. Mercury losses from these cells have been estimated to be of the order of 0.25 Kg T^{-1} of chlorine. Mercury is lost from the cell in several phases: (a) through the caustic soda product; (b) through the overall cleaning of the mercury cell room; (c) through the brine solution (d) through the hydrogen; (e) through the brine saturation. Further, the amount of mercury (at the trace level) available in caustic soda varies depending on its degree of purification. These traces of mercury can be introduced into the human food cycle if the caustic soda produced by this method is used in processing of food e.g. peeling agent for tubers and fruits, a sour-cream-butter neutralizer, a reagent in the refining process for vegetable oils and animal fats as well as a reagent to adjust alkalinity in the canned vegetable industry. The safe limit for the mercury in water is 0.5 mg L^{-1} and 0.1 μg L^{-1} for India and Europe respectively.

Mercury is usually discharged into the environment as divalent mercury or elemental mercury and which has a solubility of about 0.025–0.03 mg L^{-1} in water at 25°C. Mercurous mercury is readily precipitated by sulphide ions and these are inert. But mercury organic complexes can be attacked by bacteria to give methyl mercuric compound (<pH 7) $(CH_3\ Hg)^+$ or dimethyl mercury (>pH 7) (CH_3-Hg-CH_3). We still have much to learn about quantitative aspects of the mercury cycle. Methyl mercury can enter lake and river food chain, from water to plants to fish, to bird and to human beings, becoming concentrated at each step. Methyl mercury is particularly toxic to animals, because it can readily pass the blood–brain barrier, causing injury to cerebellum and cortex. The clinical symptoms of this damage are numbness, awkwardness or gait and blurring or vision. However, the dreaded problem of mercury pollution was tackled on a war footing in Japan with the sole objective of making it the 'zero' mercury process and it had already achieved a level of 0.5–0.6 g of mercury consumption per tonne of caustic soda. In view of this mercury menace, there is a big swing in favor of membrane or diaphragm cells for the production of chlor alkali (Ditri and Ditri 1977).

2.2.6 *Water Contamination by Corrosion*

Corrosion is most often considered to be an electrochemical process, and it contaminates the drinking water source in different ways.

Corrosion of Cementations Materials

Many drinking water distribution systems contain pollution creating materials such as cement concrete, asbestos-cement pipes and cement-asbestos-cement pipes and cement-mortar-lined steel and ductile iron pipes. There will be a possibility of asbestos fiber or sand or other inorganic materials contamination in the drinking water source by severe attack of CO_2 on cement-based pipe line materials. On the other hand, if there is no carbonate species, the leaching of the alkalinity from the cement can lead to increase in pH value of drinking water (Yu Peng and Korshin 2011; Elfstrom Broo et al. 1997).

Iron Contamination

In general failures of iron and steel pipes caused by corrosion are very rare due to the formation of protective scale on the pipes and reduce the rate of corrosion very low. But after certain period of time, corrosion will start. For example iron in contact with water, following reactions will occur,

$$\text{At anode: } Fe \rightarrow Fe^{2+} + 2e^- \tag{2.1}$$

$$\text{At cathode: } 2H_2O + 2e^- \rightarrow H_2 + 2OH^- \tag{2.2}$$

The cathodic Reaction (2.2) will generally occur slowly, but a faster alternative one will occur in the presence of oxygen and the following reaction will occur,

$$H_2O + 2e^- + {}^1\!/\!_2\, O_2 \rightarrow 2OH^- \tag{2.3}$$

For both cathodic reactions, two hydroxide ions will be produced and an alkaline condition will result near the cathode and make the water more alkaline. However, the ferrous ion can be further oxidized by dissolved oxygen and precipitate as ferric hydroxide (red mud),

$$2Fe^{2+} + 5H_2O + {}^1\!/_2\, O_2 \rightarrow 2Fe(OH)_3 + 4H^+ \quad (2.4)$$

and make the water become contamination. In neutral water dissolved oxygen is necessary for appreciable corrosion of iron. The corrosion rate will be higher as the relative motion of the water increases with respect to iron surface. Increased temperature can also increase iron corrosion. Two principal types of electrochemical corrosion cell are concern for the above corrosion process. The first results from a galvanic cell, which is due to the contact of two different metals (for example metal alloy). The other and often more important corrosion cell is the concentration cell. This cell involves a single metal, but different portions of the metal are exposed to different aqueous environments (Refait et al. 1998).

Heavy Metal Contamination

Lead may occur in drinking water by corrosion of lead plumbing or fixtures. Corrosion of plumbing is the greatest cause of concern. All type of water is corrosive to metal plumbing materials to some degree. Grounding of household electrical systems to plumbing may also exacerbate corrosion. Copper pipeline systems carrying water can be attacked by pitting corrosion. Sometimes the heavy metals come from the corrosion of internally unprotected metallic water pipelines. Heavy metals can cause a variety of adverse health effects when people are exposed to it at levels above maximum contamination level for relatively short period of time (Isaac et al. 1997).

Corrosion Due to Microorganisms

The role of microorganisms in the corrosion of metals in the pipeline systems has been recognized in recent times. Microorganisms may influence corrosion by affecting the rate of anodic or cathodic activity producing corrosive end product and making the water become contaminated. The pipe surface, joints, valves and gates provide a variety of niches for the growth of many different microorganisms that can enhance the corrosion of pipeline (Yu Peng and Korshin 2011; Elfstrom Broo et al. 1997).

Carbonate, Chloride as Comrades of Corrosion

Some of the ions like chromate and phosphate may help in preventing corrosion, the presence of other ions like carbonate, chloride, nitrate etc. in general and chloride in particular will accelerate the corrosion rate and resulting water will be contaminated with corrosion products (Yu Peng and Korshin 2011).

2.3 How Electrochemistry Is the Cure in Pollution?

There are increasing economic, social, legal, and environmental pressures to utilize the "best available technology" not entailing excessive cost and to aspire to "performance without pollution", i.e., "zero pollution processing". Electrochemical technology has an important role to play as part of an integrated approach to the avoidance of pollution, monitoring of pollution and process efficiency, cleaner processing and modern techniques for electrical energy storage and conversion (Rajeshwar and Ibanez 1997a; Genders and Weinberg 1992; Bockris 1972; Pletcher and Walsh 1993). Electrochemistry, with its unique ability to oxidize or reduce compounds at a well-controlled electrode potential or current and by just adding (at the anode) or withdrawing (at the cathode) electrons, offers many interesting possibilities in environmental pollution control. Anodic processes can be used to oxidize organic pollutants to harmless products and to remove toxic compounds from flue gases. Cathodic processes can remove heavy metal ions from wastewater solutions down to very low outlet concentrations (Rajeshwar and Ibanez 1997b; Simonsson 1997; Panizza and Cerisola 2005; Anglada et al. 2009; Chen 2004). Not only can the two electrodes of the electrochemical cell be used in purification processes, but the ion-selective membrane(s) that are often placed between the electrodes to have a selective transfer of only anions or cations can also. New electrodialytic processes using such membranes have been developed, which can solve a variety of environmental problems.

For the removal of pollutants there are several powerful tools based on principles of electrochemistry. viz,

- Electrodialytic processes
- Electrocoagulation and electroflotation
- Electro-remediation of soils
- Cathodic processes
- Anodic processes
- Electrochemical advanced oxidation processes

2.3.1 Electrodialytic Processes (EDP)

EDP is an electrochemical separation process in which ions are transferred through ion exchange membranes by means of a direct current (DC) voltage (Strathmann 1986). The principle that governs electrodialysis is an electrical potential difference across an alternating series of cation and anion exchange membranes between an anode and a cathode. The feed solution containing both positive and negative ions enters the membrane stack to which a voltage is applied, thus causing the migration of the ions toward their respective electrodes. The cation exchange membranes allow the transfer of cations but inhibit the transfer of anions. Conversely,

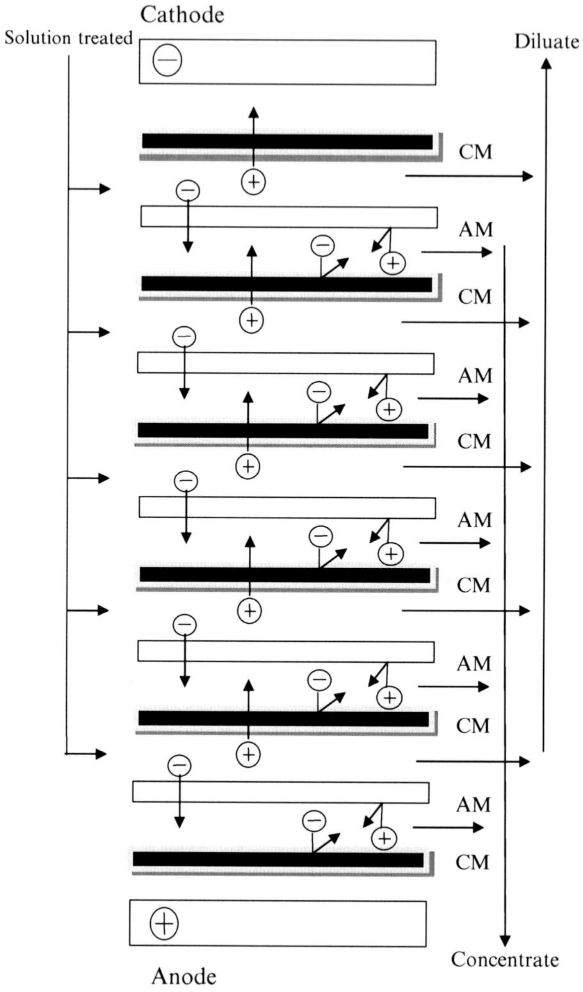

Fig. 2.1 The working principle of electrodialysis system. *AM* anionic membrane, *CM* cationic membrane

anion exchange membranes allow the transfer of anions but inhibit the transfer of cations. The result is alternating compartments containing streams of dilute ion concentration and streams rich in ion concentration exiting the stack (Fig. 2.1). The attractive characteristics of membranes used in electrodialysis applications include selectivity between ions of opposite charge, high ionic conductivity, low electrical conduction, long-term chemical stability, mechanical strength, and resistance to fouling. Polymer materials such as polystyrene, polyethylene, and polysulfone are often chosen for the membrane matrix and are often cross-linked to ensure stability.

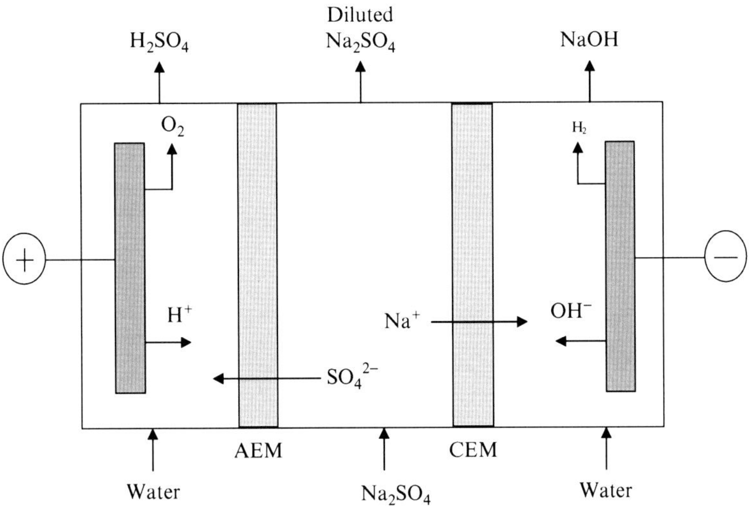

Fig. 2.2 Schematic representation of electrodialytic processes (three-compartment cell) for splitting of sodium sulfate into sodium hydroxide and sulfuric acid solutions. *AEM* anion exchange membrane, *CEM* cation exchange membrane

Fixed ionic moieties such as SO_3^-, COO^-, PO_3^{2-}, HPO^{2-}, AsO_3^{2-}, and SeO^{3-} are commonly used for cation exchange membranes, and NH^{3+}, RNH^{2+}, R_3N^+, R_2N^+, R_3P^+, R_2S^+, are common choices for anion-exchange membranes. Because of the corrosive nature of the anode compartments, electrodes are usually made of titanium and plated with precious metals.

In practice, it can be used in the separation and concentration of salts, acids, and bases from aqueous solutions, the separation of monovalent ions from multivalent ions, and the separation of ionic compounds from uncharged molecules. It can be used for either electrolyte reduction in feed streams or recovery of ions from dilute streams. Industrial applications encompass several industries and include the production of potable water from brackish water, removal of metals from wastewater, demineralization of whey, deacidification of fruit juices, and the removal of organic acids from fermentation broth. The best industrial important example is splitting of sodium sulfate solutions into sodium hydroxide and sulfuric acid solutions. The simplest design for this is to use a three-compartment cell with cationic and anionic selective membranes, as shown in Fig. 2.2. The main advantages are, very simple cell design, electrode products can be controlled, and concentrated solutions can be obtained. In addition, cation-selective membranes have a high selectivity and easy removal of metal deposit from cathode compartment is possible. The disadvantages are a low mass transfer coefficient, danger for poisoning, fouling, and slogging of membranes.

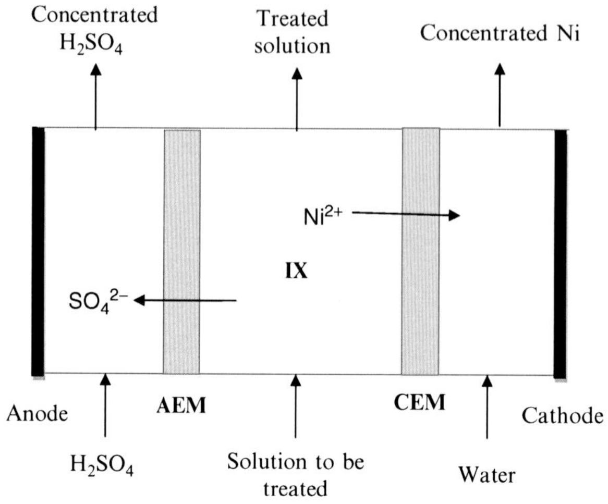

Fig. 2.3 Schematic representation of ion exchange assisted electrodialytic processes

2.3.2 *Ion Exchange Assisted Electrodialytic Processes*

Ion exchange assisted electrodialytic processes (Fig. 2.3) is a combination of electrodialysis and ion exchange technique (Rozhdetsvenska et al. 2001). This process is similar to electrodialysis with the exception that a bed of ion-exchange particles is placed in the centre (dilute) compartment. The presence of the ion-exchange bed increases the conductivity of the centre compartment and allows for improved separation. A schematic diagram is given in Fig. 2.3 for the purification of a dilute $NiSO_4$ solution. One of the best applications in industries with processes involving heavy metal electrolyte solutions; even selective removal of metal ions will be possible. The major advantages are the continuous regeneration without extra chemicals and the relatively high current efficiency. The method is capable of cleaning dilute process liquids. The disadvantages are the need of careful ion exchange resin, and choice of process parameters. In addition, the precipitation of solids lowers the efficiency of the process.

2.3.3 *Electrocoagulation and Electroflotation*

Electrocoagulation is the process of destabilising suspended, emulsified, or dissolved contaminants in an aqueous medium by introducing an electric current into the medium (Fig. 2.4) in order to produce the coagulation of colloidal particles (Vik et al. 1984; Cañizares et al. 2005; Holt et al. 2005; Lakshmanan et al. 2009; Kobya et al. 2010). In other words, electrocoagulation is the electrochemical production of destabilisation agents that brings about neutralization of electric charge for

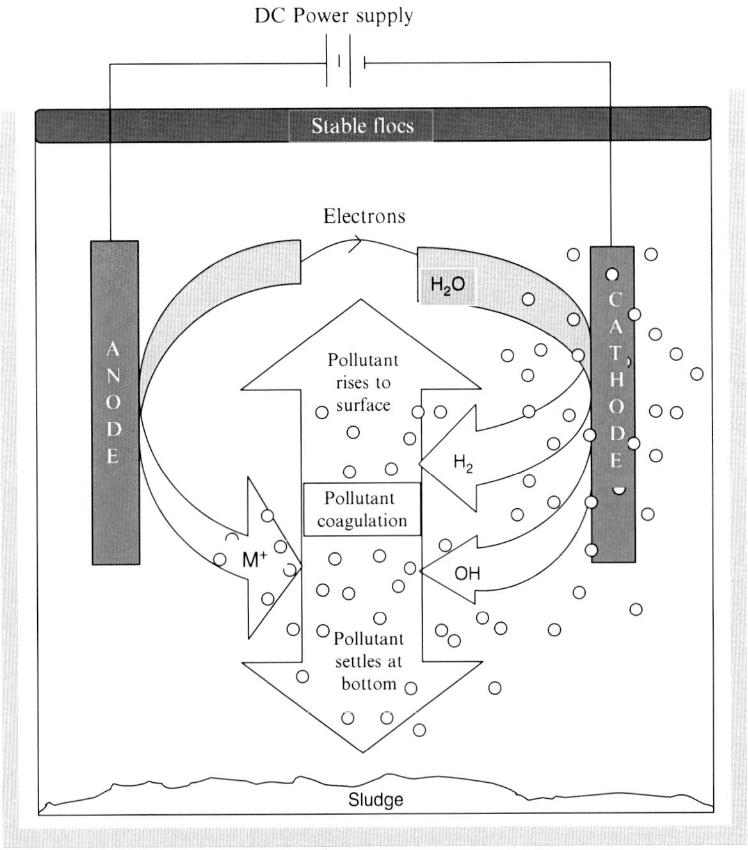

Fig. 2.4 The working principle of electrocoagulation process

removing pollutant. Once charged, the particles bond together like small magnets to form a mass. In its simplest form, an electrocoagulation unit is made up of an electrolytic cell with one anode and one cathode. The conductive metal plates/rods are, commonly known as sacrificial electrodes, made up of the same or different materials are used as the anode and cathode. Electrocoagulation was first proposed in the nineteenth century and the first electrocoagulation plant was successfully erected in 1889 at London for the treatment of sewage wastewater.

Electroflotation is a simple process that floats pollutants to the surface of a water body by tiny bubbles of hydrogen and oxygen gases generated from water electrolysis. Therefore, the electrochemical reactions at the cathode and anode are hydrogen evolution and oxygen evolution reactions, respectively. Graphite and lead dioxide are among the most common anode used in electroflotation. Dimensionally stable anodes (Ti/IrO_2, Ti/RuO_2, Ti/TiO_2-RuO_2 and Ti/Pt) are also used. In recent days highly effective (Ti/IrO_x-Sb_2O_5-SnO_2) is used and its life is predicted to be about 20 years (Shahjahan Kaisar Alam Sarkar et al. 2010).

Mechanism of Electrocoagulation

This process involves applying an electric current to sacrificial electrodes inside an electrolytic cell where the current generates a coagulating agent and gas bubbles. In detail, electrocoagulation is a technique involving the electrolytic addition of coagulating metal ions directly from sacrificial electrodes (anode). These ions coagulate with pollutants in the water, similar to the addition of coagulating chemicals such as alum and ferric chloride, and allow for easier removal of the pollutants by flotation and/or filtration. For example, in the case of aluminum, magnesium, zinc or iron electrodes, aluminum in the form of Al^{3+}, magnesium in the form of Mg^{2+}, zinc in the form of Zn^{2+} and iron in the form of Fe^{2+} will be produced by anodic dissolution during electrocoagulation process (Vasudevan et al. 2009a, b, 2011b). The dissolution reactions taking place at aluminum, magnesium, zinc and iron electrodes are given as follows,

$$Al \rightarrow Al^{3+} + 3e^- \tag{2.5}$$

$$Mg \rightarrow Mg^{2+} + 2e^- \tag{2.6}$$

$$Zn \rightarrow Zn^{2+} + 2e^- \tag{2.7}$$

$$Fe \rightarrow Fe^{2+} + 2e^- \tag{2.8}$$

Then, Al^{3+}, Mg^{2+}, Zn^{2+} and Fe^{2+} react immediately with hydroxide ions in solution to produce $Al(OH)_3$, $Mg(OH)_2$, $Zn(OH)_2$ and $Fe(OH)_2$ respectively. The reactions are as follows,

$$Al^{3+} + 3OH^- \rightarrow Al(OH)_3 \tag{2.9}$$

$$Mg^{2+} + 2OH^- \rightarrow Mg(OH)_2 \tag{2.10}$$

$$Zn^{2+} + 2OH^- \rightarrow Zn(OH)_2 \tag{2.11}$$

$$Fe^{2+} + 2OH^- \rightarrow Fe(OH)_2 \tag{2.12}$$

The metallic hydroxide was produced up to a sufficient concentration to initiate polymerization or condensation reactions. The polymerization reactions for aluminum and ferrous hydroxides are as follows,

$$Al(OH)_3 + Al(OH)_3 \rightarrow (OH)_2Al - O - Al(OH)_2 + H_2 \tag{2.13}$$

$$Fe(OH)_2 + Fe(OH)_2 \rightarrow (OH)\,Fe - O - Fe(OH) + H_2O \tag{2.14}$$

The polymeric complexes $[Al_2(O)(OH)_4]$ and $[Fe_2(O)(OH)_2]$ allows to remove pollutants from water mainly by either adsorption, surface complexation or by co-precipitation. The following reaction explains how the pollutant removed by surface complexation,

$$M - H_{(aq)} + (HO)OFe_{(S)} \rightarrow M - OFe_{(S)} + H_2O \tag{2.15}$$

where, M represents the metal pollutant act as ligand. In general, polymeric aluminum or iron complexeses can have both positive and negative charges capable of attracting the pollutants having opposite charges and remove them from water. For instance, hydrolysis of aluminum ions may generate negative and positive species (either monomeric or polymeric) such as monomeric species like $Al(OH)^{2+}$, $Al_2(OH)_2{}^{+}$, $Al_2(OH)_2{}^{4+}$, $Al(OH)_4{}^{-}$ and polymeric species like $Al_3(OH)_4{}^{5+}$, $Al_6(OH)_{15}{}^{3+}$, $Al_7(OH)_{17}{}^{4+}$, $Al_8(OH)_{20}{}^{4+}$, $Al_{13}O_4(OH)_{24}{}^{7+}$, $Al_{13}(OH)_{34}{}^{5+}$ over a wide pH range (Cañizares et al. 2005). Similarly, ferric ions generated by electrochemical oxidation of iron electrode may form both monomeric and polymeric species like $FeOH^{2+}$, $Fe_2(OH)_2{}^{4+}$, $Fe(H_2O)^{2+}$, $Fe(OH)^{4-}$, $Fe(H_2O)_3(OH)_3{}^{0}$, $Fe(H_2O)_6{}^{3+}$, $Fe(H_2O)_5(OH)^{2+}$, $Fe(H_2O)_4(OH)^{2+}$, $Fe_2(H_2O)_8(OH)_2{}^{4+}$, $Fe(H_2O)_6(OH)_4{}^{4+}$ and $Fe_2(H_2O)_6(OH)_4{}^{2+}$ over a wide pH range (Lakshmanan et al. 2009). These metallic hydroxides have a strong affinity for disperse particles. Apart from the above reactions water is also electrolyzed in a parallel reaction, producing small bubbles of oxygen at the anode,

$$2H_2O \rightarrow O_2 + 4H^+ + 4e^- \tag{2.16}$$

hydrogen at the cathode,

$$2H^+ + 2e^- \rightarrow H_2 \uparrow \tag{2.17}$$

and as a result, flotation of the coagulated particles occurs.

Electrode Materials

The materials employed in electrocoagulation processes are usually aluminum or iron. It is usually to use iron for wastewater treatment and aluminum for drinking water treatment because iron is relatively cheaper. The aluminum electrodes are also finding applications in wastewater treatment due to the high coagulation efficiency of aluminum hydroxide. But the main disadvantage in case of aluminum electrode is the residual aluminum (maximum contamination limit is 0.05–0.2 mg L^{-1}) present in the treated water due to cathodic dissolution. The excess aluminum present in the water creates health problems like cancer. To overcome this disadvantage of aluminum, in recent years alternate anodes like magnesium, zinc etc., are tried for electrocoagulation processes (Vasudevan et al. 2010a, b, c, d). When there are a significant amount of Ca^{2+} or Mg^{2+} ions in water, the cathode material is recommended to be stainless steel or galvanized iron.

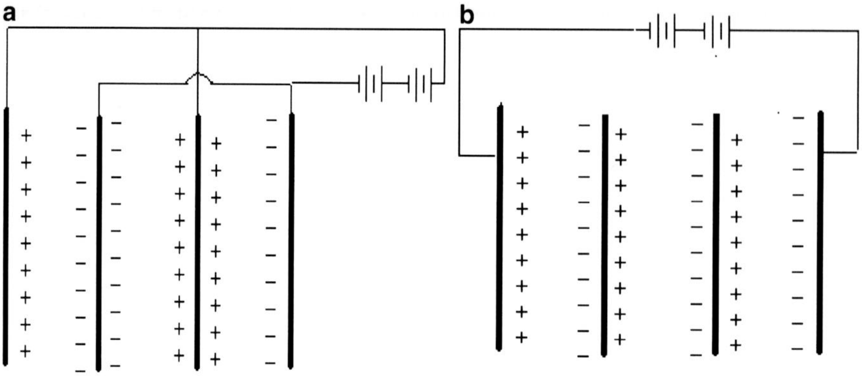

Fig. 2.5 Schematic illustrations of (**a**) Monopolar and (**b**) Bipolar configuration of cells

Electrode Configuration

There are two possible electrode configurations in the electrocoagulation process, which are monopolar and bipolar configuration as shown in the Fig. 2.5a, b. In the case of monopolar configuration, an electric potential (V) is applied between pairs of anodes and cathodes. The anodes and cathodes are arranged alternatively in the electrolytic cell pack. In this configuration, each unit/pair operates at the same voltage and the current being the sum of the individual unit currents. In the case of bipolar configuration, only two outer most electrodes physically connected to the power supply. Each electrode except for those at each end thus functions as anode on one face and as cathode on the other. Each unit is electrically in series with the others and the same current flows through every unit (Vasudevan and Lakshmi 2011, 2012b).

Effect of Operating Parameters

Current Density

Current density is an important factor which strongly influences the performance of electrocoagulation (Vasudevan and Lakshmi 2012a). The current density determines the electrocoagulant dosage rate at electrocoagulation processes. For example, in the case of aluminum, the electrochemical equivalent mass is 335.6 mg/Ah and for iron, the value is 1,041 mg/(Ah). Thus, current density should have a significant impact on pollutant removal efficiency. The amount of coagulant was determined from the Faraday law,

$$E_c = I\,t\,M/(Z\,F) \tag{2.18}$$

where I is current in A, t is the time (s), M is the molecular weight, Z is the number of electron involved in the reaction, and F is the Faraday constant (96,485.3 C/mol). The suggested current density for electrocoagulation process is 20–25 A/m^2 but the selection should be fixed with respect to other operating parameters like solution pH, temperature and flow rate.

Solution pH

It has been established that the influent pH is an important operating factor influencing the performance of electrochemical processes. The best pollutant removal is found near pH of 7. The power consumption is, however, higher at neutral pH due to the variation of conductivity. The solution pH after electrocoagulation treatment would increase for acidic influent but decrease for alkaline influent. This is one of the advantages of this process. The increase of pH at acidic condition was attributed to hydrogen evolution at cathodes. In fact, besides hydrogen evolution (Reaction 2.17), the formation of $Al(OH)_3$ near the anode (Reaction 2.9) would release H^+ leading to decrease of pH. In addition, there is also oxygen evolution reaction leading to pH decrease (Reaction 2.16). The pH decrease at alkaline conditions, is due to the formation of hydroxide precipitates with other cations and the formation of $Al(OH)_4^-$ in the solution (Vasudevan et al. 2012).

Solution Temperature

In general, all electrocoagulation units will be operated between 40°C and 60°C for better removal efficiency. But at the same time, when the temperature is too high, there is a shrink of the large pores of the electrocoagulant gel resulting in more compact flocs that are more likely to deposit on the surface of the electrode resulting low removal efficiency and higher energy consumption (Vasudevan et al. 2012).

Presence of Sodium Chloride

Sodium chloride is usually added to the water to increase the conductivity of the water or wastewater to be treated (Holt et al. 2005; Lakshmanan et al. 2009; Kobya et al. 2010). The presence of chloride ions could significantly reduce the adverse effect of other anions such as HCO_3^- and SO_4^{2-}. The presence of the HCO_3^- and SO_4^{2-} ions would lead to the precipitation of Ca^{2+} or Mg^{2+} ions that forms an insulating layer on the surface of the electrodes resulting increase in energy consumption. To overcome this difficulty, it is recommended that the addition of 15–20 % Cl^- avoids the precipitation and lower the energy consumption. Apart from the above the electrochemically generated chlorine was found to be effective in water disinfections.

Type of Current

In general, direct current (DC) is used in an electrocoagulation processes. In this case, an impermeable oxide layer may form on the cathode as well as corrosion formation on the anode due to oxidation. These prevent the effective current transport between the anode and cathode, so the removal efficiency of electrocoagulation processes declines. These disadvantages of DC have been overcome by adopting alternating current (AC) in electrocoagulation processes (Vasudevan et al. 2011a, c).

Applications

Electrocoagulation process has been widely used in textile wastewater, refractory oily wastewater, urban wastewater, restaurant wastewater, wasters loaded with phenol, laundry effluent. Likewise, it has been used to remove color, clay removal, removal of toxic metals, inorganic metals removal and drinking water treatment (Vasudevan et al. 2009a, 2010a, b, c, 2011a, b, 2012b).

2.3.4 How Electrochemistry Is the Cause and Cure for Soil Pollution?

One of the consequences of industrialization, in present case electrochemical industrialization, is that soil near industries tends to become contaminated. Such contamination occurs, for example, where a number of plants or industries like metal finishing and battery industries. These industries use inorganic and organic materials in their regular use. Metals like lead, chromium, cadmium, tin, arsenic, cyanide etc., are the very frequently used contaminants in these industries. Mismanagement allows the contaminating ions to reach the surrounding soils, where they gradually spread like a strain on a fabric. Electrochemical remediation is a promising remediation technology for soils contaminated with inorganic, organic, and mixed contaminants. In general, a direct-current electric field is imposed on the contaminated soil to extract the contaminants by the combined mechanisms of electroosmosis, electromigration, and/or electrophoresis.

Electrochemical remediation viz., electrokinetic technique (Fig. 2.6) has been widely used to remove inorganic and organic contaminants from wet solid matrix viz., soils and sludge (Yang and Liu 2001). This process involves the application of direct current across the contaminated soil or sludge through anode and cathode in the subsurface (Ko et al. 2000). The main electrode reactions are generally oxygen evolution at the anode and hydrogen evolution at the cathode. In this process the contaminants are removed by the combination of electroosmotic flow of the pollutant, electromigration of contaminants carrying ionic charges in the liquid phase and electrophoresis of the charged particles carrying contaminant on their surfaces. During electrokinetic process, the hydrogen and hydroxyl ions are

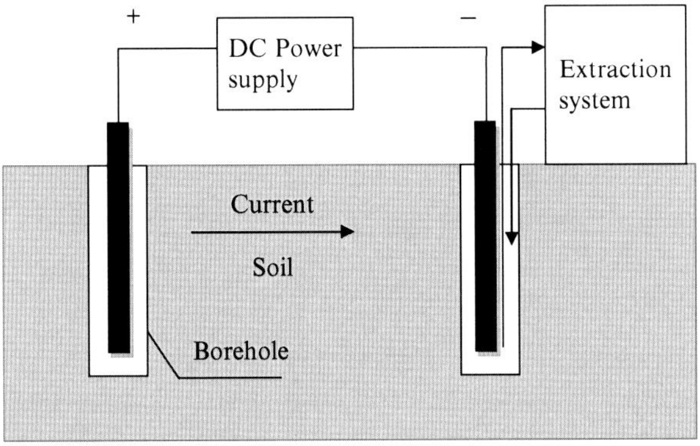

Fig. 2.6 Schematic illustration of electrochemical soil remediation

transported into soil by the applied direct-current resulting a low pH environment near the anode and a high pH environment near the cathode. The acid solution will release heavy metals and other cations that are sorbed on negatively charged clay surfaces, the cations migrating towards the cathode (Yang and Long 1999). The alkaline environment causes precipitation of most heavy metals and radionuclides, unless they form negatively charged complexes. These complexes migrate towards the anode and will form free metal ions again, when they meet the acid environment. In this way species are concentrated in the region where the two pH-environments meet and pH changes abruptly (Wang et al. 2005; Yang et al. 2005).

2.3.5 *Electrochemical Removal of Gaseous Pollutants*

Gaseous pollutants may also be removed electrochemically, provided that they are first dissolved into an electrolyte (Kreysa et al. 1994; van Velzen et al. 1990). The overall process will then generally consist of at least two steps: absorption of the gaseous species in a liquid and the subsequent electrochemical conversion of them to less harmful products.

Removal of SO_2 by Electrochemical Methods

For controlling problem of acid rains resulting essentially from sulphur dioxide emissions from fluestacks, especially in coal-fired thermal stations for power production two methods are available. In the first process, molten salt electrolysis of potassium pyrosulphate is envisaged and the second process utilizes Br_2^-.

In the first process, the cell comprises of an immobilized molten salt electrolyte viz, potassium pyropsulphate and porous gas diffusion electrodes formed a perovskite type.

The temperature of operation is 350–400°C at a current density of 5 A/dm^2. The cell voltage is <0.5 V with an energy consumption of 420 KWh/T of sulphurdioxide. The sulphur dioxide removal efficiency is 99 % However SO_3 is the byproduct gas at the cathode chamber.

This is a continuous one step operation with the following advantages.

1. Salable by product with no waste production
2. On-line temperature operation with no cooling or reheating of the flue gas.
3. Electricity is the only added reagent.

The Second process for removing sulphur dioxide from the flue gases uses anodically generated bromine from the electrolysis of hydrogen bromide in an undivided filter press reactor. The reactor consists of 34 anode of 1 m area each. The overall reaction of the process in the conversion of SO_2 and water is to form H_2SO_4

$$SO_2 + Br_2 + 2H_2O \rightarrow H_2SO_4 + 2HBr \tag{2.19}$$

The intermediate Br_2 is electrolytically regenerated in the membrane cell from HBr.

$$2HBr \rightarrow H_2 + Br_2 \tag{2.20}$$

Thus the advantage of this process is the production of the hydrogen as a byproduct in addition to value added H_2SO_4 the pilot facility treat flue gas having minimum SO_2 content of 0.016 % by volume.

Direct electrochemical oxidation creates blocking of electrodes by sulphur. Hence the new process developed consists in dissolving hydrogen suphide in sodium hydroxide and oxidizing electrochemically in two compartment cell with Nafion membrane and carbon electrodes (Current density: 10 A/dm^2) to get sulphur and hydrogen. The voltage of cell is 0.5 V and a current efficiency of 94 % is achieved.

Removal of CO_2 by Electrochemical Methods

In recent years the amount of carbon dioxide in the air has been increasing, which may adversely affect the environment in the future. Carbon dioxide is the largest contributor to the greenhouse effect. Moreover carbon dioxide is a very stable, linear molecule in which the oxygen atoms are week Lewis bases and carbon is electrophilic. In general, this carbon dioxide electroreduction is of interest as a potential component of a carbon energy cycle. The electroreduction of carbon dioxide at various metal electrodes yield many kinds of organic substances, namely CO, CH_4, C_2H_6, EtOH, other alcohols etc., four classes of metallic electrodes in aqueous electrolytes and three classes for non-aqueous media can be distinguished. Thus in aqueous solutions:

- Metallic In, Sn, Hg and Pb are selective for the production of formic acid
- Metallic Zn, Au and Ag produce carbon monoxide
- Metallic Cu exhibits a high electrocatalytic activity for the formation of hydrocarbons, aldehydes and alcohols
- Metallic Al, Ga and Group VIII elements (except Pd) show electrocatalytic activity in carbon dioxide reduction.

In non-aqueous electrolytes:

- On Pb, Tl and Hg the main product is oxalic acid
- On Cu, Ag, Au, In, Zn and Sn CO and carbonate ions are obtained
- On Ni, Pd and Pt are selective for CO formation
- On Al, Ga and Group VIII elements (except Ni, Pd and Pt) form both CO and oxalic acid.

In general, metal cathodes suffer corrosion or passivation problems that avoid its use as electrocatalysts, during long terms electrolysis. To overcome these difficulties, a number of metal transition complexes have been investigated as catalysts for the reduction of carbon dioxide.

2.3.6 *Anodic Processes*

Now-a-days, the presence of organic and inorganic pollutants are recalcitrant to treat through either chemical or biological treatments together with the stringent regulations imposed by new legislation have caused much research work to focus on waste-water treatment by electrooxidation processes. Electrochemical oxidation has been found to be an environmentally friendly technology able to mineralize completely non-biodegradable organic matter and to eliminate nitrogen species. Electrochemical oxidation of pollutants can take place through two different routes viz., (1) direct anodic oxidation, where the pollutants are destroyed at the anode surface; (2) indirect oxidation where a mediator (HClO, O_3, $H_2S_2O_8$, etc.,) is electrochemically generated to carry out the oxidation. It has to be kept in mind that during electro-oxidation of aqueous effluents, sometime both oxidation processes may coexist.

Direct Oxidation Pathways

In a direct oxidation pathway, electrooxidation of pollutants can occur directly on anodes by generating physically adsorbed "active oxygen" (adsorbed hydroxyl radicals, $^{\bullet}OH$) or chemisorbed "active oxygen" (oxygen in the oxide lattice, MO_{x+1}) (Comninellis 1994). The physically adsorbed "active oxygen" causes the complete combustion of organic compounds, and the chemisorbed "active oxygen" (MO_{x+1}) participates in the formation of selective oxidation products. In general, •OH is more effective for pollutant oxidation than oxygen in MO_{x+1}. Because oxygen evolution can also take place at the anode, high overpotentials for oxygen evolution is required in order to destroy organic pollutants to proceed with high current efficiency. Otherwise, most of the current supplied will be wasted to split water.

In direct anodic oxidation processes, the important part is the anode material. Different anode materials like granular graphite (Awad and Abuzaid 1997), planar graphite (Kannan et al. 1995), Pt or Ti/Pt (Marincic and Leitz 1978), PbO_2 (Sharifian and Kirk 1986), Ti/PbO_2 (Cossu et al. 1998), IrO_2 (Pulgarin et al. 1994), Ti/SnO_2–Sb_2O_5 (Cossu et al. 1998), Ebonex® (Chen et al. 1999), glassy carbon (Gattrell and Kirk 1990), Ti/RuO_2, Ti/Pt–Ir (Murphy et al. 1992), fiber carbon (Szpyrkowicz et al. 1994), MnO_2 (Rajalo and Petrovskaya 1996), Pt–carbon black (Boudenne and Cerclier 1999), porous carbon felt (Polcaro and Palmas 1997), reticulated vitreous carbon (Hofseth and Chapman 1999) and stainless steel (Abuzaid et al. 1999) were investigated for their suitability in the oxidation process.

Indirect Oxidation Pathways

The aim of indirect oxidation is to prevent fouling of electrode, avoiding direct electron exchange between the pollutant and the anode surface and to utilize the electrocatalytic potential of other anode materials. Therefore, the pollutants are oxidized through the mediation of some electrochemically generated redox reagents, which act as intermediaries for transfer of electrons between the electrode and the pollutant (Panizza and Cerisola 2009). Oxidation mediators can be metallic redox couples such as Ag(II/I), Ce(IV/III), Co(III/II), Fe(III/II), and Mn(III/II), or strong oxidizing chemicals such as active chlorine species, ozone, hydrogen peroxide, persulphate, percarbonate, and perphosphate.

Mediated Oxidation by Metallic Redox Couples

When metallic redox couples are applied as redox reagents the process is typically termed mediated electrochemical oxidation (MEO). Metal ions in acidic solutions are anodically oxidized from their stable oxidation state (M^{2+}) to the higher reactive oxidation state ($M^{(2+1)+}$) in which they attack the organic feed, degrading it into carbon dioxide, inorganic salts, and water:

$$M^{2+} \rightarrow M^{(2+1)+} + e^- \tag{2.21}$$

$$M^{(2+1)+} + R \rightarrow M^{2+} + CO_2 + H_2O \tag{2.22}$$

Reaction (2.22) returns the couple back to a stable state (M^{2+}) and it is then recycled through the cell in order continuously to regenerate further reaction. MEO through metallic redox reagents is mainly applied for pollution abatement in divided cells, where the anode and cathode are separated by a porous separator or an ion-exchange membrane in order to prevent reduction of the oxidative agent at the cathode. The MEO process is preferably used to treat solid waste or highly concentrated solutions with an organic content higher than 20 %, in order to avoid or limit the need for implementation of a solution-solution separation step to retain the key metallic ions from the discharged solution. For total organic oxidation, a redox couple with a high oxidation potential must be chosen as for instance the Ag(I/II) ($E^0 = 1.98$ V). Other redox couples which have been demonstrated to provide full mineralization of various organic pollutants are Co(III/II) and Ce(IV/III). Redox couples more abundant in natural waters as Fe(III/II) and Mn(III/II) are not suited for total oxidation of organics, because of their low oxidation potential ($E_0 = 0.77$ V and $E_0 = 1.56$ V respectively), but they can be used for selective partial oxidation of organic compounds (Panizza and Cerisola 2009).

Mediated Oxidation by In-Situ Generated Chemical Oxidants

Wastewater effluents containing low concentrations of organic pollutants (organic content less than 20 %) can be treated by indirect electrolysis generating in situ chemical reactants and converting them into less harmful products (Panizza and Cerisola 2009). The oxidizing chemicals can be electrogenerated either by anodic oxidation such as formation of active chlorine, ozone, and persulphate, or by cathodic reduction such as formation of hydrogen peroxide. The oxidants like hydroxyl radical ($H_2O/{}^{\bullet}OH$), ozone (O_2/O_3), peroxodisulphate ($SO_4{}^{2-}/S_2O_8{}^{2-}$), hydrogen peroxide ($H_2O/H_2O_2$), permanganate ion ($MnO_2/MnO^{4-}$), hypochlorous acid ($Cl^-/HOCl$), chlorine ($Cl^-/Cl_2$), dichromate ($Cr^{3+}/Cr_2O_7{}^{2-}$), oxygen ($H_2O/O_2$), hypochlorite ($Cl^-/ClO^-$) are used as oxidants and their respective standard reduction potentials are 2.80, 2.07, 2.01, 1.77, 1.67, 1.48, 1,36, 1.23, 1.23 and 0.81 V. Due to the natural abundance of chloride in most polluted waters, chlorine-mediated oxidation is the most widely employed in-situ generated chemical oxidant for wastewater treatment. The generation reactions are,

$$2Cl^- \rightarrow Cl_2 + 2e^- \tag{2.23}$$

$$Cl_2 + H_2O \rightarrow HOCl + H^+ + Cl^- \tag{2.24}$$

$$HOCl \leftrightarrow OCl^- + H^+ \tag{2.25}$$

The most used electrode materials for in-situ generation of active chlorine are based on platinum or on a mixture of noble metal oxides (e.g., RuO_2, TiO_2, and IrO_2) that have a good electrocatalytic properties for chlorine evolution as well as long-term mechanical and chemical stability.

Ozone is another strong chemical oxidant that can be electrochemically produced in-situ by anodic oxidation. Due to its high oxidation potential it has tremendous application in water treatment. It can be generated by,

$$3H_2O \rightarrow O_3 + 6H^+ + 6e^- \quad (2.26)$$

$$O_2 + H_2O \rightarrow O_3 + 2H^+ + 2e^- \quad (2.27)$$

Persulphate ($S_2O_8{}^{2-}$), percarbonate ($C_2O_6{}^{2-}$), and perphosphate ($P_2O_8{}^{4-}$) are various other strong oxidants produced electrochemically and used in water treatment. These oxidants are produced by anodic oxidation of the corresponding anions in the solution, but they are only generated efficiently by non-active anodes as the BDD and PbO_2 electrodes, due to the need of high oxygen overpotentials in order to reach the formation potentials.

$$2SO_4{}^{2-} \rightarrow S_2O_8{}^{2-} + 2e^- \quad (2.28)$$

$$2CO_3{}^{2-} \rightarrow C_2O_6{}^{2-} + 2e^- \quad (2.29)$$

$$2PO_4{}^{2-} \rightarrow P_2O_8{}^{4-} + 2e^- \quad (2.30)$$

Example for Anodic Process: Destroying Cyanide Waste

Destroying cyanide waste is an example for anodic oxidation for both direct and indirect oxidation pathways (Tissot and Fragniere 1994). Different cyanide compounds are commonly used in the metal finishing industry and it is absolutely essential to destroy the cyanide in the waste water disposal. One way of accomplishing the destruction of cyanide waste is to oxidize the cyanide electrolytically at high temperature (~80°C) for several days, thereby getting the harmless products of carbon dioxide and nitrogen as given by the equation:

$$2CN^- + 8OH^- \rightarrow 2CO_2 + N_2 + 4H_2O + 10e^- \quad (2.31)$$

However, in this process the complete destruction of the cyanide may be difficult especially towards the end when cyanide concentration is low. However, it is reported that the destruction of residual cyanides at low levels can be achieved by allowing sufficient treatment time (Tissot and Fragniere 1994).

In the indirect method, which is quicker, the cyanide is oxidized to harmless nitrogen and carbon dioxide by electrolyzing the waste along with sodium chloride solution of concentration 10–70 g/1 in a cell fitted with graphite anode and stainless steel cathode; the current density being 2–13 A/dm^2. The hypochlorite generated electrolytically oxidizes the CN^- as follows:

$$2CN^- + 5ClO^- + H_2O \rightarrow 2CO_2 + N_2 + 5Cl^- + 2OH^- \quad (2.32)$$

By this method, not only the cyanide but also the metal ion constituents of the wastes can be removed in one step. Complete destruction of CN^- is accomplished in the process and the heavy metal contaminants are deposited onto the cathode resulting nontoxic product for disposal.

Ozone is also used for the destruction of cyanide (Tissot and Fragniere 1994) and the ozonation of cyanide involves two separate chemical steps, oxidation of cyanide to cyanate and oxidation or hydrolysis of cyanate to nitrogen or ammonia.

$$CN^- + O_3 \rightarrow OCN^- + O_2 \tag{2.33}$$

$$OCN^- + OH^- + H_2O \rightarrow CO_3 + NH_3 \tag{2.34}$$

2.3.7 *Cathodic Processes*

Cathodic processes (Drogui et al. 2007) can be divided into two categories viz., electrodeposition and cathodic electrochemical dechlorination. Electrodeposition refers to electrochemical removal or recovery of metal contaminants from water and waste water. This technology is widely used either to remove heavy/toxic metals like Cd, Pb, Cu, Cr, etc., or recover precious metals like Pt, Au etc., from wastewater. The principle behind for the metal recovery is basically the cathodic reduction process. Metals are formed and deposited on cathode according to the following reaction,

$$M^{n+} + ne^- \rightarrow M_{(s)} \tag{2.35}$$

where 'M' represents the metal, 'n' represents the valence of the metal and '$M_{(s)}$' is the metal deposited on the cathode surface. Under some conditions, the system favors the formation of metal hydroxides instead of metal deposition due to local alkaline pH near the cathode surface. These metal hydroxides can be removed from the solutions by adsorption on the electrode as charged colloidal particles. The metal deposition is always depends on the thermodynamics and the kinetics of the system. Certainly, the ranking of the potential on the electrochemical series determines the relative reduction power of the species. For example, Cr, Pb, Cd, Zn and Cu are present in the wastewater, the metal deposition follows the electromotive force (EMF) series as: Cu > Pb > Cd > Cr > Zn. Apart from the EMF, other parametric conditions like metal concentration, electrode material and affinity existing between the pollutant influence the deposition. The main advantage of the electrodeposition is the deposited metals can be easily recycled by electrometallurgical process. The disadvantage of the process is the surface of the cathode is modified during electrodeposition and it needs additional process control.

Dechlorination of organic toxicants can occur not only by anodically but also cathodically, proved using graphite/carbon as electrode material. It was concluded that the dehalogenation results in a decreased toxicity and an increased biodegradability, thus enabling further biological treatment. It is concluded that, energy consumption and conversion rates are such that a technically and economically viable method for the detoxification of waste waters can be developed.

2.3.8 *Aerogel and Electrochemistry for Cleaning Contaminated Water*

The method called capacitive deionization (CDI), consumes only 16–32 Wh/gal. In CDI, contaminated water flows in a serpentine manner down stack of electrochemical cells. Attached to each side of the electrodes is a sheet of carbon aerogel, a highly porous material that has very large surface area. The aerogel is made by dipping a thin carbon sheet in a mixture of resorcinol and formaldehyde and then carbonizing at around 1,000°C in a vacuum or inert atmosphere. A prototype system has reduced salt concentration 100 mg/L <1 mg/L a single pass. The aerogel is regenerated by switching off the voltage, which put the ions back into solutions (Wang et al. 1993).

2.3.9 *Photo-Electrochemical Methods*

In recent years, photo-electrochemistry has led to a new and interesting possibility for treatment of pollutants from wastewater. In this case, suspensions of semiconductor particles (mostly TiO_2) can be used to harness the light with production of electrons and holes in the solid, which can destroy pollutants by means of reduction and oxidation, respectively. In this way, water containing organic, inorganic or microbiological pollutants can be effectively treated (Malato et al. 2009).

2.3.10 *Electrochemical Advanced Oxidation Processes (EAOPs)*

In the case of industrial and/or agricultural wastewater treatment, chemical oxidation is often necessary to remove organic matter (biodegradable or not) that consumes oxygen dissolved in water. The oxidations by ozone or hydrogen peroxide (H_2O_2) are methods used as complement or in competition with the activated carbon filtration or nanofiltration. But in some cases, the conventional oxidation is still inadequate and remains inefficient. Moreover, the conventional oxidation treatments as well as biological treatments prove ineffective against certain types of organic micropollutants. To eliminate these kind pollutants from water, more powerful processes, namely advanced oxidation processes (AOPs) have been developed. The AOPs are based on the in situ generation of strong oxidizing agent hydroxyl radicals ($^{\bullet}OH$) and their high oxidation power. The use of $^{\bullet}OH$ in the treatment/remediation of contaminated water is justified by a number of advantages: (i) they are not toxic (very short lifetime); (ii) they are simple to produce and use; (iii) they are very efficient to remove organic pollutants: the main feature of these powerful species is their ability to transform refractory inorganic compounds to biodegradable products; (iv) they are not corrosive to the equipment; (v) they do not induce secondary pollution: the final products of oxidation are CO_2 and H_2O, and inorganic ions (mineralization).

The term "AOPs" have been firstly defined by (Glaze et al. 1987) as "near ambient temperature and pressure water treatment processes which involve the generation of a very powerful oxidizing agent such as hydroxyl radical ($^{\bullet}OH$) in solution in sufficient quantity to effective water purification". This concept has been further extrapolated to electrochemical systems generating heterogeneous $^{\bullet}OH$ on the anode surface under the application of high current density. AOPs are applied whenever conventional oxidation techniques are insufficient, when process kinetics becomes very slow, or because contaminants are refractory to chemical oxidation in aqueous medium or partially oxidized yielding stable byproducts showing even greater toxicity than the starting pollutants. Indeed conventional oxidation techniques (use of chemical oxidants like H_2O_2, Cl_2, ClO_2, O_3, etc. . . .) prove ineffective against persistent organic pollutants (POPs) such as pesticides, polycyclic aromatic hydrocarbons (PAHs), synthetic dyes, polychlorobipenyles (PCBs), etc. (Pera-Titus et al. 2004). Contrariwise, the AOPs are able to oxidize any organic and organometallic substances until the final stage of oxidation, i.e., their transformation into CO_2, water and inorganic ions. They constitute a means to effectively treat effluents containing toxic and biorefractory pollutants. However, it must also be taken into consideration that the oxidation ability of most AOPs diminishes considerably when treating high organic matter contents (>5.0 g L^{-1}), thereby requiring the consumption of excessive amounts of expensive reactants that makes the treatment less cost affordable.

Electrochemistry constitutes one of the clean and effective ways to produce in situ hydroxyl radical ($^{\bullet}OH$), a highly strong oxidizing agent of organic matter in waters. Due to its very high standard oxidation power ($E^{o}(^{\bullet}OH/H_2O) = 2.80$ V/SHE), this radical species is able to react non-selectively with organic or organometallic pollutants yielding dehydrogenated or hydroxylated derivatives, which can be in turn completely mineralized, i.e., converted into CO_2, water and inorganic ions.

Recently, the electrochemical advanced oxidation processes (EAOPs) have received great attention by their environmental safety and compatibility (operating at mild conditions), versatility, high efficiency and amenability of automation (Brillas et al. 2009).

Anodic Oxidation

Anodic oxidation constitutes a direct way to generate hydroxyl radical by electrochemistry. This technique does not use any chemicals, hydroxyl radicals are formed by oxidation of water on high O_2 evolution overvoltage anodes (Pera-Titus et al. 2004; Martínez-Huitle and Brillas 2008; Oturan 2000a). Firstly Pt and PbO_2 anodes were investigated and on the obtained experimental results (Marselli et al. 2003) proposed the following mechanism for the oxidation of organics with concomitant oxygen evolution, assuming that both formation of $^{\bullet}OH$ and oxygen evolution take place on such anodes from the discharge of water (Reaction 2.36). Formed $^{\bullet}OH$ are heterogeneous since they are adsorbed on anode material (M) and react with organic matter (R) leading to its mineralization, i.e. transformation of R to CO_2 and H_2O (Reaction 2.37), when R does not contain heteroatom.

$$M + H_2O \rightarrow M(^{\bullet}OH) + H^+ + e^- \tag{2.36}$$

$$M(^{\bullet}OH) + R \rightarrow M + mCO_2 + nH_2O \tag{2.37}$$

Recently a new and powerful anode material, the boron-doped diamond (BDD) thin film anode appeared as an emerging anode material. This anode possesses very good properties for the electrochemical treatment of wastewaters contaminated by organic pollutants. It has a great chemical and electrochemical stability, a wide electrochemical working range. It has been shown that this new anode has an O_2 overvoltage much higher than that of conventional anodes such as Pt, PbO_2, doped SnO_2, and IrO_2 (Canizares et al. 2004). Thus the BDD anode is able to produce larger amount of $^{\bullet}OH$ physisorbed on anode surface, BDD($^{\bullet}OH$), from Reaction (2.36) that are more labile and reactive compared other anode materials, leading to a rapid and efficient destruction of organic pollutants. The effectiveness of the BDD anode was shown for the oxidation of a wide range of pollutants (Ozcan et al. 2008; Canizares et al. 2008) with almost mineralization of treated solutions. These studies were highlighted that the current efficiency is influenced by applied current density and initial pollutant concentration, low organic concentration and high applied current density favoring the mineralization degree.

Electro-Fenton Process

The second EAOPs is the electro-Fenton process in which hydroxyl radicals are generated indirectly via Fenton's reagent (mixture of H_2O_2 and iron(II)) in homogeneous medium. H_2O_2 is electrogenerated in-situ by 2-electron reduction of dissolved O_2 in acidic medium (Reaction 2.38) in presence of a catalytic amount of ferrous ions (Oturan et al. 2000). The most widely applied cathode materials for the effective H_2O_2 are based on carbonaceous materials (carbon felt, reticulated vitreous carbon (RVC), carbon sponges, carbon nanotubes (NT), or graphite).

$$O_2 + 2H^+ + 2\,e^- \rightarrow H_2O_2 \tag{2.38}$$

Compared to other chemical oxidant, H_2O_2 is a weak oxidant, but its oxidation power is enhanced in presence of Fe^{2+} ions (Fenton 1894) via Fenton's reaction:

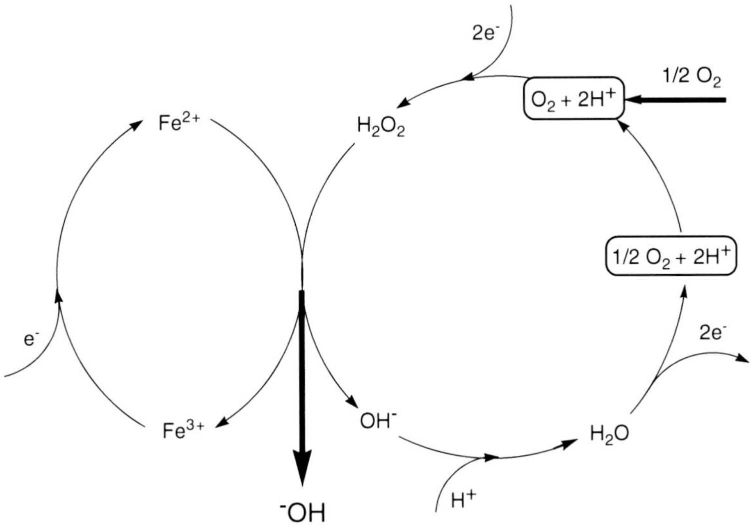

Fig. 2.7 Schematic representation of the electrocatalytic production of hydroxyl radicals by the electro-Fenton process (Oturan et al. 2000)

$$H_2O_2 + Fe^{2+} \rightarrow Fe^{3+} + OH^- + {}^{\bullet}OH \quad (2.39)$$

This enhancement is attributed to the formation of highly powerful oxidant hydroxyl radical via Reaction (2.39). Fe^{3+} formed by this reaction is reduced electrochemically at the cathode to regenerate ferrous iron in order to catalyse Fenton's reaction:

$$Fe^{3+} + e^- \rightarrow Fe^{2+} \quad (2.40)$$

The simultaneous generation of H_2O_2 and regeneration of ferrous iron at the cathode allow continuous and catalytic formation of hydroxyl radicals that react on organic pollutants by hydrogen atom abstraction or addition reactions:

$$\text{Organic pollutants} + {}^{\bullet}OH \rightarrow \text{oxidation intermediates} \quad (2.41a)$$

$$\text{Intermediates} + {}^{\bullet}OH \rightarrow\rightarrow\rightarrow CO_2 + H_2O + \text{inorganic ions} \quad (2.41b)$$

When the anode is Pt, the O_2 needed for the Reaction (2.38) to produce H_2O_2 is formed on the anode by oxidation of water (Oturan et al. 2001). Thus, the electro-Fenton process constitutes an overall catalytic system. Both catalytically cycle taking place during this system for the continuous regeneration of Fe^{2+} and H_2O_2 in order to produce hydroxyl radicals via Fenton reaction are schematized in Fig. 2.7.

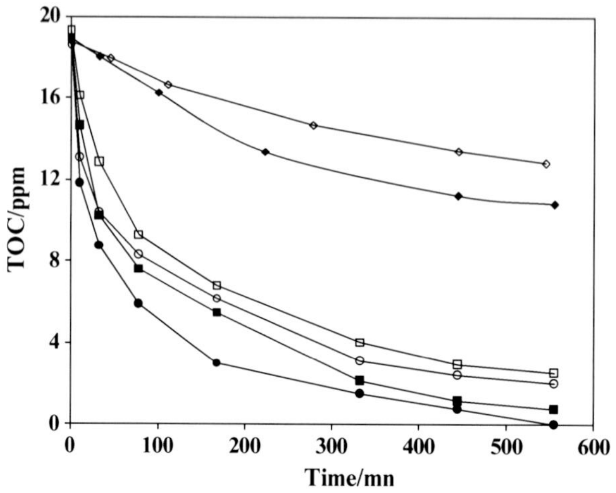

Fig. 2.8 Effect of pH and medium on TOC removal efficiency during degradation of 0.2 mM methyl parathion (insecticide) aqueous solution by electro-Fenton process at I = 150 mA. pH: 1 (◇); 3 (○) and 4 (□) in H_2SO_4 medium; pH: 1 (♦); 3 (•) and 4 (■) in $HClO_4$ medium (Diagne et al. 2007). *TOC* total organic carbon

Effect of Operating Parameter on the Process Efficiency

The electro-Fenton process is governed by a number of operating parameters, the most important being solution pH, applied current, catalyst concentration, medium (supporting electrolyte) and initial organic content.

Effect of Solution pH and Medium

The solution pH is one of key parameter in electro-Fenton process. Now the optimal value of this parameter is well known. The value of pH 3 was suggested firstly (Sun and Pignatello 1993) and confirmed thereafter by other groups (Diagne et al. 2007). This value is valid for the Fenton process and all related process including electro-Fenton process (Fig. 2.8). However electro-Fenton process can be carried out more or less effectively in the range of 2.5–3.5. It is suggested that iron ions (catalyst) are lost from the solution by precipitation for $pH > 4$. For $pH < 2.5$ the efficiency of the process decreases; the lower effectiveness can be attributed to the formation of the peroxonium ion ($H_3O_2^+$) from the following reaction (Feng et al. 2003) making H_2O_2 electrophilic, enhancing its stability and reducing its reactivity with Fe^{2+} in the Fenton reaction:

$$H_2O_2 + H^+ \rightarrow H_3O_2^+ \tag{2.42}$$

The lower pH values promote also the waste Reaction (2.43) and decreasing process efficiency (Tang and Huang 1996):

$$H_2O_2 + {}^{\bullet}OH \rightarrow HO_2{}^{\bullet} + H_2O \quad (2.43)$$

Figure 2.8 shows also the effect of the supporting electrolyte. $HClO_4$ is not a complexant for iron ions. When the electrolyte anions acts as complexing agent (such as $SO_4{}^{2-}$ or Cl^-) of iron ions, the free iron ion concentration decreases in the solution as in the case of H_2SO_4, leading to a decrease in the process efficiency.

Effect of Applied Current

The applied current (or current density) is another key parameter of electro-Fenton process, because the formation rate of H_2O_2 (Reaction 2.38) as well as the regeneration rate of ferrous iron (Reaction 2.40), and consequently the generation rate of $^{\bullet}OH$ are governed by the value of the applied current. In a general way, the process efficiency increases with applied current till a given value, since more $^{\bullet}OH$ are generated. Then the increase of applied current does not results better results because of enhancement of wasting reactions such as H_2 evolution at the cathode (Fig. 2.9) so the optimal current value should be determined experimentally for degradation kinetics or for mineralization efficiency before each trial (Özcan et al. 2010). As shown on Fig. 2.9a, the total disappearance of 0.5 mM (96 ppm) clopyralid is reached at 20 min under 300 and 500 mA applied current. Thus the optimal current value can be taken as 300 mA under operating conditions of Fig. 2.9. The Fig. 2.9b indicates the effect of initial organic matter on the mineralization efficiency. At the same experimental conditions the almost mineralization of 1.5 mM (288 mg L^{-1}) clopyralid solution was reached at 120 min while that of 3 mM (576 ppm) solution needed more longer time of 240 min at 300 mA applied current.

Effect of Catalyst Concentration

The catalyst concentration becomes an influent parameter particularly in the case of ferrous (Fe^{2+}) or ferric (Fe^{3+}) iron as catalyst. As shown on the Fig. 2.10 for the case of phenylurea herbicide fenuron, the degradation rate increases by decreasing the catalyst (Fe^{3+}) concentration from 0.1 to 1.0 mM (Oturan et al. 2010). The treatment time needed for complete destruction of 0.2 mM fenuron is three fold higher in the case of 1.0 mM catalyst concentration compared to 0.1 mM. This behavior can be explained by the parasitic reaction between Fe^{2+} and $^{\bullet}OH$.

$$Fe^{2+} + {}^{\bullet}OH \rightarrow Fe^{3+} + OH^- \quad (2.44)$$

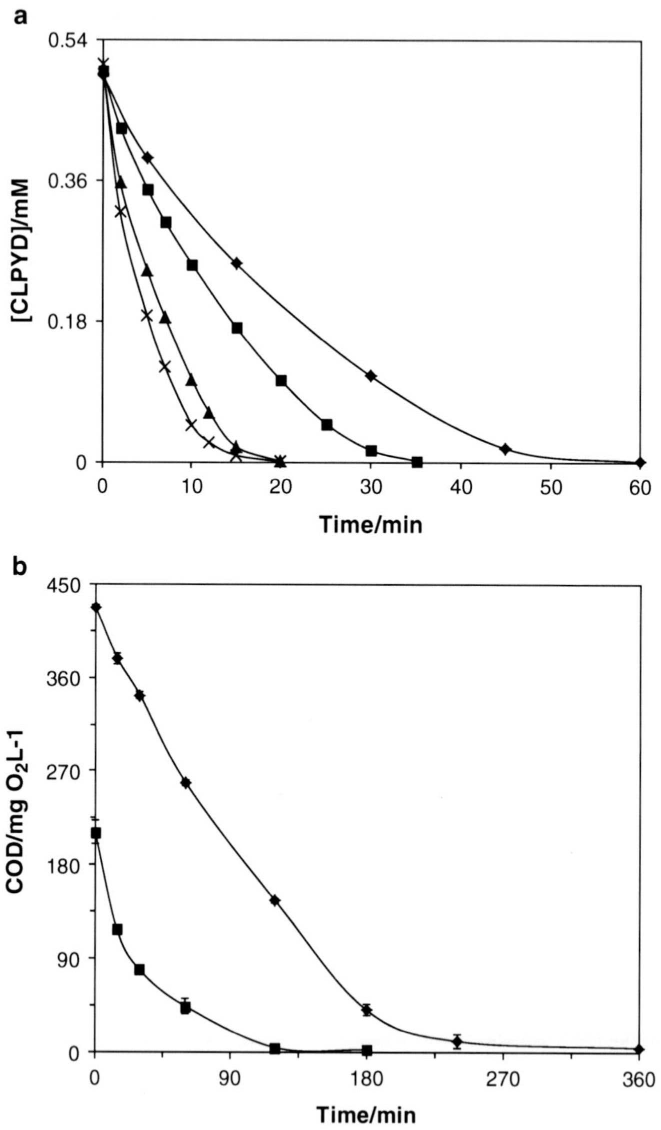

Fig. 2.9 Oxidative degradation/mineralization of 0.5 mM Clopyralid (systemic herbicide) aqueous solution by electro-Fenton process. (**a**) effect of applied current on degradation kinetics, I: 60 (♦), 100 (■), 300 (▲) and 500 (×) mA, and (**b**) Time course of COD removal for initial concentrations of 1.5 mM (■) and 3.0 (♦) mM aqueous clopyralid solutions at I = 300 mA. Operating conditions: [Na_2SO_4]: 0.05 M, [Fe^{3+}]: 0.1 mM, V: 0.15 L, pH: 3, T: 35 °C

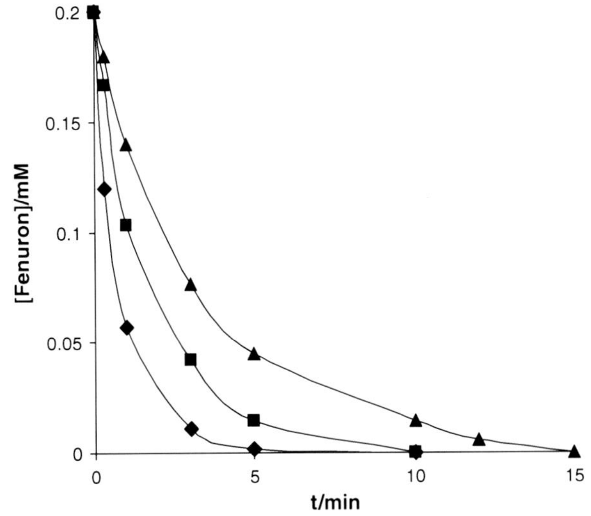

Fig. 2.10 Effect of the catalyst (Fe^{3+}) concentration on the degradation kinetics of 0.20 mM fenuron (phenylurea herbicide) solution during the electro-Fenton process at 100 mA and room temperature [243]. C_0 (mM): 0.2 (♦); 0.5 (■); 1 (▲)

Reaction (2.44) becomes competitive with organic pollutants to be treated for $^{\bullet}OH$ in the case of high ferrous iron concentration. In the case of ferric iron as catalyst, it is quickly reduced at the cathode to ferrous iron, so a high initial concentration of Fe^{3+} results in higher Fe^{2+} concentration in the solution leading to decrease of the process efficiency.

Application to the Removal of Organic Micropollutants from Water

Electro-Fenton process was successfully applied to removal of organic pollutants from water with high oxidation and/or mineralization efficiency, mainly by Oturan's and Brillas' groups. The mineralization of several organic pollutants such as pesticide active ingredients (Oturan et al. 1999; Özcan et al. 2008), pesticide commercial formulations (Oturan and Oturan 2005), synthetic dyes (Guivarch et al. 2003; Sirés et al. 2008; Marco and Oturan 2011), pharmaceuticals (Dirany et al. 2012), industrial pollutants (Oturan et al. 2000), landfill leachates (Trabelsi et al. 2012) etc., was thoroughly studied with almost mineralization efficiency in each case, showing that the electro-Fenton process can be constitutes an alternative when conventional treatment processes remain inefficient.

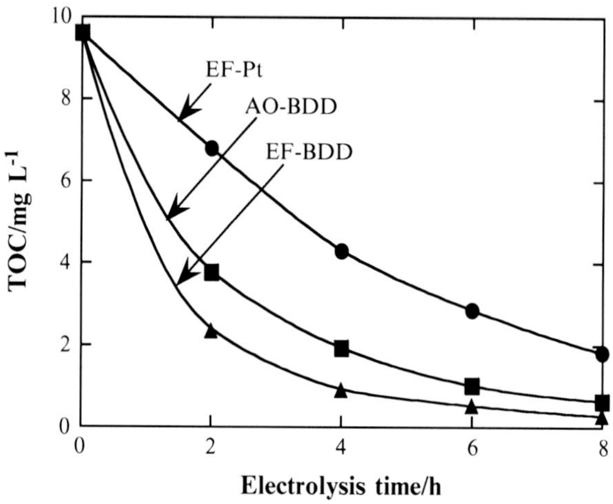

Fig. 2.11 TOC abatement in function of the electrolysis during the mineralization of 0.1 mM atrazine solution by electro-Fenton process using Pt or BDD and a carbon-felt cathode. *AO-BDD* anodic oxidation with BDD anode, *EF-Pt* electro-Fenton process with Pt anode, *EF-BDD* electro-Fenton process with BDD anode (Following the reference Oturan et al. 2012)

The performance of electro-Fenton process can be significantly enhanced by the replacement of the classical anode Pt by the emergent anode BDD (boron doped diamond). This configuration offers two main advantages: (i) generation of supplementary heterogeneous BDD($^{\bullet}$OH) by Reaction (2.36) in addition those produced homogeneously in bulk solution, (ii) high oxidizing power of BDD($^{\bullet}$OH) than others M($^{\bullet}$OH) since they are physisorbed, and (iii) direct oxidation of organics because of high oxidation window (about 3 V). A striking example of the performance of electro-Fenton with BDD anode (Oturan et al. 2012) concerning the mineralization of herbicide atrazine. A large variety of AOPs have been applied to remove atrazine from aqueous medium for avoiding its negative impact on ecosystems and human health. However, chemical, photochemical and photocatalytic AOPs yielded the persistent cyanuric acid (2,4,6-trihydroxy-1,3,5-triazine) as predominant by-product (De Laat et al. 1999). These studies reached to 40–60 % mineralization degree corresponding to the mineralization of side chains of atrazine. The anodic oxidation with a Pt anode converted atrazine into cyanuric acid (Mamián et al. 2009), whereas AO with a DSA anode allowed 46 % of total organic carbon (TOC) decay (Malpass et al. 2006). On the other hand, the electro-Fenton with BDD anode allowed not only almost total mineralization (97 %) of atrazine (Fig. 2.11), but also a high mineralization degree (90 % TOC removal) of cyanuric acid (Fig. 2.12) which is considered long ago as the end-products of atrazine mineralization and defined as refractory to $^{\bullet}$OH produced by AOPs. As can be seen on Fig. 2.12, electro-Fenton process with Pt anode leads only 4 % of TOC removal, evidencing the interest to use of BDD anode in electro-Fenton process.

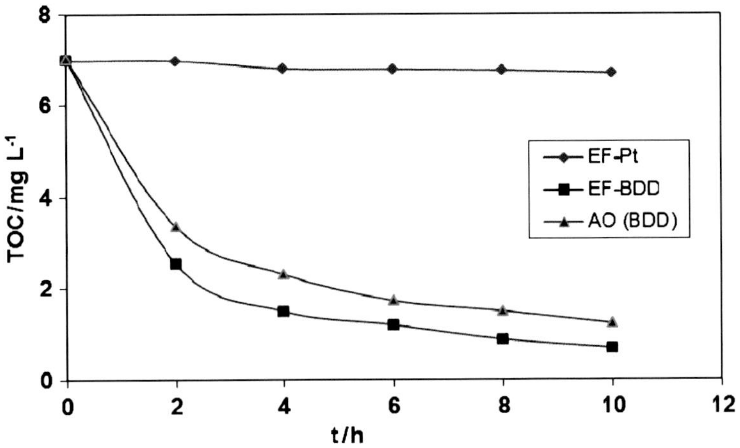

Fig. 2.12 Mineralization of cyanuric acid by anodic oxidation with BDD anode and electro-Fenton using Pt (EF-Pt) and BDD (EF-BDD) anodes. [cyanuric acid]0 = 0.2 mM (25.8 ppm), I = 1 A

Integrated Processes: Coupling of Electro-Fenton with Other AOPs

In order to enhance the efficiency of electro-Fenton process its coupling with other AOPs has been carried out. One of these integrated processes is the sonoelectro-Fenton in which the solution electrochemical reactor is simultaneously irradiated by ultrasounds (Oturan et al. 2008). Compared to the electro-Fenton sole, a significant synergetic effect was obtained by this coupling due to additional effect of enhancement of mass transfer rate to the electrode.

The most developed integrated process is the photoelectron-Fenton in which the electro-Fenton reactor is irradiated by an UV lamp (Ricardo et al. 2012). The interest is double: Firstly the generation of supplementary $^{\bullet}OH$ by irradiation of iron(III) $Fe(OH)^{2+}$ which is the predominant form at pH 3 (Reaction 2.45) and H_2O_2 (Reaction 2.46).

$$Fe(OH)^{2+} + hv \rightarrow Fe^{2+} + {}^{\bullet}OH \tag{2.45}$$

$$H_2O_2 + hv \rightarrow 2\,{}^{\bullet}OH \tag{2.46}$$

And secondly, the destruction of Fe(III)-carboxylic acid complexes (Reaction 2.47) resulting in ferrous ion regeneration and also in a significant decay in solution TOC content of the system due to the decarboxylation reaction (Oturan et al. 2011):

$$[Fe(RCO_2)]^{2+} + hv \rightarrow Fe^{2+} + CO_2 + R^{\bullet} \tag{2.47}$$

Recently solar photoelectro-Fenton process has attracted much attention (Ruiz et al. 2011). In this process, the electrical energy needed by electro-Fenton reactor was produced from solar energy in addition of natural solar irradiation of the solution. The process is particularly interesting for countries benefiting strong sunlight.

Peroxi-coagulation (or peroxi-electrocoagulation) constitutes another integrated method. This process involve an electro-Fenton like process using a sacrificial iron anode in order to generate iron(II) ions (Boye et al. 2003; Brillas et al. 2003) that react with catholically generated H_2O_2 following Reaction (2.38). Then Fenton reaction (Reaction 2.39) takes place and produces Fe^{3+} ions that induce the coagulation. Thus the pollutants are removed either by oxidation with $^{\bullet}OH$ produced from Fenton reaction (Reaction 2.39) and by coagulation.

2.4 Conclusion

Although electrochemistry, on one hand is the cause for pollution by the discharge of pollutants from electrochemical industries, its successful application to combat pollution supersedes the former and becomes really a cure in pollution prevention. Electrochemical technologies are vital and enabling discipline in many areas of environmental treatment, including: (a) clean synthesis, (b) monitoring of process efficiency and pollutants, (c) removal of contaminants, (d) recycling of pollutants, (e) water sterilization, and (f) clean energy conversion. The field of environmental electrochemistry has witnessed many progresses and has demonstrated many successes, as evidenced by the increasing literature. In this review, a variety of selected electrochemical processes for environmental protection of water, soil and air have been presented through electrodialytic processes, ion exchange assisted electrodialytic processes, electrocoagulation and electroflotation, electro remediation of soils, electrochemical removal of gaseous pollutants, anodic and cathodic processes, aerogel and electrochemistry for cleaning contaminated water, photo-electrochemical methods and electrochemical advanced oxidation processes. Finally, due to the specific advantages in a number of applications, electrochemical processes for the treatment of waste and prevention of pollution will find increasing acceptance in future developments.

Acknowledgments The author wishes to express their gratitude to the Director, Central Electrochemical Research Institute, Karaikudi, for publishing this article.

References

Abuzaid NS, Al-Hamouz Z, Bukhari AA, Essa MH (1999) Electrochemical treatment of nitrite using stainless steel electrodes. Water Air Soil Pollut 109:429–442. doi: 10.1155/2010/232378

Anglada A, Urtiaga A, Ortiz I (2009) Contributions of electrochemical oxidation to wastewater treatment: fundamentals and review of applications. J Chem Technol Biotechnol 84:1747–1755. doi: 10.1002/jctb.2214

Awad YM, Abuzaid NS (1997) Electrochemical treatment of phenolic wastewater: efficiency, design considerations and economic evaluation. J Environ Sci Health A 32:1393–1414. doi: 10.1080/10934529709376617

Ayoob S, Gupta AK (2006) Fluoride in drinking water: a review on the status and stress effects. Crit Rev Environ Sci Technol 36:433–487

Ball P (2008) Water: water – an enduring mystery. Nature 452:291–292. doi: 10.1038/452291a

Bockris JOM (1972) Electrochemistry of cleaner environments. Plenum Press, New York

Bockris JOM, Nagy Z (1974) Electrochemistry for ecologists. Plenum Press, New York

Boudenne JL, Cerclier O (1999) Performance of carbon black-slurry electrodes for 4- chlorophenol oxidation. Water Res 33:494–504. doi: 10.1016/S0043-1354(98)00242-5

Boye B, Dieng MM, Brillas E (2003) Electrochemical degradation of 2,4,5-trichlorophenoxyacetic acid in aqueous medium by peroxi-coagulation. Effect of pH and UV light. Electrochim Acta 48:781–790. doi: 10.1016/S0013-4686(02)00747-8

Brillas E, Boye B, Banos MA, Calpe JC, Garrido JA (2003) Electrochemical degradation of chlorophenoxy and chlorobenzoic herbicides in acidic aqueous medium by the peroxi-coagulation method. Chemosphere 51:227–235. doi: 10.1016/S0045-6535(02)00836-6

Brillas E, Sirés I, Oturan MA (2009) Electro-Fenton process and related electrochemical technologies based on Fenton's reaction chemistry. Chem Rev 109:6570–6631. doi: 10.1021/cr900136g

Canizares P, Saez C, Lobato J, Rodrigo MA (2004) Electrochemical treatment of 2,4- dinitrophenol aqueous wastes using boron-doped diamond anodes. Electrochim Acta 49:4641–4650. doi: 10.1016/j.electacta.2004.05.019

Cañizares P, Carmona M, Lobato J, Martínez F, Rodrigo MA (2005) Electrocoagulation of aluminum electrodes in electrocoagulation processes. Ind Eng Chem Res 44:4178–4185. doi: 10.1021/ie048858a

Canizares P, Paz R, Saez C, Rodrigo MA (2008) Electrochemical oxidation of alcohols and carboxylic acids with diamond anodes – a comparison with other advanced oxidation processes. Electrochim Acta 53:2144–2153. doi: 10.1016/j.electacta.2007.09.022

Chakraborti D, Ghorai SK, Das B, Pal A, Nayak B, Shah BA (2009) Arsenic exposure through groundwater to the rural and urban population in the Allahabad-Kanpur track in the upper Ganga plain. J Environ Monit 11:1455–1459. doi: 10.1039/b914858m

Chen G (2004) Electrochemical technologies in wastewater treatment. Sep Purif Technol 38:11–41. doi: 10.1016/j.seppur.2003.10.006

Chen G, Betterton EA, Arnold RG (1999) Electrolytic oxidation of trichloroethylene using a ceramic anode. J Appl Electrochem 29:961–970. doi: 10.1023/A:1003541706456

Chen L, Xu Z, Liu M, Huang Y, Fan R, Su Y, Hu G, Peng X, Peng X (2012) Lead exposure assessment from study near a lead-acid battery factory in China. Sci Total Environ 429:191–198. doi: org/10.1016/j.scitotenv.2012.04.015

Comninellis C (1994) Electrocatalysis in the electrochemical conversion/combustion of organic pollutants for waste water treatment. Electrochim Acta 39:1857–1862. doi: org/10.1016/0013-4686(94)85175-1

Cossu R, Polcaro AM, Lavagnolo MC, Mascia M, Palmas S, Renoldi F (1998) Electrochemical treatment of landfill leachate: oxidation at Ti/PbO_2 and Ti/SnO_2 anodes. Environ Sci Technol 32:3570–3573. doi: 10.1021/es971094o

Dagogo DI (1994) Air and blood lead levels in a battery factory. Sci Total Environ 152:269–273. doi: org/10.1016/0048-9697(94)90317-4

De Laat J, Gallard H, Ancelin A, Legube B (1999) Comparative study of the oxidation of atrazine and acetone by H2O2/UV, Fe(III)/UV, Fe(III)/H2O2/UV and Fe(II) or Fe(III)/H2O2. Chemosphere 39:2693–2706. doi: 10.1016/S0045-6535(99)00204-0

Diagne M, Oturan N, Oturan MA (2007) Removal of methyl parathion from water by electrochemically generated Fenton's reagent. Chemosphere 66:841–848. doi: 10.1016/j.chemosphere.2006.06.033

Dirany A, Sirés I, Oturan N, Özcan A, Oturan MA (2012) Electrochemical treatment of the antibiotic sulfachloropyridazine: kinetics, reactions pathways, and toxicity evaluation. Environ Sci Technol 46:4074–4082. doi: 10.1021/es204621q

Ditri PA, Ditri FM (1977) Mercury contamination- a human tragedy. Wiley, New York
Drogen J, Passek L (1965) Continuous electrolytic destruction of cyanide waste. Plat Surf Finish 18:310–313
Drogui P, Blais JF, Mercier G (2007) Review of electrochemical technologies for environmental applications. Recent Patents Eng 1:257–272
Elfstrom Broo A, Berghult B, Hedberg T (1997) Copper corrosion in drinking water distribution systems – the influence of water quality. Corros Sci 39:1119–1132. doi: PII: SOOlo-938X(97)00026-7
Feng J, Hu X, Yue PL, Zhu HY, Lu CQ (2003) Degradation of azo-dye Orange II by a photoassisted Fenton reaction. Ind Eng Chem Res 42:2058–2066. doi: 10.1021/ie0207010
Fenton HJH (1894) Oxidation of tartaric acid in presence of iron. J Chem Soc 65:889–910
Friberg L (1974) Cadmium in the environment, 2nd edn., Chap. 2. CRC Press, Cleveland
Gattrell M, Kirk DW (1990) The electrochemical oxidation of aqueous phenol at a glassy carbon electrode. Can J Chem Eng 68:997–1003. doi: 10.1002/cjce.5450680615
Genders JD, Weinberg NL (1992) Electrochemistry for a cleaner environment. The Electrochemistry, New York
Glaze WH, Kang JW, Chapin DH (1987) The chemistry of water treatment processes involving ozone, hydrogen peroxide, and ultraviolet radiation. Ozone Sci Eng 9:335–352
Guivarch E, Trévin S, Lahitte C, Oturan MA (2003) Degradation of diuron by the electro-Fenton process. Environ Chem Lett 1:39–44. doi: 10.1007/s10311-002-0017-0
Haiao CY, Wu HD, Lai JS, Kuo HW (2001) A longitudinal study of the effects of long-term exposure to lead among lead battery factory workers in Taiwan (1989–1999). Sci Total Environ 279:151–158. doi: 10.1016/S0048-9697(01)00762-8
Hofseth CS, Chapman TW (1999) Electrochemical destruction of dilute cyanide by copper-catalyzed oxidation in a flow-through porous electrode. J Electrochem Soc 146:199–207. doi: 10.1149/1.1391587
Holt PK, Barton GW, Mitchell CA (2005) The future for electrocoagulation as a localised water treatment technology. Chemosphere 59:355–367. doi: org/10.1016/j.chemosphere.2004.10.023
Isaac RA, Gil L, Cooperman AN, Hulme K, Eddy B, Ruiz M, Jacobson K, Larson C, Pancorbo OC (1997) Corrosion in drinking water distribution systems: a major contributor of copper and lead to wastewaters and effluents. Environ Sci Technol 31:3198–3203. doi: S0013-936X(97)00185-5 CCC
Kannan N, Sivadurai NS, Berchmans LJ, Vijayavalli R (1995) Removal of phenolic compounds by electrooxidation method. J Environ Sci Health A 30:2185–2203. doi: 10.1080/10934529509376331
Ko SO, Schlautmann MA, Carraway ER (2000) Cyclodextrin-enhanced electrokinetic removal of phenanthrene from a model clay soil. Environ Sci Technol 34:1535–1541. doi: 10.1021/es990223t
Kobya M, Demirbas E, Dedeli A, Sensoy MT (2010) Treatment of rinse water from zinc phosphate coating by batch and continuous electrocoagulation processes. J Hazard Mater 173:326–334. doi: org/10.1016/j.jhazmat.2009.08.092
Kreysa G, Juttner K (1994) In: Lapique F, Storck A, Wragg AA (eds) Electrochemical engineering, energy. Plenum Press, New York, p 255
Kumar R, Singh RD, Sharma KD (2005) Water resources of India. Curr Sci 89:794–811
Lakshmanan D, Clifford DA, Samanta G (2009) Ferrous and ferric ion generation during iron electrocoagulation. Environ Sci Technol 3:3853–3859. doi: 10.1021/es8036669
Malato S, Fernandez-Ibanez P, Maldonado MI, Blanco J, Gernjak W (2009) Decontamination and disinfection of water by solar photocatalysis: recent overview and trends. Catal Today 147:1–59. doi: 10.1016/j.cattod.2009.06.018
Malpass GRP, Miwa DW, Machado SAS, Olivi P, Motheo AJ (2006) Oxidation of the pesticide atrazine at DSA (R) electrodes. J Hazard Mater B137:565–572. doi: 10.1016/j.jhazmat.2006.02.045

Mamián M, Torres W, Larmat FE (2009) Electrochemical degradation of atrazine in aqueous solution at a platinum electrode. Port Electrochim Acta 27:371–379. doi: 0872–1904

Marco P, Oturan MA (2011) Degradation of alizarin red by electro-Fenton process using a graphite-felt cathode. Electrochim Acta 56:7084–7087. doi: 10.1016/j.electacta.2011.05.105

Marincic L, Leitz FB (1978) Electro-oxidation of ammonia in waste water. J Appl Electrochem 8:333–345. doi: 10.1007/BF00612687

Mark Shannon A, Bohn W, Elimelech M, Georgiadis G, Marinas J, Mayes M (2008) Science and technology for water purification in the coming decades. Nature 452:301–310. doi: 10.1038/nature06599

Marselli B, Garcia-Gomez J, Michaud PA, Rodrigo MA, Comninellis C (2003) Electrogeneration of hydroxyl radicals on boron-doped diamond electrodes. J Electrochem Soc 150:D79–D83. doi: 10.1149/1.1553790

Martínez-Huitle CA, Brillas E (2008) Electrochemical alternatives for drinking water disinfection. Angew Chem Int Ed 47:1998–2005

Montgomery MA, Elimelech M (2007) Water and sanitation in developing countries: including health in the equation. Environ Sci Technol 41:17–24. doi: 10.1021/es072435t

Murphy OJ, Hitchens GD, Kaba L, Verostko CE (1992) Direct electrochemical oxidation of organics for wastewater treatment. Water Res 26:443–451. doi: 10.1016/0043-1354(92)90044-5

Nagendran R, Parthasaradhy NV (1966) Present trends in the treatment of brass buttons. Trans SAEST 13:6–12

Oturan MA (2000a) An ecologically effective water treatment technique using electrochemically generated hydroxyl radicals for in situ destruction of organic pollutants: application to herbicide 2,4-D. J Appl Electrochem 30:475–482

Oturan N, Oturan MA (2005) Degradation of three pesticides used in viticulture by electrogenerated Fenton's reagent. Agron Sustain Dev 25:267–270. doi: 10.1051/agro:2005005

Oturan MA, Aaron JJ, Oturan N, Pinson J (1999) Degradation of chlorophenoxyacid herbicides in aqueous media, using a novel electrochemical method. Pestic Sci 55:558–562. doi: 10.1002/(SICI)1096-9063(199905)55:5<558::AID-PS968>3.3.CO;2–8

Oturan MA, Peiroten J, Chartrin P, Acher AJ (2000) Complete destruction of p-nitrophenol in aqueous medium by electro-Fenton method. Environ Sci Technol 34:3474–3479. doi: 10.1021/es990901b

Oturan MA, Oturan N, Lahitte C, Trevin S (2001) Production of hydroxyl radicals by electrochemically assisted Fenton's reagent – application to the mineralization of an organic micropollutant, pentachlorophenol. J Electroanal Chem 507:96–102. doi: 10.1016/S0022-0728(01)00369-2

Oturan MA, Sires I, Oturan N, Perocheau S, Laborde JL, Trevin S (2008) Sonoelectro- Fenton process: a novel hybrid technique for the destruction of organic pollutants in water. J Electroanal Chem 624:329–332. doi: 10.1016/j.jelechem.2008.08.005

Oturan MA, Edelahi MC, Oturan N, El Kacemi K, Aaron JJ (2010) Kinetics of oxidative degradation/mineralization pathways of the phenylurea herbicides diuron, monuron and fenuron in water during application of the electro-Fenton process. Appl Catal B Environ 97:82–89. doi: 10.1016/j.apcatb.2010.03.026

Oturan MA, Oturan N, Edelahi MC, Podvorica FI, El Kacemi K (2011) Oxidative degradation of herbicide diuron in aqueous medium by Fenton's reaction based advanced oxidation processes. Chem Eng J 171:127–135

Oturan N, Brillas E, Oturan MA (2012) Unprecedented total mineralization of atrazine and cyanuric acid by anodic oxidation and electro-Fenton with a boron-doped diamond anode. Environ Chem Lett 10:165–170. doi: 10.1007/s10311-011-0337-z

Özcan A, Şahin Y, Oturan MA (2008) Removal of propham from water by using electro- Fenton technology: kinetics and mechanism. Chemosphere 73:737–744. doi: 10.1016/j.chemosphere.2008.06.027

Özcan A, Sahin Y, Koparal AS, Oturan MA (2008) Propham mineralization in aqueous medium by anodic oxidation using boron-doped diamond anode: influence of experimental param-

eters on degradation kinetics and mineralization efficiency. Water Res 42:2889–2898. doi: 10.1016/j.watres.2008.02.027
Özcan A, Oturan N, Şahin Y, Oturan MA (2010) Electro-Fenton treatment of aqueous clopyralid solutions. Int J Environ Anal Chem 90:478–486. doi: 10.1080/03067310903096011
Panizza M, Cerisola G (2005) Application of diamond electrodes to electrochemical processes. Electrochim Acta 51:191–199. doi: 10.1016/j.electacta.2005.04.023
Panizza M, Cerisola G (2009) Direct and mediated anodic oxidation of organic pollutants. Chem Rev 109:6541–6569. doi: 10.1021/cr9001319
Pera-Titus M, García-Molina V, Baños MA, Giménez J, Esplugas S (2004) Degradation of chlorophenols by means of advanced oxidation processes: a general review. Appl Catal B Environ 47:219–256. doi: 10.1016/j.apcatb.2003.09.01
Pletcher D, Walsh FC (1993) Industrial electrochemistry, 2nd edn. Blackie, London
Polcaro MA, Palmas S (1997) Electrochemical oxidation of chlorophenols. Ind Eng Chem Res 36:1791–1798. doi: 10.1021/ie960557g
Pulgarin C, Adler N, Peringer P, Comninellis C (1994) Electrochemical detoxification of a 1,4-benzoquinone solution in wastewater treatment. Water Res 28:887–893, org/10.1016/0043-1354(94)90095-7
Rajalo G, Petrovskaya T (1996) Elective electrochemical oxidation of sulphides in tannery wastewater. Environ Technol 17:605–612. doi: 10.1080/09593331708616424
Rajeshwar K, Ibanez JG (1997a) Environmental electrochemistry: fundamentals and applications in pollution abatement. Academic, London
Rajeshwar K, Ibanez J (1997b) Environmental Electrochemistry. Academic, San Diego
Refait PH, Abdelmoula M, Genin JMR (1998) Mechanisms of formation and structure of green rust one in aqueous corrosion of iron in the presence of chloride ions. Corros Sci 40:1547–1560. doi: PII] S9909!827X"87#99955!2
Ricardo S, Brillas E, Sires I (2012) Finding the best Fe^{2+}/Cu^{2+} combination for the solar photoelectro-Fenton treatment of simulated wastewater containing the industrial textile dye Disperse Blue 3. Appl Catal B Environ 115:107–116. doi: 10.1016/j.apcatb.2011.12.026
Roy Chowdhury T, Basu GK, Mandal BK, Biswas BK, Chowdhury UK, Chanda CR, Lodh D, Roy SL, Saha KC, Roy S, Kabir S, Zaman QQ, Chakraborti D (1999) Arsenic poisoning in the Ganges delta. Nature 401:545–546. doi: 10.1038/44056
Rozhdetsvenska L, Monzie I, Chanel S, Mahmoud A, Muhr L, Grévillot G, Belyakov V, Lapicque F (2001) Ion exchange-assisted electrodialysis for treatment of dilute copper-containing wastes. Chemie Ingenieur Technik 73:761. doi: 10.1002/1522-2640(200106)
Ruiz EJ, Hernandez-Ramirez A, Peralta-Hernandez JM, Arias C, Brillas E (2011) Application of solarphotoelectro-Fenton technology to azo dyes mineralization: effect of current density, Fe^{2+} and dye concentrations. Chem Eng J 171:385–392. doi: 10.1016/j.cej.2011.03.004
Shahjahan Kaisar Alam Sarkar Md, Evans GM, Donne SW (2010) Bubble size measurement in electroflotation. Minerals Eng 23:1058–1065. doi: org/10.1016/j.mineng.2010.08.015
Sharifian H, Kirk DW (1986) Electrochemical oxidation of phenol. J Electrochem Soc 133:921–924. doi: 10.1149/1.2108763
Simonsson D (1997) Electrochemistry for a cleaner environment. Chem Soc Rev 26:181–189. doi: 10.1039/CS9972600181
Sirés I, Oturan N, Guivarch E, Oturan MA (2008) Efficient removal of triphenylmethane dyes from aqueous medium by in situ electrogenerated Fenton's reagent at carbon-felt cathode. Chemosphere 72:592–600. doi: 10.1016/j.chemosphere.2008.03.010
Srianujata S (1998) Lead-the toxic metal to stay with human. J Toxicol Sci 23:237–240. doi: doi.org/10.2131/jts.23.SupplementII_237
Strathmann H (1986) Electrodialysis. In: Bungay PM, Lonsdale HK, de Pinho MND (eds) Synthetic membranes: science, engineering, and applications. Reidel, Dordrecht/Holland
Sun Y, Pignatello JJ (1993) Photochemical reactions involved in the total mineralization of 2,4-D by Fe3+/H2O2/UV. Environ Sci Technol 27:304–1696

Szpyrkowicz L, Naumczyk J, Zilio-Grandi F (1994) Application of electrochemical process for tannery wastewater treatment. Toxicol Environ Chem 44:189–202. doi: 10.1080/02772249409358057

Tang WZ, Huang CP (1996) 2,4-dichlorophenol oxidation kinetics by Fenton's reagent. Environ Technol 17:1371–1378. doi: 10.1080/09593331708616506

Tissot PT, Fragniere KM (1994) Anodic oxidation of cyanide on a reticulated three- dimensional electrodes. J Appl Electrochem 24:509–512. doi: 10.1007/BF00249850

Trabelsi S, Oturan N, Bellakhal N, Oturan MA (2012) Application of Doehlert matrix to determine the optimal conditions for landfill leachate treatment by electro-Fenton process. J Mater Environ Sci 3:426–433. doi: 10.1016/j.cej.2011.03.072, 2028–2508

van Velzen D, Langenkamp H, Moryouseff A (1990) HBr electrolysis in the Ispara Mark 13A flue gas desulphurization process: electrolysis in a DEM cell. J Appl Electrochem 20:60–68. doi: 10.1007/BF01012472

Vasudevan S, Lakshmi J (2011) Effects of alternating and direct current in electrocoagulation process on the removal of cadmium from water – a novel approach. Sep Purif Technol 80:643–651. doi: 10.1016/j.seppur.2011.06.027

Vasudevan S, Lakshmi J (2012a) Process conditions and kinetics for the removal of copper from water by electrocoagulation. Environ Eng Sci 29:563–572. doi: 10.1089/ees.2010.0429

Vasudevan S, Lakshmi J (2012b) Electrochemical removal of boron from water: adsorption and thermodynamic studies. Can J Chem Eng 90:1017–1026. doi: 10.1002/cjce.20585

Vasudevan S, Jayaraj J, Lakshmi J, Sozhan G (2009a) Removal of iron from drinking water by electrocoagulation: adsorption and kinetics studies. Korean J Chem Eng 26:1058–1064. doi: 10.2478/s11814-009-0176-9

Vasudevan S, Lakshmi J, Sozhan G (2009b) Studies on a Mg-Al-Zn alloy as an anode for the removal of fluoride from drinking water in an electrocoagulation process. Clean 37:372–378. doi: 10.1002/clen.200900031

Vasudevan S, Lakshmi J, Vanathi R (2010a) Electrochemical coagulation for chromium removal: process optimization, kinetics, isotherms and sludge characterization. Clean 38:9–16. doi: 10.1002/clen.200900169

Vasudevan S, Epron F, Lakshmi J, Ravichandran S, Mohan S, Sozhan G (2010b) Removal of NO3 – from drinking water by electrocoagulation – an alternate approach. Clean 38:225–229. doi: 10.1002/clen.200900226

Vasudevan S, Lakshmi S, Sozhan G (2010c) Studies relating to removal of arsenate by electrochemical coagulation: optimization, kinetics, coagulant characterization. Sep Sci Technol 45:1313–1325. doi: 10.1080/01496391003775949

Vasudevan S, Lakshmi J, Packiyam M (2010d) Electrocoagulation studies on removal of cadmium using magnesium electrode. J Appl Electrochem 40:2023–2032. doi: 10.1007/s10800-010-0182-y

Vasudevan S, Suresh Kannan B, Lakshmi J, Mohanraj S, Sozhan G (2011a) Effects of alternating and direct current in electrocoagulation process on the removal of fluoride from water. J Chem Technol Biotechnol 86:428–436. doi: 10.1002/jctb.2534

Vasudevan S, Lakshmi J, Sozhan G (2011b) Studies on the Al–Zn–In-alloy as anode material for the removal of chromium from drinking water in electrocoagulation process. Desalination 275:260–268. doi: 10.1016/j.desal.2011.03.011

Vasudevan S, Lakshmi J, Sozhan G (2011c) Effects of alternating and direct current in electrocoagulation process on the removal of cadmium from water. J Hazard Mater 192:26–34. doi: 10.1016/j.jhazmat.2011.04.081

Vasudevan S, Lakshmi J, Sozhan G (2012) Electrocoagulation studies on the removal of copper from water using mild steel electrode. Water Environ Res 84:209–219. doi: 10.2175/106143011X13225991083640

Vik EA, Carlson DA, Eikum AS, Gjessin ET (1984) Electrocoagulation of potable water. Water Res 18:1355–1360. doi: org/10.1016/0043-1354(84)90003-4

Wang J, Angnes L, Tobias H, Roesner RA, Hong KC, Glass RS, Kong FM, Pekala RW (1993) Carbon aerogel composite electrodes. Anal Chem 65:2300–2303. doi: 10.1021/ac00065a022

Wang JD, Soong WT, Chao KY, Hwang YH, Jang CS (1998) Occupational and environmental lead poisoning: case study of a battery recycling smelter in Taiwan. J Toxicol Sci 23:241–245. doi: org/10.2131/jts.23.SupplementII_241

Wang JY, Zhang DS, Stabnikova O, Jay JH (2005) Evaluation of electrokinetic removal of heavy metals from sewage sludge. J Hazard Mater B124:139–146. doi: org/10.1016/j.jhazmat.2005.04.036

Weiner R (1963) Effluent treatment in the metal finishing industry. Robert Draper Ltd, Teddington

Yang CCG, Liu YC (2001) Remediation of TCE contaminated soils by in situ EK-Fenton process. J Hazard Mater B85:317–331. doi: org/10.1016/S0304-3894(01)00288-6

Yang GCC, Long YW (1999) Removal and degradation of phenol in a saturated flow by in-situ electrokinetic remediation and Fenton-like process. J Hazard Mater 69:259–271. doi: org/10.1016/S0304-3894(99)00059-X

Yang L, Nakhla G, Bassi A (2005) Electro-kinetic dewatering of oily sludges. J Hazard Mater B125:130–140. doi: org/10.1016/j.jhazmat.2005.05.040

Yu Peng C, Korshin GV (2011) Speciation of trace inorganic contaminants in corrosion scales and deposits formed in drinking water distribution systems. Water Res 45:5553–5563. doi: 10.1016/j.watres.2011.08.017

Chapter 3
Heterogeneous Photocatalysis for Pharmaceutical Wastewater Treatment

Devagi Kanakaraju, Beverley D. Glass, and Michael Oelgemöller

Abstract Environmentally sustainable solutions for wastewater management including the improvement of water quality and water recycling are considered key priority areas globally. The challenges facing our water resources are unprecedented due to the presence of organic and inorganic pollutants derived from numerous anthropogenic activities. This situation has been further complicated by emerging persistent contaminants such as pharmaceuticals which possess low biodegradability and resistance to chemical and biological treatments. Excretion of pharmaceuticals and their metabolites via human waste, improper disposal and veterinary applications is recognized as the principal sources of pharmaceuticals ending up in various compartments of the environment. Heterogeneous photocatalysis using semiconductor titanium dioxide (TiO_2) has proven to be a promising treatment method for the degradation of pharmaceuticals.

Here, we review recent research concerning the application of TiO_2 photocatalysis for the removal of selected pharmaceuticals belonging to different therapeutic drug classes. These classes of pharmaceuticals were chosen based on their environmental prevalence and potential adverse effects. The highlighted conclusions from this review are that (1) TiO_2 photocatalysis may play a major role in the degradation of pharmaceuticals; (2) various factors such as catalyst loading, initial concentrations, and water matrices significantly influence both the degradation rate and kinetics of degradation; (3) mineralization remains incomplete, despite complete abatement of the parent pharmaceutical due to the formation of stable by-products; (4) structures or number of intermediates formed differ due to the variation of photocatalytic experimental parameters, and (5) laboratory-scale experiments with artificial pharmaceutical solutions, in particular single compounds, are more common than pilot-scale or real wastewater samples. The main conclusion is

D. Kanakaraju • B.D. Glass • M. Oelgemöller (✉)
School of Pharmacy and Molecular Sciences, James Cook University, Townsville, QLD 4811, Australia
e-mail: kdevagi@frst.unimas.my; beverley.glass@jcu.edu.au; michael.oelgemoeller@jcu.edu.au

E. Lichtfouse et al. (eds.), *Green Materials for Energy, Products and Depollution*, Environmental Chemistry for a Sustainable World 3, DOI 10.1007/978-94-007-6836-9_3,

that the use of heterogeneous photocatalysis can be considered a state-of-the-art pharmaceutical wastewater treatment. Further studies are needed to optimize the operating conditions for maximum degradation of wastewater containing multiple pharmaceuticals under realistic conditions and on industrial scales.

Keywords Pharmaceutical pollutants • TiO_2 photocatalysis • Wastewater • Antibiotic • Analgesic • Drugs • Advanced oxidation processes

Contents

Abbreviations

AAS	Atomic absorption spectrometry
AMX	Amoxicillin
AN	Analgesic
AOP	Advanced oxidation processes
API	Active pharmaceutical ingredient
BOD	Biochemical oxygen demand
CBZ	Carbamazepine
CIP	Ciprofloxacin
COD	Chemical oxygen demand
CPC	Compound parabolic collector
DCF	Diclofenac
DO	Dissolved oxygen
DOC	Dissolved organic carbon
DOM	Dissolved organic matter
E_g	Band gap energy
ESI-MS	Electrospray ionization mass spectrometry
FIA	Flow injection analysis
FQ	Fluoroquinolone
GC-MS	Gas chromatography mass spectrometry
HPLC	High performance liquid chromatography

IBP	Ibuprofen
IC	Ion chromatography
ICP-AES	Inductively coupled plasma atomic emission spectroscopy
IUPAC	International union of pure and applied chemistry
LC/ESI-MS-MS	Liquid chromatography-electrospray ionization tandem mass spectrometry
LC/ESI-TOF-MS	Liquid chromatography-electrospray ionization time-of-flight mass spectrometry
LC-MS	Liquid chromatography mass spectrometry
LED	Light emitting diode
L-H	Langmuir-Hinshelwood
Log K_{ow}	Octanol/water partition coefficient
MOX	Moxifloxacin
NHE	Normal hydrogen electrode
NOM	Natural organic matter
NOR	Norfloxacin
NPX	Naproxen
NSAIDs	Non-steroidal anti-inflammatory drugs
OTC	Oxytetracycline
OX	Oxolinic acid
QTOF-MS	Quadrupole time of light-mass spectrometry
SMX	Sulfamethoxazole
$t_{1/2}$	Half-life time
TC	Tetracycline
TOC	Total organic carbon
TOF-MS	Time-of-flight mass spectrometry
UPLC-ESI/MS	Ultra performance liquid chromatography-electrospray mass spectrometry
USEPA	United States environmental protection agency
UV	Ultraviolet
WWTP	Wastewater treatment plant

3.1 Introduction: Advanced Oxidation Processes

Since water is essential to life, it is not surprising that there are a number of issues related to this essential resource. Poor sanitation, water scarcity, deteriorating water quality, waterborne diseases and lack of clean water supply are all posing global challenges due to the rising demand by an increasing world population. Further pressure on water resources has resulted from the introduction of emerging recalcitrant contaminants or xenobiotics, such as pharmaceuticals, endocrine disruptors, surfactants and personal care products. These substances display much more complex characteristics and resistance to degradation. Numerous studies highlighting the prevalence of pharmaceuticals in water environments has increased

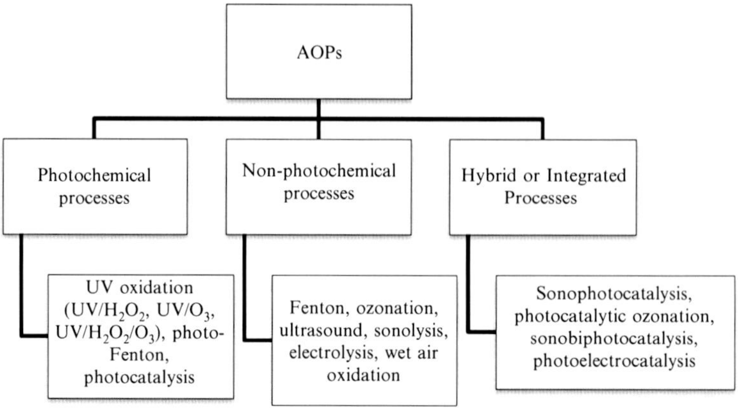

Fig. 3.1 Types of destructive advanced oxidation processes (AOPs) for wastewater purification

significantly since 2000, Heberer (2002a), Tixier et al. (2003), Fent et al. (2006), Choi et al. (2008) and Miège et al. (2009). This development has raised widespread concern of poorly known potential effects of pharmaceutical compounds and their metabolites on human and aquatic organisms, despite occurring in trace quantities ranging from ng/L to μg/L, Kümmerer (2009) and Mompelat et al. (2009).

Currently existing sewage treatment technologies can only be considered as a compromise for emerging micropollutants, although they have proven efficient for microbial, carbon, heavy metals and nitrogen removal. Complex pharmaceuticals with various physical and chemical properties in wastewater, lead to inadequate removal by existing wastewater treatment technologies which are not designed to handle this specific class of pollutants, Suárez et al. (2008). Incomplete removal of pharmaceutical residues clearly supports the urgent need for innovative technologies that can deal with their presence, either by improving existing or engaging in alternative technologies. Advanced Oxidation Processes (AOPs) are regarded to be appropriate to degrade pollutants which are known to be non-biodegradable or have low biodegradability, persistent, and possess high chemical stability. AOPs for water and wastewater treatment include photolysis and photocatalysis, ozonation, Fenton and photo-Fenton, ultrasound radiation, sonolysis, electrochemical oxidation, and wet air oxidation. These advanced technologies can be grouped into photochemical, non-photochemical and hybrid techniques (Fig. 3.1).

In principle, all AOPs are characterized by a common chemical feature known as reactive oxygen species which can react with non-biodegradable or recalcitrant compounds in water or wastewater, Dalrymple et al. (2007). Reactive oxygen or free radical species are strong oxidants required to initiate AOPs in order to mineralize pollutants to simpler and nontoxic molecules. These free radical species are based on atoms or molecules consisting of one or more unpaired electrons such as the hydroxyl radical ($HO^{\bullet}$), superoxide anion radical ($O_2^{\bullet -}$), hydroperoxyl radical ($HO_2^{\bullet}$) or alkoxyl radical ($RO^{\bullet}$). Of these, the $HO^{\bullet}$ has attracted the most attention. The appealing characteristics of $HO^\bullet$ radicals are its non-selective nature, high oxidation potential (2.8 V) when compared to other oxidants and its ability to react with a wide range of contaminants without any additives, with rate constants normally in the order of 10^6–10^9 mol L^{-1} s^{-1}, Andreozzi et al. (1999). When $HO^\bullet$ radicals reacts with organic molecules, they can either abstract a hydrogen atom or add to multiple bonds yielding oxidized intermediates or, in the case of complete mineralization, produce carbon dioxide, water and inorganic acids (3.1).

$$HO^\bullet + \text{Pollutant} \rightarrow CO_2 + H_2O + \text{inorganic acids} \quad (3.1)$$

During an ideal treatment by AOPs, contaminants are structurally altered forming smaller and more biodegradable compounds until complete mineralization is achieved. For the application of AOPs in wastewater treatment, the objectives can be either to (i) increase the biodegradability of wastewater before applying conventional biological process, (ii) reduce the level of toxicity and micropollutants in the effluent or (iii) disinfect the wastewater instead of applying traditional disinfection methods such as chlorination, which is known to generate carcinogenic and mutagenic by-products such as trihalomethanes and haloacetic acids, Rizzo (2011). AOPs have been proven to be efficient for the treatment of industrial wastewater; however, due to the high chemical oxygen demand (COD) levels in industrial wastewater, AOPs are only considered efficient for a low COD level of <5 g/L. When dealing with high levels of COD, techniques such as incineration or wet oxidation are preferred, Malato et al. (2002). Semiconductor material mediated photocatalysis in particular with titanium dioxide (TiO_2), is distinctive compared to other AOPs for the removal of persistent pollutants from wastewaters.

3.2 Heterogeneous Photocatalysis

The IUPAC defines photocatalysis as a "change in the rate of a chemical reaction or its initiation under the action of ultraviolet, visible or infrared radiation in the presence of a substance, the photocatalyst that absorbs light quanta and is involved in the chemical transformation of the reaction partners", Braslavsky (2007). Heterogeneous photocatalysis may be also termed as a process which uses a semiconductor metal oxide as catalyst and oxygen as an oxidizing agent, Andreozzi et al. (1999).

In 1972, Fujishima and Honda discovered the possibility of water splitting by means of a photoelectrochemical cell consisting of a rutile TiO_2 photoanode and a Pt counter electrode, Fujishima et al. (2008). This discovery opened ways for new applications such as air purification, self-cleaning surfaces application, organic synthesis, disinfection, and anti-cancer therapy. An interesting application of this breakthrough is in water purification. Water purification by means of illuminating TiO_2 was first proposed by Frank and Bard for cyanide and sulfite removal,

Fujishima et al. (2008). Since then, degradation of various organic target compounds such as dyes, pesticides and pharmaceuticals has been demonstrated.

A conventional heterogeneous photocatalysis process can be divided into five individual steps, Herrmann (1999):

i. Diffusion of the reactants from the bulk phase to the surface of the catalyst;
ii. Adsorption of at least one of the reactants;
iii. Reaction in the adsorbed phase;
iv. Desorption of the product(s);
v. Removal of the product(s) from the interface region.

An ideal photocatalyst should be chemically and biologically inert, photoactive, photostable, inexpensive, non-toxic, and should be excited with visible and near/or ultraviolet (UV) light. Despite the existence of various chalcogenide semiconductor photocatalysts (oxides and sulfides) such as zinc oxide (ZnO), zinc sulfide (ZnS), ferric oxide (Fe_2O_3), cadmium sulfide (CdS), cerium dioxide (CeO_2), tungsten trioxide (WO_3), tin dioxide (SnO_2) and titanium dioxide (TiO_2), none of these fulfils all the characteristics of an ideal photocatalyst. The choice of catalyst in photocatalytic studies is frequently narrowed to TiO_2, CdS and ZnO. Most studies involve TiO_2 as a semiconductor candidate despite its limitations, such as a low efficiency and a narrow light response factor, Leary and Westwood (2011). CdS and ZnO are known to undergo self-oxidation, which leads to lower photoactivity, and also the release of potentially dangerous metals such as Cd^{2+}, which makes these photocatalysts unattractive, Fox and Dulay (1993). Notably, ZnO has been reported to be a better photocatalyst in photocatalytic degradation than TiO_2 due to its broader absorption within the solar spectrum, Elmolla and Chaudhuri (2010a).

3.2.1 Mechanisms of TiO_2 Photocatalysis

Mechanistic processes of TiO_2 induced photocatalytic degradation of organic pollutants have been well described in the literature, Fox and Dulay (1993), Hoffmann et al. (1995), Gaya and Abdullah (2008), Chong et al. (2010) and Augugliaro et al. (2012). Photocatalysis occurs due to absorption of a photon with sufficient energy either equal or higher than the band gap energy (E_g) of the catalyst. The E_g represents the difference between the valence band (vB) and the conduction band (cB) of TiO_2. The vB and cB energies of TiO_2 are about +3.1 eV (vs. NHE) and −0.1 eV (vs. NHE), respectively. Consequently, the E_g between the vB and cB is 3.2 eV for anatase TiO_2. When TiO_2 is excited with UV light ($\lambda < 380$ nm for anatase TiO_2 which is more common or $\lambda < 400$ nm for rutile TiO_2), an electron (e^-) is promoted from the vB to the cB of TiO_2 generating a hole (h^+) in the vB and an electron (e^-) in the cB (3.2). The photogenerated e^--h^+ pair or charge carriers can either migrate to the surface of the TiO_2, where they can actively perform their roles in oxidation-reduction reactions with the adsorbed pollutant (P) or recombine in the

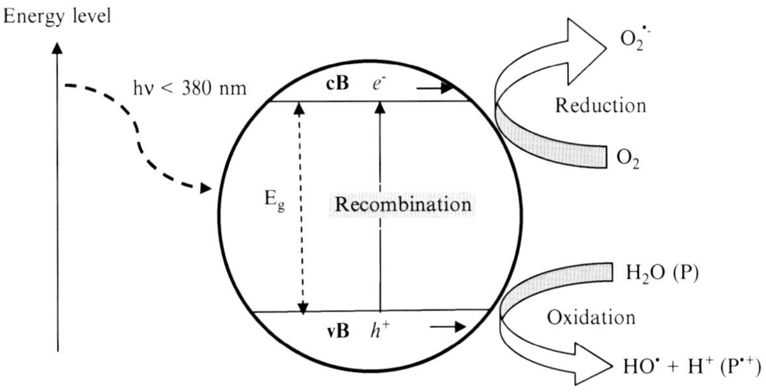

Fig. 3.2 Photogenerated e^--h^+ pair on the TiO_2 surface and oxidation-reduction reactions

bulk or on the surface of the TiO_2. The schematic representation of photogeneration of the e^--h^+ pair and of the subsequent oxidation-reduction is presented in Fig. 3.2.

$$\text{Excitation of TiO}_2\text{with photon } (\lambda \leq 380\text{ nm}): \ TiO_2 + h\nu \rightarrow e^-{}_{cB} + h^+{}_{vB} \quad (3.2)$$

A series of oxidation-reduction reactions may occur during TiO_2 photocatalysis (3.3, 3.4, 3.5, 3.6, 3.7, 3.8, 3.9, 3.10, and 3.11). A generated h^+ can react with an adsorbed water molecule, which is an essential process in TiO_2 photocatalysis, or reacts with OH^- anions to form powerful $HO^•$ radicals (3.3 and 3.4). The $HO^•$ radicals can subsequently oxidize water pollutants P to complete mineralization (3.5). When molecular oxygen is available, it is adsorbed onto the surface of TiO_2 and can scavenge an electron to form the superoxide anion radical ($O_2{}^{•-}$) (3.6).

$$h^+{}_{vB} + H_2O_{ads} \rightarrow HO^\bullet{}_{ads} + H^+ \quad (3.3)$$

$$h^+{}_{vB} + OH^-{}_{ad} \rightarrow HO^\bullet{}_{ads} \quad (3.4)$$

$$HO^\bullet{}_{ads} + P \rightarrow \rightarrow \rightarrow H_2O + CO_2 \quad (3.5)$$

$$(O_2)_{ads} + e^- \rightarrow O_2{}^{\bullet -} \quad (3.6)$$

Holes can also directly oxidize pollutants P by electron transfer (3.7 and 3.8).

$$h^+{}_{vB} + P_{ads} \rightarrow P_{ads}{}^{\bullet +} \quad (3.7)$$

$$HO^\bullet{}_{ads} + P_{ads} \rightarrow \text{intermediates/degradation products} \quad (3.8)$$

The photogenerated e^--h^+ pairs can undergo rapid recombination within nano seconds in the absence of electron scavengers, such as oxygen, releasing heat without favouring any reactions (3.9, 3.10, and 3.11). The $e^{-}{}_{TR}$ and $h^{+}{}_{TR}$ (3.9, 3.10, and 3.11) represent the surface trap valence band electron and conduction band hole, respectively.

$$\text{Charge carrier trapping}: \; e^{-}{}_{cB} \rightarrow e^{-}{}_{TR} \tag{3.9}$$

$$\text{Charge carrier trapping}: \; h^{+}{}_{vB} \rightarrow h^{+}{}_{TR} \tag{3.10}$$

$$e^{-}\text{-}h^{+}\text{recombination}: \; e^{-}{}_{TR} + h^{+}{}_{vB}\left(h^{+}{}_{TR}\right) \rightarrow e^{-}{}_{cB} + \text{heat} \tag{3.11}$$

Wavelengths between 300 and 400 nm (near UV range) are of interest in TiO_2 photocatalysis, which are provided either by artificial UV lamps or by a small section of the solar spectrum. Sunlight-induced photocatalytic processes have been the characteristic domain of TiO_2 as a photocatalyst candidate. Demonstration-scale solar treatments of emerging contaminants using compound parabolic collectors (CPC) underpin this, Pérez-Estrada et al. (2005) and Miránda-García et al. (2011). However, a limitation of solar photocatalysis is the poor overlap of the solar spectrum with the absorption spectrum of TiO_2 (<5 %). Thus, only 5 % of solar irradiation comprises the UV light that can be harvested by TiO_2 photocatalysis. TiO_2 doping with non-metals, metal-ion implantation, co-doping and sensitizing TiO_2 with dyes has been applied to address this limitation and to improve the TiO_2 photocatalytic efficiencies, Kumar and Devi (2011). The main objective of employing doping is to increase absorption in the visible region as it induces an optical response change i.e. decrease in band gap. One major drawback of doping is significant increase in the cost of the photocatalyst due to expense of ion implantation facilities, Zhang et al. (2009).

3.3 Titanium Dioxide: Structure and Properties

One most prevalent application of TiO_2 is as a white pigment in paints, plastics, paper, fibres, foods, pharmaceuticals, and personal care products mainly due to its light scattering properties and high refractive index. Worldwide production of TiO_2 powder has been reported as five million tons in 2005, with projections up to a 10 % increase by 2015, Skocaj et al. (2011). Additionally, production of nano-sized TiO_2 which permits more applications such as anti-fogging and self-cleaning coatings has also increased from 2,000 t in 2005 to 5,000 t in 2010, Weir et al. (2012). However, the concern on safety and health aspects which arise from the usage of nano-TiO_2 needs to be addressed. A recent study was conducted to assess the amount of TiO_2 in various food, personal care products and pharmaceuticals and predict adverse effects of TiO_2 nanoparticles on the environment, Weir et al. (2012).

TiO_2 exists in three crystalline forms namely anatase, rutile and brookite with the anatase form more commonly used as an active photocatalyst than pure rutile phase. In contrast, brookite TiO_2 is not commonly applied in photocatalysis. The anatase form tends to be the most photoactive and the most stable form at low temperatures (<700 °C). Anatase and brookite are however, thermodynamically metastable and they can be irreversibly converted to the rutile form, the most stable form at high temperatures, Silva and Faria (2009). Surface and structural properties such as crystal structure, surface area, particle size distribution, porosity, band gap and surface hydroxyl density control the efficiency of TiO_2 as photocatalyst. Of these, crystallinity and specific surface area have great effect on the photocatalytic activity of TiO_2.

The more common application of pure anatase compared to the pure rutile TiO_2 is attributed to a higher density of superficial hydroxyl groups, which leads to an improved capacity of anatase to adsorb oxygen, a larger specific surface area compared to pure rutile samples, and a lower recombination velocity of e^{-}-h^{+} pairs in anatase, Hoffmann et al. (1995) and Achilleos et al. (2010a). Combinations of anatase and rutile have demonstrated better photocatalytic activity due to the promotion of charge pair separation and inhibition of e^{-}-h^{+} recombination, Pelaez et al. (2012).

The most prominent TiO_2 photocatalyst, Degussa P25 (now known as AEROXIDE® TiO_2 P25) contains 80:20 anatase to rutile weight ratio and has demonstrated good performance in photocatalytic applications due to its large surface area of 50 m^2/g, with a particle size range of 20–30 nm. Although TiO_2 P25 has been widely used as a benchmark photocatalyst, its effectiveness has been limited by poor light absorption in the visible region, as a result of its large band gap.

The sol–gel method has been one of the widely used synthetic methods for the preparation of TiO_2 nanocrystalline forms. Several synthetic pathways are available for the preparation of the photocatalyst from precursor chemicals, for example, titanium alkoxide or titanium halogenide. The precursors are normally calcined at very high temperatures to obtain the desired crystal form and strong adhesion to surface materials or solid support such as activated carbon, zeolite, silica and glass. Hydroxyl groups from the catalyst surface and the support can react upon loss of a molecule of water, creating an oxygen bond thus increasing the adherence of the catalyst to the support during calcinations, Shan et al. (2010). The rate of hydrolysis is difficult to control due to the affinity of the TiO_2 precursors for water leading to poor modification of TiO_2's intrinsic properties such as surface structure, structural properties and porosity. Nitric acid, hydrochloric acid and complexing reagents such as oxalate and citrate have also been commonly used in the sol–gel synthesis. Weak acids such as acetic acid also exhibit excellent control of the hydrolysis rate of TiO_2, as they maintain pH due to the chelating effects of anions formed (e.g. acetate ion), generation of pH buffer, and stabilization of the sol, Shamaila et al. (2010).

3.4 Pharmaceuticals in the Environment

The United States Environmental Protection Agency's (USEPA) priority pollutants such as volatile organics, semivolatile, pesticides and polyvinylchloride biphenyls, metals and inorganics have been the main focus of water pollution over the past

30 years. However, in recent years, concerns in water pollution have shifted towards emerging contaminants. In particular, pharmaceutical and personal care products have become pollutants of interest due to their notable detection and potential detrimental impact on aquatic environments, Khetan and Collins (2007). Pharmaceuticals, classified as environmental emerging contaminants because of their endocrine-disrupting properties are either of natural origin or are produced synthetically, Oller et al. (2011).

The term 'pharmaceutical' or active pharmaceutical ingredient (API) for the purpose of this review refers to chemicals used for diagnosis, treatment, alteration, and prevention of disease, health conditions or functions of the human body and also includes veterinary pharmaceuticals and illicit or recreational drugs, Daughton and Ternes (1999). APIs are complex molecules with various functionalities and physicochemical and biological properties, balanced in terms of hydrophilicity and lipophilicity with molecular weights ranging from 200 to 1,000 Da, Kümmerer (2009). Pharmaceuticals are generally designed with high stability for their intended effects on humans and are eventually metabolised in the body by biochemical processes. They can be eliminated from the human body either after being partially or completely converted to water soluble metabolites or, in some cases, without being metabolized, Khan and Ongerth (2004). Their classifications are generally based on their therapeutic uses and those of environmental interest can be categorized into eight groups, namely non-steroidal anti-inflammatory drugs, antibiotics, beta-blockers, antiepileptics, blood lipid-lowering agents, antidepressants, hormones and antihistamines, Khetan and Collins (2007).

Inappropriate disposal and excretion of consumed medications by households, pharmaceutical manufacturers, hospitals, pharmacies and animal farming or aquaculture have been identified as the major sources of pharmaceuticals either as the parent drug or their metabolites in water courses (Fig. 3.3). Wastewater treatment plants (WWTPs) are thus a major source of the presence of this class of micropollutants in the environment. Various studies have reported on the discrepancies in efficiency and variation of WWTPs in the removal of pharmaceuticals due to their stability and differences in physicochemical properties. In fact, studies have also confirmed the detection of pharmaceuticals in effluents and influents from WWTP samples collected, in trace concentrations, Rodil et al. (2012).

A significant contribution of pharmaceuticals in surface water also comes from inappropriate disposal of expired or unused medications from households. Occurrence of pharmaceuticals in groundwater is related to a leaching process from landfills and soil and from sludge or manure spreading. Groundwater, being the extraction source of drinking water, has also been affected by pharmaceutical pollution, which can cause long term accumulation of trace amounts in humans. Finally, the accumulation of pharmaceuticals can be also expected in aquatic and terrestrial organisms via the food chain, Hoeger et al. (2005) and Deegan et al. (2011).

Monitoring studies conducted thus far have verified the occurrence of pharmaceutical in trace amounts in surface waters, groundwater and also sewage effluents, Andreozzi et al. (2003), Carballa et al. (2004), Lishman et al. (2006) and Moldovan

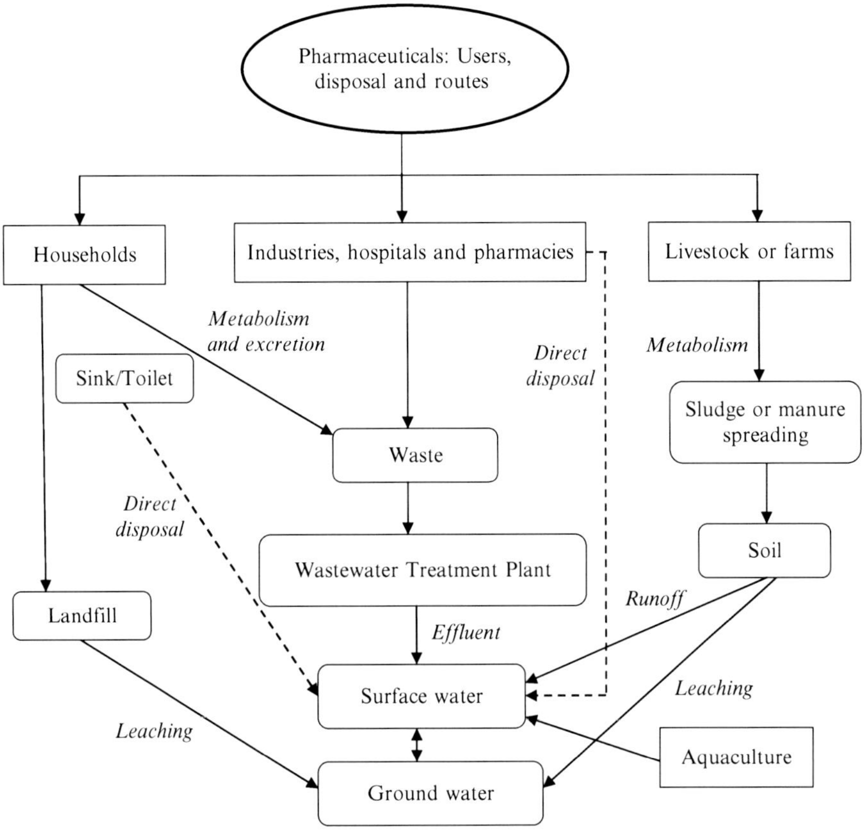

Fig. 3.3 Sources and routes of pharmaceuticals in water (Modified from Nikolaou et al. (2007) and Mompelat et al. (2009))

(2006). Moreover, concentrations of pharmaceuticals in the water environment have also been established based on their therapeutic use, Nikolaou et al. (2007), Pal et al. (2010) and Ziylan and Ince (2011). The occurrence of pharmaceutical residues in drinking and tap water has also been reported, Ternes et al. (2002) and Rodil et al. (2012). A study reported levels of numerous pharmaceuticals and endocrine disruptors in the finished water of 19 full-scale water utilities in the United States, Benotti et al. (2009).

Once released into the environment, APIs are exposed to different natural conditions and resulting transformations, which is not only affected by their physicochemical properties, but also their structure. These changes create further complexities and concerns regarding the identity, toxicity and potential risk of any new products, which may be formed.

Although studies are generally focused on a wide spectrum of pharmaceuticals, those frequently prescribed have gained importance amongst researchers.

Table 3.1 Relative annual per capita consumption (in mg/cap/y) of selected pharmaceuticals in Australia and some European countries, data taken from Carballa et al. (2008)

Pharmaceuticals	Australia (1998)	France (1998)	Germany (2001)	Spain (2003)	Sweden (2005)
Ibuprofen	720.6	2,841.0	1,553.4	6,391.2	7,864.3
Naproxen	1,159.9	646.2	n.a	986.1	n.a
Diclofenac	222.8	254.7	594.7	747.7	375.9
Sulfamethoxazole	n.a	382.9	571.4	294.0	160.4
Trimethoprim	136.8	n.a	n.a	85.6	n.a
Metoprolol	313.5	n.a	631.1	53.2	n.a
Carbamazepine	506.3	601.7	946.6	463.0	820.2
Diazepam	371.7	6.8	5.3	20.8	19.8
Bezafibrate	n.a	589.7	315.5	92.6	66.7

n.a not available

For example, acidic pharmaceuticals such as ibuprofen, ketoprofen, bezafibrate, and naproxen have been frequently monitored in receiving waters and WWTPs, Lindqvist et al. (2005) and Wang et al. (2010). Selected therapeutic classes such as antibiotics, NSAIDs and analgesics have also been reviewed, Kümmerer (2009) and Ziylan and Ince (2011). Accurate statistics on sales and consumption of pharmaceuticals present a challenge, due to difficulty in monitoring of prescription medication, in addition to non-prescription medicines, which can be obtained over-the counter or through the internet. Statistics on global consumptions and production of pharmaceuticals are not available in the literature. In addition, about 3,000 new APIs are introduced each year as a result of advancements in medicine, Beausse (2004).

The annual per capita consumption of selected pharmaceuticals in some European countries and Australia are presented in Table 3.1. The consumption patterns vary according to countries and demand or popularity of a drug. In Australia, 262 million prescriptions were dispensed via community pharmacies in 2008 which accounts for about 12 prescriptions per person, AIHW (2011). This figure represents an increase of 24 % over prescriptions dispensed in 2001. Among the mostly prescribed medicines are cholesterol-lowering drugs, antibiotics and drugs for the treatment of high blood pressure. Amoxicillin (3.9 %), an antibiotic, emerged in the top 10 most commonly prescribed medicines in 2009–2010 followed by paracetamol (3.2 %), an analgesic and cephalexin (3.2 %), a cephalosporin antibiotic. The presences of amoxicillin and cephalexin were confirmed in WWTP influents and surface water (ng/L) in a monitoring study of 28 antibiotics in watersheds of south-east Queensland. None of these drugs was, however, detected in drinking water, Watkinson et al. (2009).

Although the fate of many pharmaceuticals released into the environment remains unknown, three possible scenarios have been proposed: (i) the pharmaceutical is mineralized to carbon dioxide, water, and other inorganic ions or acids such nitrate, sulfate and phosphate ions, (ii) the pharmaceutical is not readily degradable and will be retained in sludge due to its lipophilic nature or (iii) the pharmaceutical will be metabolized to a more hydrophilic form than the parent compound, Halling-Sørensen et al. (1998) and Klavarioti et al. (2009).

The main concern from the presence of pharmaceuticals and their metabolites is the potential adverse effects that they may pose on human health and to other living organisms, which thus far remains unknown and unclear. Due to their complex physicochemical properties as a result of mixtures of various pharmaceuticals and other pollutants in the environment, advanced treatment such as TiO_2 photocatalysis undoubtedly can be applied for their degradation and a substantial amount of work has been published in this field, Abellán et al. (2007), Yurdakal et al. (2007), Giraldo et al. (2010) and Tong et al. (2012).

3.5 Degradation, Mineralization and Transformation Pathways of Pharmaceuticals Through Heterogeneous Photocatalysis

Positive developments in the photooxidation efficiency of TiO_2 based photocatalysis (UV/TiO_2) of pharmaceuticals have lead to an increasing interest in its application for the removal of micropollutants in spiked wastewaters and also real wastewater such as wastewater effluents, river water and drinking water. Determination of optimal operating conditions to yield the highest performance in terms of degradation and mineralization has been investigated in various heterogeneous photocatalytic studies on pharmaceuticals. The disappearance of the parent API over the treatment time has been the primary focus. However, studies have reported that kinetic assessment of APIs during UV/TiO_2 degradation can often be complex, due to the formation of various transformation products. Interferences between the disappearance of the parent API and newly formed transformation products may also occur. The non-selective nature of hydroxyl radicals contributes to this complexity, Oller et al. (2011).

Another general observation is that while high removal rates of the parent API can be accomplished by UV/TiO_2, mineralization remains incomplete in most cases. The degree of mineralization also tends to vary according to the nature of the water matrices. In general, a solution of the API prepared in the laboratory with distilled water or deionized water generally yields higher mineralization compared to that of doped effluent or influent samples from WWTPs, groundwater or river water.

The interference provided by the presence of organic matter such as humic and fulvic acids and radical scavengers such as carbonate species, HCO_3^- and CO_3^{2-} cannot be ruled out, as these species retard photocatalysis by competing for active sites and radiation attenuation. Organic matter present in water can be classified into two major classes namely dissolved organic matter and particulate organic matter. The existence of organic matter can be naturally occurring organic matter, such as humic and fulvic acids, or that derived from anthropogenic activities, Oppenländer (2003).

The ultimate aim of TiO_2 photocatalysis (and also for other AOPs) is threefold, to degrade pollutants, to reduce toxicity and to enhance the biodegradability of the treated wastewater. Global parameters such as total organic carbon (TOC), COD, biochemical oxygen demand (BOD), dissolved organic carbon (DOC) and BOD/COD ratio are also often measured. The ratio of BOD_5/COD is commonly used in this context and ratio values of above 0.4 are regarded as highly biodegradable in wastewater, Abellán et al. (2007). Toxicity assays are used to indicate the toxicity of photocatalytically treated wastewater and various types of bioassays based on microorganisms, plants and algae, invertebrates and fish have been applied to measure the response, Rizzo (2011).

In terms of the API selection, it is either driven by their consumption which can correlate to the high probability of detection in the environment or due to an existing gap of information on the degradation rate of the particular API. Frequently investigated operational parameters include catalyst load, initial concentration, type of photocatalysts, pH of the solution, wavelength and light intensity. Other operational parameters for enhancement of the degradation rate by the addition of co-oxidants and the effect of the concentration of dissolved oxygen (DO) have also been incorporated in various TiO_2 photocatalysis studies. These variables are chosen in order to establish their relationship with respect to the API degradation. However, it is also widely accepted that the design and geometry of the photoreactor dictates the optimal degradation rate of APIs, Malato et al. (2009) and Friedmann et al. (2010).

3.5.1 Non-steroidal Anti-inflammatory and Analgesic Drugs

Non-steroidal anti-inflammatory drugs (NSAIDs) primarily reduce inflammation, while analgesic (AN) drugs are widely used to relieve pain, Ziylan and Ince (2011). Examples of NSAIDs include naproxen, ketoprofen, diclofenac, fenoprofen, indomethacin, and ibuprofen, whereas ANs are acetaminophen (paracetamol), acetylsalicylic acid (aspirin) and the opioid analgesics such as morphine. Chemical structures of selected NSAIDs and ANs are presented in Table 3.2. One of the most prominent characteristics of NSAIDs is the carboxylic acid moiety which contributes to both its activity and side effects.

Photocatalytic degradation of NSAIDs and ANs has been conducted by various researchers primarily aiming to explore the degradation kinetics and degradation efficiency of APIs under selected experimental conditions. Among the most frequently studied NSAIDs and ANs are diclofenac, ketoprofen, ibuprofen, naproxen and paracetamol. Photocatalytic studies of NSAIDs and ANs are summarized in Table 3.3 and detailed discussions of selected studies are provided to highlight both the significant findings and differences in the experimental conditions.

Table 3.2 Chemical structures of selected non-steroidal anti-inflammatory drugs (NSAID) and analgesics

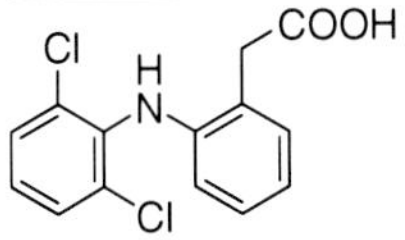	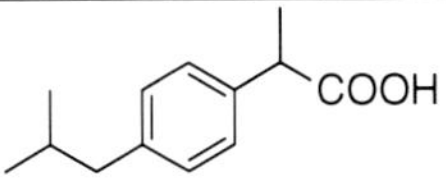	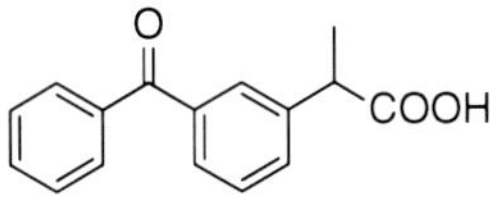
Diclofenac ($C_{14}H_{11}C_{12}NO_2$)	Ibuprofen($C_{13}H_{18}O_2$)	Ketoprofen ($C_{16}H_{14}O_3$)
2-(2-(2,6-dichloro-phenylamino)phenyl) acetic acid	(*RS*)-2-(4-(2-methylpropyl)phenyl) propanoic acid	(*RS*)-2-(3 benzoyl-phenyl) propanoic acid
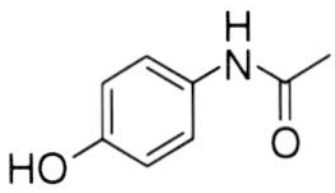	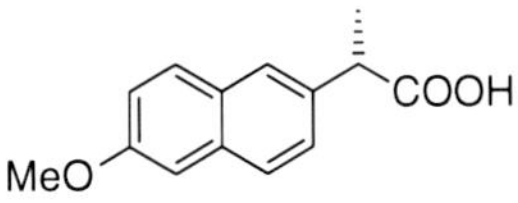	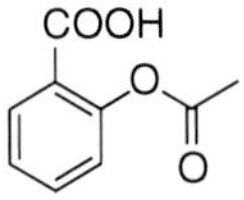
Acetaminophen ($C_8H_9NO_2$)	Naproxen ($C_{14}H_{14}O_3$)	Acetylsalicylic acid($C_9H_8O_4$)
N-(4-hydroxyphenyl) acetamide	(+)-(*S*)-2-(6-methoxynaphthalen-2-yl) propanoic acid	2-acetoxybenzoic acid

Diclofenac

Diclofenac (DCF) has been identified as one of the most recognized pharmaceuticals in the aquatic environment as a result of its elevated use worldwide, Heberer (2002b). The global consumption of DCF was reported to be 940 t per year with a daily dose of 100 mg. These numbers clearly demonstrate its significance as an environmentally important pharmaceutical, Zhang et al. (2008a). Threats on fish such as rainbow trout, and birds, specifically the vulture populations, have been reported, Oaks et al. (2004). NSAIDs such as DCF (and NPX) with a partition coefficient (log K_{ow}) greater than 3 are generally capable of bioaccumulating in the tissues of organisms, Rizzo et al. (2009a). The concentration of DCF has been reported to be less than 10 ng/L in a drinking water sample from tap water in Berlin and ranged between 0.14 and 1.48 μg/L in WWTPs effluents and surface water, Zhang et al. (2008a). As for the removal efficiency by WWTPs, there has been a wide range varying from 0 % to 80 % with typical ranges between 21 % and 40 %, which clearly indicates insufficient removal of this pharmaceutical, Zhang et al. (2008a).

One of the most comprehensive studies of the effect of UV/TiO_2 on DCF to date was performed by Calza and his group (2006). The study included the identification of intermediates formed, mineralization and also degradation rate controlling parameters. The photocatalytic efficiency increased with the amount of TiO_2, but high loadings (0.8–0.9 g/L) showed no positive effect on the degradation due to scattering of light. The optimum combination of TiO_2 loading/initial

Table 3.3 Highlights of the heterogeneous photocatalytic studies on non-steroidal anti-inflammatory drugs and analgesics

Compound(s)	Water matrix	Experimental features	Analytical methods	Findings	Reference
Diclofenac					
	Distilled water	Solar simulator suntest CPS + (Xe arc lamp, 1,500 W, 750 W/m^2) Initial concentration: 0.76–9.24 mg/L TiO_2 P25 loading: 104–896 mg/L	HPLC, TOC, LC-MS, IC, toxicity measurement on bacterium, *Vibrio fischeri*	Factorial design was used to investigate effect of TiO_2 loading and DCF initial concentration on reaction rate. Toxicity and TOC decreased. Complete DCF elimination after 1 h of irradiation	Calza et al. (2006)
	Milli-Q water	Black light fluorescent lamp (125 W, 300 and 420 nm, intensity 4.7×10^{-7} Einstein/s) Initial concentration: 5–80 mg/L TiO_2 P25 loading: 0.2-1.6 g/L	UV–vis (254 nm, 276 nm), COD, bioassays on *Daphnia magna*, *Pseudokirchneriella subcapitata*, *Artemia salina*	Degradation data fitted pseudo-first order kinetic model. High DCF concentrations (40 and 80 mg/L) and 1.6 g TiO_2/L followed second-order kinetic model	Rizzo et al. (2009a)
	Deionized water, groundwater and treated municipal wastewater	UV-A lamp (9 W, 350–400 nm, intensity 3.37×10^{-6} Einstein/s) Initial concentration: 5–20 mg/L TiO_2 loading: 50–800 mg/L Type of TiO_2: TiO_2 Degussa P25, Hombikat UV 100, Aldrich, Tronox AK-1, Tronox TRHP-2, Tronox TR Temperature: 23 °C H_2O_2 concentration: 0.07–1.4 mM	UV (276 nm), TOC, toxicity measurement with *Daphnia magna*	TiO_2 Degussa P25 performed best for DCF decomposition which was affected by water matrix and increased initial concentration. Irradiated solution, in particular after 2 h irradiation, showed high toxicity	Achilleos et al. (2010a)

	Milli-Q water	UV (254 nm; low pressure Hg lamp; photon flux: 8.33×10^{-8} Einstein/s) and near UV-vis (366 nm; medium pressure Hg lamp; photon flux: 2.38×10^{-6} Einstein/s) Initial concentration: 8 mg/L TiO_2 loading: 0.1–2 g/L Type of TiO_2: TiO_2 Degussa P25, synthesized TiO_2 anatase and rutile, and composite multi-walled carbon nanotube Gas flow rate: 200 mL/min O_2/Ar (50 % vol of oxygen) DO % (v/v): 21–100 H_2O_2 concentration: 3 and 5 mM pH: 6	HPLC, UV-vis, TOC, LC-MS	Optimum degradation was obtained with synthesized anatase (0.5 g/L) and 50 % O_2 (v/v) under UV irradiation with rate constant 0.9 min^{-1}	Martínez et al. (2011)
	Urban wastewater and Milli-Q water	High pressure Hg lamp ($\lambda = 313$ nm; intensity 3×10^{-5} Einstein/s) Initial concentration: 30–80 mg/L TiO_2 loading: 0.5–2.5 g/L Gas flow rate: 30–50 L/h Ozone concentration: 5–30 mg/L pH: 7	HPLC, COD, BOD, dissolved ozone, TOC	Concentrations of ozone gas and initial DCF concentrations greatly affected DCF and TOC removals	Aguinaco et al. (2012)

(continued)

Table 3.3 (continued)

Compound(s)	Water matrix	Experimental features	Analytical methods	Findings	Reference
	Milli-Q water	High pressure Hg lamp (313 nm; intensity 3×10^{-5} Einstein/s) Initial concentration: 10^{-4} M TiO_2 P25 loading: 1.5 g/L Temperature: 20 °C Oxygen flow rate: 30 L/h Ozone gas concentration: 10 mg/L pH: 5, 7 and 9 (buffering conditions)	HPLC, TOC, dissolved ozone	Efficiency of TiO_2 photolytic systems (O_2/UV-A/TiO_2, O_3/UV-A/TiO_2, UV-A/TiO_2) and single ozonation were compared. DCF was completely removed under photocatalytic process and TOC removed up to 80 % which was higher than single ozonation	Garciá-Araya et al. (2010)
Diclofenac mixture					
Diclofenac (amoxicillin, carba-mazepine)	Distilled water, urban wastewater effluent	Black light fluorescent lamp (125 W, 300 and 420 nm, intensity 4.7×10^{-7} Einstein/s) Initial concentration: 2.5 mg/L TiO_2 P25 loading: 0.2–0.8 g/L	UV–vis (DCF: 276 nm), TOC, bioassays on *Daphnia magna*, *Pseudokirchneriella subcapitata*, *Lepidium sativum*	DCF degradation was examined as single compound and in mixtures with amoxicillin and carbamazepine. Degradation rate was faster in spiked distilled water compared to real wastewater	Rizzo et al. (2009b)
Naproxen					
	Milli-Q water	Solar simulator (Xe lamp, 1 kW) Initial concentration: 2–1,000 mg/L TiO_2 P25 loading: 0.1–1 g/L Temperature: 20–40 °C DO concentration: 40 ± 2 mg/L Volumetric rate: 0.1–0.4 L/min	HPLC, DOC, COD, BOD_5, ESI-MS, toxicity assay on *Vibrio fischeri*	Degradation of NPX mainly contributed by photolysis although photocatalysis was more efficient for DOC removal. BOD_5/COD ratios were found to be below 0.05 suggesting no-biodegradable product formed during photocatalysis. The photoproducts formed through decarboxylation showed a high cytotoxicity	Méndez-Arriaga et al. (2008a)

Naproxen mixture					
Naproxen (Diclofenac and ibuprofen)	Milli-Q water	Solar simulator (Xe-OP lamp, 1 kW; photon flux: 6.9 μEinstein/s) Initial concentration: 200 ppm TiO_2 loading: 0.1–2 g/L Temperature: 20, 30 and 40 °C Circulation flow rate: 0.1, 0.2 and 0.4 mL/min DO concentration: 8, 40, 8 (initial)-2 (final) mg/L	HPLC, TOC, BOD_5, COD, DO, ESI/MS, toxicity assay on *Vibro fisheri*	Increase in TiO_2 loading showed no improvement on degradation rate and the maximum TiO_2 loading was 0.1 g/L. Flow rate had no effect on degradation rate while increase in temperature to 40 °C yielded 99 % conversion	Méndez-Arriaga et al. (2008b)
Naproxen (carbamazepine, clofibric acid, phenazone, ofloxacin, furosemide, ranitidine)	Ultrapure water	Medium pressure Hg lamp (125 W) Initial concentration: 5 mg/L TiO_2 P25 loading: 1 g/L Temperature: 30 °C Oxygen concentration: 22 mg/L pH: 3 and 11	UV-vis (NPX: 230 nm)	Adsorption of NPX increased with decreasing pH. NPX of 5 mg/L decreased with increasing radiation time and in accordance to pseudo-first order kinetics in experiments conducted in batch photoreactor without membrane	Molinari et al. (2006)
Ibuprofen					
Solar photocatalysis	Water type not provided	Three solar pilot plant reactors (Two CPCs and one axis-parabolic collector) Initial concentration: 20–200 mg/L TiO_2 loading: 0.1–1.0 g/L H_2O_2 concentration: 900 and 1,028 mg/L	HPLC, TOC, BOD_5, COD	IBP and TOC removal increased proportionally with TiO_2 concentration and addition of H_2O_2 accelerates the degradation process. Solar photocatalysis was efficient to degrade IBP	Méndez-Arriaga et al. (2009a)

(continued)

Table 3.3 (continued)

Compound(s)	Water matrix	Experimental features	Analytical methods	Findings	Reference
Hybrid method	Milli-Q water	Initial concentration: 0.024–0.044 mM TiO_2 P25 concentration: 10 mg/L Temperature: 25 °C H_2O_2 concentration: 2.15 μmol/L/min	HPLC, DOC	Individual and hybrid processes of Fenton, sonolysis, and photocatalysis were compared. Combination of TiO_2/Fe^{2+}/sonolysis yielded faster degradation for DOC removal (98 %) and complete degradation of IBP in 240 min	Méndez-Arriaga et al. (2009b)
Hybrid method	Milli-Q water	Initial concentration: 0.02–0.12 mM TiO_2 P25 loading: 1 g/L Temperature: 25 ± 2 °C Ultrasonic power: 16, 35 and 55 mW/mL Fe^{3+} concentration: 0.045 mM	HPLC, UV–vis, ESI/MS, TOC	Compared to sonolysis and photocatalysis, addition of Fe^{3+} and TiO_2 showed better degradation of IBP. Combination of UV, TiO_2 and ultrasound produced highest degradation rate of 108×10^{-7} min^{-1}	Madhavan et al. (2010)
Ibuprofen mixture					
Ibuprofen (and carba-mazepine)	Milli-Q water and wastewater from WWTP	Solar simulator (1,000 W Phillip Xe lamp, intensity 272.3 W/m^2) and UV-A lamp (9 W Radium lamp, photon flux 3.37×10^{-6} Einstein/s) Initial concentration: 5–20 mg/L TiO_2 loading: 50–3,000 mg/L Type of TiO_2: TiO_2 Degussa P25, Hombikat UV 100, Aldrich, Tronox AK-1, Tronox TRHP-2, Tronox TR H_2O_2 concentration: 0.07–1.4 mM pH: 3–10	UV (IBP: 220 nm), DOC	Simulated solar and UV-A light were efficient on IBP degradation in particular in the presence of TiO_2 Degussa P25	Achilleos et al. (2010b)

Ibuprofen (diclofenac and naproxen)	Milli-Q water	Solar simulator (Xe-OP lamp, 1 kW; photon flux: 6.9 μEinstein/s) Initial concentration: 200 ppm TiO_2 loading: 1 g/L Temperature: 20, 30 and 40 °C Circulation flow rate: 0.1 and 0.4 mL/min DO concentration: 8, 40, 8 (initial)-2 (final) mg/L	HPLC, TOC, BOD_5, COD, DO, ESI/MS, toxicity assay on *Vibro fisheri*	DO influenced the degradation of IBP and the optimum TiO_2 loading for maximum degradation was 1 g/L	Méndez-Arriaga et al. (2008b)
Paracetamol					
	Milli-Q water	Black light blue UV-A (8 W) and UV-C (15 W) Initial concentration: 2.0–16.0 μM TiO_2 loading: 0.04–5.0 g/L Temperature: 26 °C Oxygen concentration: 8.2–36.3 mg/L pH: 3.5–11.0	HPLC, TOC, GCMS	Rapid paracetamol degradation and mineralization occurred in the presence of TiO_2 under UV-C irradiation. Also, direct photolysis under UV-C radiation was more effective compared to that of UV-A	Yang et al. (2008)
	Milli-Q water	UV lamp (254 nm) Initial concentration: 4.0 mM TiO_2 loading: 0.4 g/L Temperature: 26 ± 2 °C H_2O_2 concentration: 48 mM Oxygen concentration: 8.2 mg/L (saturated) or 1.3–37.1 mg/L	HPLC, GCMS, IC	$HO^{\bullet}$ radical played an important part in the degradation of paracetamol with a second-order rate constant ($1.7 \times 10^9\ M^{-1}\ s^{-1}$)	Yang et al. (2009)

(continued)

Table 3.3 (continued)

Compound(s)	Water matrix	Experimental features	Analytical methods	Findings	Reference
	Double-distilled water	Metal halide lamp (250 W) Initial concentration: 20–200 μM TiO_2 loading: 0.25–1.0 g/L Temperature: 15 °C pH: 3.0–11.0	HPLC, GCMS	Initial concentration of 100 μM was degraded by 95 % in the presence of 1.0 g/L TiO_2 within 100 min of irradiation. TiO_2 photocatalysis is efficient on paracetamol degradation without generating any toxic compound other than the parent drug	Zhang et al. (2008b)
	Milli-Q water	UV-A/LED (365 nm) Light intensity: 1–4 mW/cm^2 Initial concentration: 100–385 ppb TiO_2 P25 loading: 0.005–0.06 g/L Temperature: 25 ± 2 °C	RP-LC, LC-MS-MS	Addition of TiO_2 accelerates the paracetamol degradation and complete conversion was obtained within 20 min. Smaller amounts of TiO_2 and stronger UV light helps for more light to penetrate into sample	Xiong and Hu (2012)

concentration, 0.6 g/L/8.17 mg/L yielded the highest degradation rate with a residual DCF percentage of 0.4 % after 30 min of irradiation. Longer irradiation times of up to 2 h were required to achieve complete mineralization. The toxicity of irradiated solutions was monitored with a microtox bioassay (*Vibrio fischeri*) and showed a decrease in the inhibition percentage to less than 1 % in 2 h, thus demonstrating the efficiency of photocatalytic detoxification. A total of 11 transformation products with four products with similar m/z ratios of 312 were generated from hydroxylation of DCF. This was attributed to the non-selective nature of the $HO^{\bullet}$ radical. Hydroxy and bishydroxy-DCF derivatives were formed with subsequent formation of other chloro and hydroxy-phenol derivatives.

UV/TiO_2 degradation of DCF alone and in mixture solutions was examined in two studies by Rizzo et al. (2009a, b). For the investigation of DCF as a single compound in Milli-Q water, the authors investigated two operational parameters under oxygen supply, the initial concentration (5–80 mg/L) and loading of TiO_2 Aeroxide P25 (0.2–1.6 g/L) with a batch reactor. All degradation data followed pseudo-first order kinetics, but at higher initial concentrations, 40 mg/L and 80 mg/L and 1.6 g TiO_2/L, DCF degradation was reported to follow second-order kinetics. COD removal increased from 78 % to 85 % proportionally with TiO_2 loading from 0.4 to 0.8 g/L and a DCF initial concentration of 15 mg/L, but declined when DCF concentrations were increased to 40–80 mg/L, Rizzo et al. (2009a).

The degradation of DCF in the presence of two other APIs, namely amoxicillin (AMX) and carbamazepine (CBZ), was investigated in a subsequent study by the same group, Rizzo et al. (2009b). In this study, spiked samples in Milli-Q water and wastewater samples collected from a WWTP in Italy downstream of the biological process were used. The experimental set-up was similar to that in their previous study. Photocatalytic oxidation displayed higher removal from the sample mixture prepared using pure water after 30 min of irradiation, than the sample mixture in real wastewater. This difference is attributed to the existence of other oxidizable species and radical scavengers such as carbonates in the wastewater samples, which are able to compete with the APIs during the photocatalytic treatment. The following Eqs. (3.12) and (3.13) show the scavenging effects of carbonate species on hydroxyl radicals:

$$HCO_3^{-} + HO^{\bullet} \rightarrow H_2O + CO_3^{-\bullet} \quad k = 8.5 \times 10^6\,M^{-1}\,s^{-1} \tag{3.12}$$

$$CO_3^{2-} + HO^{\bullet} \rightarrow OH^{-} + CO_3^{-\bullet} \quad k = 4.0 \times 10^8\,M^{-1}\,s^{-1} \tag{3.13}$$

TOC removal of up to 80 % after 2 h and a rapid increase after 30 min in the sample mixtures in pure water was observed. However, TOC $t_{1/2}$ (half-life time) was found to be significantly higher in the sample mixtures in both water matrices compared to that of the single compound samples investigated.

UV-A photocatalytic irradiation was applied to the decomposition of DCF with different types of TiO_2, water matrices, H_2O_2 concentrations, TiO_2 loadings and

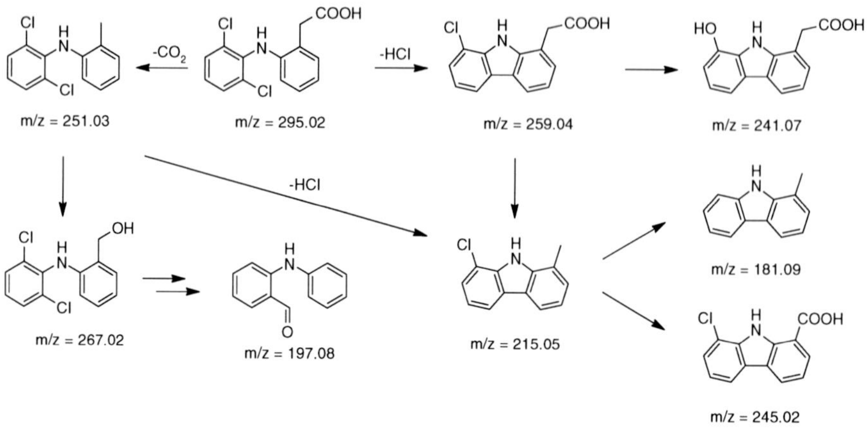

Scheme 3.1 Photocatalytic degradation pathways and degradants of diclofenac (DCF), Martínez et al. (2011)

initial concentrations investigated, Achilleos et al. (2010a). An immersion-well type reactor equipped with a UV-A lamp (350–400 nm, 9 W) was used in this study. All experiments were conducted at a nominal pH of 6 with continuous oxygen flow. Among six TiO_2 samples tested, the TiO_2 Degussa P25 experiment showed the highest conversion of 85 % (initial concentration 10 mg/L and TiO_2 250 mg/L) after 4 h of irradiation due to the better photocatalytic activity of anatase. Nevertheless, the P25 TiO_2 loading resulted in no improvement of DCF conversion and in all cases only 85 % conversion was attained. The results demonstrated that UV-A/TiO_2 treatment is effective for DCF abatement as well as mineralization. Another major finding was that photocatalytically treated 10 mg/L DCF (500 mg/L TiO_2) samples taken after 120 min were highly toxic to the freshwater species *Daphnia magna.*

An attempt was made to compare the effect of the irradiation source (near UV–vis and UV) on the photocatalytic degradation of DCF with commercial TiO_2 P25 and synthesized TiO_2, Martínez et al. (2011). This study highlighted that UV radiation (254 nm) was found to be superior to near UV–vis (366 nm) for DCF photocatalytic degradation. The optimum conditions for DCF removal included 0.5 g/L of synthesized anatase in the presence of 50 % (v/v) DO. In terms of mineralization, irradiation of aqueous DCF in the presence of 1 g/L TiO_2 P25 with near UV–vis caused about 40 % reduction in TOC in 60 min where the concentration declined from 112 mg/L to 68.6 mg/L. Eight photoproducts were identified regardless of reaction conditions. The main photoproduct formed was identified as a monohalogenated carbazole with a m/z of 259 g/mol as a result of photocyclization of DCF which was also in accordance with a DCF photolysis study, Agüera et al. (2005). Another important pathway, decarboxylation generated two other intermediates corresponding to a m/z of 251 g/mol and 215 g/mol (Scheme 3.1), respectively.

More recently, the combination of heterogeneous photocatalysis with other AOPs for DCF degradation has been studied. The effect of photocatalytic ozonation (O_3/UV-A/TiO_2) on DCF was assessed, Aguinaco et al. (2012). Irradiation was performed using a high pressure Hg lamp which was placed in the centre of a borosilicate-type glass photoreactor. Commercial TiO_2 Degussa P25 and ozone generated from pure oxygen were used. One interesting finding was that the wastewater sample of DCF showed no differences in terms of degradation rate and TOC removal for a DCF concentration of 30 mg/L, whereas differences were observed when Milli-Q water was used under the same conditions indicating a significant effect of the water matrix. The study reported negligible differences when investigating TiO_2 loadings (0.5–2.5 g/L) on 30 mg/L DCF and in all cases complete degradation was achieved in 5.5 min.

From the results above on the UV/TiO_2 degradation of DCF, it is evident that attempts have been made to correlate selected operational parameters with DCF degradation in pure water and also wastewater. It can be concluded that the choice of initial concentration of DCF, TiO_2 loading and irradiation time differs between the studies. However, there is no single study which exclusively covers all the operating parameters such as pH (from acidic to basic) and TiO_2 loading (low to high range). The effects of radiant flux and wavelength are not clearly understood despite the fact that UV-A type of irradiation has been chosen for the majority of the studies reviewed.

Naproxen

The occurrence of naproxen (NPX) has been reported both in water and wastewater. A recent review has also indicated its continuous presence in the water bodies, Ziylan and Ince (2011). Concentrations of NPX in WWTP effluents and surface water ranged between 0.1–2.6 μg/L and 0.01–0.1 μg/L respectively, Boyd et al. (2005). However, there is still a considerable lack of information on its photocatalytic degradation compared to its direct photolysis which has thus far been performed with artificial UV light and direct sunlight, Felis et al. (2007) and Packer et al. (2003). NPX photodegradation products were found to be more toxic than the parent compound based on a toxicity studies conducted using *Daphnia magna* and *Vibrio fischeri*, Isidori et al. (2005).

Only one study has been conducted on the photocatalytic and photolytic degradation of NPX, Méndez-Arriaga et al. (2008a). In this study, 3 h of photolysis and photocatalysis with a solar simulator produced 90 % and 40 % NPX removal, respectively. Lower percentage removal during the photocatalytic process resulted from either low adsorption of NPX onto the TiO_2 surface or possible recombination and deactivation of the $HO^{\bullet}$ radical. A higher mineralization (20 %) level was however achieved with photocatalysis compared to photolysis (5 %). Initial degradation rate of NPX increased with TiO_2 loading (0.1–1 g/L) for 0.8 mM/L NPX. The by-products formed from the 180 min of treatment were identified

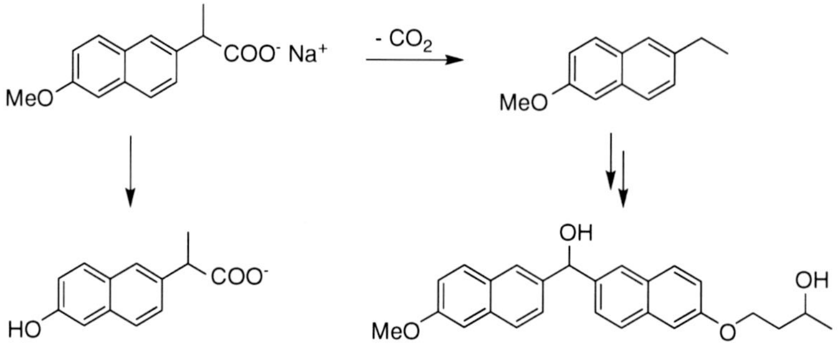

Scheme 3.2 The proposed degradation pathway of naproxen (NPX) by TiO_2 photocatalysis, Méndez-Arriaga et al. (2008a)

by means of liquid chromatography-electrospray ionization-time of flight mass spectrometry (LC/ESI-TOF-MS). Two major pathways including demethylation and decarboxylation were thus suggested to occur on the photocatalytic treatment of NPX (Scheme 3.2).

In another study, the photocatalytic degradation of NPX was evaluated together with two other NSAIDs, namely DCF and ibuprofen (IBP), Méndez-Arriaga et al. (2008b). A Duran tubular photoreactor was placed in a solar simulator and irradiated with a Xe-OP lamp (290–400 nm; photon flux 6.9 μEinstein/s). A maximum degradation of NPX was obtained at a TiO_2 loading of 0.1 g/L for 200 mg/L NPX (30 °C and flow rate of 0.2 mL/min) with a kinetic rate constant of 7.0×10^{-3} min^{-1}. An increase in TiO_2 (0.1–1 g/L) did not improve the degradation rate. NPX degradation and TOC conversion however showed a slight increase at 40 °C. The reaction rate for each NSAID did not vary significantly. Biodegradability (BOD_5/COD) showed no improvement after photocatalytic treatment of NPX, implying that the by-products formed affect post-biological treatment.

Another study was carried out to evaluate the feasibility of commercial nanofiltration membranes and TiO_2 P25 Degussa for NPX and six other pharmaceuticals (furosemide, rantinide, ofloxacine, phenazone, carbamazepine and clofibric acid) degradation, Molinari et al. (2006). For irradiation purposes, a medium pressure Hg lamp (125 W) was used under continuous oxygen bubbling. NPX showed a 95 % adsorption at pH 3 and only 3 % at pH 11 due to high solubility. Degradation of NPX was only performed in a batch photoreactor without any membrane. A comparison with the membrane photoreactor for NPX removal was thus not possible. In a batch photoreactor, the concentration of NPX decreased with increasing irradiation time following pseudo-first order kinetics. The rate constants for 5 mg/L NPX at pH 3 and pH 11 were 7.86×10^{-2} and 4.91×10^{-1} min^{-1}, respectively.

Although NPX is known to undergo direct photolysis under natural sunlight, Packer et al. (2003), various aspects of the efficiency of photocatalytic NPX degradation remain to be investigated. Further research needs to be undertaken in order to gain sufficient knowledge on the degradation kinetics and toxicity outcome of photocatalytically treated samples, in addition to the by-products generated.

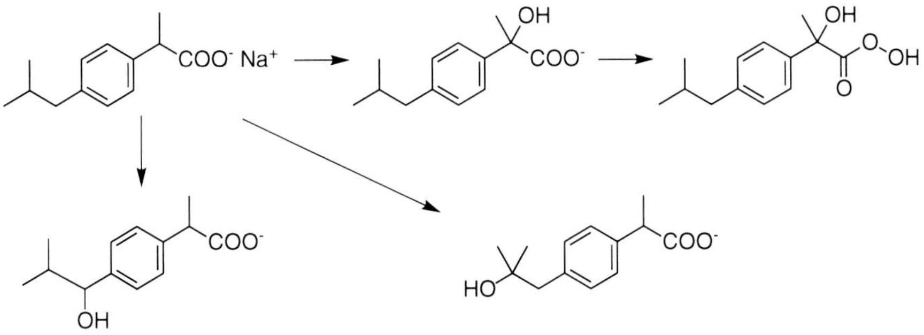

Scheme 3.3 Photocatalytic degradation pathways of ibuprofen (IBP), Méndez-Arriaga et al. (2008b)

Ibuprofen

As for the other NSAIDs, ibuprofen (IBP) has been found in water, wastewater and wastewater effluents due to only partial removal in WWTPs. For example, the concentrations of IBP and one of its metabolites was recorded to be more than 500 ng/L in the effluent of a primary WWTP, Sabri et al. (2012). Elimination of IBP in WWTPs varies between 75 % and 90 %, Tixier et al. (2003). The most prominent metabolites of IBP, namely carboxy and/or hydroxy IBP, have been reported to be present after biological treatment as toxic by-products, Mozia and Morawski (2012). Despite being one of the most consumed pharmaceuticals, there has been little attention devoted to the photocatalytic oxidation of IBP thus far.

A systematic study of IBP (and CBZ) photocatalytic degradation and mineralization spiked in Milli-Q and wastewater samples was performed under UV-A and simulated solar irradiation, Achilleos et al. (2010b). The study concluded that heterogeneous photocatalysis with both, UV-A and solar irradiation, was effective for IBP removal in the presence of TiO_2 in particular Degussa P25. The study proposed that this treatment can be employed as a post-secondary treatment in WWTPs.

UV/TiO_2 degradation of IBP and two other NSAIDs, DCF and NPX, revealed that IBP can be completely removed, Méndez-Arriaga et al. (2008b). A maximum conversion of 200 mg/L IBP was achieved in the presence of 1 g/L TiO_2 after 240 min of irradiation with a first-order kinetic rate constant of 9.1×10^{-3} min^{-1}. A low initial concentration of IBP (25 ppm) resulted in maximum degradation within 60 min and 50 % mineralization. Further investigation showed that biodegradability (BOD_5/COD index) of IBP was enhanced in the presence of excess DO. The study thus proposed the feasibility of post-biological treatment. Electrospray ionization/mass spectrometry (ESI/MS) analysis at the end of 4 h of photocatalytic treatment revealed that hydroxylation or $HO^{\bullet}$ attack on the propanoic moiety or/and isobutyl side chain is the dominant pathway for IBP degradation. Demethylation and decarboxylation processes resulted in the formation of other organic acids such as propionic, formic or hydropropionic acid or their sodium salts (Scheme 3.3).

Solar photocatalysis has proven to be an effective method for IBP degradation based on a study conducted in three solar pilot plants, Méndez-Arriaga et al. (2009a). TOC and IBP removal were enhanced in the presence of H_2O_2. In contrast, no improvement was found on inclusion of $Na_2S_2O_8$ as the pH decreased up to 4.5 which correlates to the pK_a of IBP (4.5) and thus leads to precipitation. Improvement in the biodegradability (BOD_5/COD) of the treated solution suggested coupling potential with biological treatment.

Effective photocatalytic degradation rates have also been reported when hybrid AOP methods were applied to IBP, Méndez-Arriaga et al. (2009b) and Madhavan et al. (2010). One study confirmed that AOP hybrid processes (sonophotocatalysis with TiO_2, sonophoto-Fenton and sonophotobicatalysis with TiO_2 and Fe^{2+}) improved the degree of mineralization of IBP in parallel with IBP removal. The combination of TiO_2/Fe^{2+}/sonolysis turned out to be the most beneficial application, owing to its highest mineralization (98 %), Méndez-Arriaga et al. (2009b). The combination of UV and ultrasound with TiO_2 (UV/TiO_2/ultrasound) enhanced the degradation of 18.5 mg/L IBP (in 1 g/L TiO_2 P25) from 61 % (photocatalysis only) to 85 % after 15 min of irradiation, Madhavan et al. (2010).

Acetaminophen

Acetaminophen commonly known as paracetamol is the API in a variety of over-the-counter analgesic and antipyretic drugs. This drug is extensively used around the globe. In England, it was used in more than 400 t in 2000 while the export quantity from China was estimated to be 15,348 t in 2001, Zhang et al. (2012). Its occurrence in levels up to 6 ppb and 10 ppb, respectively, in natural waters, in European sewage treatment plant effluents and in the United States has been reported. However, only few studies deal with the removal of paracetamol by means of UV/TiO_2.

A study dedicated to explore the effect of various operating parameters on photocatalytic degradation for paracetamol in Milli-Q water was reported in 2008, Yang et al. (2008). Comparing UV-A/TiO_2 and UV-C/TiO_2 irradiation, the latter was found to be more effective to degrade 4.0 mM (0.4 g/L TiO_2) of the parent compound as well as reducing TOC to about 60 % in 300 min. Increasing TiO_2 loading from 0.04 g/L to 5.0 g/L caused the degradation rate to increase from 4.9 ± 0.5 to $14.7 \pm 1.7 \times 10^{-3}$ min^{-1} until it levelled off to $13.7 \pm 1.2 \times 10^{-3}$ min^{-1} at 0.8 g/L. Insignificant effects were found at 7.0 g/L due to scattering phenomena. Increase in pH from 3.5 to 9.5 also increased the degradation rates from 11.0 to 16.5×10^{-3} min^{-1}. However, a higher pH of 11 significantly slowed down the degradation rate due to repulsion between negatively charged TiO_2 and paracetamol. A similar pH effect was also observed by Zhang et al. (2008b). Addition of oxygen showed a significant impact on paracetamol degradation compared to argon as the former inhibits e^--h^+ recombination.

In order to establish the nature of the most active species (e_{cb}^-, h_{vb}^+, $HO^\bullet$, $HO_2^\bullet$, $O_2^{\bullet-}$ and H_2O_2) for paracetamol degradation, a subsequent study was undertaken by Yang et al. (2009) using UV-A irradiation. It was proposed that the $HO^\bullet$ radical plays a prominent role in the photocatalytic degradation of paracetamol.

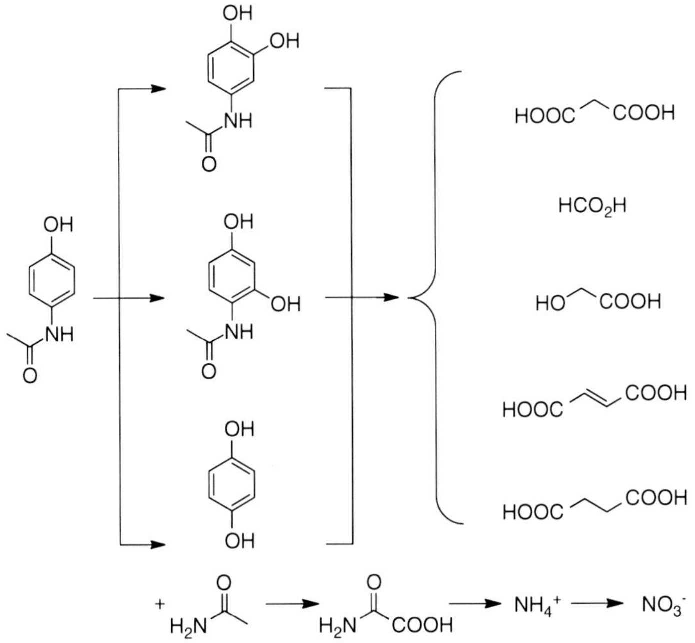

Scheme 3.4 Photocatalytic degradation pathways and degradants of paracetamol, Yang et al. (2008, 2009)

Both studies identified 11 common intermediates, including aromatic compounds, carboxylic acids, and nitrogen containing aliphatic compounds from photocatalytic degradation of paracetamol (Scheme 3.4). Two additional inorganic intermediates, namely ammonium and nitrate, were proposed as a result of successive mineralization of the intermediates. The degradation is initiated by hydroxylation through $HO^{\bullet}$ addition onto the aromatic ring, which subsequently results in further oxidation to form carboxylic acid derivatives, Yang et al. (2009).

Recent advances in the photocatalytic degradation of paracetamol revealed that UV-A/LEDs (Light Emitting Diode) can be used instead of the traditional Hg lamps. Conventional lamps are known for their high energy cost and the incorporation of toxic mercury, Xiong and Hu (2012). Negligible amounts of paracetamol were removed with UV-A/LED photolysis. In contrast, TiO_2 P25 caused a significant level of degradation with complete degradation within 20 min. An increase in light intensity from 1 to 4 mW/cm^2 significantly enhanced the degradation rate of paracetamol. One important advantage demonstrated in this study is the possibility to employ low average amount of TiO_2 (0.01 g/L) due to the high light intensity from LED lamps. This finding has important implications for pursuing the applications of UV-A/LED light sources in water and wastewater treatment.

Table 3.4 Classifications of antibiotics (Adapted from Homem and Santos (2011))

Classes	Features of structures	Chemical structure
Penicillins	Thiazolidine ring connected to a ß-lactam ring, to which a side chain is attached	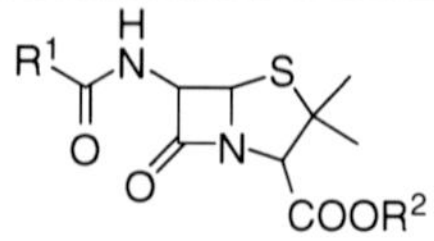
Imidazoles	Heterocyclic compounds of 5 member di-unsaturated ring structure with two nitrogen atoms at nonadjacent positions	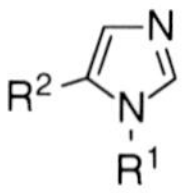
Quinolones	Two fused rings with a carboxylic acid and a ketone group. When R^4 is F, then it is a fluoroquinolone	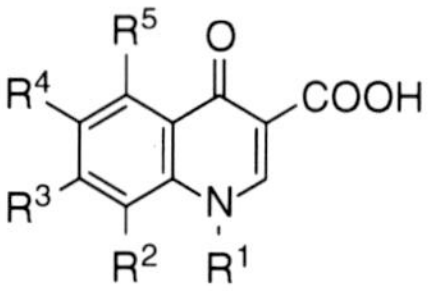
Sulfonamides	Sulfonyl group connected to amine group	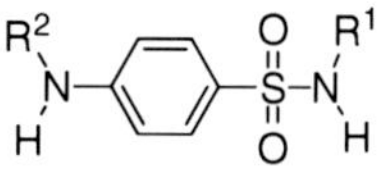
Tetracyclines	Consists of an octahydrotetracene ring skeleton and four fused rings	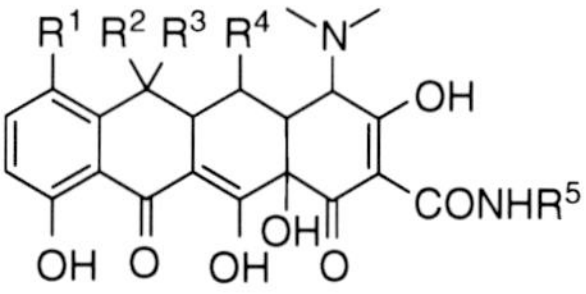
Trimethoprim	Diaminopyrimidine, analogue of the pteridine moiety of folic acid	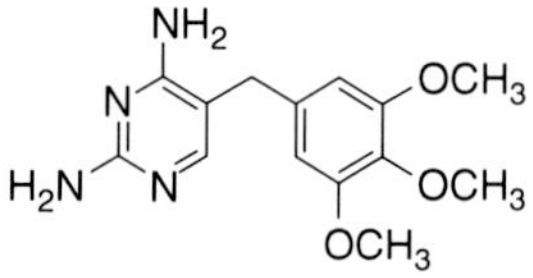

3.5.2 *Antibiotics*

In recent years, there has been an increasing amount of literature on the photocatalytic treatment of antibiotics, due to their extensive use in both humans and animals as agents to prevent and treat microbial infections. Classifications of the major antibiotics are presented in Table 3.4, according to importance attached to their chemical structure and mechanism of action.

The annual consumption of antibiotics for human and veterinary usage is estimated to be about 100,000–200,000 t, Xu et al. (2007). The first detection of antibiotics in river water was reported in 1982 in England, where macrolides, tetracycline (TC) and sulphonamides at levels of 1 μg/L were found, Homem and Santos (2011). Since then, numerous studies have reported the presence of antibiotics in water courses around the world, Watkinson et al. (2009). The polar and non-volatile nature of antibiotics contributes significantly to their accumulation in the environment due to their persistence either as the parent compound or metabolites after conventional water treatment. One of the major effects of the continuous accumulation is the emergence of antibiotic resistance strains of bacteria, Elmolla and Chaudhuri (2010a). Accumulation of small quantities of antibiotics in receiving water over a long term has been the main concern, as they may have adverse effects on both humans and aquatic organisms.

Studies conducted to assess the potential of UV/TiO_2 degradation on four important classes of antibiotics namely β-lactam antibiotics, quinolones, tetracyclines and sulphonamides are summarized in Table 3.5.

β-Lactam Antibiotics

Amoxicillin (AMX) ($C_{16}H_{19}N_3O_5S$) has been one of the most popular candidates amongst the β-lactam antibiotics selected for UV/TiO_2 treatment. This API has three pK_as namely 2.7, 7.5 and 9.6, Dimitrakopoulou et al. (2012). In water, AMX shows absorption up to 290 nm (Fig. 3.4). Oral consumption of 500 mg of AMX has resulted in 86 ± 8 % of AMX unchanged in human urine within 2 h of consumption, Martins et al. (2009). Excretion of unchanged AMX due to a slow rate of metabolism in humans has led to its discharge into the environment. AMX has also been detected in effluents from drug manufacturing facilities with levels reaching several hundreds of milligrams per liter, Mavronikola et al. (2009). The possibilities of AMX elimination by means of UV/TiO_2 have been investigated by several research groups, Martins et al. (2009) and Dimitrakopoulou et al. (2012).

AMX spiked ultrapure water and secondary effluent water were used in a UV-A/TiO_2 degradation comparative study, Dimitrakopoulou et al. (2012). Of the various catalysts tested, TiO_2 Degussa P25 (250 mg/L) was found to be the most efficient TiO_2 catalyst, yielding complete degradation after 25 min and 93 % mineralization within 90 min of irradiation for 10 mg/L of AMX. Under the same conditions, TiO_2 (Anatase) required at least 45 min to cause a complete removal. The other six TiO_2 catalysts needed much longer reaction times and in most cases lead to incomplete degradation. In addition, the degrees of mineralization failed to rise above 75 % after 90 min of irradiation. Direct photolysis with a UV-A light source had little effect on AMX degradation (only 4 %) due to its poor absorbance within the emission spectrum of the irradiation source. The comparison between ultrapure water and secondary effluent showed that the latter hindered the degradation due to the presence of organic matter and scavenging effects of bicarbonates and chlorides.

AMX spiked hospital wastewater samples yielded complete degradation with TiO_2 Degussa P25 photocatalysis (800 mg/L at pH 4 and 30 °C) after 30 min of irradiation while only 85 % degradation was achieved using the photo-Fenton process after an extended period of 60 min, Martins et al. (2009). The lower degradation in the latter case was explained by the pH conditions as it has a significant effect on the treatment. At pH 4 and beyond, the formation of Fe^{2+} hydroxide complexes and Fe^{3+} species retard the production of $HO^•$. In contrast, the

Table 3.5 Highlights of heterogeneous photocatalytic studies on antibiotics

Compound(s)	Water matrix	Experimental features	Analytical methods	Findings	Reference
Amoxicillin					
	Ultrapure water and treated wastewater (secondary effluent)	9 W Radium Ralutec Initial concentration: 2.5–30 mg/L TiO_2 loading: 100–750 mg/L Type of TiO_2: TiO_2 Degussa P25 (21 nm), Hombikat UV 100 (5 nm), Millennium PC50 (20–30 nm), Millennium PC100 (15–25 nm), Millennium PC105 (15–25 nm), Millennium PC500 (5–10 nm), Tronox AK1 (20 nm), Aldrich Anatase AA (15 nm) pH: 5 and 7.5	TOC, HPLC, toxicity on *Escherichia coli*, *Klebsiella pneumonia* and *Enterococcus faecalis*	UV-A/TiO_2 is an effective abatement method for AMX in water in particular with TiO_2 Degussa P25. Antibiotic activity of irradiated solution showed that *Escherichia coli* and *Klebsiella pneumonia* were highly resistant to AMX up to 25 mg/L	Dimitrakopoulou et al. (2012)
	Hospital wastewater	Medium pressure Hg lamp (125 W; intensity: 401 W/m^2) TiO_2 P25 loading: 400–1,200 mg/L Temperature: 20–40 °C pH: 3–11	HPLC, COD, toxicity on *Artemia salina*	Response surface methodology was used to study the combined effect of pH and TiO_2 loading on COD reduction and AMX photocatalysis. Maximum COD removal was achieved with 800 mg/L TiO_2 and pH 3. Toxicity inhibition of 46.3 % was obtained	Martins et al. (2009)

	Water type not provided	15 W low pressure luminescent Hg lamp (UV, 365 nm) and 15 W daylight fluorescent lamp (Vis, 0.6 mW/cm^2) Solar radiation (16 mW/cm^2) Initial concentration: 20–100 mg/L TiO_2 concentration: 1 g/L Temperature: 20 ± 1 °C pH: 3–9	UV–vis (230 nm), IC, UPLC-ESI/MS, QTOF/MS, COD	Carbon and iron doped titania and Degussa P25 was tested for AMX degradation under artificial visible light and solar radiation. Doped titania yielded comparable degradation with that of Degussa P25 under UV light	Klauson et al. (2010)
Amoxicillin mixtures					
Amoxicillin (ampicillin and cloxacillin)	Distilled water	UV lamp (6 W, 365 nm) Initial concentration: 104 mg/L TiO_2 (Anatase) loading: 0.5–2.0 g/L ZnO loading: 0.2–2.0 g/L Temperature: 22 ± 2 °C pH: 3–11 H_2O_2 concentration: 50–300 mg/L	HPLC, COD, BOD_5, DO, DOC	All treatments degraded the antibiotics and also improved biodegradability except for UV/ZnO which did not show any effect on the latter. Photo-Fenton showed highest rate constant (0.029 min^{-1}) and was most cost-effective	Elmolla and Chaudhuri et al. (2010b)
Amoxicillin (ampicillin and cloxacillin)	Distilled water	UV lamp (6 W, 365 nm) Initial concentration: 104 mg/L TiO_2 (Anatase) loading: 0.5–2.0 g/L Temperature: 22 ± 2 °C pH: 3–11 H_2O_2 concentration: 50–300 mg/L	HPLC, COD, BOD, DO, DOC	Addition of H_2O_2 at pH 5 with TiO_2 0.1 g/L led to complete degradation of all antibiotics. Mineralization was also observed through DOC removal and formation of nitrate, ammonia and sulphate	Elmolla and Chaudhuri et al. (2010c)

(continued)

Table 3.5 (continued)

Compound(s)	Water matrix	Experimental features	Analytical methods	Findings	Reference
Amoxicillin (Diclofenac)	Deionized water	Solar simulator (250–765 W/m^2) Initial concentration: 200 mg/L TiO_2 loading: 100 mg/L	HPLC, UV-vis	Degradation of AMX was more efficient in the presence of TiO_2 where about 70 % removal was achieved after 6 h of irradiation at 400 W/m^2	Kockler et al. (2012)
Sulfonamide and other sulfa antibiotics					
Sulfamethoxazole	Milli-Q water	Solar simulator Xe lamp (1,000 W) Initial concentration: 100 ppm TiO_2 concentration: 0.1–2.0 g/L Temperature: 25 ± 0.1 °C Oxygen concentration: 8 mg/L Initial pH: 5 (not buffered) pH: 2, 5, 7 and 11	HPLC, TOC, BOD_5/COD, FIA, IC, LCMS	Increase in TiO_2 amount improved SMX degradation and TOC removal. Change in pH showed more effect on TOC removal than that on the degradation of SMX	Abellán et al. (2007)
Sulfamethoxazole (and trimethoprim)	Milli-Q water	Solar simulator (Xe lamp, 1,500 W) Initial concentration: 100 mg/L TiO_2 loading: 0.1–2.0 g/L Temperature: 25 °C	HPLC, TOC, BOD, COD	Increase in the amount of TiO_2 has great influence on the reaction rate. The optimum catalyst concentration was between 0.5 and 1.0 g TiO_2/L for SMX degradation	Abellán et al. (2009)
Sulfamethoxazole	Reverse-osmosis water	Hg-Ar (Germicidal UV-C) lamp Initial concentration: 3–12 mg/L TiO_2 P25 loading: 0.5 g/L Temperature: 25 ± 0.1 °C DO concentration: 8.8 mg/L Pure oxygen concentration: 42 mg/L	HPLC, COD, DO, pH, UV–vis, LCMS, toxicity test on *Daphnia magna*	Kinetic rate constant decreased with increasing SMX concentration. DO did not favour removal of SMX	Nasuhoglu et al. (2011)

Sulfamethoxazole	Ultrapure water, municipal wastewater and ground water	UV-A Radium lamp (9 W) Initial concentration: 2.5–30 mg/L TiO_2 loading: 100–750 mg/L Type of TiO_2: TiO_2 Degussa P25, Hombikat UV 100, Millennium PC50, Millennium PC100, Millennium PC105, Millennium PC500 Temperature: 25 °C pH: 4–8	HPLC, TOC	Degradation of SMX was described by L-H model. Excess DO showed large effect on TOC removal rather compared to SMX degradation. Initial reaction rate for SMX decomposition increased from 0.428 mg/(L × min) to 1.510 mg/(L × min) when SMX concentration was increased from 2.5 mg/L to 20 mg/L	Xekoukoulotakis et al. (2011)
Sulfachlorpyridazine, sulfapyridine and sulfisoxazole	Milli-Q water	High pressure Hg lamp (125 W) Initial concentration: 50–200 μM TiO_2 P25 loading: 0.25–3.0 g/L Initial pH: 3–11 Radical quencher: isopropanol and potassium iodine	TOC, HPLC, HPLC/MS/MS	All three sulfa drugs, sulfachlorpyridazine, sulfapyridine and sulfisoxazole can be effectively degraded in aqueous system with removals of 85.2 %, 92.5 % and 85.0 % after 1 h of irradiation respectively. Effect of pH varied between the drugs	Yang et al. (2010)
Sulfamethoxazole and sulfonamide analogues (sulfisoxazole, sulfathiazole, sulfamethizole)	Deionized water	Xe arc lamp (450 W) Initial concentration: 5–480 μM TiO_2 P25 loading: 0.01–1.0 g/L Hombikat UV-100 loading: 0.1 g/L Temperature: 25 °C pH: 3–11	HPLC, LC/MS/MS, DOC	UV-A/TiO_2 photocatalysis is very effective for degrading sulfonamides particularly in natural waters exhibiting either alkaline pH or low concentrations of NOM, or both conditions	Hu et al. (2007)

(continued)

Table 3.5 (continued)

Compound(s)	Water matrix	Experimental features	Analytical methods	Findings	Reference
Quinolones					
Oxolinic acid	Water type not provided	Annular reactor with black light (36 W) Initial concentration: 18 ppm TiO_2: 1 g/L (suspension) and 8.79 g for immobilized form pH: 9	Spectrophotometer, TOC, COD, antibacterial activity with *Escherichia coli*	Treated solutions showed no antibiotic activity. TiO_2 coated on sintered glass cylinder efficient to remove OX acid	Palominos et al. (2008)
Oxolinic acid	Ultrapure water	Black lamp (14 W/m^2; 365 nm) Initial concentration: 20 mg/L TiO_2 P25 loading: 0.2–1.5 g/L pH: 7.5–11	LC with fluorescence detector, HPLC-MS DOC, COD, antimicrobial activity on *Escherichia coli*	Experimental conditions of pH 7.5 and 1.0 g/L of TiO_2 favoured between 80 % and 100 % of OX acid degradation. About 20 % of OX acid was adsorbed on TiO_2 under dark conditions	Giraldo et al. (2010)
Ciprofloxacin	Double distilled water	High pressure Hg lamp (0.38 mW/cm^2) Initial concentration: 100 μM TiO_2 P25 loading: 1.5 g/L Temperature: 25 °C pH: 3.0–11.0	HPLC, LC/MS/MS	Oxidative and reductive processes are important for CIP degradation. Hydroxyl and azide radicals can react with CIP at high rate constants, $(2.15 \pm 0.10) \times 10^{10}$ and $(2.90 \pm 0.12) \times 10^{10}$ m^{-1} s^{-1} respectively	An et al. (2010)

Ciprofloxacin	Deionized water	O_3 free Xe arc lamp (450 W) Initial concentration: 100 μM TiO_2 and Hombikat UV 100 loading: 0.5 g/L Temperature: 25 °C pH: 6	HPLC, HPLC/UV/FLD, LC/MS/MS, antibacterial activity with *Escherichia coli*	Degradation rate of CIP under comparable solution conditions (100 mM CIP, 0 or 0.5 g/L TiO_2, pH 6, 25 °C) followed the trend UV-A/TiO_2 > Vis-TiO_2 > UV-A	Paul et al. (2010)
Ciprofloxacin	Hospital wastewater	Medium pressure Hg lamp (125 W) Initial concentration: 200 μg/L TiO_2 P25 loading: 400 mg TiO_2/700 mL	LC/FLD, UV–vis, COD	Heterogeneous photocatalysis and ozonation were compared for CIP degradation in hospital wastewater. Both treatments produced complete degradation in 60 min and degradation products generated were similar	Vasconcelos et al. (2009)
Moxifloxacin	Deionized water	UV-A (485 μW/cm^2) Initial concentration: 12.5–124.6 μM TiO_2 P25 loading: 0.25–8 g/L Temperature: 25–65 °C Oxygen concentration: 0–100 % pH: 7 Stirring speeds: 2.3, 7.9 and 13.2 rps Effect of radical inhibitors: isopropanol and potassium iodide	HPLC	Degradation kinetics was affected by the variables investigated. Oxidative holes play more important role than hydroxyl radicals as it led to 63 % MOX degradation	Van Doorslaer et al. (2012)

(continued)

Table 3.5 (continued)

Compound(s)	Water matrix	Experimental features	Analytical methods	Findings	Reference
Norfloxacin	Double distilled water	Medium pressure Hg lamp (125 W) Initial concentration: 0.15–0.50 mM TiO_2 loading: 1 g/L Type of TiO_2: Degussa P25, Hombikat UV100, PC500 H_2O_2 concentration: 10 mM pH: 4.2, 6.3, 8.1 and 10.4	UV–vis, TOC	TiO_2 Degussa P25 was found to be more efficient than other catalysts. Degradation rate increased with initial concentrations at optimum pH 10.4 from 0.009×10^{-3} mol L^{-1} min^{-1} (0.15 mM) to 0.017×10^{-3} mol L^{-1} min^{-1} (0.50 mM)	Haque and Muneer (2007)
Levofloxacin	Reverse osmosis water	Hg-Ar Germicidal UV-C lamp (254 nm; intensity: $1.3 \times 10^{-3} \pm 0.3$ Einstein/min/L) Initial concentration: 20 mg/L TiO_2 P25 loading: 0.05–0.5 g/L	HPLC, COD, antibacterial activity on *Escherichia coli* and *Pseudomonas fluorescens*	First study to report photocatalysis and ozonation under UV-C radiation for levofloxacin removal. About 97 % levofloxacin was removed after 2 h under photocatalysis. No antibacterial activity found under both treatments	Nasuhoglu et al. (2012)
Fluoroquinolone mixtures					
Ofloxacin norfloxacin, ciprofloxacin and enrofloxacin	Distilled water	Solar irradiation (800 W Xe lamp) Initial concentration: 0.028 mM (ofloxacin), 0.031 mM (norfloxacin), 0.027 mM (ciprofloxacin) and 0.028 mM (enrofloxacin) TiO_2 loading: 0.1–1.5 g/L pH: 2–9 H_2O_2 concentration: 41.25–330 mg/L	HPLC, TOC, antibacterial activity with *Bacillus subtilis*	Photocatalytic reactions of the four FQs could be described by a pseudo-first order kinetic model. Complete removal of all FQs at 0.5 g/L TiO_2 and 82.5 mg/L of H_2O_2 at pH 6 after 90 min of irradiation	Li et al. (2012)

Ciprofloxacin, danofloxacin, enrofloxacin, levofloxacin, marbofloxacin and moxifloxacin	River water	Natural sunlight (290–470 W/m^2) Phosphorus-covered low pressure Hg lamp (12 W/m^2, 310 nm) Initial concentration: 20–50 μg/L TiO_2 P25 loading: 0.5 g/L	HPLC with fluorescence detector, ESI-MS/MS	About 15 min sufficient to remove the FQs under natural sunlight	Sturini et al. (2012)
Ciprofloxacin and moxifloxacin	Deionized water	UV-A (485 $\mu W/cm^2$) and UV-C (389 $\mu W/cm^2$) Initial concentration MOX: 37.4 μM Initial concentration CIP: 45.3 μM TiO_2 P25 loading: 0.5 g/L Temperature: 25 °C pH: 3–10	HPLC	The highest removal rates were obtained at pH 7 in the presence of TiO_2 (0.5 g/L) as a photocatalyst were as follows: k_1, CIP, UV-A = 0.137 min^{-1}, k_1, CIP, UV-C = 0.163 min^{-1}, k_1, MOX, UV-A = 0.227 min^{-1}, k_1, MOX, UV-C = 0.236 min^{-1}	Van Doorslaer et al. (2011)
Tetracyclines					
Tetracycline	Water type not provided	Solar simulator (300–800 nm, Xe lamp, 250 W/m^2) TiO_2 P25 loading: 0.5–1.5 g/L ZnO loading: 0.2–1.5 g/L pH range for TiO_2: 3–10 pH range for ZnO: 6–11	Spectrophotometer, TOC, AAS, antibacterial activity with *Staphylococcus aureus*	Irradiation in the presence of catalyst completely removes the solution's antibacterial activity. The catalytic activity of both catalysts indicates that ZnO presents a slightly higher oxidation rates than TiO_2	Palominos et al. (2009)

(continued)

Table 3.5 (continued)

Compound(s)	Water matrix	Experimental features	Analytical methods	Findings	Reference
Tetracycline	Deionized water	UV lamp (125 W), UV-A lamp (black light, 160 W), solarium device (6 W × 20 W lamps, 300–400 nm) Initial concentration: 40 mg/L TiO_2 P25 loading: 0.5 and 1 g/L Oxygen gas flow: 200 mL/min	HPLC, TOC, UV–vis, antibacterial activity on *Staphylococcus aureus* strain	In the absence of TiO_2, no degradation was observed. Degradation of TC up to 50 % was obtained after 10, 20 and 120 min under UV lamp, solarium device and UV-A lamp respectively	Reyes et al. (2006)
Tetracycline	Water type not provided	Medium pressure Hg lamp (700 W) Initial concentration: 67 mg/L TiO_2 loading: 30 mg/L	HPLC, DOC, LC-ESI/MS/MS, toxicity on *pseudomonas*	Toxicity test carried out on *Pseudomonas* showed decrease in toxicity during tetracycline photocatalysis. Irradiation of 90 min is not sufficient to generate biodegradable by-products	Mboula et al. (2012)
Oxytetracyline	Deionized water	Solar simulator (Xe-OP lamp 1 kW) and CPC reactor Initial concentration: 20 mg/L TiO_2 P25 loading: 0.1 and 0.5 g/L pH: 3, 9, 11 and free initial pH 4.4	HPLC, TOC, BOD, COD, UV-vis, LC-MSD-TOF, ESI-MS, toxicity test on *Vibrio fischeri*	Combination of 0.5 g/L TiO_2 at nominal pH (pH ~ 4.4) showed complete removal after 40 min of irradiation with 7.5 kJ/L of UV dose	Pereira et al. (2011)

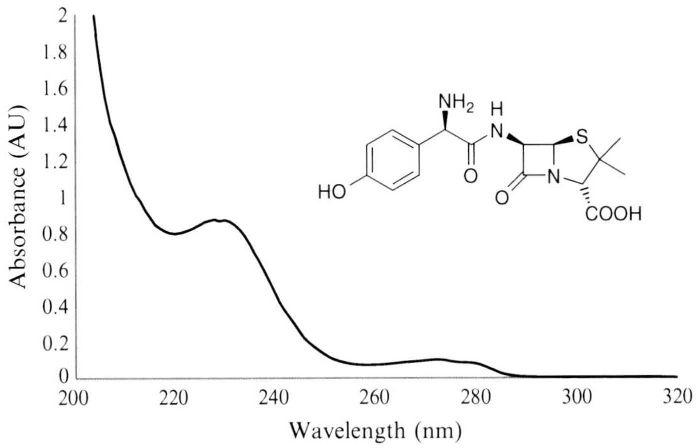

Fig. 3.4 Absorbance spectrum of amoxicillin in water

percentage COD removal was higher for photo-Fenton than for photocatalysis with 64.6 % versus 44 %, respectively. Toxicity inhibition studies with *Artemia salina* showed that both treatments were efficient in removing toxicity and reductions of 43.5 % and 46.3 % were obtained in photo-Fenton and UV/TiO_2 processes, respectively.

Another study compared the efficiency of four AOPs, namely Fenton, photo-Fenton, TiO_2 photocatalysis (UV/TiO_2 and UV/TiO_2/H_2O_2) and ZnO photocatalysis (UV/ZnO) on AMX (104 mg/L) in distilled water in combination with two other antibiotics, ampicillin (105 mg/L) and cloxacillin (103 mg/L) under UV-A irradiation (365 nm), Elmolla and Chaudhuri (2010b). Based on the pseudo-first order rate constants under optimum operating conditions, photo-Fenton demonstrated the highest rate constant of 0.029 min^{-1}, followed by Fenton with 0.0144 min^{-1} with $t_{1/2}$ of 49.5 min and 69.3 min, respectively. Although the performances of UV-A/TiO_2/H_2O_2 and UV/ZnO were similar in terms of the reported rate constants of 0.0005 min^{-1} and 0.00056 min^{-1}, respectively, the rate constant for UV/TiO_2 was not reported therefore preventing a meaningful comparison. With the exception of UV-A/ZnO, biodegradability (BOD_5/COD) was improved for all treatment methods investigated.

A separate study reported the efficacy of TiO_2 (Anatase with purity >99 %) on AMX (and ampicillin and cloxacillin) degradation in distilled water, irradiated with UV-A lamp (365 nm, 6 W) which was further enhanced by the addition of H_2O_2, Elmolla and Chaudhuri (2010c). These results, somewhat contradictory with those reported by Martins et al. (2009), may be due to differences in the photocatalytic activity of the different TiO_2 used, the irradiation source, addition of H_2O_2 and also the water matrix. None of these studies described the degradation products formed during the photocatalytic treatment.

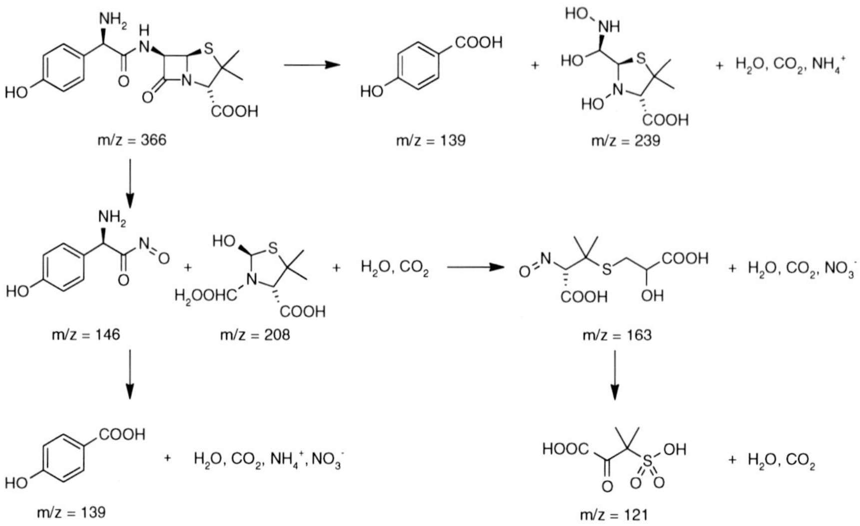

Scheme 3.5 Proposed photocatalytic degradation pathway for amoxicillin (AMX) (10 mg/L AMX; pH 6; 20 °C), Klauson et al. (2010)

A comparison between photocatalysis using doped and undoped titania with artificial UV-A and sunlight showed that that latter produced a three times faster degradation for AMX than artificial UV, Klauson et al. (2010). The improved efficiency of 2 h of solar (16 mW/cm^2) photocatalytic illumination with TiO_2 Degussa P25 over 6 h irradiation with artificial UV-A light (365 nm, 0.5 mW/cm^2) for AMX was apparent. The faster degradation occurred at pH 6 (20 °C) and a concentration of 50 mg/L and yielded the highest photocatalytic efficiency of 80 %. Doped catalysts with Fe and C also showed comparable efficiencies to that of Degussa P25. With artificial UV light, COD was removed to a percentage between 10 % and 40 % in all experiments. This study revealed the superiority of solar photocatalysis for AMX conversion. Photocatalytic degradation pathways were proposed to occur for different AMX concentrations, 10, 25 and 100 mg/L (pH 6, 20 °C) at three different reaction rates of control. The majority of the degradation products differed under these three reaction conditions, with only a few common products identified. Only small amounts of ammonia, nitrate and sulphate were detected, which implies that the heteroatoms N and S remained in the organic by-products. *p*-Hydroxybenzoic acid (m/z 139) was a common degradation product found under all photocatalytic conditions. It is proposed to form through the fragmentation of the peptide bond, which is in close proximity to the aromatic ring in AMX. Scheme 3.5 shows the proposed degradation pathway for AMX.

It can be concluded that artificial UV and solar photocatalysis with TiO_2 facilitated efficient AMX degradation. ZnO was also effective, albeit only one study used this photocatalyst. As a result of this, more research is needed to evaluate the

Table 3.6 Selected sulfonamides and their structures

Common name	-R
Sulfathiazole	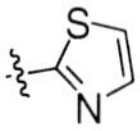
Sulfisoxazole	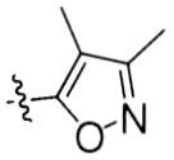
Sulfapyridine	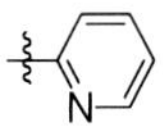
Sulfamethoxazole	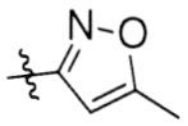
Sulfachloropyridazine	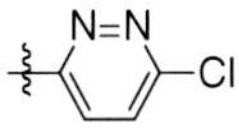

potential of ZnO for AMX degradation by optimizing the parameters influencing its degradation as this catalyst has higher photocatalytic activity than TiO_2. There are also similarities expressed between studies where the UV-A type of irradiation was sufficient to cause AMX degradation.

Sulfonamide and Other Sulfa Antibiotics

Sulfonamides and sulfanilamides are derived from sulfanilic acid and their common features are the sulfanilamide group and a five or six member heterocyclic ring (shown as R in the general structure) and are represented in Table 3.4. Selected sulphonamides, with their common names and the nature of their R groups are shown in Table 3.6. Sulfonamides and sulfanilamide derivatives are antibacterial agents used both in animals and humans for the treatment of infections. Almost 80 % of all sulfonamides have been detected in the environment, which could easily infiltrate groundwater, and have shown high toxicity to microorganisms, algae and plants, Baran et al. (2011).

Sulfamethoxazole (SMX) ($C_{10}H_{11}N_3O_3S$), being the most commonly used member of the group of sulfonamide antibiotics, is prescribed for urinary tract infections and has been detected in the environment. This synthetic antibiotic is used either alone or in combination with trimethoprim. SMX has been detected in plant effluents at concentrations ranging between 0.01 and 2.0 μg/L in countries such as Germany, Spain and Italy and in surface waters between 0.03 and 0.48 ug/L, Trovó et al. (2009). SMX has two pK_a values, $pK_{a1} = 1.6$ and $pK_{a2} = 5.7$, and thus exists predominantly in the cationic form at $pH < 1.6$ and in the anionic form at $pH > 5.7$. There have been reports on UV/TiO_2 treatment of this pharmaceutical using various lamp sources, Abellán et al. 2009 and Nasuhoglu et al. (2011).

SMX has been reported to be susceptible to direct photolysis with a Xe lamp resulting in significant degradation, but poor mineralization, Abellán et al. (2007, 2009). In water, SMX is capable of absorbing light up to 310 nm. When irradiated

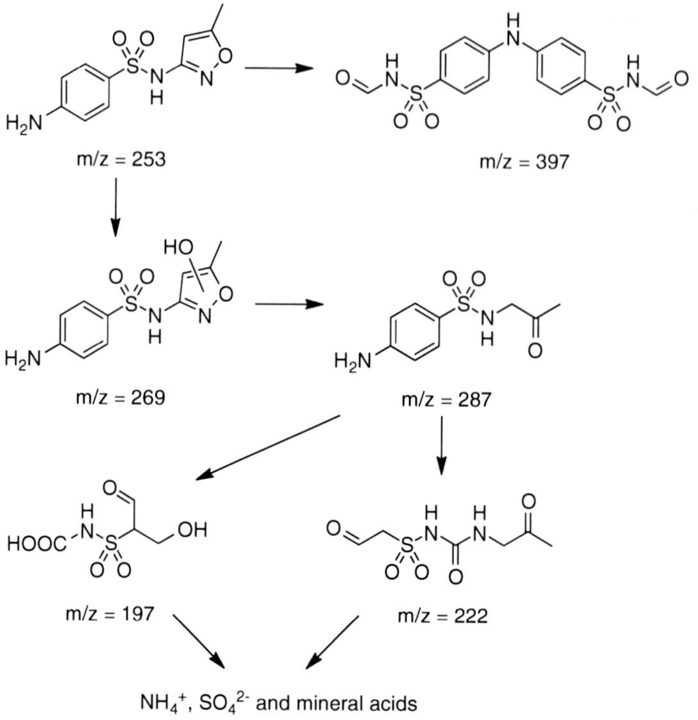

Scheme 3.6 Degradation pathways of sulfamethoxazole (SMX) with TiO_2 photocatalysis, Abellán et al. (2007)

without the use of a filter and photocatalyst, SMX degraded up to 80 % but only resulted in a reduction of 14 % of TOC, Abellán et al. (2007). In contrast, the photocatalytic degradation of 100 ppm SMX (irradiation of 6 h) in the presence of 1.0 g/L TiO_2 yielded 88 % reduction in SMX. With 2.0 g/L TiO_2, the degradation marginally increased to 91 %. The optimum TiO_2 P25 concentration was found between 0.5 and 1.0 g/L. Variations in pH (2–11) resulted in very minor changes and the degradation remained almost constant at 85 %, with a very small increase at lower pHs. A total of five intermediates were identified using mass spectrometry. Of these, the intermediate with a molecular weight of m/z = 397 was found to be the major product and was assigned to a dimer product of the parent SMX. Other intermediates possessed molecular weights of 222 in ES positive ion mode and 197, 269 and 287 in ES negative ion mode, Abellán et al. (2007) (Scheme 3.6).

UV-C irradiation (254 nm) was employed to study the photolytic and photocatalytic degradation of SMX in pure water, Nasuhoglu et al. (2011). Irradiation was carried out in a cylindrical acrylic photoreactor equipped with an Hg-Ar lamp source. The results suggest that photolysis is an important removal mode for SMX (12 mg/L) which corroborates the findings of Abellán et al. (2007). The high tendency for direct photolysis was explained based on the matching of the SMX absorbance with the highest intensity emission of the UV-C lamp. UV-C radiation yielded complete degradation of 12 mg/L SMX within 10 min, while UV/TiO_2 required 30 min to achieve a similar level of degradation. However, mineralization measured as COD was more efficient on photocatalysis treatment with 87 % removal, compared to only 24 % removal upon photolysis. Photoproducts generated as a result of both treatments were reported to be more toxic than the parent SMX, as demonstrated by the acute toxicity *Daphnia magna* testing.

SMX was reported to undergo UV-A (350–400 nm) induced photocatalytic degradation with TiO_2 Degussa P25, Xekoukoulotakis et al. (2011). A 9 W lamp (photon flux 2.81×10^{-4} Einstein/min) in an immersion well batch photoreactor was used under continuous oxygen sparging. Complete degradation of 10 mg/L SMX with 250 mg/L TiO_2 P25 was achieved in 30 min. Likewise, a TOC reduction of up to 90 % was observed. The photocatalytic degradation at different initial concentrations (2.5–30 mg/L) followed the Langmuir-Hinshelwood (L-H) kinetic model.

The behaviour of three sulfa pharmaceuticals, namely sulfachlorpyridazine, sulfapyridine and sulfisoxazole with similar structures towards UV/TiO_2 treatment were investigated by Yang et al. (2010). A Pyrex reactor with a high pressure Hg lamp ($\lambda_{max} = 365$ nm) was chosen. Photocatalytic treatment proved generally effective on these sulfa drugs, with removal efficiencies varying between 85.2 % and 92.5 % after 60 min of irradiation. TOC was concomitantly removed in high percentages (81.5–90.8 %) within 240 min with all three drugs showing complete mineralization to CO_2, H_2O and inorganic ions. Photocatalytic degradation rate constants of these three sulfa drugs also increased with the amount of TiO_2 from 0.25 to 3 g/L. It was proposed that h^+ and $HO^\bullet$ play the most important role in the disappearance of sulfa pharmaceuticals. Remarkably, different photocatalytic degradation rate constants for the three drugs were obtained at different pH values despite their similarity in structure. This was attributed to changes in their zwitterionic nature in different pH environments. Various intermediates formed rapidly within 15 or 20 min of irradiation and then slowly disappeared. The two major pathways for photocatalytic degradation of the sulfa drugs were hydroxylation and cleavage of the S-N bond by the photohole (Scheme 3.7).

Another study employed a UV-A light source to determine the degradation of SMX and a few other sulphonamides, namely sulfamethizole, sulfathiazole and sulfisoxazole, Hu et al. (2007). The effects of natural organic matter (NOM) and bicarbonate ion on the degradation of SMX were investigated. The occurrences of NOM and bicarbonate ions in water are known to impede the degradation of compounds by scavenging $HO^\bullet$. However, the results showed that an increase in bicarbonate concentrations generated a higher degradation rate for SMX. This was attributed to considerably higher HCO_3^- and CO_3^{2-} concentrations compared to SMX, which is likely to induce a stronger interaction with Ti (IV) than between SMX and Ti (IV). As a result, these ions efficiently scavenged the adsorbed $HO^\bullet$ radicals, which in turn reduced the recombination of the charge carriers. The $CO_3^{-\bullet}$ radicals generated were proposed to react with SMX in the bulk rather than on the

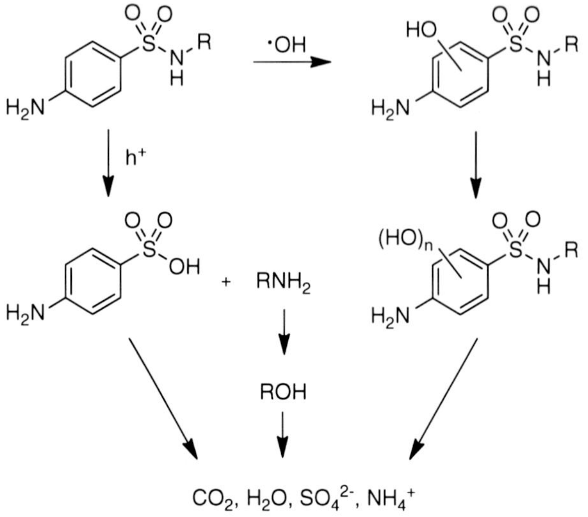

Scheme 3.7 Photocatalytic degradation pathways of sulfa pharmaceuticals, Yang et al. (2010)

surface because they are more stable and posses longer lifetimes. In contrast, an increase in NOM concentrations (2–20 mg/L) inhibited the SMX degradation, in particular at higher concentrations of NOM.

Quinolones and Fluoroquinolones

Quinolones are synthetic antibacterial agents which are widely used in both humans and livestock. The basic structure of quinolones contains two fused rings with a carboxylic acid and a ketone group (Table 3.4). First and second generation quinolones are active against gram-negative bacteria, while the third and fourth generations have extended activity against gram-positive bacteria as well, Nasuhoglu et al. (2012). Fluoroquinolones (FQs) have attracted substantial attention due to their incomplete metabolism and frequent detection in the environment. A FQ compound contains a fluorine substituent in position R^4 of the quionolone ring as shown in Table 3.4.

FQs are excreted to between 20 % and 80 % and their persistence has led to detection in surface waters, municipal wastewaters and tertiary treated effluents, Michael et al. (2010). As with other synthetic antibiotics, FQs are not fully metabolized and thus excreted unchanged. The persistent nature of FQs is associated with the quinolone ring, which results in high chemical stability and contributes additionally to their resistance towards hydrolysis, Sturini et al. (2010). Only a few studies dealing with the degradation of FQs by TiO_2 mediated photocatalytic treatment have been reported thus far. In general, incomplete mineralization of FQs results in the generation of a range of intermediates which still contain the quinolone ring structure and as a result continue to exhibit antibacterial activity.

The performance of immobilized TiO_2 integrated into an annular reactor was tested with black light (360 nm) using 18 ppm oxolinic acid (OX) at pH 9 under oxygenated conditions. A comparison was made with a 1 g/L titania suspension, using the same reactor. The coated titania was capable of removing OX acid to over 90 % after 60 min, while TOC was reduced by 50 % after 100 min of irradiation. In suspension, 90 % removal was achieved after 20 min whereas total removal took 40 min. The superior performance of the suspended TiO_2 was explained by the superior adsorption, which subsequently increased the reaction rate, Palominos et al. (2008).

A systematic evaluation of OX acid removal in ultrapure water was conducted by Giraldo et al. (2010). The optimum conditions for complete degradation of 20 mg/L OX acid after 60 min of irradiation were with 1.0 g/L TiO_2 and pH 7.5. COD was reduced by 50 % and the degradation followed pseudo-first order kinetics. The study also revealed that OX acid intermediates, which were identified by high performance liquid chromatography-mass spectrometry (HPLC/MS) exhibited lower toxicity than the parent compound in addition to a loss of antimicrobial activity based on an *Escherichia coli* assay.

UV-A/TiO_2 is also an effective technology for oxidation of ciprofloxacin (CIP) in pure water, An et al. (2010) and Paul et al. (2010). UV-A and Vis light sources were employed to study the photolytic and photocatalytic treatment of CIP, Paul et al. (2010). Based on the pseudo-first order rate constants, the efficiency of CIP degradation increased in the order UV-A/TiO_2 > Vis/TiO_2 > UV-A only. Degradation of CIP by UV-A/TiO_2 is proposed to occur via three mechanisms: direct UV-A photolysis, charge transfer and charge separation involving the semiconductor. Various piperazine ring transformation products were formed upon UV-A/TiO_2 and Vis/TiO_2 treatment and their proposed structures are shown in Scheme 3.8. Energy efficiencies for the three treatment protocols were compared and the UV-A/TiO_2 turned out to be the most energy efficient process.

Degradation of 100 μM CIP and a short $t_{1/2}$ of 1.9–10.9 min were furthermore achieved at different pH ranges (3.0–11.0) using 1.5 g TiO_2/L Degussa P25 and UV-A irradiation (365 nm) emitted from a high pressure Hg lamp. The degradations followed pseudo-first order kinetics and the rate constant increased from $0.06 \pm 0.01\ min^{-1}$ at pH 3 to $0.38 \pm 0.01\ min^{-1}$ at pH 9. At a higher pH value of 11, the degradation rate decreased to $0.07 \pm 0.01\ min^{-1}$ due to the repulsive effect between the negatively charged compound and the negatively charged TiO_2 surface. Addition of hydroxyl radicals to CIP and photohole attack were established as the two predominant reaction pathways, An et al. (2010).

Photocatalytic degradation of four FQs, namely ofloxacin, norfloxacin, ciprofloxacin and enrofloxacin, was examined simultaneously using a solar simulator, Li et al. (2012). Alkaline conditions (pH 9) resulted in almost complete degradation (98–100%) for all four FQs after 150 min of irradiation. Likewise, TOC was reduced by 58 %. The threshold TiO_2 load was found to be 0.5 g/L at pH 9 to achieve 96–100% of degradation of the four FQs. An increase in the TiO_2 load from 0.1 g/L to 1.5 g/L did not produce a direct correlation with TOC removal. The effects of pH changes on these four FQs were rather complex, due

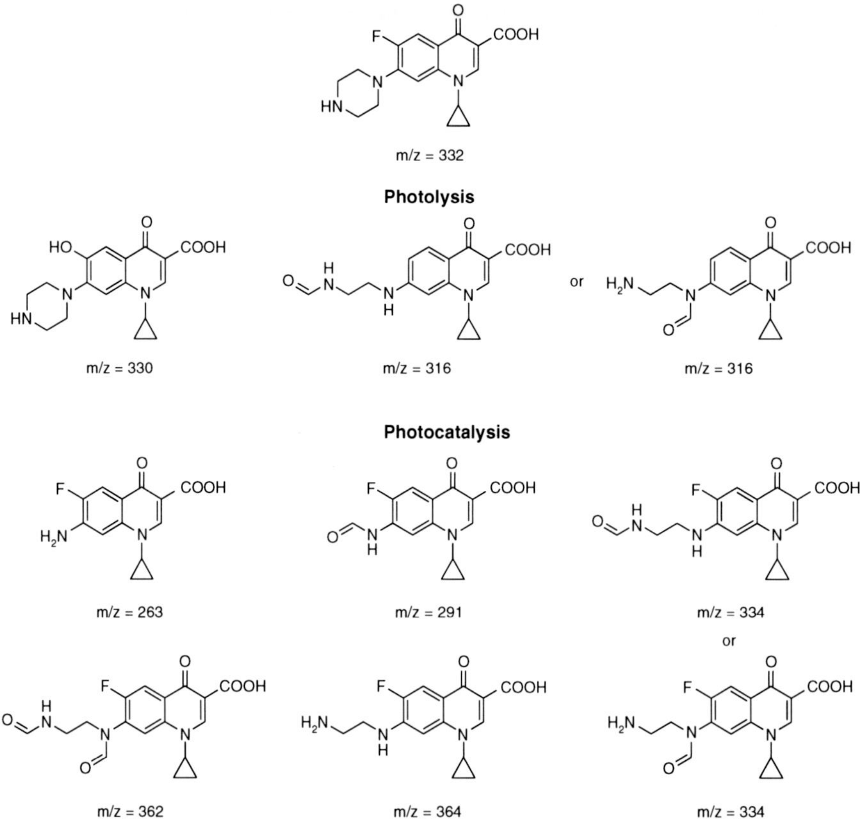

Scheme 3.8 Possible degradation products of ciprofloxacin (CIP) from photolysis and TiO_2 photocatalysis, Paul et al. (2010)

Table 3.7 Molecular formulas and pK_a values of four fluoroquinolones (FQ), Li et al. (2012)

	Ofloxacin	Norfloxacin	Ciprofloxacin	Enrofloxacin
Chemical formula	$C_{18}H_{20}FN_3O_4$	$C_{16}H_{18}FN_3O_3$	$C_{17}H_{18}FN_3O_3$	$C_{19}H_{22}FN_3O_3$
pK_a	pK_{a1}: 6.05	pK_{a1}: 6.20	pK_{a1}: 6.00	pK_{a1}: 6.03
	pK_{a2}: 8.11	pK_{a2}: 8.70	pK_{a2}: 8.80	pK_{a2}: 8.70

to different ionization states of the catalyst and the substrates (two pK_a values for each). In particular, these FQs can either exist in their cationic ($pH < pK_{a1}$), or anionic ($pH > pK_{a2}$) or in zwitterionic form ($pK_{a1} < pH < pK_{a2}$) (Table 3.7) due to the presence of a carboxylic acid group in the quinolone structure and an additional amine group from the piperazinyl ring.

A third generation FQ, moxifloxacin (MOX) was subjected to UV/TiO_2 and the influence of various parameters was examined using a batch photoreactor equipped with a UV-A lamp (300–440 nm, 485 $\mu W/cm^2$), Van Doorslaer et al. (2012). A catalyst loading of 5 g/L, temperature of 25 °C and air sparging at 60 mL/min were found optimal for removing 37.4 μM of MOX. The initial degradation rate and the $t_{1/2}$ were determined to be 16.2 ± 0.3 μM/min and 1.6 min, respectively. A stirring speed of 13.2 rps produced a higher degradation rate. The TiO_2 P25 loading ranged from 0.1 to 5.0 g/L.

An independent study highlighted that TiO_2 Degussa P25 was more effective than two other common photocatalysts, Hombikat UV 100 and PC 500 for the degradation of 0.25 mM of norfloxacin (NOR) (pH 6.3) with 1 g/L of photocatalyst, Haque and Muneer (2007). Almost complete degradation was achieved in 80 min of irradiation from a medium pressure Hg lamp (125 W) under oxygen purging. Likewise, the TOC content was depleted as a function of irradiation time. The efficiency of the photocatalytic degradation of NOR was subsequently correlated with pH, substrate concentration, catalyst concentration, and the addition of electron acceptor, H_2O_2. The results revealed that addition of 10 mM of H_2O_2 significantly enhanced the degradation rate and mineralization of 0.25 mM NOR (pH 6.3 and 1 g/L TiO_2 P25).

A recent study was conducted with unfiltered river water to explore the photocatalytic decomposition of six FQs, namely ciprofloxacin, danofloxacin, enrofloxacin, levofloxacin, marbofloxacin and moxifloxacin, under natural sunlight. Compared to direct photolysis under natural sunlight, all of the tested drugs except CIP underwent faster degradation with degradation rate constants ranging from 0.22 to 2.78 min^{-1} in the presence of TiO_2 (0.5 g/L). The results demonstrated the effectiveness of solar TiO_2 photocatalysis for the removal of FQs from river water matrix despite the presence of other non-target matrix components, Sturini et al. (2012). CIP in hospital wastewater was also satisfactorily removed with UV/TiO_2 using a medium pressure Hg lamp. For comparison, ozonation was used as an alternative AOP. Both processes generated similar degradation products and oxidation of the piperazine group was found to be the major degradation pathway, Vasconcelos et al. (2009).

The results presented for quinolones and FQs are significant and demonstrate the potential removal from wastewaters not only for single compounds but also for mixtures.

Tetracyclines

A comparison of photocatalytic degradation and mineralization of oxytetracycline (OTC) in a laboratory and at pilot-scale was reported by Pereira et al. (2011). Laboratory scale experiments with a solar simulator showed that photocatalysis of 20 mg/L OTC with 0.5 g/L TiO_2 at free initial pH produced 95 % degradation. In contrast, only 36 % degradation was achieved upon direct photolysis after 60 min of irradiation. Solar photocatalysis in a CPC pilot plant reactor showed complete

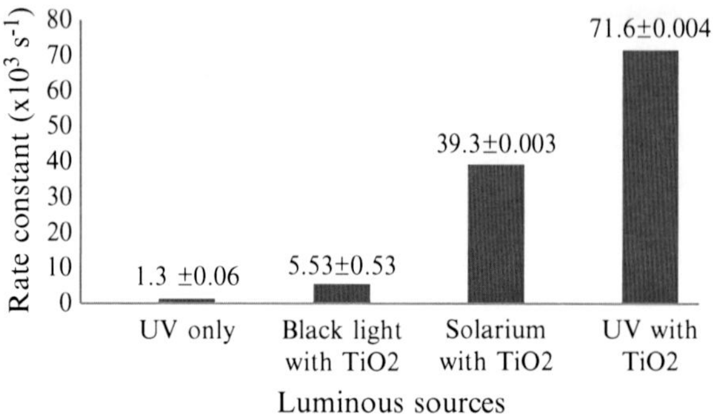

Fig. 3.5 Comparison of pseudo-first order constant for tetracycline (TC) degradation (TC: 40 mg/L; TiO_2 P25: 0.5 g/L) (Data taken from Reyes et al. (2006))

removal of OTC with energy consumption 10 times less than solar photolysis which was also performed by means of CPC reactor. Solar photocatalysis in the pilot plant reactor yielded 80 % mineralization. All experiments followed pseudo-first order kinetics, but the rate constants increased from the photolytic to photocatalytic operation modes. Solar photocatalysis in a CPC reactor produced the highest pseudo-first order rate constant of 2.63 ± 0.03 L/kJ, indicating the efficiency of such a photoreactor to harvest and utilize solar photons.

Several studies involving tetracycline (TC) have been carried out to determine its photocatalytic degradation, Reyes et al. (2006), Palominos et al. (2009) and Mboula et al. (2012). Catalyst performance and biodegradability or antibacterial efficiency arising from the treated solutions differed for all studies reported. Using a multivariate and response surface method approach, the optimum degradation rate for each catalyst was investigated. ZnO exhibited a slightly higher oxidation rate than TiO_2 P25. The optimum oxidation conditions for TiO_2 were established as 1.5 g/L and at pH 8.7. For ZnO, the most appropriate conditions were 1.0 g/L and pH 11, Palominos et al. (2009).

The TC disappearance over the irradiation time, at different pH values and with three different light sources, namely UV lamp ($\lambda > 254$ nm), UV-A lamp ($\lambda = 365$ nm) and a solarium device ($\lambda = 300$–400 nm), was carried out by Reyes et al. (2006). Based on the pseudo-first order rate constants, direct photolysis was regarded negligible for TC. In contrast, faster degradation rates were achieved with the solarium and UV light in the presence of TiO_2 (Fig. 3.5). These findings were supported by the high irradiance intensities measured from the UV lamp (1,210 $\mu W/cm^2$ at 365 nm) and the solarium lamp (1,980 $\mu W/cm^2$ at 365 nm), respectively. Rapid disappearance of TC and subsequent complete mineralization was achieved after 2 h of irradiation.

3.5.3 Antiepileptics

The most commonly investigated antiepileptic is the dibenzazepine derivative, carbamazepine (CBZ) (Fig. 3.6). CBZ is also used as sedative for numerous mental

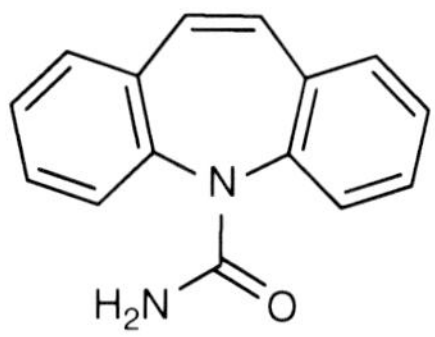

Fig. 3.6 Chemical structure of carbamazepine ($C_{15}H_{12}N_2O$) (5*H*-dibenzo [*b*,*f*]azepine-5-carboxamide)

disorders, can cause toxic effects in the liver and is a haematopoietic. Only 72 % of CBZ is absorbed when orally administered, while 28 % remains unchanged and is finally excreted through faeces. CBZ has been known to persist in the environment and studies have reported the removal efficiency of CBZ in WWTPs as below 10 %, Andreozzi et al. (2002) and Zhang et al. (2008a). Thus far, CBZ has been detected in various water compartments around the world. In Germany, CBZ was detected in surface water samples at a concentration up to 1,075 ng/L, Heberer (2002b) and also in groundwater and rivers in the USA, Canada and Germany, Khetan and Collins (2007). These observations clearly confirm the CBZ presence in the aquatic environment. Although the effects of CBZ are not clearly known or confirmed, its toxic effect on aquatic organisms has been already published. There is also evidence of specific chronic effects of CBZ on the oligochaete *Chiromus*, Oetken et al. (2005). These facts show that this antiepileptic needs to be removed with effective remediation technology. In this context, photocatalysis has been shown to be effective for CBZ degradation. A summary of UV/TiO_2 studies of CBZ is tabulated in Table 3.8.

There has been an insufficient amount of information available pertaining to photocatalytic degradation of CBZ. Photocatalytic degradation of CBZ has however been examined in deionized water, Dai et al. (2012) and Im et al. (2012), real wastewater samples from WWTPs, Achilleos et al. (2010b) and formulated hospital wastewater, Chong and Jin (2012). In general, there are some discrepancies in the photocatalytic efficiency on CBZ.

A comparative study on the effect of different on CBZ (4.2 μM) degradation efficiencies showed the following trend: UV/Fenton (86.9 %) > UV/TiO_2 (70.4 %) > Fenton (67.8 %) > UV/H_2O_2 (40.6 %) > UV (14.2 %). On the basis of operating costs, the Fenton process was the most cost-effective in comparison with the other four AOPs (Table 3.9), Dai et al. (2012). As shown in Table 3.9, UV/TiO_2 and UV can be considered as costly approaches for wastewater treatment. However the authors reported that this statement can be challenged on the grounds that the cost of photocatalytic treatment can be reduced by applying sunlight as the irradiation source.

The removal efficiency by UV-A (350–400 nm) and solar irradiation provided by a solar simulator on CBZ was investigated along with another API, IBP, Achilleos et al. (2010b). Greater degradation was attained with UV-A irradiation than with artificial solar radiation for both APIs. Degradation of CBZ in pure water was found to be more dependent on changes in the loading of Degussa P25 TiO_2 compared to IBP. Conversion of 74 % CBZ was achieved after 120 min of reaction with 100 mg/L

Table 3.8 Highlights of heterogeneous photocatalytic studies on antiepileptics

Compound(s)	Water matrix	Experimental features	Analytical methods	Findings	Reference
Carbamazepine					
	Deionized water	Medium pressure Hg lamp (30 W) Initial concentration: 4.2 μM TiO_2 loading: 0.5–2.0 g/L Temperature: 25 °C pH: 6.5	HPLC	Comparison of UV/TiO_2 with UV, UV/H_2O_2 and UV/Fenton showed the UV/TiO_2 yielded 70.4 % degradation which was slightly lower than UV/Fenton at 86.9 % degradation. All processes followed pseudo-first order kinetics	Dai et al. (2012)
	Hospital wastewater	Annular slurry photoreactor UV-A black light (8 W) Initial concentration: 5,000 μg/L TiO_2 Anatase particles	HPLC, COD	Photodegradation at high concentration of CBZ in synthetic hospital wastewater was fitted to the L-H kinetics model	Chong and Jin (2012)
	Milli-Q water	UV-A lamp: 2.17 mW/cm^2 UV-C lamp: 3.56 mW/cm^2 Initial concentration: 0.021 mM TiO_2 loading: 0.5 g/L pH: 5.4	HPLC, IC, TOC	UV-A and UV-C irradiation efficient to degrade CBZ	Im et al. (2012)
Carbamazepine mixtures					
Carbamazepine (and ibuprofen)	Milli-Q water and wastewater from WWTP	Solar simulator (1,000 W Phillip Xe lamp) and UV-A lamp (9 W Radium lamp) Initial concentration: 5–20 mg/L TiO_2 loading: 50–3,000 mg/L Type of TiO_2: Degussa P25, Hombikat UV100, Aldrich,	UV-vis (CBZ: 284 nm), DOC	Degradation under UV-A irradiation in pure water was sensitive to TiO_2 P25 loading. Solar and UV-A photocatalysis appears to be efficient on CBZ degradation	Achilleos et al. (2010b)

		Tronox AK-1, Tronox TRHP-2, Tronox TR H_2O_2 concentration: 0.07–1.4 mM pH: 3–10			
Carbamazepine (Clofibric acid, iomeprol)	Milli-Q water	Solar simulator Xe short arc lamp (1,000 W) (cut-off 296 nm) Initial concentration: 1–5.5 mg/L TiO_2 loading (P25 and Hombikat UV100): 500 mg/L	UV–vis, HPLC, DOC	Degradation of CBZ dependent on the concentration of NOM as it greatly influences its kinetics	Doll and Frimmel (2005a)
Carbamazepine (clofibric acid, iomeprol)	Ultrapure water	Solar simulator Xe short-arc lamp (1,000 W) Initial concentration: 4.3 mg/L TiO_2 and Hombikat UV100 loading: 0.1 g/L Temperature: 20 ± 2 °C	HPLC, DOC, ICP-AES, HPLC/TQMS	Pseudo-first order degradation rate of CBZ for TiO_2 P25 and UV100 were 4.7×10^{-3} and 0.13×10^{-3} s^{-1}	Doll and Frimmel (2005b)
Carbamazepine (Clofibric acid, iomeprol, iopromide)	Milli-Q water	Solar simulator Xe short arc lamp (1,000 W) Initial concentration: 1.0 and 4.3 mg/L TiO_2 P25 loading: 0.01–1,000 mg/L Temperature: 20 ± 2 °C Hombikat UV100 loading: 0.1–1,000 mg/L	HPLC, DOC, ICP-AES, HPLC/ESMS, DOC	Degradation of CBZ was faster with TiO_2 P25 due to higher adsorption	Doll and Frimmel (2004)

Table 3.9 Comparison of operating costs of advanced oxidation processes (AOP) studied (Data taken from Dai et al. (2012))

Process	Treatment cost (US$/kg)
UV	210.4
UV/H_2O_2	42.6
Fenton	5.4
UV/Fenton	26.7
UV/TiO_2	70.2

P25 under UV-A irradiation, but it decreased to 35 % upon solar irradiation, which was attributed to completely different reactor set-ups. Wastewater samples spiked with 10 mg/L CBZ had a detrimental effect on DOC removal, due to the presence of scavengers of $HO^•$ radicals and naturally occurring DOC in the raw wastewater sample. Nevertheless, the study proved that UV-A and solar irradiation can be applied to CBZ removal (and also for IBP). Direct photolysis with a UV-A lamp produced negligible degradation and DOC removal for CBZ.

The results from another study confirmed that direct photolysis supplied by a UV-C irradiation source is not effective for CBZ elimination, Im et al. (2012). An important contribution of the study is that UV-C irradiation increased the removal efficiency compared to that of UV-A irradiation employed on the 0.021 mM CBZ in the presence of 0.5 g/L TiO_2 P25. Addition of radical scavengers decreased the removal rate, implying that $HO^•$ has a crucial role to play in the photocatalytic degradation of this compound. Addition of oxygen also enhanced the mineralization.

A TiO_2 nanofiber was evaluated as a pre-treatment option and for improvement of biodegradability of 5,000 μg/L CBZ in synthetic hospital wastewater, Chong and Jin (2012). This application was found to be efficient for CBZ abatement in synthetic hospital wastewater and also for COD removal. High CBZ degradation rate of 48.33 μg/L min was obtained, which also resulted in 40 % COD reduction after 4 h of treatment. The study highlighted the fact that the TiO_2 based system has a good potential as a sustainable pre-treatment system for hospital wastewater.

Doll and Frimmel (2004, 2005a, b) conducted multiple photocatalytic studies on CBZ under various conditions, which revealed that TiO_2 photocatalysis is a promising abatement method for CBZ degradation. When a comparison was made with Hombikat UV100, CBZ was degraded more efficiently with TiO_2 P25 under simulated solar irradiation (1,000 W), Doll and Frimmel (2004). In the subsequent study, photocatalytic degradation of CBZ was investigated with two types of TiO_2 photocatalysts in the presence and absence of NOM under simulated solar UV light, Doll and Frimmel (2005a). It was noted that different concentrations of NOM had different effects on the kinetics of CBZ degradation. In general, NOM retarded the photocatalytic degradation rate of CBZ. This was attributed to competitive inhibition as NOM scavenged the holes. The research group also suggested possible photocatalytic degradation products for CBZ, Doll and Frimmel (2005b) (Scheme 3.9).

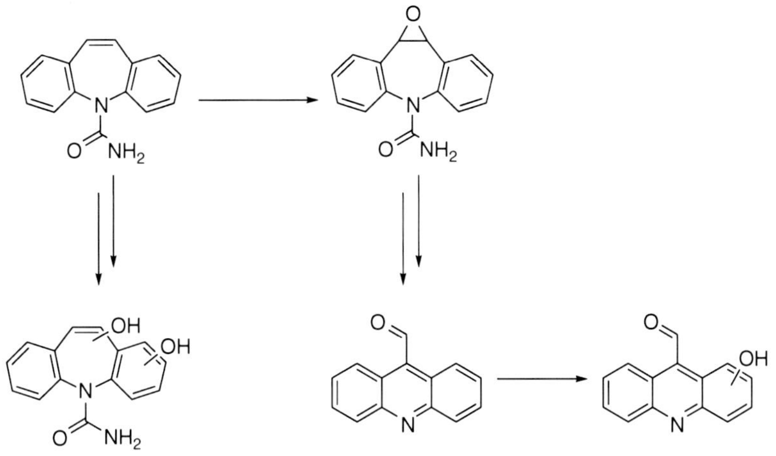

Scheme 3.9 Photocatalytic degradation pathways of carbamazepine (CBZ), Doll and Frimmel (2005b)

3.6 Conclusion

A typical UV/TiO_2 investigation is shown in Fig. 3.7. After one or more model compounds have been chosen, the next step is the photocatalytic experimental set-up or the choice of the photoreactor and light source, which will have a significant effect on the outcome of the degradation rate, Legrini et al. (1993). There are various factors that need to be taken into consideration with respect to photoreactors, namely the material it is made of, shape, geometry, radiation path lengths, and the cooling apparatus. Similar issues have been highlighted in a review which has commented on this lack of information, when explaining photoproducts formed from the photocatalytic and photolytic studies, Fatta-Kassinos et al. (2011a).

When considering light sources, artificial light sources have been employed in most studies. These can be grouped into monochromatic light sources such as low pressure Hg lamps and polychromatic light source namely medium pressure Hg lamps and high pressure Hg lamps. Other irradiation devices include Xe, Xe-arc, Hg and Ra lamps. Solar simulators have been commonly applied to simulate solar irradiation. LED based light sources are also emerging as an alternative light source, due to advantages which include no mercury waste, longer life time compared to Hg based lamps and improved flexibility in terms of reactor design, Landgraf (2001) and Chen et al. (2007). Very few studies have made a comparison between the lamp's spectral domain and the absorption spectrum of the reactant. Photon flux, intensity and irradiance emitted from the light source also need to be clearly stated as these values determine the rate of photons emitted, which has a direct correlation to the degradation rate. Selection of a radiation source (shape and dimension) is also greatly influenced by the type of photoreactor.

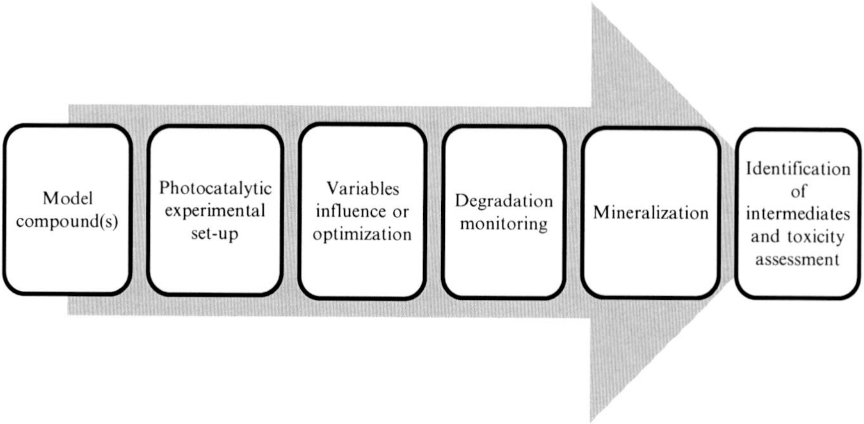

Fig. 3.7 Typical steps involved in heterogeneous photocatalytic degradation

Reaction rates are known to be dependent on operation conditions. Optimization of these variables thus allows for maximum degradation. Statistical approaches such as multivariate analysis and response surface methodology have also been applied to establish the optimal degradation rate. One of the most commonly investigated and important factors is the TiO_2 loading. It has been shown that the degradation rate is not always proportional to the catalyst load, but an optimum load needs to be determined. Difficulty arises when comparing studies for a particular compound of interest. In general, optimal catalyst load is influenced by the photoreactor design including its diameter and source of irradiation. Commercially available TiO_2 such as Degussa P25, Hombikat UV 100, PC 500 and Aldrich have been used in various photocatalytic degradation studies of pharmaceuticals. Among them, TiO_2 Degussa P25 has demonstrated its superiority in removing various APIs. A concentration below the optimum catalyst loading is typically chosen for further investigation, due to the fact that catalyst deposition might occur during the experiment. For example, 0.4 g/L TiO_2 was chosen instead of 0.5 g/L which gave the best results for metoprolol and propranolol degradation, Romero et al. (2011).

Initial concentrations of APIs applied in photocatalytic studies tend to be higher than the environmental level. Furthermore, in most cases single pharmaceuticals are investigated despite the fact that real wastewaters contain complex mixtures of pharmaceuticals. Although most studies have demonstrated efficiency for API in spiked distilled water or ultrapure water, real wastewater samples are rarely used. The presence of radical scavengers such as HCO_3^-, CO_3^{2-} ions and NOM in real wastewater typically impacts on degradation efficiencies. Radical scavengers or inhibitors are frequently studied to determine the role of reactive intermediates, e.g. the $HO^\bullet$ radical, $O_2^{\bullet -}$ anion and holes in the photocatalytic degradation. Isopropanol, benzoquinone and iodide ions are commonly used for this purpose.

In general, laboratory scale experiments are more common than pilot scale operations due to a more controllable environment. There is usually no direct correlation between the effectiveness of photodegradation in the laboratory with that under real environmental conditions. Conducting photodegradation studies on pilot scales will provide more realistic environmental conditions such as season, latitude, and hardness of water. These are generally not encountered when working in the laboratory. Trials using solar energy have been reported for pharmaceutical degradation, although the results thus far vary between compounds of interest. Commercial and industrial scale applications of solar energy have been established in a few European countries, the USA and Canada.

In the subsequent step, degradation monitoring of pharmaceuticals is generally carried out by using HPLC detection or UV–vis spectrophotometry. The latter has been reported to be a fast and useful tool of monitoring although it is not the state-of-the-art technique for API detection. The kinetics of degradation in terms of percentage API removed or API disappearances versus irradiation time or concentration change versus irradiation time (C/C_0 vs time) are commonly reported. The L-H model has been often applied to describe photocatalytic mineralization of pharmaceutical degradation.

Apart from determining the removal rate of APIs with photocatalytic treatment, the degree of mineralization is a critical parameter in heterogeneous photocatalysis. DOC has been frequently measured as an indication of the formation of CO_2. In most cases, complete degradation of the pharmaceutical does not correspond directly with the mineralization rate mainly due to due to the formation of more stable compounds or transformation products during the degradation. A non-biodegradable fraction frequently remains in the treated solution. Mineralization can be also determined by measuring the formation of inorganic ions such as chlorine, sulphate and nitrate.

Although the ultimate goal of heterogeneous photocatalysis is to achieve complete mineralization of parent compounds, it is also important to evaluate the by-products generated to ensure that there are no toxic or more persistent compounds produced during the treatment. This will contradict the ultimate goal of the application of such an AOP. However, photodegradation products are not commonly studied. A few possible reasons for this are: (i) difficulties in separating and identifying a large number of these transformation products formed; (ii) lack of or non-existent analytical standards to determine the identity of these transformation products; and (iii) requirement for more than one analytical technique or sample preparation technique due to their diverse physicochemical properties, Agüera et al. (2005) and Fatta-Kassinos et al. (2011b). Moreover, studies have also assessed the toxicity of the photocatalytically treated samples by applying toxicity assays on various microorganisms and invertebrates.

In conclusion, despite most promising findings reported, the lack of compatibility in terms of degradation schemes and photoproducts formed, removal rates and degree of mineralization of APIs highlights the complex behaviour of the various APIs towards UV/TiO_2. Meaningful comparison between studies for scaling-up and application to real wastewater treatment at a pilot scale is limited by all these factors. Additional studies are thus urgently required to eliminate these shortcomings.

Acknowledgements The authors thank James Cook University for financial support (FAIG award 2009, GRS awards 2011 and 2012). DK thanks the Malaysian Government for a University Doctorate Training Award.

References

Abellán MN, Bayarri B, Giménez J, Costa J (2007) Photocatalytic degradation of sulfamethoxazole in aqueous suspension of TiO_2. Appl Catal B Environ 74(3–4):233–241. doi:10.1016/j.apcatb.2007.02.017

Abellán MN, Giménez J, Esplugas S (2009) Photocatalytic degradation of antibiotics: the case of sulfamethoxazole and trimethoprim. Catal Today 144(1–2):131–136. doi:10.1016/j.cattod.2009.01.051

Achilleos A, Hapeshi E, Xekoukoulotakis NP, Mantzavinos D, Fatta-Kassinos D (2010a) Factors affecting diclofenac decomposition in water by UV-A/TiO_2 photocatalysis. Chem Eng J 161(1–2):53–59. doi:10.1016/j.cej.2010.04.020

Achilleos A, Hapeshi E, Xekoukoulotakis NP, Mantzavinos D, Fatta-Kassinos D (2010b) UV-A and solar photodegradation of ibuprofen and carbamazepine catalyzed by TiO_2. Sep Sci Technol 45(11):1564–1570. doi:10.1080/01496395.2010.487463

Agüera A, Estrada LAP, Ferrer I, Thurman EM, Malato S, Fernández-Alba AR (2005) Application of time-of-flight mass spectrometry to the analysis of phototransformation products of diclofenac in water under natural sunlight. J Mass Spectrom 40(7):908–915. doi:10.1002/jms.867

Aguinaco A, Beltrán FJ, García-Araya JF, Oropesa A (2012) Photocatalytic ozonation to remove the pharmaceutical diclofenac from water: influence of variables. Chem Eng J 189–190:275–282. doi:10.1016/j.cej.2012.02.072

AIHW Australian Institute of Health and Welfare (2011) Drugs in Australia 2010: tobacco, alcohol and other drugs, vol 27, Drug statistics series. Australian Institute of Health and Welfare, Canberra, Cat. no. PHE 154. ISBN 978-1-74249-230-8

An TC, Yang H, Li GY, Song WH, Cooper WJ, Nie XP (2010) Kinetics and mechanism of advanced oxidation processes (AOPs) in degradation of ciprofloxacin in water. Appl Catal B Environ 94(3–4):288–294. doi:10.1016/j.apcatb.2009.12.002

Andreozzi R, Caprio V, Insola A, Marotta R (1999) Advanced oxidation processes (AOP) for water purification and recovery. Catal Today 53(1):51–59. doi:10.1016/s0920-5861(99)00102-9

Andreozzi R, Marotta R, Pinto G, Pollio A (2002) Carbamazepine in water: persistence in the environment, ozonation treatment and preliminary assessment on algal toxicity. Water Res 36(11):2869–2877. doi:10.1016/s0043-1354(01)00500-0

Andreozzi R, Marotta R, Paxeus N (2003) Pharmaceuticals in STP effluents and their solar photodegradation in aquatic environment. Chemosphere 50(10):1319–1330. doi:10.1016/s0045-6535(02)00769-5

Augugliaro V, Bellardita M, Loddo V, Palmisano G, Plamisano L, Yurdakal S (2012) Overview of oxidation mechanisms of organic compounds by TiO_2 in heterogeneous photocatalysis. J Photochem Photobiol C Photochem Rev 13(3):224–245. doi:10.1016/j.jphotochemrev.2012.04.003

Baran W, Adamek E, Ziemiańska J, Sobczak A (2011) Effects of the presence of sulfonamides in the environment and their influence on human health. J Hazard Mater 196:1–15. doi:10.1016/j.jhazmat.2011.08.082

Beausse J (2004) Selected drugs in solid matrices: a review of environmental determination, occurrence and properties of principal substances. Trac Trends Anal Chem 23(10–11):753–761. doi:10.1016/j.trac.2004.08.005

Benotti MJ, Trenholm RA, Vanderford BJ, Holady JC, Stanford BD, Snyder SA (2009) Pharmaceuticals and endocrine disrupting compounds in US drinking water. Environ Sci Technol 43(3):597–603. doi:10.1021/es801845a

Boyd GR, Zhang SY, Grimm DA (2005) Naproxen removal from water by chlorination and biofilm processes. Water Res 39(4):668–676. doi:10.1016/j.watres.2004.11.013

Braslavsky SE (2007) Glossary of terms used in photochemistry, 3rd edition. Pure Appl Chem 79(3):293–465. doi:10.1351/pac200779030293

Calza P, Sakkas VA, Medana C, Baiocchi C, Dimou A, Pelizzetti E, Albanis T (2006) Photocatalytic degradation study of diclofenac over aqueous TiO_2 suspensions. Appl Catal B Environ 67(3–4):197–205. doi:10.1016/j.apcatb.2006.04.021

Carballa M, Omil F, Lema JM, Llompart M, Garcia-Jares C, Rodriguez I, Gomez M, Ternes T (2004) Behavior of pharmaceuticals, cosmetics and hormones in a sewage treatment plant. Water Res 38(12):2918–2926. doi:10.1016/j.watres.2004.03.029

Carballa M, Omil F, Lema JM (2008) Comparison of predicted and measured concentrations of selected pharmaceuticals, fragrances and hormones in Spanish sewage. Chemosphere 72(8):1118–1123. doi:10.1016/j.chemosphere.2008.04.034

Chen HW, Ku Y, Irawan A (2007) Photodecomposition of o-cresol by UV-LED/TiO_2 process with controlled periodic illumination. Chemosphere 69(2):184–190. doi:10.1016/j.chemosphere.2007.04.051

Choi K, Kim Y, Park J, Park CK, Kim M, Kim HS, Kim P (2008) Seasonal variations of several pharmaceutical residues in surface water and sewage treatment plants of Han River, Korea. Sci Total Environ 405(1–3):120–128. doi:10.1016/j.scitotenv.2008.06.038

Chong MN, Jin B (2012) Photocatalytic treatment of high concentration carbamazepine in synthetic hospital wastewater. J Hazard Mater 199:135–142. doi:10.1016/j.jhazmat.2011.10.067

Chong MN, Jin B, Chow CWK, Saint C (2010) Recent developments in photocatalytic water treatment technology: a review. Water Res 44(10):2997–3027. doi:10.1016/j.watres.2010.02.039

Dai CM, Zhou XF, Zhang YL, Duan YP, Qiang ZM, Zhang TC (2012) Comparative study of the degradation of carbamazepine in water by advanced oxidation processes. Environ Technol 33(10):1101–1109. doi:10.1080/09593330.2011.610359

Dalrymple OK, Yeh DH, Trotz MA (2007) Removing pharmaceuticals and endocrine-disrupting compounds from wastewater by photocatalysis. J Chem Technol Biotechnol 82(2):121–134. doi:10.1002/jctb.1657

Daughton CG, Ternes TA (1999) Pharmaceuticals and personal care products in the environment: agents of subtle change? Environ Health Perspect 107(6):907–938. doi:10.2307/3434573

Deegan AM, Shaik B, Nolan K, Urell K, Oelgemöller M, Tobin J, Morrissey A (2011) Treatment options for wastewater effluents from pharmaceutical companies. Int J Environ Sci Technol 8(3):649–666

Dimitrakopoulou D, Rethemiotaki I, Frontistis Z, Xekoukoulotakis NP, Venieri D, Mantzavinos D (2012) Degradation, mineralization and antibiotic inactivation of amoxicillin by UV-A/TiO_2 photocatalysis. J Environ Manage 98:168–174. doi:10.1016/j.jenvman.2012.01.010

Doll TE, Frimmel FH (2004) Kinetic study of photocatalytic degradation of carbamazepine, clofibric acid, iomeprol and iopromide assisted by different TiO_2 materials – determination of intermediates and reaction pathways. Water Res 38(4):955–964. doi:10.1016/j.watres.2003.11.009

Doll TE, Frimmel FH (2005a) Photocatalytic degradation of carbamazepine, clofibric acid and iomeprol with P25 and Hombikat UV100 in the presence of natural organic matter (NOM) and other organic water constituents. Water Res 39(2–3):403–411. doi:10.1016/j.watres.2004.09.016

Doll TE, Frimmel FH (2005b) Removal of selected persistent organic pollutants by heterogeneous photocatalysis in water. Catal Today 101(3–4):195–202. doi:10.1016/j.cattod.2005.03.005

Elmolla ES, Chaudhuri M (2010a) Degradation of amoxicillin, ampicillin and cloxacillin antibiotics in aqueous solution by the UV/ZnO photocatalytic process. J Hazard Mater 173(1–3):445–449. doi:10.1016/j.jhazmat.2009.08.104

Elmolla ES, Chaudhuri M (2010b) Comparison of different advanced oxidation processes for treatment of antibiotic aqueous solution. Desalination 256(1–3):43–47. doi:10.1016/j.desal.2010.02.019

Elmolla ES, Chaudhuri M (2010c) Photocatalytic degradation of amoxicillin, ampicillin and cloxacillin antibiotics in aqueous solution using UV/TiO_2 and UV/H_2O_2/TiO_2 photocatalysis. Desalination 252(1–3):46–52. doi:10.1016/j.desal.2009.11.003

Fatta-Kassinos D, Vasquez MI, Kummerer K (2011a) Transformation products of pharmaceuticals in surface waters and wastewater formed during photolysis and advanced oxidation processes – degradation, elucidation of byproducts and assessment of their biological potency. Chemosphere 85(5):693–709. doi:10.1016/j.chemosphere.2011.06.082

Fatta-Kassinos D, Meric S, Nikolaou A (2011b) Pharmaceutical residues in environmental waters and wastewater: current state of knowledge and future research. Anal Bioanal Chem 399(1):251–275. doi:10.1007/s00216-010-4300-9

Felis E, Marciocha D, Surmacz-Gorska J, Miksch K (2007) Photochemical degradation of naproxen in the aquatic environment. Water Sci Technol 55(12):281–286. doi:10.2166/wsat.2007.417

Fent K, Weston AA, Caminada D (2006) Ecotoxicology of human pharmaceuticals. Aquat Toxicol 76(2):122–159. doi:10.1016/j.aquatox.2005.09.009

Fox MA, Dulay MT (1993) Heterogeneous photocatalysis. Chem Rev 93(1):341–357. doi:10.1021/cr00017a016

Friedmann D, Mendive C, Bahnemann D (2010) TiO_2 for water treatment: parameters affecting the kinetics and mechanisms of photocatalysis. Appl Catal B Environ 99(3–4):398–406. doi:10.1016/j.apcatb.2010.05.014

Fujishima A, Zhang XT, Tryk DA (2008) TiO_2 photocatalysis and related surface phenomena. Surf Sci Rep 63(12):515–582. doi:10.1016/j.surfrep.2008.10.001

Garciá-Araya JF, Beltran FJ, Aguinaco A (2010) Diclofenac removal from water by ozone and photolytic TiO_2 catalysed processes. J Chem Technol Biotechnol 85(6):798–804. doi:10.1002/jctb.2363

Gaya UI, Abdullah AH (2008) Heterogeneous photocatalytic degradation of organic contaminants over titanium dioxide: a review of fundamentals, progress and problems. J Photochem Photobiol C Photochem Rev 9(1):1–12. doi:10.1016/j.jphotochemrev.2007.12.003

Giraldo AL, Penuela GA, Torres-Palma RA, Pino NJ, Palominos RA, Mansilla HD (2010) Degradation of the antibiotic oxolinic acid by photocatalysis with TiO_2 in suspension. Water Res 44(18):5158–5167. doi:10.1016/j.watres.2010.05.011

Halling-Sørensen B, Nielsen SN, Lanzky PF, Ingerslev F, Lützhøft HCH, Jørgensen SE (1998) Occurrence, fate and effects of pharmaceutical substances in the environment – a review. Chemosphere 36(2):357–394. doi:10.1016/S0045-6535(97)00354-8, DOI:10.1016/S0045-6535%2897%2900354-8

Haque MM, Muneer M (2007) Photodegradation of norfloxacin in aqueous suspensions of titanium dioxide. J Hazard Mater 145(1–2):51–57. doi:10.1016/j.jhazmat.2006.10.086

Heberer T (2002a) Tracking persistent pharmaceutical residues from municipal sewage to drinking water. J Hydrol 266(3–4):175–189. doi:10.1016/s0022-1694(02)00165-8

Heberer T (2002b) Occurrence, fate and removal of pharmaceutical residues in aquatic environment: a review of recent research data. Toxicol Lett 131(1–2):5–17. doi:10.1016/s0378-4274(02)00041-3

Herrmann JM (1999) Heterogeneous photocatalysis: fundamentals and applications to the removal of various types of aqueous pollutants. Catal Today 53(1):115–129. doi:10.1016/S0920-5861(99)00107-8

Hoeger B, Kollner B, Dietrich DR, Hitzfeld B (2005) Water-borne diclofenac affects kidney and gill integrity and selected immune parameters in brown trout (Salmo trutta f. fario). Aquat Toxicol 75(1):53–64. doi:10.1016/j.aquatox.2005.07.006

Hoffmann MR, Martin ST, Choi WY, Bahnemann DW (1995) Environmental applications of semiconductor photocatalysis. Chem Rev 95(1):69–96. doi:10.1021/cr00033a004

Homem V, Santos L (2011) Degradation and removal methods of antibiotics from aqueous matrices – a review. J Environ Manage 92(10):2304–2347. doi:10.1016/j.jenvman.2011.05.023

Hu LH, Flanders PM, Miller PL, Strathmann TJ (2007) Oxidation of sulfamethoxazole and related antimicrobial agents by TiO_2 photocatalysis. Water Res 41(12):2612–2626. doi:10.1016/j.watres.2007.02.026

Im JK, Son HS, Kang YM, Zoh KD (2012) Carbamazepine degradation by photolysis and titanium dioxide photocatalysis. Water Environ Res 84(7):554–561. doi:10.2175/106143012x13373550427273

Isidori M, Lavorgna M, Nardelli A, Parrella A, Previtera L, Rubino M (2005) Ecotoxicity of naproxen and its phototransformation products. Sci Total Environ 348(1–3):93–101. doi:10.1016/j.scitotenv.2004.12.068

Khan SJ, Ongerth JE (2004) Modelling of pharmaceutical residues in Australian sewage by quantities of use and fugacity calculations. Chemosphere 54(3):355–367. doi:10.1016/j.chemosphere.2003.07.001

Khetan SK, Collins TJ (2007) Human pharmaceuticals in the aquatic environment: a challenge to green chemistry. Chem Rev 107(6):2319–2364. doi:10.1021/cr020441w

Klauson D, Babkina J, Stepanova K, Krichevskaya M, Preis S (2010) Aqueous photocatalytic oxidation of amoxicillin. Catal Today 151(1–2):39–45. doi:10.1016/j.cattod.2010.01.015

Klavarioti M, Mantzavinos D, Kassinos D (2009) Removal of residual pharmaceuticals from aqueous systems by advanced oxidation processes. Environ Int 35(2):402–417. doi:10.1016/j.envint.2008.07.009

Kockler J, Kanakaraju D, Glass B, Oelgemöller M (2012) Photochemical and photocatalytic degradation of diclofenac and amoxicillin using natural and simulated sunlight. J Sustain Sci Manage 7(1):23–29, ISSN 1823–8556

Kumar SG, Devi LG (2011) Review on modified TiO_2 photocatalysis under UV/visible light: selected results and related mechanisms on interfacial charge carrier transfer dynamics. J Phys Chem A 115(46):13211–13241. doi:10.1021/jp204364a

Kümmerer K (2009) The presence of pharmaceuticals in the environment due to human use-present knowledge and future challenges. J Environ Manage 90(8):2354–2366. doi:10.1016/j.jenvman.2009.01.023

Landgraf S (2001) Application of semiconductor light sources for investigations of photochemical reactions. Spectrochim Acta A Mol Biomol Spectrosc 57(10):2029–2048. doi:10.1016/s1386-1425(01)00502-9

Leary R, Westwood A (2011) Carbonaceous nanomaterials for the enhancement of TiO_2 photocatalysis. Carbon 49(3):741–772. doi:10.1016/j.carbon.2010.10.010

Legrini O, Oliveros E, Braun AM (1993) Photochemical processes for water treatment. Chem Rev 93(2):671–698. doi:10.1021/cr00018a003

Li WH, Guo CS, Su S, Xu J (2012) Photodegradation of four fluoroquinolone compounds by titanium dioxide under simulated solar light irradiation. J Chem Technol Biotechnol 87(5):643–650. doi:10.1002/jctb.2759

Lindqvist N, Tuhkanen T, Kronberg L (2005) Occurrence of acidic pharmaceuticals in raw and treated sewages and in receiving waters. Water Res 39(11):2219–2228. doi:10.1016/j.watres.2005.04.003

Lishman L, Smyth SA, Sarafin K, Kleywegt S, Toito J, Peart T, Lee B, Servos M, Beland M, Seto P (2006) Occurrence and reductions of pharmaceuticals and personal care products and estrogen by municipal wastewater treatment plants in Ontario, Canada. Sci Total Environ 367(2–3):544–558. doi:10.1016/j.scitotenv.2006.03.021

Madhavan J, Grieser F, Ashokkumar M (2010) Combined advanced oxidation processes for the synergistic degradation of ibuprofen in aqueous environments. J Hazard Mater 178(1–3):202–208. doi:10.1016/j.jhazmat.2010.01.064

Malato S, Blanco J, Vidal A, Richter C (2002) Photocatalysis with solar energy at a pilot-plant scale: an overview. Appl Catal B Environ 37(1):1–15. doi:10.1016/S0926-3373(01)00315-0

Malato S, Fernandez-Ibanez P, Maldonado MI, Blanco J, Gernjak W (2009) Decontamination and disinfection of water by solar photocatalysis: recent overview and trends. Catal Today 147(1):1–59. doi:10.1016/j.cattod.2009.06.018

Martínez C, Canle M, Fernandez MI, Santaballa JA, Faria J (2011) Aqueous degradation of diclofenac by heterogeneous photocatalysis using nanostructured materials. Appl Catal B Environ 107(1–2):110–118. doi:10.1016/j.apcatb.2011.07.003

Martins AF, Mayer F, Confortin EC, Frank CD (2009) A study of photocatalytic processes involving the degradation of the organic load and amoxicillin in hospital wastewater. Clean Soil Air Water 37(4–5):365–371. doi:10.1002/clen.200900022

Mavronikola C, Demetriou M, Hapeshi E, Partassides D, Michael C, Mantzavinos D, Kassinos D (2009) Mineralisation of the antibiotic amoxicillin in pure and surface waters by artificial UVA and sunlight-induced Fenton oxidation. J Chem Technol Biotechnol 84(8):1211–1217. doi:10.1002/jctb.2159

Mboula VM, Hequet V, Gru Y, Colin R, Andres Y (2012) Assessment of the efficiency of photocatalysis on tetracycline biodegradation. J Hazard Mater 209:355–364. doi:10.1016/j.jhazmat.2012.01.032

Méndez-Arriaga F, Giménez J, Esplugas S (2008a) Photolysis and TiO_2 photocatalytic treatment of naproxen: degradation, mineralization, intermediates and toxicity. J Adv Oxid Technol 11(3):435–444

Méndez-Arriaga F, Esplugas S, Giménez J (2008b) Photocatalytic degradation of non-steroidal anti-inflammatory drugs with TiO_2 and simulated solar irradiation. Water Res 42(3):585–594. doi:10.1016/j.watres.2007.08.002

Méndez-Arriaga F, Maldonado MI, Giménez J, Esplugas S, Malato S (2009a) Abatement of ibuprofen by solar photocatalysis process: enhancement and scale up. Catal Today 144(1–2):112–116. doi:10.1016/j.cattod.2009.01.028

Méndez-Arriaga F, Torres-Palma RA, Petrier C, Esplugas S, Giménez J, Pulgarin C (2009b) Mineralization enhancement of a recalcitrant pharmaceutical pollutant in water by advanced oxidation hybrid processes. Water Res 43(16):3984–3991. doi:10.1016/j.watres.2009.06.059

Michael I, Hapeshi E, Michael C, Fatta-Kassinos D (2010) Solar Fenton and solar TiO_2 catalytic treatment of ofloxacin in secondary treated effluents: evaluation of operational and kinetic parameters. Water Res 44(18):5450–5462. doi:10.1016/j.watres.2010.06.053

Miège C, Choubert JM, Ribeiro L, Eusèbe M, Coquery M (2009) Fate of pharmaceuticals and personal care products in wastewater treatment plants – conception of a database and first results. Environ Pollut 157(5):1721–1726. doi:10.1016/j.envpol.2008.11.045

Miránda-García N, Suarez S, Sanchez B, Coronado JM, Malato S, Maldonado MI (2011) Photocatalytic degradation of emerging contaminants in municipal wastewater treatment plant effluents using immobilized TiO_2 in a solar pilot plant. Appl Catal B Environ 103(3–4):294–301. doi:10.1016/j.apcatb.2011.01.030

Moldovan Z (2006) Occurrences of pharmaceuticals and personal care products as micropollutants in rivers from Romania. Chemosphere 64(11):1808–1817. doi:10.1016/j.chemosphere.2006.02.003

Molinari R, Pirillo F, Loddo V, Palmisano L (2006) Heterogeneous photocatalytic degradation of pharmaceuticals in water by using polycrystalline TiO_2 and a nanofiltration membrane reactor. Catal Today 118(1–2):205–213. doi:10.1016/j.cattod.2005.11.091

Mompelat S, Le Bot B, Thomas O (2009) Occurrence and fate of pharmaceutical products and by-products, from resource to drinking water. Environ Int 35(5):803–814. doi:10.1016/j.envint.2008.10.008

Mozia S, Morawski AW (2012) The performance of a hybrid photocatalysis-MD system for the treatment of tap water contaminated with ibuprofen. Catal Today 193(1):213–220. doi:10.1016/j.cattod.2012.03.016

Nasuhoglu D, Yargeau V, Berk D (2011) Photo-removal of sulfamethoxazole (SMX) by photolytic and photocatalytic processes in a batch reactor under UV-C radiation (lambda (max) = 254 nm). J Hazard Mater 186(1):67–75. doi:10.1016/j.jhazmat.2010.10.080

Nasuhoglu D, Rodayan A, Berk D, Yargeau V (2012) Removal of the antibiotic levofloxacin (LEVO) in water by ozonation and TiO_2 photocatalysis. Chem Eng J 189–190:41–48. doi:10.1016/j.cej.2012.02.016

Nikolaou A, Meric S, Fatta D (2007) Occurrence patterns of pharmaceuticals in water and wastewater environments. Anal Bioanal Chem 387(4):1225–1234. doi:10.1007/s00216-006-1035-8

Oaks JL, Gilbert M, Virani MZ, Watson RT, Meteyer CU, Rideout BA, Shivaprasad HL, Ahmed S, Chaudhry MJI, Arshad M, Mahmood S, Ali A, Khan AA (2004) Diclofenac residues as the cause of vulture population decline in Pakistan. Nature 427(6975):630–633. doi:10.1038/nature02317

Oetken M, Nentwig G, Löffler D, Ternes D, Oehlmann J (2005) Effects of pharmaceuticals on aquatic vertebrates. Part I. The antiepileptic drug carbamazepine. Arch Environ Contam Toxicol 49:353–361. doi:10.1007/s00244-004-0211-0

Oller I, Malato S, Sanchez-Perez JA (2011) Combination of advanced oxidation processes and biological treatments for wastewater decontamination – a review. Sci Total Environ 409(20):4141–4166. doi:10.1016/j.scitotenv.2010.08.061

Oppenländer T (2003) Photochemical purification of water and air. Advanced oxidation processes (AOPs): principles, reaction mechanisms, reactor concepts. Wiley-VCH, Weinheim, pp 101–121. ISBN 3-527-30563-7

Packer JL, Werner JJ, Latch DE, McNeill K, Arnold WA (2003) Photochemical fate of pharmaceuticals in the environment: naproxen, diclofenac, clofibric acid and ibuprofen. Aquat Sci 65(4):342–351. doi:10.1007/s00027-003-0671-8

Pal A, Gin KYH, Lin AYC, Reinhard M (2010) Impacts of emerging organic contaminants on freshwater resources: review of recent occurrences, sources, fate and effects. Sci Total Environ 408(24):6062–6069. doi:10.1016/j.scitotenv.2010.09.026

Palominos RA, Mora A, Mondaca MA, Perez-Moya M, Mansilla HD (2008) Oxolinic acid photo-oxidation using immobilized TiO_2. J Hazard Mater 158(2–3):460–464. doi:10.1016/j.jhazmat.2008.01.117

Palominos RA, Mondaca MA, Giraldo A, Penuela G, Perez-Moya M, Mansilla HD (2009) Photocatalytic oxidation of the antibiotic tetracycline on TiO_2 and ZnO suspensions. Catal Today 144(1–2):100–105. doi:10.1016/j.cattod.2008.12.031

Paul T, Dodd MC, Strathmann TJ (2010) Photolytic and photocatalytic decomposition of aqueous ciprofloxacin: transformation products and residual antibacterial activity. Water Res 44(10):3121–3132. doi:10.1016/j.watres.2010.03.002

Pelaez M, Nolan NT, Pillai SC, Seery MK, Falaras P, Kontos AG, Dunlop PSM, Hamilton JWJ, Byrne JA, O'Shea K, Entezari MH, Dionysiou DD (2012) A review on the visible light active titanium dioxide photocatalysts for environmental applications. Appl Catal B Environ 125:331–349. doi:10.1016/j.apcatb.2012.05.036

Pereira JHOS, Vilar VJP, Borges MT, Gonzalez O, Esplugas S, Boaventura RAR (2011) Photocatalytic degradation of oxytetracycline using TiO_2 under natural and simulated solar radiation. Sol Energ 85(11):2732–2740. doi:10.1016/j.solener.2011.08.012

Pérez-Estrada LA, Maldonado MI, Gernjak W, Agüera A, Fernández-Alba AR, Ballesteros MM, Malato S (2005) Decomposition of diclofenac by solar driven photocatalysis at pilot plant scale. Catal Today 101(3–4):219–226. doi:10.1016/j.cattod.2005.03.013

Reyes C, Fernandez J, Freer J, Mondaca MA, Zaror C, Malato S, Mansilla HD (2006) Degradation and inactivation of tetracycline by TiO_2 photocatalysis. J Photochem Photobiol A Chem 184(1–2):141–146. doi:10.1016/j.jphotochem.2006.04.007

Rizzo L (2011) Bioassays as a tool for evaluating advanced oxidation processes in water and wastewater treatment. Water Res 45(15):4311–4340. doi:10.1016/j.watres.2011.05.035

Rizzo L, Meric S, Kassinos D, Guida M, Russo F, Belgiorno V (2009a) Degradation of diclofenac by TiO_2 photocatalysis: UV absorbance kinetics and process evaluation through a set of toxicity bioassays. Water Res 43(4):979–988. doi:10.1016/j.watres.2008.11.040

Rizzo L, Meric S, Guida M, Kassinos D, Belgiorno V (2009b) Heterogeneous photocatalytic degradation kinetics and detoxification of an urban wastewater treatment plant effluent contaminated with pharmaceuticals. Water Res 43(16):4070–4078. doi:10.1016/j.watres.2009.06.046

Rodil R, Quintana JB, Concha-Grana E, Lopez-Mahia P, Muniategui-Lorenzo S, Prada-Rodriguez D (2012) Emerging pollutants in sewage, surface and drinking water in Galicia (NW Spain). Chemosphere 86(10):1040–1049. doi:10.1016/j.chemosphere.2011.11.053

Romero V, De la Cruz N, Dantas RF, Gimenez PMJ, Esplugas S (2011) Photocatalytic treatment of metoprolol and propranolol. Catal Today 161(1):115–120. doi:10.1016/j.cattod.2010.09.026

Sabri N, Hanna K, Yargeau G (2012) Chemical oxidation of ibuprofen in the presence of iron species at near neutral pH. Sci Total Environ 427–428:382–389. doi:10.1016/j.scitotenv.2012.04.034

Shamaila S, Sajjad AKL, Chen F, Zhang JL (2010) Synthesis and characterization of mesoporous-TiO_2 with enhanced photocatalytic activity for the degradation of chloro-phenol. Mater Res Bull 45(10):1375–1382. doi:10.1016/j.materresbull.2010.06.047

Shan AY, Ghazi TIM, Rashid SA (2010) Immobilisation of titanium dioxide onto supporting materials in heterogeneous photocatalysis: a review. Appl Catal A Gen 389(1–2):1–8. doi:10.1016/j.apcata.2010.08.053

Silva CG, Faria JL (2009) Anatase vs. rutile efficiency on the photocatalytic degradation of clofibric acid under near UV to visible irradiation. Photochem Photobiol Sci 8(5):705–711. doi:10.1039/b817364h

Skocaj M, Filipic M, Petkovic J, Novak S (2011) Titanium dioxide in our everyday life; is it safe? Radiol Oncol 45(4):227–247. doi:10.2478/v10019-011-0037-0

Sturini M, Speltini A, Maraschi F, Profumo A, Pretali L, Fasani E, Albini A (2010) Photochemical degradation of marbofloxacin and enrofloxacin in natural waters. Environ Sci Technol 44(12):4564–4569. doi:10.1021/es100278n

Sturini M, Speltini A, Maraschi F, Profumo A, Pretali L, Irastorza EA, Fasani E, Albini A (2012) Photolytic and photocatalytic degradation of fluoroquinolones in untreated river water under natural sunlight. Appl Catal B Environ 119:32–39. doi:10.1016/j.apcatb.2012.02.008

Suárez S, Carballa M, Omil F, Lema JM (2008) How are pharmaceutical and personal care products (PPCPs) removed from urban wastewaters? Rev Environ Sci Biotechnol 7:125–138. doi:10.1007/s11157-008-9130-2

Ternes TA, Meisenheimer M, McDowell D, Sacher F, Brauch HJ, Gulde BH, Preuss G, Wilme U, Seibert NZ (2002) Removal of pharmaceuticals during drinking water treatment. Environ Sci Technol 36(17):3855–3863. doi:10.1021/es015757k

Tixier C, Singer HP, Oellers S, Muller SR (2003) Occurrence and fate of carbamazepine, clofibric acid, diclofenac, ibuprofen, ketoprofen, and naproxen in surface waters. Environ Sci Technol 37(6):1061–1068. doi:10.1021/es025834r

Tong AYC, Braund R, Warren DS, Peake BM (2012) TiO_2-assisted photodegradation of pharmaceuticals – a review. Cent Eur J Chem 10(4):989–1027. doi:10.2478/s11532-012-0049-7

Trovó AG, Nogueira RFP, Aguera A, Fernandez-Alba AR, Sirtori C, Malato S (2009) Degradation of sulfamethoxazole in water by solar photo-Fenton. Chemical and toxicological evaluation. Water Res 43(16):3922–3931. doi:10.1016/j.watres.2009.04.006

Van Doorslaer X, Demeestere K, Heynderickx PM, Van Langenhove H, Dewulf J (2011) UVA and UVC induced photolytic and photocatalytic degradation of aqueous ciprofloxacin and moxifloxacin: reaction kinetics and role of adsorption. Appl Catal B Environ 101(3–4):540–547. doi:10.1016/j.apcatb.2010.10.027

Van Doorslaer X, Heynderickx PM, Demeestere K, Debevere K, Van Langenhove H, Dewulf J (2012) TiO_2 mediated heterogeneous photocatalytic degradation of moxifloxacin: operational variables and scavenger study. Appl Catal B Environ 111–112:150–156. doi:10.1016/j.apcatb.2011.09.029

Vasconcelos TG, Kummerer K, Henriques DM, Martins AF (2009) Ciprofloxacin in hospital effluent: degradation by ozone and photoprocesses. J Hazard Mater 169(1–3):1154–1158. doi:10.1016/j.jhazmat.2009.03.143

Wang L, Ying GG, Zhao JL, Yang XB, Chen F, Tao R, Liu S, Zhou LJ (2010) Occurrence and risk assessment of acidic pharmaceuticals in the Yellow River, Hai River and Liao River of North China. Sci Total Environ 408(16):3139–3147. doi:10.1016/j.scitotenv.2010.04.047

Watkinson AJ, Murby EJ, Kolpin DW, Constanzo SD (2009) The occurrence of antibiotics in an urban watershed: from wastewater to drinking water. Sci Total Environ 407(8):2711–2723. doi:10.1016/j.scitotenv.2008.11.059

Weir A, Westerhoff P, Fabricius L, Hristovski K, von Goetz N (2012) Titanium dioxide nanoparticles in food and personal care products. Environ Sci Technol 46(4):2242–2250. doi:10.1021/es204168d

Xekoukoulotakis NP, Drosou C, Brebou C, Chatzisymeon E, Hapeshi E, Fatta-Kassinos D, Mantzavinos D (2011) Kinetics of UV-A/TiO_2 photocatalytic degradation and mineralization of the antibiotic sulfamethoxazole in aqueous matrices. Catal Today 161(1):163–168. doi:10.1016/j.cattod.2010.09.027

Xiong P, Hu JY (2012) Degradation of acetaminophen by UV-A/LED/TiO_2 process. Sep Purif Technol 91(SI):89–95. doi:10.1016/j.seppur.2011.11.012

Xu WH, Zhang G, Zou SC, Li XD, Liu YC (2007) Determination of selected antibiotics in the Victoria Harbour and the Pearl River, South China using high performance liquid chromatography electrospray ionization tandem mass spectrometry. Environ Pollut 145(3):672–679. doi:10.1016/j.envpol.2006.05.038

Yang L, Yu LE, Ray MB (2008) Degradation of paracetamol in aqueous solutions by TiO_2 photocatalysis. Water Res 42(13):3480–3488. doi:10.1016/j.watres.2008.04.023

Yang LM, Yu LE, Ray MB (2009) Photocatalytic oxidation of paracetamol: dominant reactants, intermediates, and reaction mechanisms. Environ Sci Technol 43(2):460–465. doi:10.1021/es8020099

Yang H, Li GY, An TC, Gao YP, Fu JM (2010) Photocatalytic degradation kinetics and mechanism of environmental pharmaceuticals in aqueous suspension of TiO_2: a case of sulfa drugs. Catal Today 153(3–4):200–207. doi:10.1016/j.cattod.2010.02.068

Yurdakal S, Loddo V, Augugliaro V, Berber H, Palmisano G, Palmisano L (2007) Photodegradation of pharmaceutical drugs in aqueous TiO_2 suspensions: mechanism and kinetics. Catal Today 129(1–2):9–15. doi:10.1016/j.cattod.2007.06.044

Zhang YJ, Geissen SU, Gal C (2008a) Carbamazepine and diclofenac: removal in wastewater treatment plants and occurrence in water bodies. Chemosphere 73(8):1151–1161. doi:10.1016/j.chemosphere.2008.07.086

Zhang X, Wu F, Wu XW, Chen PY, Deng NS (2008b) Photodegradation of acetaminophen in TiO_2 suspended solution. J Hazard Mater 157(2–3):300–307. doi:10.1016/j.jhazmat.2007.12.098

Zhang W, Zou LD, Wang LZ (2009) Photocatalytic TiO_2/adsorbent nanocomposites prepared via wet chemical impregnation for wastewater treatment: a review. Appl Catal A Gen 371(1–2):1–9. doi:10.1016/j.apcata.2009.09.038

Zhang HH, Cao BP, Liu WP, Lin KD, Feng J (2012) Oxidative removal of acetaminophen using zero valent aluminum-acid system: efficacy, influencing factors, and reaction mechanism. J Environ Sci 24(2):314–319. doi:10.1016/s1001-0742(11)60769-9

Ziylan A, Ince NH (2011) The occurrence and fate of anti-inflammatory and analgesic pharmaceuticals in sewage and fresh water: treatability by conventional and non-conventional processes. J Hazard Mater 187(1–3):24–36. doi:10.1016/j.jhazmat.2011.01.057

Chapter 4
Water Depollution Using Ferrites Photocatalysts

Virender K. Sharma, Chun He, Ruey-an Doong, and Dionysios D. Dionysiou

Abstract The presence of organic pollutants and pathogenic microorganisms in water has become an increasing concern throughout the world. Heterogeneous photocatalytic technologies have been applied to control the organic pollutants and microorganisms in water. Development of narrow band-gap photocatalysts which function in the visible light remains a challenge in the wastewater treatment processes. Spinel ferrites has attracted a remarkable attention because of a relatively narrow band gap of about 2.0 eV, which has considerable photo-response in the visible light region. This chapter reviews recent advances in ferrites and the application of visible light photocatalysts to the remediation of contaminants such as H_2S, phenols, and dyes in water. Recent development in synthesis and characterization of ferrite and hybrid ferrites with other semiconductors is reviewed. The applications of ferrites in photocatalytic conversion of visible solar energy to generate e^-/h^+, which in turn produce reactive oxygen species through redox processes for the degradation of the pollutants in water, are demonstrated. We discuss the enhanced visible photocatalytic activity of ferrites by doping with metals and combing with

V.K. Sharma (✉)
Center of Ferrate Excellence and Chemistry Department, Florida Institute of Technology, 150 West University Boulevard, Florida 32901, USA
e-mail: vsharma@fit.edu

C. He
School of Environmental Science and Engineering, Sun Yat-sen University, Guangzhou 510275, China

R.-a. Doong
Department of Biomedical Engineering and Environmental Sciences, National Tsing Hua University, Hsinchu 30013, Taiwan

D.D. Dionysiou
Environmental Engineering and Science Program, 705 Engineering Research Center, University of Cincinnati, Cincinnati, OH 45221-0012, USA

E. Lichtfouse et al. (eds.), *Green Materials for Energy, Products and Depollution*, Environmental Chemistry for a Sustainable World 3, DOI 10.1007/978-94-007-6836-9_4,

other photocatalysts. Moreover, the addition of H_2O_2 to ferrite either in dark or visible light irradiation indicates the enhanced degradation efficiency for organic pollutants.

Keywords Water pollution • Photocatalyst • ferrite • H_2S • Phenols • Dyes • H_2O_2 • Surfactant • Agrochemical • Glyphosate

Contents

4.1 Introduction

Freshwater is a precious resource on earth and is critical to sustain life. In many regions of the world, daily need of water is not met. According to World Health Organization (WHO), more than 880 million people in the world do not have access to potable water (WHO 2010). Water contamination caused death to 1.8 million children from diarrhea every year (WHO 2010). Water scarcity can also affect ecosystem as numerous species might not be able to cope with decrease in availability of freshwater. One of the greatest challenges in this century is to provide access to clean water. The development of nanotechnology in the past decade offers prospects of meeting challenges of safe and sustainable water demand (Di Paola et al. 2012; Qu et al. 2012). Furthermore, a combination of nanotechnology and solar energy may lead to innovative water purification technologies.

One of the well studied technologies to clean water is the photocatalytic remediation (Di Paola et al. 2012; Rajeshwar 2011; Rajeshwar et al. 2012). The photocatalytic processes on bare semiconductor titanium dioxide (TiO_2) have been studied for several decades, but TiO_2 is only active under UV irradiation ($\lambda < 400$ nm) (Abe 2011; Chen et al. 2011; Hoffmann et al. 1995; Li and Liu 2011; Serpone et al. 2012; Tao et al. 2011). The solar spectrum has very small fraction of incoming light in the UV region (ca. 4 %). A use of all the UV light would result in only 2 % solar conversion efficiency (Abe 2011). Comparatively, the portion of visible light ($400 < \lambda < 800$ nm) in the solar spectrum is much more abundant (ca. 46 %). A use of visible light up to 600 nm increases the efficiency to 16 % and further improvement to 32 % is achievable if visible light use is extended to 800 nm (Kubacka et al. 2012; Linic et al. 2011; Paracchino et al. 2011; Serpone and Emeline 2012).

Currently, the efficiency for degradation of pollutants under visible-light irradiation is still low due to the fast charge recombination and backward reactions of photocatalysts (Chen et al. 2011). The efficient visible-light-driven photocatalysts requires closing the band gaps to harvest visible light in the long-wavelength regions as well as to improve the separation of photogenerated electrons (e^-) and holes (h^+) efficiently. Dopants into TiO_2 have been added to narrow the band gap in order for TiO_2 to be appropriate for absorbing visible light (Kamat 2012; Kubacka et al. 2012; Serpone et al. 2009). However, doped TiO_2 materials under visible light have lower chemical activity of surface active centers and decreased photoactivities compared to those formed under UV light irradiation (Serpone and Emeline 2012). Moreover, commonly used dopants are rare, expensive and/or toxic metals, which may not fulfill the principle of sustainable chemistry. Identification of suitable candidates capable of degrading pollutants under visible light is essential to provide clean remediation processes. The objective of this article is to provide information on the progress made in remediation of water using environmentally benign visible-light active iron-based oxides, ferrites.

Ferrites have a molecular formula $M\text{-}Fe_2O_4$ in which Fe_2O_3 is combined with a metal oxide (M-O). Ferrites may provide an alternative to TiO_2 because they have shown to be efficient in the visible light region (Han et al. 2007). Metals used in ferrites include Ca^{2+}, Zn^{2+}, Mg^{2+}, Ni^{2+}, Co^{2+}, and Mn^{2+} (Benko and Koffyberg 1986; Borse et al. 2008; Dom et al. 2011; Han et al. 2007; Tamaura et al. 1999). These ferrites, when prepared with a spinel structure, show promising photocatalytic activity (Kim et al. 2009). Spinel structures have the general formula AB_2X_4, where, in the case of ferrites, A is a metal, B is Fe^{3+} and X is oxygen (Burdett et al. 1982). As shown in Fig. 4.1 for example for $ZnFe_2O_4$, the spinel structure usually consists of both tetrahedral and octahedral sites, where the metal atoms occupy one-eighth of the tetrahedral sites and Fe^{3+} occupies one-half of the octahedral sites (Degueldre et al. 2009). The lattice parameter a and the cation-oxide distance R_{ZnO} and R_{FeO} in the octahedral and tetrahedral substructures were related by Eq. 4.1 (Degueldre et al. 2009)

$$A = 2(2)^{1/2}[R_{FeO}\cos(\psi/2) + R_{ZnO}\sin(\varphi/2)] \quad (4.1)$$

The tetrahedron internal angle (φ) and the octahedron equatorial angle (ψ) were identified as 109.471° and 95.375° for O-Zn-O and Fe-O-Fe, respectively.

The ferrites of Zn^{2+}, Ca^{2+}, and Mg^{2+} seem to be favorable as they are more relatively environmentally friendly. The energy diagram of these ferrites is shown in Fig. 4.2 (Dom et al. 2011). The band gaps of ~2.0 eV show the potential of ferrites to absorb visible light absorption and also the possibility for degrading pollutants in water (Casbeer et al. 2012; Hou et al. 2010; Ida et al. 2010; Li et al. 2011; Su et al. 2012).

The present chapter first describes briefly the synthesis and characterization of ferrites, followed by application of ferrites to degrade contaminants. Most of studies in literature have been performed on $ZnFe_2O_4$ and hence examples are presented using this form of ferrite.

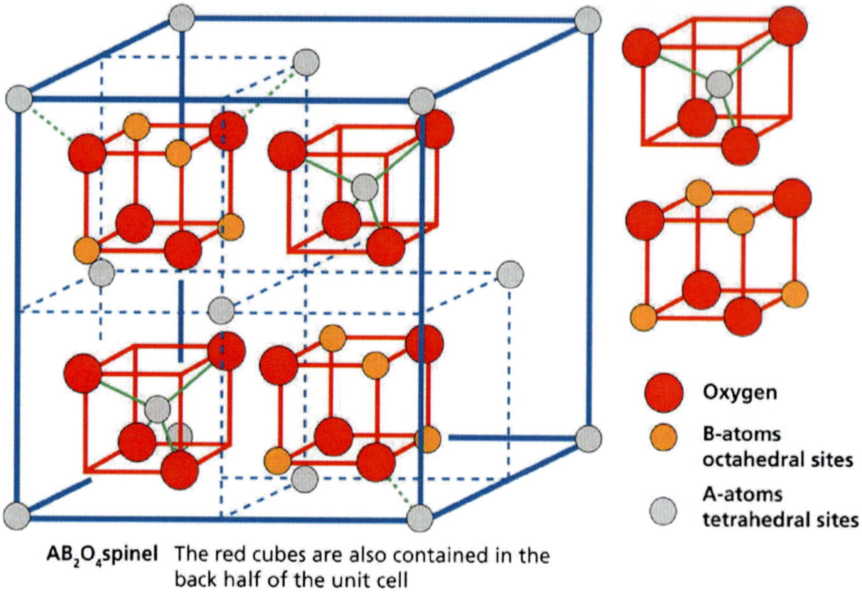

Fig. 4.1 Schematic description of the crystallographic structure for bulk $ZnFe_2O_4$. Occupied tetrahedral site in spinel sub-cell, Zn is *gray* and O is in *red*. Occupied octahedral site in spinel sub-cell, Fe is *yellow*, and O is in *red*. The arrangement in one unit cell with 3D succession of octahedral and tetrahedral sub-cells (Adapted from (Degueldre et al. 2009) with permission from Elsevier Ltd.)

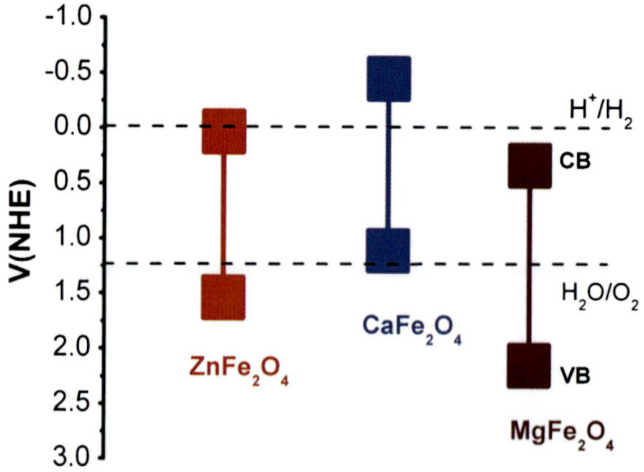

Fig. 4.2 Schematic diagram of MFe_2O_4 (M: Mg, Ca, Zn) showing the feasibility of materials as visible light photocatalysts. The thickness of the band edge (CB-*conduction*/VB *valence band*) indicates the possible variation in the value depending on various physico–chemical parameters of electrolyte and environmental conditions viz. *p*H, temperature, concentration etc. and there possible effects on the Fermi energy (Adapted from (Dom et al. 2011) with the permission of Elsevier Ltd.)

4.2 Synthesis of Ferrites

4.2.1 Spinel Ferrites

A number of methods used in synthesis of spinel ferrites are shown in Fig. 4.3. These method include thermal, sol-gel and citrate, co-precipitation, and solid-state reactions (Casbeer et al. 2012; Dom et al. 2011; Hou et al. 2011; Jadhav et al. 2012; Li et al. 2011; Pardeshi and Pawar 2011; Pradeep et al. 2011; Salunkhe et al. 2012; Su et al. 2012; Zhang et al. 2010a). In synthesis of ferrites, salts of Fe(III) and M(II) were used as precursors. Synthesis methods have recently been summarized (Casbeer et al. 2012). Highly visible-light active $ZnFe_2O_4$ nanotube array has also been prepared using sol-gel methods (Li et al. 2011). Microwave sintering process may be advantageous due to its shorter synthesis time compared to conventional methods (Dom et al. 2011).

4.2.2 Composite Ferrites

In composite ferrites, the emphasis has been on the coupling of titania and zinc ferrite through different approaches. Ferrite is sensitive to visible light while titania has high photoactivity (Chen et al. 2010; Cheng et al. 2004) and composites showed increased photocatalytic activity (Tan et al. 2012). Sol-gel and hydrothermal methods were successfully used to prepare either titania doped ferrite or ferrite

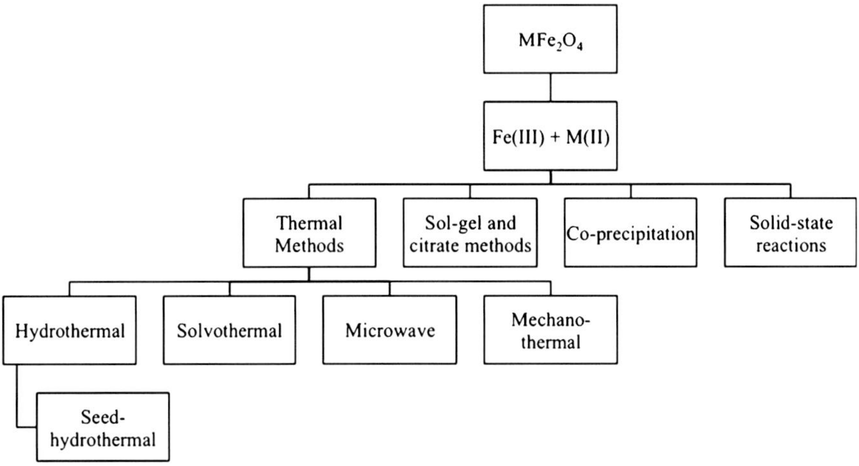

Fig. 4.3 Preparation methods of ferrites using Fe(III) and M(II) salts as precursors (Adapted from (Casbeer et al. 2012) with the permission of Elsevier Ltd)

doped titnia (Cheng et al. 2004; Liu et al. 2004; Moreira et al. 2012). More recently, TiO_2-$ZnFe_2O_4$ with an intermediate layer of alumina has been prepared by a multistep wet chemical method (Hankare et al. 2011). Other methods include the liquid catalytic phase transformation at low temperature and the two-step electrochemical processes (Hou et al. 2010; Shihong et al. 2009). Nanocomposites of $ZnFe_2O_4$ nanoplates and Ag nanoparticles and $ZnFe_2O_4$/multi-walled carbon nanotubes (MWCNTs) have also been prepared (Cao et al. 2011; Chen et al. 2010).

4.3 Characterization

Figure 4.4 shows the optical properties of ferrites, analyzed by UV–vis absorption spectroscopy. Ferrites have strong absorption in the visible range of 400–700 nm (Dom et al. 2011; Liu et al. 2009; Subramanian et al. 2004; Valenzuela et al. 2002). This is significant because of performing photocatalytic activity under visible light. Comparatively, TiO_2 does not show any significant absorption in the visible region. The electron excitation from the O-2p level into the Fe 3d level for ferrites may be causing absorption in the visible region (Lv et al. 2010). The two persistent absorption bands of spinel structures of nano $ZnFe_2O_4$ at $v_1 = 545$ and $v_2 = 292\ cm^{-1}$ in the Fourier Transform Infrared (FTIR) spectrum were observed (Pradeep et al. 2011). These bands could be related to the oxygen-metal cation complex presence in the tetrahedral and octahedral sites of the ferrite. Other surface techniques have also been applied to characterize ferrite, which include X-ray diffraction (XRD), X-ray spectroscopy (XAS), transmission electron microscopy (TEM), scanning electron microscopy (SEM), neutron diffraction analysis, and

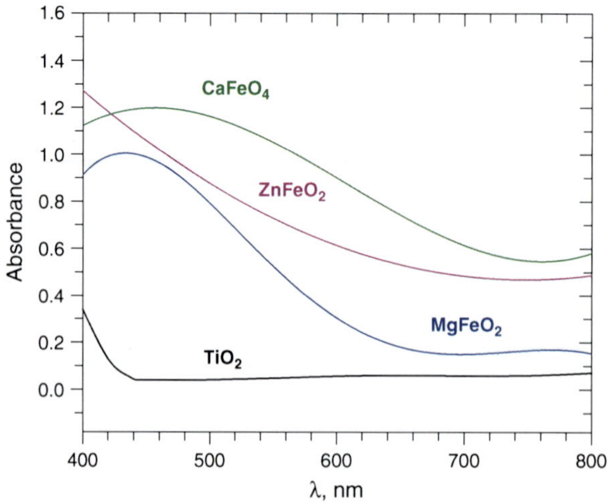

Fig. 4.4 Visible light spectra of metal ferrites (Adapted from (Dom et al. 2011; Liu et al. 2009; Subramanian et al. 2004; Valenzuela et al. 2002))

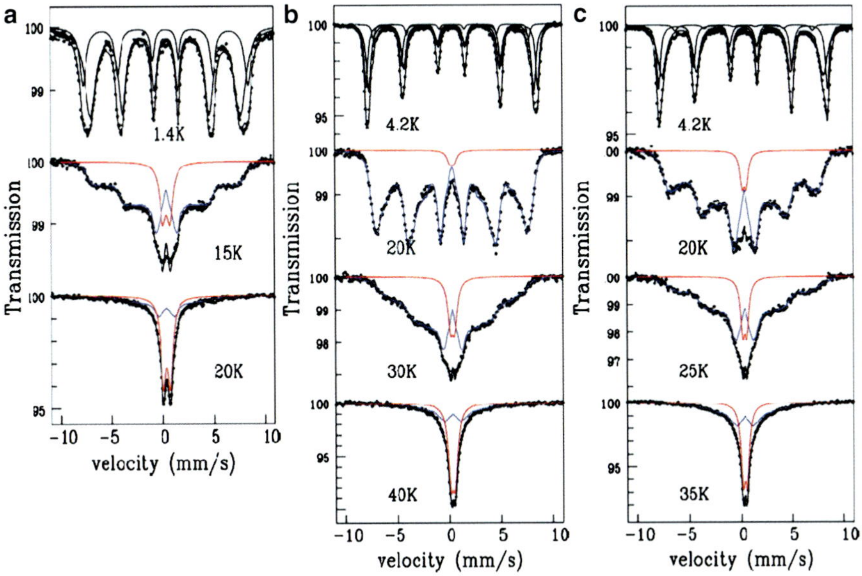

Fig. 4.5 Mössbauer spectra obtained at different temperatures for Z3 (**a**), Z11 (**b**), and Z19 (**c**) samples (Adapted from (Blanco-Gutiérrez et al. 2011) with the permission of the American Chemical Society)

Brunauer-Emmett-Teller (BET) surface area (Blanco-Gutierrez et al. 2011; Hankare et al. 2011; Hou et al. 2011; Li et al. 2011; Moreira et al. 2012; Nilsen et al. 2007).

Mossbauer spectroscopy has also been used to characterize $ZnFe_2O_4$ samples (Blanco-Gutiérrez et al. 2011). The zero-field Mossbauer spectra of samples, prepared differently are presented in Fig. 4.5 (Blanco-Gutiérrez et al. 2011). Samples Z3 and Z19 were synthesized using solvothermal method while Z11 was prepared by means of sol-gel method. All samples had different treatment time while temperature was ranged from 160 °C to 200 °C. Treatment times were 2 and 288 h for Z3 and Z19 samples, respectively. Sample Z11 was treated for 24 h. The mean particle sizes of samples were 3, 11, and 19 nm for Z3, Z11, and Z19, respectively. Spectra of samples at 1.4 and 4.2 K had two broadened sextets in which suggested the presence of Fe^{3+}. The non-Lorentzian shape of lines gave mean hypefine parameters as 47.2, 50.2, and 49.9 T for Z3, Z11, and Z19, respectively. The characteristic doublets appeared with increase in temperature (Fig. 4.5). This is indicative of superparamagnetic behavior of particles of such samples.

4.4 Oxidation by Ferrites

The absorption of light by semiconductor photocatalysis results in several processes (Fig. 4.6) (Teoh et al. 2012). In the first step, formation of electron–hole (e-h^+) occurs through bandgap excitation. A number of reactions can take place in absence

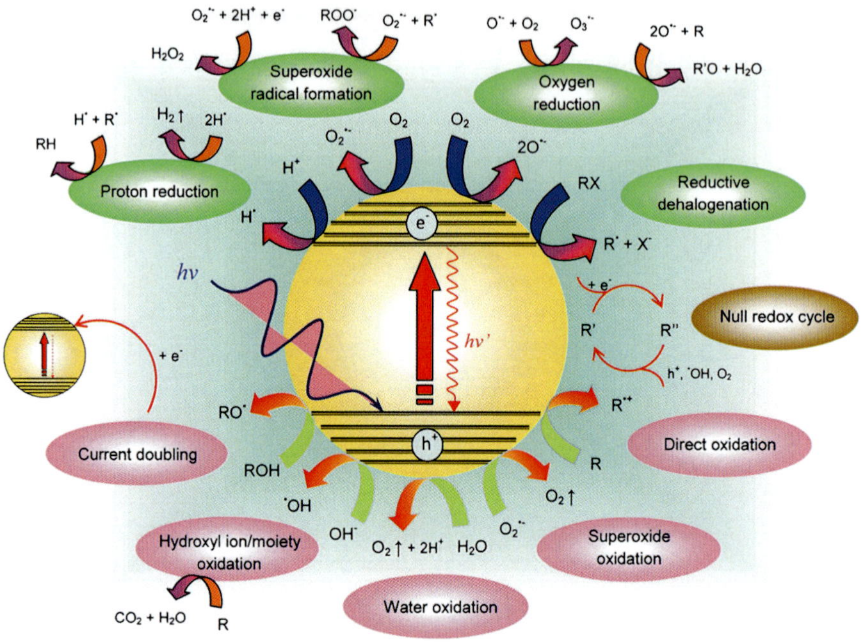

Fig. 4.6 Possible reaction pathways arising from the excitation of photocatalyst (Adapted from (Teoh et al. 2012) with the permission of American Chemical Society)

and presence of contaminants in water (Fig. 4.6). Both electron and hole can diffuse to the surfaces of semiconductors and reduce and oxidize the adsorbed contaminants, respectively (Fig. 4.6). Other possibility is the recombination of electron and hole, which decreases the efficiency of the photocatalysts to react with contaminants in water. Hydroxyl radicals ($^{\bullet}OH$) have been suggested to be dominated species to oxidize contaminants. Hydroxyl radicals are efficient in abstracting hydrogen atom and attaching to electron-rich moieties (Park and Choi 2005; Turchi and Ollis 1990). Superoxide species ($O_2^{\bullet -}$) are generated by the reaction of O_2 with electron (Gerischer and Heller 1991; Schwitzgebel et al. 1995). Other reactive oxygen species such as singlet oxygen ($^1O_2$), hydrogen peroxide (H_2O_2), and hydroperoxyl radicals ($HO_2^{\bullet}$) may also be produced and be involved in the photocatalytic oxidation reactions. In the next section, applications of ferrites (e.g., $ZnFe_2O_4$) as a photocatlysts in degrading contaminants are presented.

4.4.1 Degradation of Contaminants

Ferrites have been useful in decontamination of inorganic compounds and disinfection (Li et al. 2008; Liu et al. 1996; Rana et al. 2005; Rawat et al. 2007; Zhang et al. 2010b; Zhao et al. 2010). Examples include photodehydrogenation

Table 4.1 Degradation of phenols by ferrites

Contaminant	Catalyst	Irradiation	Reference
Phenol	Zn-Al ferrite	No	Xu et al. (2007)
4-Chlorophenol	$ZnFe_2O_4$-modified TiO_2 nanotube array electrode	Visible light	Hou et al. (2011)
4-Chlorophenol	$ZnFe_2O_4$ nanotube	Visible light	Li et al. (2011)
Phenol	$TiO_2/CoFe_2O_4$ composite	UV-visible light	Li et al. (2012)
2,4-Dichlorophenol	N-doped titania supported *on* $SrFe_{12}O_{19}$	Sunlight	Aziz et al. (2012b)
2,4-Dichlorophenol	TiO_2 nanocomposites/SiO_2 coating supported on $NiFe_2O_4$	Sunlight	Aziz et al. (2012a)
Phenol o-Nitrophenol p-Nitrophenol Picric acid	Multi-walled carbon nanotube supported on $NiFe_2O_4$	UV light	Xiong et al. (2012)

of H_2S using $ZnFe_2O_4$ and inactivation of *Escherichia coli* by composite ferrites, $Ag/MgFe_2O_4$, $Ag/Ni_2Fe_2O_4$, Ag/Zn_2FeO_4, and $Ag/CoFe_2O_4$. Ferrites have also shown effectiveness in degrading aliphatic compounds such as methanol, ethanol, isopropanol, acetaldehyde, oxalic acid, and butenes (Gibson and Hightower 1976; Manova et al. 2004, 2011; Shchukin et al. 2004; Tsoncheva et al. 2010). Most of studies on the remediation by ferrites are on the oxidation of phenols and dyes and are discussed below.

Phenols

A few studies on the oxidation of phenols by ferrites have been performed and are summarized in Table 4.1 (Aziz et al. 2012b; Hou et al. 2010; Li et al. 2012, 2011; Valenzuela et al. 2002; Xu et al. 2007). The concentrations of phenols in these studies were generally from 10 to 50 mg l^{-1}. These studies have demonstrated that phenols could be removed successfully using ferrite alone and composite ferrites. The current focus is on the composite ferrites to enhance the photocatalytic oxidation of phenols through synergistic effects. The presence of ferrites in the TiO_2-ferrite composite not only increases absorbance of visible light, but also increases absorbance in the UV light in order to assist photooxidation of phenol (Li et al. 2012, 2011). Furthermore, composites could be magnetically separated (Aziz et al. 2012a, b; Li et al. 2012). The photocatalytic oxidation of phenols using a magnetically recyclable photocatalyst, multi-walled carbon nanotubes (MWNT) supported $NiFe_2O_4$ is shown in Fig. 4.7 (Xiong et al. 2012). Figure 4.7a shows the degradation of phenol using $NiFe_2O_4$/MWNT nanocomposites having different MWNT content. The degradation of phenols increased with increasing amount of MWNT in the nanocomposites, however, the degradation rate had no such relationship. The degradation of phenol under UV radiation followed pseudo-first-order kinetics and obtained rate constant (k) as a function of content of MWNT are

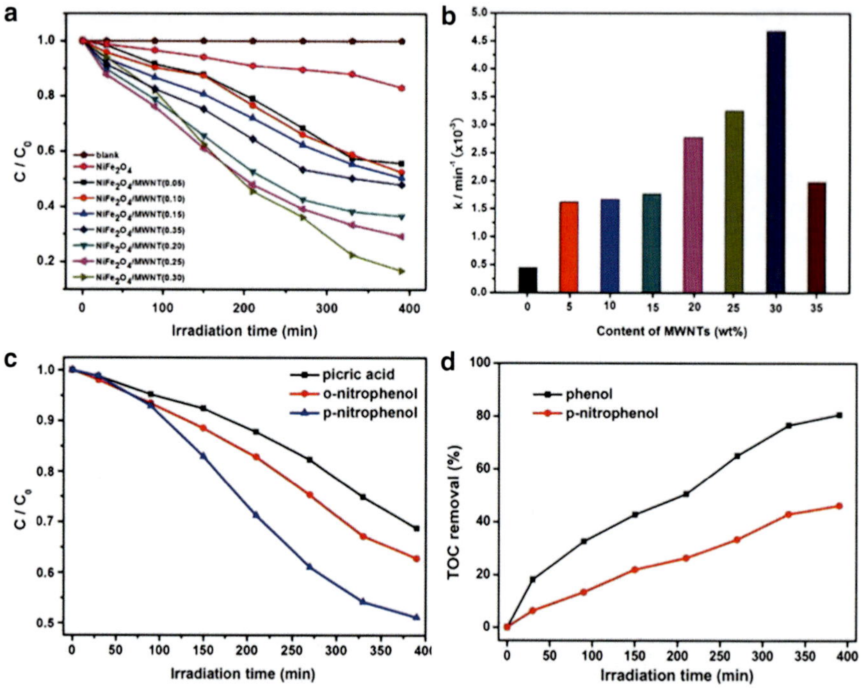

Fig. 4.7 (**a**) Photocatalytic degradation of phenol by $NiFe_2O_4$/multi-walled carbon nanotubes (MWNT) photocatalysts with differing MWNT contents under UV light irradiation; (**b**) The pseudo-first-order rate constant *k* as a function of MWNT content; (**c**) Photocatalytic degradation of nitrophenols (o-nitrophenol, p-nitrophenol, picric acid) by $NiFe_2O_4$/MWNT(0.30) nanocomposite photocatalysts under UV light irradiation; (**d**) Evolution of TOC removal of phenol and p-nitrophenol with irradiation time (Adapted from (Xiong et al. 2012) with the permission of Elsevier Ltd.)

shown in Fig 4.7b. $NiFe_2O_4$/MWNT nanocomposite with 30 wt % MWNT exhibited the best photocatalytic activity. Degradation of picric acid, o-nitrophenol, and p-nitrophenol are shown in Fig. 4.7c. Most of the phenol degraded significantly under UV light in 400 min. This study also determined the total organic carbon (TOC) for the degradation of phenol and p-nitrophenol (Fig. 4.7d). Removals of TOC were ~80 % and ~40 % for phenol and p-nitrophenol, respectively. Incomplete removal of TOC suggests that the oxidation of phenol and p-nitrophenol resulted in simple organic compounds in the $NiFe_2O_4$/MWNT/UV radiation (Xiong et al. 2012).

Dyes

Numerous studies on the oxidation of degradation of dyes by ferrites alone and composite ferrites have been carried out (Baldrian et al. 2006; Casbeer et al. 2012; Li et al. 2011; Moreira et al. 2012). Dyes studied include methyl orange (MO),

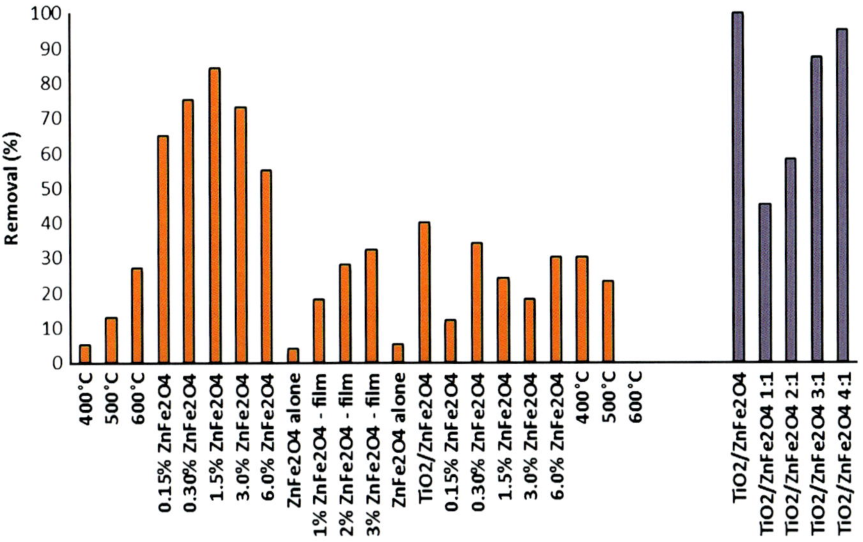

Fig. 4.8 Degradation of methyl orange (MO) and rhodamine B (RhB) dyes by $TiO_2/ZnFe_2O_4$ composite photocatalysts. All listed reactions contain the composite unless otherwise noted. MO solutions irradiated with visible light, while RhB solutions are irradiated with UV light (Adapted from (Casbeer et al. 2012) with the permission of Elsevier Ltd.)

methylene blue (MB), rhodamine B (RhB), bromophenol blue, Chicago sky blue, eosine yellow, evans blue, naphthol blue black, phenol red, poly B-411, and reactive orange 16. The concentration levels of dyes were usually in the range of 50–500 mg L^{-1}. In a recent work, results of the studies have been summarized (Casbeer et al. 2012). Ferrites as photocatalyst alone were effective in degrading dyes, but the efficiency of degradation was enhanced when ferrites were used as composite photocatalysts. This is shown in Fig. 4.8. The degradation of RhB by ferrites was examined by UV light while visible light was applied in irradiating the MO dye. The temperature of ferrite synthesis influenced the degradation of RhB. Increase in the amount of TiO_2 in the composite ferrites increased the degradation efficiency of dyes (Fig. 4.8). Decrease in crystal size of ferrites also increased the degradation efficiency of dyes (Fan et al. 2009).

In a recent work, the photocatalytic degradation of acid orange II (AOII) was studied by combining $ZnFe_2O_4$ with H_2O_2 under visible light ($\lambda > 400$ nm) (Su et al. 2012). Significantly direct photolysis of AOII by visible light was difficult, but oxidation of AOII was possible due to generation of $^{\bullet}OH$ radicals in the system (Fig. 4.9). The Fenton-type reaction can produce $^{\bullet}OH$ radicals (Path A). The formation of $^{\bullet}OH$ radicals is also due to oxidation of water by holes on the surface of the ferrite (Path B). The capture of electron by H_2O_2 gives $^{\bullet}OH$ radicals (Path C). The Path C is advantageous because it decreases the recombination of electron and hole and hence enhance the formation of $^{\bullet}OH$ radicals. The mechanism of the reactions was examined by adding scavengers of reactive species, oxalate,

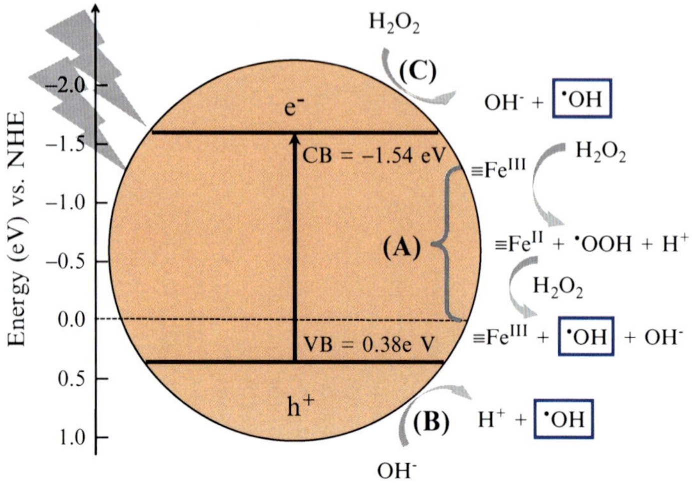

Fig. 4.9 A hypothetical scheme for the generation of OH radical in H_2O_2-$ZnFe_2O_4$-visible light system (Adapted from (Su et al. 2012) with the permission of Elsevier Ltd.)

iso-propoanol, Cr(VI), and KI, into the system. Degradation rate did not show any influence by adding oxalate. This suggests that oxidation of AOII by hole was less likely to be involved because hole is highly reactive with oxalate. Cr(VI) can react rapidly with electron and the degradation rate decreases in the presence of Cr(VI), which suggest that electron formed at the surface of the ferrite is involved. Both iso-propoanol and KI are good scavengers of $^{\bullet}OH$ radicals and degradation of AOI decreased very significantly. This confirmed the dominant role of the $^{\bullet}OH$ radicals in degradation of AOII. Basically, Paths, A, B, and C enhanced the degradation of AOII (Su et al. 2012). Degradation of AOII in the $ZnFe_2O_4/H_2O_2$/visible light was almost complete within 2 h. The amount of $ZnFe_2O_4$ and the concentration of H_2O_2 influenced the degradation of AOII.

4.5 Conclusion

$ZnFe_2O_4$-mediated photocatalytic oxidation of organic contaminations is a promising alternative technology for remediation of contaminants in water treatment. Pioneering works indicate that $ZnFe_2O_4$ possesses a relatively low visible-light-driven activity for organic pollutants, however, the technology is a potentially economical and benign process due to its capability to absorb the visible light of solar energy, stability against photo- and chemical- corrosion, low cost, and non-toxicity. In this case, it is still a challenge to explore the highly-efficient modified ferrites for remediation of contaminants under visible light. The size and morphology control of ferrites is one of the most important approaches that can enhance the photocatalytic activity. Compared with a traditional synthesized method, i.e. co-precipitation and sol-gel techniques, resulting in large particles and a broad size distribution, some innovative methods should be developed which can synthesize smaller size $ZnFe_2O_4$. An alternative promising approach is to modify ferrites with an appropriate matching-band-gap-energy semiconductor in a way that the photogenerated e^-/h^+ can be transferred between ferrites and matching-band-gap-energy semiconductor. This effective charge suppresses the combination of the photoinduced electrons and holes, which is beneficial to improve the photocatalytic efficiency. Further improvement in the modification of ferrites and ferrite composite for photocatalytic degradation of organic pollutants is still needed.

Acknowledgements V.K. Sharma and D.D. Dionysiou acknowledge support from the National Science Foundation grant (CBET 1236331) for ferrite research. C. He wishes to thank the National Natural Science Foundation of China (No. 20877025), National Natural Science Foundation of Guangdong Province (No. S2011010001836) and the Fundamental Research Funds for the Central Universities (No. 09lgpy20).

References

Abe R (2011) Development of a new system for photocatalytic water splitting into H_2 and O_2 under visible light irradiation. Bull Chem Soc Jpn 84:1000–1030

Aziz AA, Cheng CK, Ibrahim S, Matheswaran M, Saravanan P (2012a) Visible light improved, photocatalytic activity of magnetically separable titania nanocomposite. Chem Eng J 183:349–356

Aziz AA, Yong KS, Ibrahim S, Pichiah S (2012b) Enhanced magnetic separation and photocatalytic activity of nitrogen doped titania photocatalyst supported on strontium ferrite. J Hazard Mater 199–200:143–150

Baldrian P, Merhautová V, Gabriel J, Nerud F, Stopka P, Hrubý M, Beneš MJ (2006) Decolorization of synthetic dyes by hydrogen peroxide with heterogeneous catalysis by mixed iron oxides. Appl Catal Environ 66:258–264

Benko FA, Koffyberg FP (1986) The effect of defects on some photoelectrochemical properties of semiconducting $MgFe_2O_4$. Mater Res Bull 21:1183–1188

Blanco-Gutierrez V, Climent-Pascual E, Torralvo-Fernandez MJ, Saez-Puche R, Fernandez-Diaz MT (2011) Neutron diffraction study and superparamagnetic behavior of $ZnFe_2O_4$ nanoparticles obtained with different conditions. J Solid State Chem 184:1608–1613

Blanco-Gutiérrez V, Jiménez-Villacorta F, Bonville P, Torralvo-Fernández MJ, Sáez-Puche R (2011) X-ray absorption spectroscopy and Mössbauer spectroscopy studies of superparamagnetic $ZnFe_2O_4$ nanoparticles. J Phys Chem C 115:1627–1634

Borse PH, Jun H, Choi SH, Hong SJ, Lee JS (2008) Phase and photoelectrochemical behavior of solution-processed Fe_2O_3 nanocrystals for oxidation of water under solar light. Appl Phys Lett 93

Burdett JK, Price GD, Price SL (1982) Role of the crystal-field theory in determining the structures of spinels. J Am Chem Soc 104:92–95

Cao J, Kako T, Li P, Ouyang S, Ye J (2011) Fabrication of p-type $CaFe_2O_4$ nanofilms for photoelectrochemical hydrogen generation. Electrochem Commun 13:275–278

Casbeer E, Sharma VK, Li X (2012) Synthesis and photocatalytic activity of ferrites under visible light: a review. Sep Purif Technol 87:1–14

Chen C-H, Liang Y-H, Zhang W-D (2010) $ZnFe_2O_4$/MWCNTs composite with enhanced photocatalytic activity under visible-light irradiation. J Alloys Compd 501:168–172
Chen X, Liu L, Yu PY, Mao SS (2011) Increasing solar absorption for photocatalysis with black hydrogenated titanium dioxide nanocrystals. Science 331:746–750
Cheng P, Li W, Zhou T, Jin Y, Gu M (2004) Physical and photocatalytic properties of zinc ferrite doped titania under visible light irradiation. J Photochem Photobiol A 168:97–101
Degueldre C, Kuri G, Borca CN, Grolimund D (2009) X-ray micro- fluorescence, diffraction and absorption spectroscopy for local structure investigation of a radioactive zinc ferrite deposit. Corros Sci 51:1690–1695
Di Paola A, García-López E, Marcì G, Palmisano L (2012) A survey of photocatalytic materials for environmental remediation. J Hazard Mater 211–212:3–29
Dom R, Subasri R, Radha K, Borse PH (2011) Synthesis of solar active nanocrystalline ferrite, MFe_2O_4 (M: Ca, Zn, Mg) photocatalyst by microwave irradiation. Solid State Commun 151:470–473
Fan G, Gu Z, Yang L, Li F (2009) Nanocrystalline zinc ferrite photocatalysts formed using the colloid mill and hydrothermal technique. Chem Eng J 155:534–541
Gerischer H, Heller A (1991) The role of oxygen in photooxidation of organic molecules on semiconductor particles. J Phys Chem 95:5261–5267
Gibson MA, Hightower JW (1976) Oxidative dehydrogenation of butenes over magnesium ferrite kinetic and mechanistic studies. J Catal 41:420–430
Han SB, Kang TB, Joo OS, Jung KD (2007) Water splitting for hydrogen production with ferrites. Sol Energ 81:623–628
Hankare PP, Patil RP, Jadhav AV, Garadkar KM, Sasikala R (2011) Enhanced photocatalytic degradation of methyl red and thymol blue using titania-alumina-zinc ferrite nanocomposite. Appl Catal Environ 107:333–339
Hoffmann MR, Martin ST, Choi W, Bahnemann DW (1995) Environmental applications of semiconductor photocatalysis. Chem Rev 95:69–96
Hou X, Feng J, Liu X, Ren Y, Fan Z, Wei T, Meng J, Zhang M (2011) Synthesis of 3D porous ferromagnetic $NiFe_2O_4$ and using as novel adsorbent to treat wastewater. J Colloid Interface Sci 362:477–485
Hou Y, Li X, Zhao Q, Quan X, Chen G (2010) Electrochemically assisted photocatalytic degradation of 4-chlorophenol by $ZnFe_2O_4$-modified TiO_2 nanotube array electrode under visible light irradiation. Environ Sci Technol 44:5098–5103
Ida S, Yamada K, Matsunaga T, Hagiwara H, Matsumoto Y, Ishihara T (2010) Preparation of p-type $CaFe_2O_4$ photocathodes for producing hydrogen from water. J Am Chem Soc 132:17343–17345
Jadhav SV, Jinka KM, Bajaj HC (2012) Nanosized sulfated zinc ferrite as catalyst for the synthesis of nopol and other fine chemicals. Catal Today 198:98–105
Kamat PV (2012) Manipulation of charge transfer across semiconductor interface. A criterion that cannot be ignored in photocatalyst design. J Phys Chem Lett 3:663–672
Kim HG, Borse PH, Jang JS, Jeong ED, Jung O, Suh YJ, Lee JS (2009) Fabrication of $CaFe_2O_4/MgFe_2O_4$ bulk heterojunction for enhanced visible light photocatalysis. Chem Commun 39:5889–5891
Kubacka A, Fernández-García M, Colón G (2012) Advanced nanoarchitectures for solar photocatalytic applications. Chem Rev 112:1555–1614
Li C-J, Wang JN, Wang B, Gong JR, Lin Z (2012) Direct formation of reusable $TiO_2/CoFe_2O_4$ heterogeneous photocatalytic fibers via two-spinneret electrospinning. J Nanosci Nanotechnol 12:2496–2502
Li S, Wang E, Tian C, Mao B, Kang Z, Li Q, Sun G (2008) Jingle-bell-shaped ferrite hollow sphere with a noble metal core: simple synthesis and their magnetic and antibacterial properties. J Solid State Chem 181:1650–1658
Li X, Hou Y, Zhao Q, Teng W, Hu X, Chen G (2011) Capability of novel $ZnFe_2O_4$ nanotube arrays for visible-light induced degradation of 4-chlorophenol. Chemosphere 82:581–586

Li Y, Liu Z (2011) Particle size, shape and activity for photocatalysis on titania anatase nanoparticles in aqueous surroundings. J Am Chem Soc 133:15743–15752

Linic S, Christopher P, Ingram DB (2011) Plasmonic-metal nanostructures for efficient conversion of solar to chemical energy. Nat Mater 10:911–921

Liu G-G, Zhang X-Z, Xu Y-J, Niu X-S, Zheng L-Q, Ding X-J (2004) Effect of $ZnFe_2O_4$ doping on the photocatalytic activity of TiO_2. Chemosphere 55:1287–1291

Liu J, Lu G, He H, Tan H, Xu T, Xu K (1996) Studies on photocatalytic activity of zinc ferrite catalysts syntedesized by shock waves. Mater Res Bull 31:1049–1056

Liu Z, Zhao Z, Miyauchi M (2009) Efficient visible light active $CaFe_2O_4/WO_3$ based composite photocatalysts: effect of interfacial modification. J Phys Chem C 113:17132–17137

Lv H, Ma L, Zeng P, Ke D, Peng T (2010) Synthesis of floriated $ZnFe_2O_4$ with porous nanorod structures and its photocatalytic hydrogen production under visible light. J Mater Chem 20:3665–3672

Manova E, Tsoncheva T, Paneva D, Mitov I, Tenchev K, Petrov L (2004) Mechanochemically synthesized nano-dimensional iron-cobalt spinel oxides as catalysts for methanol decomposition. Appl Catal Gen 277:119–127

Manova E, Tsoncheva T, Paneva D, Popova M, Velinov N, Kunev B, Tenchev K, Mitov I (2011) Nanosized copper ferrite materials: mechanochemical synthesis and characterization. J Solid State Chem 184:1153–1158

Moreira E, Fraga LA, Mendonça MH, Monteiro OC (2012) Synthesis, optical, and photocatalytic properties of a new visible-light-active $ZnFe_2O_4$-TiO_2 nanocomposite material. J Nanopart Res 14:1–10

Nilsen MH, Nordhei C, Ramstad AL, Nicholson DG, Poliakoff M, Cabanas A (2007) XAS (XANES and EXAFS) investigations of nanoparticulate ferrites synthesized continuously in near critical and supercritical water. J Phys Chem C 111:6252–6262

Paracchino A, Laporte V, Sivula K, Grätzel M, Thimsen E (2011) Highly active oxide photocathode for photoelectrochemical water reduction. Nat Mater 10:456–461

Pardeshi SK, Pawar RY (2011) $SrFe_2O_4$ complex oxide an effective and environmentally benign catalyst for selective oxidation of styrene. J Mol Catal A Chem 334:35–43

Park H, Choi W (2005) Photocatalytic conversion of benzene to phenol using modified TiO_2 and polyoxometalates. Catal Today 101:291–297

Pradeep A, Priyadharsini P, Chandrasekaran G (2011) Structural, magnetic and electrical properties of nanocrystalline zinc ferrite. J Alloys Compd 509:3917–3923

Qu X, Brame J, Li Q, Alvarez PJJ (2012) Nanotechnology for a safe and sustainable water supply: enabling integrated water treatment and reuse. Acc Chem Res 45. doi: 10.1021/ar300029v

Rajeshwar K (2011) Solar energy conversion and environmental remediation using inorganic semiconductor-liquid interfaces: the road traveled and the way forward. J Phys Chem Lett 2:1301–1309

Rajeshwar K, De Tacconi NR, Timmaji HK (2012) New-generation oxide semiconductors for solar energy conversion and environmental remediation. J Nano Res 17:185–191

Rana S, Rawat J, Misra RDK (2005) Anti-microbial active composite nanoparticles with magnetic core and photocatalytic shell: TiO_2-$NiFe_2O_4$ biomaterial system. Acta Biomater 1:691–703

Rawat J, Rana S, Srivastava R, Misra RDK (2007) Antimicrobial activity of composite nanoparticles consisting of titania photocatalytic shell and nickel ferrite magnetic core. Mater Sci Eng C 27:540–545

Salunkhe AB, Khot VM, Phadatare MR, Pawar SH (2012) Combustion synthesis of cobalt ferrite nanoparticles – influence of fuel to oxidizer ratio. J Alloys Compd 514:91–96

Schwitzgebel J, Ekerdt JG, Gerischer H, Heller A (1995) Role of the oxygen molecule and of the photogenerated electron in TiO_2-photocatalyzed air oxidation reactions. J Phys Chem 99:5633–5638

Serpone N, Emeline AV (2012) Semiconductor photocatalysis – past, present, and future outlook. J Phys Chem Lett 3:673–677

Serpone N, Emeline AV, Horikoshi S (2009) Photocatalysis and solar energy conversion (chemical aspects). Photochemistry 37:300–361

Serpone N, Emeline AV, Horikoshi S, Kuznetsov VN, Ryabchuk VK (2012) On the genesis of heterogeneous photocatalysis: a brief historical perspective in the period 1910 to the mid-1980s. Photochem Photobiol Sci 11:1121–1150

Shchukin DG, Ustinovich EA, Sviridov DV, Kulak AI (2004) Titanium and iron oxide-based magnetic photocatalysts for oxidation of organic compounds and sulfur dioxide. High Energ Chem 38:167–173

Shihong X, Daolun F, Wenfeng S (2009) Preparations and photocatalytic properties of visible-light-active zinc ferrite-doped TiO_2 photocatalyst. J Phys Chem C 113:2463–2467

Su M, He C, Sharma VK, Abou Asi M, Xia D, Li X-Z, Deng H, Xiong Y (2012) Mesoporous zinc ferrite: synthesis, characterization, and photocatalytic activity with H_2O_2/visible light. J Hazard Mater 211–212:95–103

Subramanian V, Wolf EE, Kamat PV (2004) Catalysis with TiO_2/Gold nanocomposites. Effect of metal particle size on the fermi level eqilibration. J Am Chem Soc 126:4943–4950

Tamaura Y, Ueda Y, Matsunami J, Hasegawa N, Nezuka M, Sano T, Tsuji M (1999) Solar hydrogen production by using ferrites. Sol Energ 65:55–57

Tan D, Bi D, Shi P, Xu S (2012) Preparation and photocatalytic property of TiO_2/NiFe $_2O_4$ composite photocatalysts. Adv Mater Res 518–523:775–779

Tao J, Luttrell T, Batzill M (2011) A two-dimensional phase of TiO_2 with a reduced bandgap. Nat Chem 3:296–300

Teoh WY, Scott JA, Amal R (2012) Progress in heterogeneous photocatalysis: from classical radical chemistry to engineering nanomaterials and solar reactors. J Phys Chem Lett 3:629–639

Tsoncheva T, Manova E, Velinov N, Paneva D, Popova M, Kunev B, Tenchev K, Mitov I (2010) Thermally synthesized nanosized copper ferrites as catalysts for environment protection. Catal Commun 12:105–109

Turchi CS, Ollis DF (1990) Photocatalytic degradation of organic water contaminants: mechanisms involving hydroxyl radical attack. J Catal 122:178–192

Valenzuela MA, Bosch P, Jimenez-Becerrill J, Quiroz O, Paez AI (2002) Preparation, characterization and photocatalytic activity of ZnO, Fe_2O_3 and $ZnFe_2O_4$. J Photochem Photobiol A 148:177–182

WHO (2010) UNICEF Progress on sanitation and drinking-water 2010 update

Xiong P, Fu Y, Wang L, Wang X (2012) Multi-walled carbon nanotubes supported nickel ferrite: a magnetically recyclable photocatalyst with high photocatalytic activity on degradation of phenols. Chem Eng J 195–196:149–157

Xu A, Yang M, Qiao R, Du H, Sun C (2007) Activity and leaching features of zinc-aluminum ferrites in catalytic wet oxidation of phenol. J Hazard Mater 147:449–456

Zhang G-Y, Sun Y-Q, Gao D-Z, Xu Y-Y (2010a) Quasi-cube $ZnFe_2O_4$ nanocrystals: hydrothermal synthesis and photocatalytic activity with TiO_2 (Degussa P25) as nanocomposite. Mater Res Bull 45:755–760

Zhang S, Niu H, Cai Y, Zhao X, Shi Y (2010b) Arsenite and arsenate adsorption on coprecipitated bimetal oxide magnetic nanomaterials: $MnFe_2O_4$ and $CoFe_2O_4$. Chem Eng J 158:599–607

Zhao L, Li X, Zhao Q, Qu Z, Yuan D, Liu S, Hu X, Chen G (2010) Synthesis, characterization and adsorptive performance of $MgFe_2O_4$ nanospheres for SO_2 removal. J Hazard Mater 184:704–709

Chapter 5
Bioindicators of Toxic Metals

Slavka Stankovic and Ana R. Stankovic

Abstract Growing social concern about environmental quality has been observed in recent years, on global and local scales. As the world's population continues to grow, it becomes imperative to understand the dynamic interactions between human activities and the environment. With the growth of all aspects of the activities of modern man, warning signs appeared that the sources of the basic environmental components, air, water, and land are not unlimited. The advent of an increasing number of contaminants reduces the power of self-purification of these media, and degradation of nature and biosphere continues infinitely, consequently acting on the man.

Toxic heavy metals in air, soil, and water are global problems that are a growing threat to the environment. There are hundreds of sources of heavy metals pollution. Metals in the environment arise from natural sources directly or indirectly from human activities such as rapid industrialization, urbanization, and other anthropogenic sources. They are potential hazards to aquatic, animal, and human life because of their toxicity and bioaccumulative and nonbiodegradable nature. Non essential metals such as Hg, Cd, Cr, Pb, and As are toxic in their chemically combined forms as well as the elemental form. The major impacts of pollution can be stated as health problems with exposed human populations and ecosystems contamination. Those problems have been a cause of increasing public concern throughout the world.

During the last two decades the interest in using bioindicators as monitoring tools to assess environmental pollution by toxic metals has steadily increased. Bioindicators are flora and fauna members or groups of them which are used indirectly to measure the levels of metal contaminants in their environment, systematically collected and analyzed to identify potential health hazards to biota and humans. In

S. Stankovic (✉) • A.R. Stankovic
Faculty of Technology and Metallurgy, University of Belgrade,
Karnegijeva 4, Belgrade 11000, Serbia
e-mail: slavka@tmf.bg.ac.rs

E. Lichtfouse et al. (eds.), *Green Materials for Energy, Products and Depollution*, Environmental Chemistry for a Sustainable World 3, DOI 10.1007/978-94-007-6836-9_5,

practice, microbial systems, fungi, animal and plants are used as bioindicators to formulate conclusions regarding the environmental conditions. Similarly, different kind of living organisms, plants and animals, and even humans, can be used as bioindicators for environmental pollution by toxic metals because of their ability to absorb them from air, water, sediment, soil, and food chain.

Problems concerning toxic metals are very wide and encompass all parts of the nature. In this review the recent studies of organisms used as a toxic metals bioindicators in air, water, and soil ecosystems are presented. Also, the related toxicity, storage or defense mechanisms of living organisms and ecological effects of heavy metals pollution will be considered.

Keywords Heavy metal • Bioindicator • Soil • Water • Air • Pollution • Hg • Cd • Cr • Pb • As

Contents

Abbreviations

BLB	Bioluminescent Bacteria
Bw	Body Weight
CAFE	Clean Air for Europe
CAT	Catalase

COM	European Commission
d.w.	Dry Weight
DNA	Deoxyribonucleic Acid
EC	European Commission
EEA	European Environment Agency
EEC	European Economic Community
Eh	Soil Redox Potential
EPA	Environmental Protection Agency
EU	European Union
FAO	Food and Agriculture Organization of the United Nation
GPX	Glutathione Peroxidase
GSH	Glutathione
GST	Glutathione S-transferases
HBM	Human Biomonitoring
IARC	International Agency for Research on Cancer
ICES	International Council for the Exploration of the Sea
LP	Lipid Peroxidation
MAP	Mediterranean Action Plan
MED POL	Marine Pollution Assessment and Control Component of MAP
MeHg	Methylmercury
MG	Methylglyoxal
MSFD	Marine Strategy Framework Directive
MT	Metallothionein
NCM	Nordic Council of Ministers
OECD	Organization for Economic Cooperation and Development
OSPAR	Oslo/Paris Convention (for the Protection of the Marine Environment of the North-East Atlantic)
PC	Phytochelatin
PTWI	Provisional Tolerable Weekly Intake
RNA	Ribonucleic Acid
ROS	Reactive Oxygen Species
SOD	Superoxide Dismutase
SPCS	Soil Pollution Control Site
T	Tonne
t/yr	Tonne per Year
TRI	Toxic Release Inventory
UK	United Kingdom
UNEP/MAP	United Nations Environment Programme/Mediterranean Action Plan
USA, US	United States of America, United States
USEPA	United States Environmental Protection Agency
WFD	Water Framework Directive
WHO	World Health Organization
wt%	Weight Percent
ww	Wet Weight

5.1 Introduction

Increasing industrialization and urbanization have been accompanied throughout the world by the extraction and distribution of mineral substances from their natural deposits. The pressure from the activities of the urban population is intense, as anthropogenic emissions of potentially toxic trace metals have accelerated considerably.

Metals are in the environment occurring in varying concentrations in parent rock, soil, water, air and all biological matter. In natural systems, potentially toxic heavy metals can originate from rocks, ore minerals, volcanoes, Szyczewski et al. (2009), and weathering releases metals during soil formation and transports them to surface and/or aquifer waters, Siegel (2002). Natural metal loading is aggravated by anthropogenic sources. A wide spectrum of anthropogenic sources include smelting of metallic ores, industrial fabrication and commercial application of metals, traffic, agro-chemicals pesticides as well as burning of fossil fuels. Anthropogenic impacts of toxic metals are related mostly to the mining, extraction and refining stages and can be the cause of substantial air, water and soil pollution, Norgate et al. (2007).

Metal uptake by organisms occurs through diffusion as well as ingestion of food and particulate matter, Szyczewski et al. (2009). The metals serving as micronutrients in living organisms usually occur in trace amounts that are precisely defined for each species. Both their deficiency and high excess badly affect living organisms. They, also directly or indirectly, throughout air, water and food (plants, animals and humans) get into human bodies. The excessive content of metals in the human body may in many ways affect the body and psychophysical development Stankovic et al. (2011a), Jovic et al. (2012), Stankovic and Jovic (2012). Therefore, the content of heavy metals needs to be known not only in water, air, soil and sediment, but also in biological samples: plants, animals, and on the end, in the human population.

Traces of heavy metals in living organisms are not just a reflection of human activity. Together with essential nutrients, living organisms also take up heavy metals and their compounds and can accumulate them. Metals fall into one of two categories, essential and non-essential. Essential metals are required for the optimal functioning of biological and biochemical processes in organisms; non-essential elements have no known biological function in organisms and exert their toxicity by competing with essential metals for active enzyme or membrane protein sites, Stankovic et al. (2011a). Even metals that are biologically essential have the potential to be harmful to humans and other living organisms at high levels of exposure, Hu (2002).

In the world today there are different methods of monitoring the concentration of pollutants, mostly as instrumental physicochemical measurements. The use of living organisms as indicators for environmental stability has long been widely recognized. Furthermore, exposure can be measured by either levels or effects. Plants and animals are commonly known as toxic metals bioindicators. Higher plants, animals, alga, fungi, bacteria and lichen have been used as bioindicators in air, soil and water pollution surveys over the past few decades.

Bioindicators are a good tool for monitoring the state of the environment. The widespread development and application of bioindicators has occurred primarily since the 1960s, Holt and Miller (2011), using various animals like birds, mollusks and mammals. Bioindicators are living organisms that are particularly sensitive to the environment, and provide information about ecosystem health. Indicator species respond well both, to the presence or absence of other species as well as the presence of pollutants.

The application of living organisms has many advantages over standard methods of chemical analysis of trace metals in air and water. For example, in water samples the concentration of metals are often below the detection limit of the instrument that is used, Morillo et al. (2005), while the concentration of metals in water living organisms are significantly higher (102–106 times) compared to the water, and allow determination of biologically available elements and/or metal compounds in aqueous media, Roméo et al. (2005). Hence, using live organisms as analysis samples instead of sampling directly air and water reduces the possibility of measuring mistakes.

Various aquatic organisms occur in rivers, lakes, seas and marines potentially useful as metal pollutant bioindicators, including fish, shellfish, oyster, mussels, Jovic et al. (2011), Joksimovic et al. (2011a), Stankovic et al. (2011a), Stankovic and Jovic (2012), clams, aquatic animals and aquatic plants and macroalgae, Joksimovic et al. (2011b), Akcali and Kucuksezgin (2011), Rybak et al. (2012), Luy et al. (2012).

Metal content in bioindicators depends not only on the concentration of metals in air, water, soil and sediment, but also depends on environmental conditions, biological factors of the organisms themselves. It is therefore necessary to monitor the impact of all these factors in this complex system.

Heavy metals are released into the environment – the air, surface water, soil and hence into the groundwater and crops: once in the environment, they do not disappear, but accumulate in soils, sediments and biomass.

Although the adverse effects of heavy metals on human health have long been known, exposure of humans to heavy metals unfortunately continues, especially in less developed countries. The main threats to human health from heavy metals are associated with exposure to Pb, Cd, Hg and As, Schwartz et al. (2011), Stankovic et al. (2011a), Stankovic and Jovic (2012).

One of the largest problems associated with the persistence of toxic metals is the potential for bioaccumulation and biomagnification causing heavier exposure for some organisms than toxic metals present in the environment alone. In this review bioindicators of toxic metals will be presented in biotic material of the following ecosystems: air, water, and soil/sediment.

5.2 Metals and Environmental Pollution

5.2.1 Toxic Metals

Toxic metals are redistributed in the biosphere and dispersed in the air, soil, water, plants and animals throughout bioaccumulation, and consequently, in human beings through food chain bio-magnification causing chronic ailments. Environmental problems and toxic metal pollutions are serious problem in many parts of the world. Much has been done globally in the past few decades to improve human health related to environmental pollutants: on the Earth Summit in Rio de Janeiro, Brazil in 1992 at the World Conference on Environment and Development was convened to address urgent problems of environmental protection and socio-economic development. The world leaders adopted Agenda 21, a plan for achieving sustainable development in the twenty-first century.

Heavy metals normally occur in nature and some are essential to life, but can become toxic through accumulation in organisms. Only a few of the many metals present in the environment are essential and necessary in very small amounts to all living organisms, so-called micronutrients. Essential micronutrients are involved in the functional activities of living organisms. Pb, Cd, Cr, Cu, Zn, Ni, As and Hg are the most common heavy metal pollutants, and Hg, Pb and Cd are of the greatest concern because of their ability to transport long distances in the atmosphere. Dissolved Cu can affect lower trophic levels such as phytoplankton. Pb, Cd, and Hg bioacumulate through the food-chain posing a toxic risk to species higher in the food chain and to humans, Stankovic et al. (2011a), Stankovic and Jovic (2012).

For *heavy metals* many different definitions have been proposed – some based on density, some on atomic number or atomic weight, and some on chemical properties or toxicity. There is an alternative term *toxic metal*, for which no consensus of exact definition exists either. Depending on context, heavy metals can include elements lighter than carbon and can exclude some of the heaviest metals, Duffus (2002). Heavy metals occur naturally in the ecosystem with large variations in concentration. In modern times, anthropogenic sources of heavy metals pollutions have been introduced to the ecosystem. The classification of elements from the periodic table, according to the toxicity and their uptake, where easily exposed to organisms or not, is presented in Table 5.1.

Table 5.1 Classification of elements according to toxicity and their uptake, Wood (1974)

Not critical	Toxic, partially dissolved or easily exposed	Very toxic and easily exposed
Na, C, F, K, S, Sr, H, Cl	Ti, Ga, Hf, Rh, Nb, Ir	Be, As, Au, Cu, Pd, Pb
P, Li, Mg, Al, O, Br, Si	La, Zr, Os, Ta, Ru, Re,	Co, Se, Hg, Zn, Ag, Sb
Fe, Rb, Ca, N	W	Ni, Te, Tl, Pt, Sn, Cd, Bi

Metals fall into one of two categories, essential and non-essential. Essential metal micronutrients, such as Cu, Zn, Fe, Mn, Co, Mo, Cr and Se, are required in small quantities, a few mg or μg per day, and Ca, Mg, Na, P and S, are also required, but in larger quantities, 100 mg or more per day, for the optimal functioning of biological and biochemical processes in humans. Non-essential elements, such as Cd, Pb, Hg and As, have no known biological function in organisms and exert their toxicity by competing with essential metals for active enzyme or membrane protein sites, Valavanidis et al. (2006), Zhou et al. (2008).

Very toxic metals (Table 5.1) are potential hazards to aquatic, animal, and human life because of their toxicity and bioaccumulative and nonbiodegradable nature. Nonessential metals such as Hg, Cd, Pb, As, and Sb are toxic in their chemically combined forms as well as the elemental form, Ahmad (2006), Peralta-Videa et al. (2009), Tangahu et al. (2011). The elements can interact and be interdependent. The strongest toxic properties are characteristic for inorganic metals compounds, which dissociate well and are easily soluble because they can easily penetrate through cell membranes and get into organisms, Szyczewski et al. (2009).

Together with essential nutrients, living organisms also take up toxic metal compounds and can concentrate them. Not all the traces of toxic metals in plants and animals are the results of human activity. The circulation and migration of metals in the natural environment are related to rock decay, volcano eruptions, evaporation of oceans, forest fires and soil formation processes too, Szyczewski et al. (2009). Some metals arise through the absorption processes in organisms of naturally occurring elements in soil components. Purely theoretically every 1,000 kg of "normal" soil contains 200 g Cr, 80 g Ni, 16 g Pb, 0.5 g Hg and 0.2 g Cd, CAOBISCO (1996). Therefore it is not always easy to assign a definite cause for increased toxic metal content.

Toxic metals are commonly considered as simulators or inhibiting factors of life processes, due to which they may appear toxic for living organisms. This depends on their concentration, ability to form complexes and degree of oxidation.

The excessive content of metals in living organisms and in the human body may in many ways affect the body and psychophysical development, Szyczewski et al. (2009). As certain heavy metals such as Pb, Cd, As and Hg have been recognised to be potentially toxic within specific limiting values, a considerable potential hazard exists for human nutrition.

5.2.2 *Toxic Metals Production and Emission*

Metal pollution has increased with the technological progress of human society. The Industrial Revolution brought an exponential increase in the intensity of metal emissions released, both, in absolute masses and in the number and type of toxic metal compounds: about 90 % of mine outputs of Cd, Cu, Zn, Ni, and Pb were consumed in the last century, Nriagu (1996), Fig. 5.1. World mining

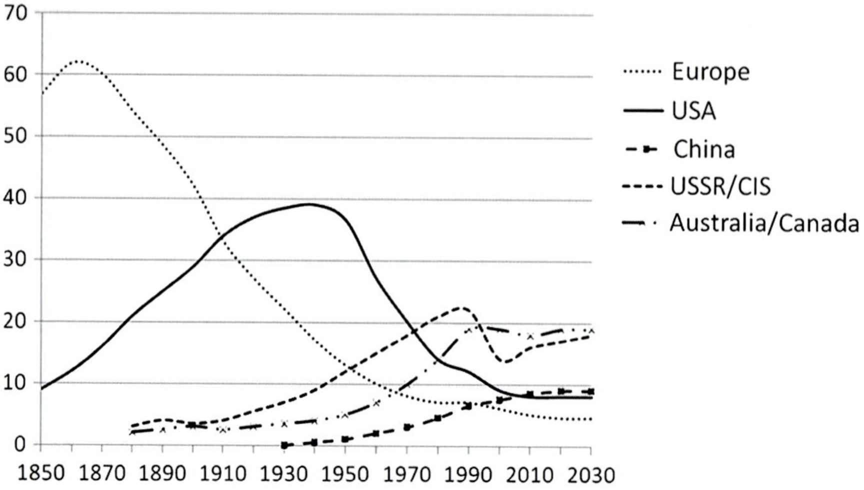

Fig. 5.1 Distribution of global mine production among regions in %, Östensson (2006)

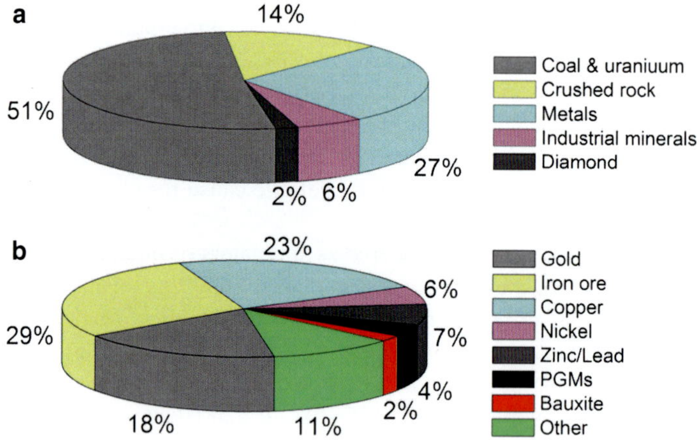

Fig. 5.2 (**a**) Value of world mining activities (**b**) world metals value at mine

activities (Fig. 5.2a) are known to release significant amounts of toxic metals into the surrounding environment. Some toxic metals, including Hg, As, and Pb, are frequently associated with mining. These elements are present at low levels in soil, rock and water, but the process of world metals mining may release quantities harmful to the health of people and the ecosystem.

The cumulative releases of heavy metals into our environment are indeed massive and overwhelmed the natural biochemical cycles of the ecosystems. The levels of the toxic metals such as Hg, Pb, Cd, Cu, As, Zn, and Cr are increasing cumulatively in the environment due to atmospheric emissions from human activities. For example,

Hg levels in Minnesota, Wisconsin, Canada, and Sweden were found to be increasing in sediments, soils, and fish at rates between 2 % and 5 % per year. The four major industrial sources of total metal emissions as reported by USEPA Toxic Release Inventory (EPA TRI 2002–2005) include the electric power industry, the primary metals industry, hazardous waste disposal facilities and world metal mining operations, Östensson (2006) (Fig. 5.2b).

The toxicity impacts mostly occur in the mining, smelting and refining stage via emissions of contaminants in the ores, the fossil fuels or of the auxiliary materials used in these processes. In waste management, metals may end up in landfills and there form a source of slow pollution. Also, metals often occur as ore contaminants in fossil fuels or phosphate rock but also in other metals, and are emitted to the environment together with their host material. Metals are elements not degradable: once in the environment, they do not disappear and their environmental impact is dominant: they tend to accumulate in soils, sediments and biomass, der Voet et al. (2000).

The sources of anthropogenic contamination or pollution of the environment by heavy metals include different branches of industry, the power industry, transport, municipal waste, and fertilizers used to fertilize soil, Szyczewski et al. (2009). The heavy metals from these sources are dispersed in the environment and they contaminate air, water and soil. They, directly or indirectly, throughout air, water, plants and animals get into human bodies. As already mentioned, the metals classified as heavy metals are: Cu, Co, Cr, Cd, Fe, Zn, Pb, Sn, Hg, Mn, Ni, Mo, V and W, Szyczewski et al. (2009). Within the group of heavy metals one can distinguish both, the elements essential for living organisms (microelements) and the elements whose physiological role is unknown, and thus they are "inactive" towards plants, animals and people. The metals serving as microelements in living organisms usually occur in trace amounts that are precisely defined for each species. Both, their deficiency and excess badly affect living organisms, Szyczewski et al. (2009).

The most common heavy metal environmental pollutants are Cu, Cd, Cr, As, Ni, Pb, and Hg, especially in locations where metals tend to accumulate, der Voet et al. (2000). However, metals can become biologically unavailable via geological routes or via soil processes. Also, their hazardous potential can be reduced by transformations in nature, but also in technological processes to a less harmful state. The free-metal ionic activity may be more important in producing metal toxicity than the total concentration of a metal.

Potentially toxic metals follow natural pathways and cycles through the biosphere. Terrestrial and aquatic life forms can suffer short-or long term perturbations if these pathways or cycles are interrupted by natural events or impacted by human activities. Mobilization of suspect metals, resulting in their bioavailability and hence access to a food web, can impact the integrity of a web. Conversely, immobilization of specific metals can have a negative impact if it causes a deficiency of an essential micronutrient within an ecosystem.

The chemical form of a potentially toxic metal determines its bioavailability to a food web. When a heavy metal enters a food web, organisms can react to its bioavailability in different ways. Some organisms may discriminate against the uptake of one or more potentially toxic metals. Others may incorporate the metal in their soft or hard parts in proportion to the concentration in the growth environments, excreting any excess. Still other organisms may be tolerant of heavy metals and will accumulate concentrations greatly in excess amounts in a growth environment without any damage, Siegel (2002). Currently microorganisms or plants which exhibit hyper accumulation can be used to remove heavy metals from water and soils by concentrating them in their bio matter, Stojanović et al. (2012).

Trace element pollution results from various sources, mostly: (i) industrial wastewaters such as Hg from chlor-alkali plants, (ii) mining and smelter wastes, As and Cd, (iii) urban run-off, particularly Pb, (iv) agricultural run-off (where Cu is still used as a pesticide), (v) atmospheric deposition, and (vi) leaching from solid waste dumps. Essentially, the toxic metals have become a focus of public interest since analytical techniques have made it possible to detect them even in very small traces, Szyczewski et al. (2009).

Mercury (Hg)

Mercury is an element that occurs naturally in the environment. However, human activity has significantly changed its cycling in recent centuries. The main natural sources of Hg are diffusion from the Earth's mantle though the lithosphere, evaporation from the sea surface, geothermal and volcanic activity WHO (2007).

The largest anthropogenic source of Hg on a global scale is the combustion of coal and other fossil fuels. Others sources include metal and cement production, forest fires and waste disposal.

Hg, one of the most toxic trace metals, is spread to the atmosphere at significant amounts. On a global scale, the estimated natural emission of Hg represents about one-third of the total, and anthropogenic emissions represent about two-thirds, WHO (2007). Total values of global Hg emission to the air from both anthropogenic and natural sources vary from 4,000 to 9,230 t/year, UNEP (2008) and it is in inorganic form that can be converted biologically to methylmercury (MeHg) in soil and water. MeHg bioaccumulates and enters the human body readily via the dietary route, Stankovic et al. (2011a, b), Stankovic and Jovic (2012).

This element is used in many industries, including pharmaceutical and chloralkali manufacturing industries, electrical engineering, mining, metallurgy and agriculture. Annually, anthropogenic pollution reaches approximately 2,000 t of this metal, most of which comes from burning coal, milling of nonferrous metals and gold mining by amalgamation, Kalisinska et al. (2012). In Europe, the majority of Hg emitted into the environment comes from the combustion of solid fuels, mainly coal (47 %), cement production (13 %), Fe milling (5 %), Zn and Pb milling (7 %) and incineration of waste (5 %), Kalisinska et al. (2012). According to official emission data the total Hg anthropogenic emissions in Europe was 413 t/year in 1990 and 195 t/year in 2003, WHO (2007).

The total concentrations of Hg in natural waters are normally very low, below 1 ng/L. The concentration range for mercury in drinking-water is, in the average 25 ng/L. The WHO guideline value is 1 μg/L, WHO (2007). Representative values for dissolved total mercury are: open ocean 0.5–3 ng/L; coastal seawater 2–15 ng/L; freshwater lakes and rivers 1–3 ng/L. In the European topsoils, Hg concentrations range from 10 to 160 μg/kg, reaching a median value of 40 μg/kg, WHO (2007).

Arsenic (As)

Arsenic is one of the few elements for which a large proportion of the world-production was 30,453 t in 1995 and 40,000 t in 2008 and ends more or less in the environment, Reimann et al. (2009). This is due to its wide-spread use in pesticides, fungicides, herbicides and insecticides and especially as a wood preservative. Further uses are in the glass and electronics industry, in metal treatment, alloys and bronze, in the production of dyes and colours, and in some pharmaceuticals. Another important source of anthropogenic As emissions is the burning of coal and oil. The mean As concentration of coal is given as 10 mg/kg, Reimann et al. (2009), but coals containing up to 3.5 wt% As have been reported from China. Severe health problems in populations living close to power plants burning such coal are well documented, Bencko (1997).

The total annual anthropogenic As emissions were estimated as 28,000–54,000 t/year, but in this range is not included natural As emissions to the atmosphere, including volcanism, forest fires and soil dust. For example, significant amounts of As species are released through hydrothermal activities in Yellowstone National Park. Such sources have not previously been assessed nor their contribution to the global As cycle, Reimann et al. (2009). Plants generally show low As concentrations, much lower than in the supporting soils and the suggested value for the world reference plant of As is 0.1 mg/kg, Nagajyoti et al. (2010). This is clear indication that the majority of plants have mechanisms for avoiding As uptake, but mosses and lichenes consistently show higher As values than other terrestrial plants, Reimann et al. (2009), Serbula et al. (2012).

The most carcinogenic of all substances named in current drinking water regulations is As. Values up to 3,200 μg/L for As have been reported from in India and Bangladesh water wells. The drinking water action limit for As was quite recently lowered to 10 μg/L (from 50 μg/L). A lower As limit of 2 μg/L was suggested for a better safety margin but rejected due to the financial implications, Reimann et al. (2009). Concentrations of As in air is generally very low and ranges from 0.4 to 30 ng/m^3; in seawater from 1.0 to 8.0 μg/L, while in marine sediments As concentration ranges from 1.0 to 60.0 mg/kg, depending on the sediment kind. The world-average value for As in soil is 5 mg/kg, but varies considerably among geographic regions, Reimann et al. (2009). The toxicity of As has been well known at least since Roman times, when arsenolite (As_2O_3) was often used as a poison. Trivalent (As^{3+}) compounds are usually more toxic than As^{5+} compounds, Tangahu et al. (2011), Stankovic et al. (2011a).

Lead (Pb)

The current annual worldwide production of Pb is approximately 5.4 million tonnes and continues to rise. Pb is used in the automobile batteries production (6 %), while the remainder is used in the production of pigments, glazes, solder, plastics, cable sheathing, ammunition, weights, gasoline additive, and a variety of other products, WHO (2007). Pb from soil tends to concentrate in root vegetables and leafy green vegetables (onion and spinach, respectively). Individuals will absorb more Pb in their food if their diets are deficient in Ca, Fe, or Zn, Hu (2002). In general, ingestion of Pb through food and water is the major exposure pathway for Pb in humans.

According to Krystofova et al. (2009) levels of Pb in the environment are not stable and vary according to industrial production, urbanization, climate changes and many other factors. The levels of Pb in the environment vary between 4 and 20 mg/g of dust. Uncontaminated waters contain Pb in concentrations ranging from 0.001 to 0.06 mg/L, while in seawater to 0.03 μg/L, Bardi (2010). Levels of Pb in soils are between 5 and 30 mg/kg. Pb is released into the atmosphere from natural and anthropogenic sources. Natural emissions are from wind resuspension and from sea salt, volcanoes, forest fires and biogenic sources. Major anthropogenic emission sources of Pb on a global scale include the combustion of fossil fuels, from traffic, non-ferrous metal production and iron and steel production, WHO (2007). Pb levels in air in Europe fell by 50–70 % between 1990 and 2003 because of restrictions on the use of leaded petrol, WHO (2007).

The natural Pb content in soil typically ranges from below 10 mg/kg soil up to 30 mg/kg soil. In Europe, Pb concentrations in topsoils are spatially heterogeneous and vary from below 10 mg/kg up to >70 mg/kg. The median value is estimated to be 22.6 mg/kg. In addition to atmospheric deposition, agricultural practices are a source of Pb input to soils from mineral and organic fertilizers and other sources, WHO (2007). Pb is incorporated into several crops through absorption, by the roots, from soil and through direct deposition on plant surfaces. The Pb levels in various food crops amounted to 2–136 mg/kg for grains and cereals, 5–649 mg/kg for vegetables and 6–73 mg/kg for fats and oils, WHO (2007). The uptake of soil-borne Pb by vegetables is very low. This explains the very low Pb concentrations in root vegetables, seeds and fruit. In leafy vegetables, the accumulation of airborne Pb largely exceeds the soil-borne part taken up via the roots. Airborne Pb is mainly accumulated at the leaf surface and can be removed to a large extent by thorough washing of the crop. There is, however, a relationship between the Pb content of the soil and the Pb content of the atmospheric deposition, WHO (2007).

The EU ambient air quality guideline for Pb is 0.5 μg/m^3 and in the immediate vicinity of specific industrial sources, the value was 1 μg/m^3 until 2010, WHO (2007). With regard to drinking-water, Pb contamination of the water can occur from, for example, industrial discharges or highway run-offs. Furthermore, drinking-water can be contaminated with Pb from lead pipes, lead-soldered copper pipes, lead-containing brass joints in plastic pipes, or from other parts of the water system. In particular, soft water has the potential for dissolving Pb. The limit value for Pb in drinking-water in the EU is currently 25 μg/L and will be reduced to 10 μg/L by 2013, WHO (2007).

Within the EU, the maximum permissible level of Pb in foods ranges from 0.02 mg/kg ww for milk and milk supplements for infants up to 1 mg/kg for mussels, Stankovic et al. (2011b). A PTWI of 25 μg/kg bw/week has been established for all age groups by the Joint FAO/WHO Expert Committee on Food Additives, Stankovic et al. (2011b). Pb accumulates in bone and bone-Pb concentrations are associated with the Pb concentration in the whole blood and plasma, WHO (2007).

Cadmium (Cd)

Cd is released to the biosphere from both natural and anthropogenic sources. The major natural sources for mobilisations of Cd from the earth's crust are volcanoes and weathering of rocks, than airborne soil particles, sea spray, biogenic material and forest fires. The total Cd emission to air from natural sources is estimated at about 150–2,600 t, WHO (2007). The weathering of rocks releases Cd to soils and aquatic systems. Within the biosphere Cd is translocated by different processes.

The anthropogenic sources of Cd include non-ferrous metal production, stationary fossil fuel combustion, waste incineration, iron and steel production, and cement production, WHO (2007). World Cd mine production was 19,700 t in the year 2000, 73 % was for automobile batteries. In the US, combustion of coal and oil is assumed to be responsible for around 76 % of the total emission of Cd to the atmosphere. The global Cd production increased with a factor of four from 1950 to 1990 and in the recent decade the production has slightly decreased, NCM (2003). Cd emissions in Europe constituted 485 t in 1990 and 257 t in 2003. Emissions from EU countries have decreased by 50 %. In European countries the dominant sources of Cd are atmospheric deposition and commercial phosphate fertilizers, OECD (1994). Global emission of Cd to air until the year 2000 was closed to 3,000 t/year, NCM (2003).

On the basis of the Cd contents in surface soils from many parts of the world, the average value lies between 0.07 and 1.1 mg/kg; values above 0.5 mg/kg usually reflect anthropogenic inputs, WHO (2007). There are three main anthropogenic sources of terrestrial Cd: atmospheric deposition, agricultural application of phosphate fertilizers, and use of municipal sewage sludge as a fertilizer on agricultural soils. It has been reported that 90 % of Cd in the soil remains in the 15 cm of the top. Accumulation of Cd in the soil depends on the soil properties: generally clay soils retain more Cd than sandy soils, WHO (2007). Cd levels of up to 5 mg/kg have been reported in sediments from river and lakes, and from 0.03 to 1 mg/kg in marine sediments. The average Cd content of seawater is about 5–20 ng/L in open seas, while concentrations of 80–250 ng/L has been reported in French and Norwegian coastal zones. A concentration measured in European rivers roughly varies from 10 to 100 ng/L, OSPAR (2002).

Drinking-water generally contains low Cd levels, and a value of 1 μg/l or less is often assumed to be a representative content in most situations. Thus, Cd exposure from drinking-water is relatively unimportant compared with the dietary contribution, WHO (2007). Drinking-water contains very low concentrations of Cd, usually between 0.01 and 1 μg/L. The average daily intake of water is about 2 L per person, WHO (2007).

Chromium (Cr)

Cr (VI) exists in soils naturally and is the sixth most abundant element in the Earth's crust. It is also present as a result of human practices, and mostly from those practices associated with industry. Cr occurs naturally in Irish agricultural soils in concentrations between 5 and 250 mg/kg and in various soil types ranging from 1.04 to 3,015 mg/kg worldwide. Cr has been extensively shown to induce general environmental toxicity, as well as more specific effects of an acute and chronic nature such as neurotoxicity, dermatoxicity, genotoxicity, carcinogeneticity, and immunotoxicity. It is believed that Cr inflicts most damage during reduction of Cr (VI) to Cr (III), a process considered to be initiated in the cell by glutathione, Boyle and Kakouli-Duarte (2008). Cr is found in all phases of the environment, including air, water and soil. Naturally occurring in soil, Cr ranges from 10 to 50 mg/kg. In fresh water, Cr concentrations generally range from 0.1 to 117 mg/L, whereas values for seawater range from 0.2 to 50 μg/L. In the atmosphere Cr concentration vary widely, from background concentrations of 0.0012 $\mu g/m^3$ in the air sample, Shanker et al. (2005).

5.3 Metal Pollution

5.3.1 Air

Air pollution is certainly not a new phenomenon and the recognition thereof was marked as early as during the 1930s catastrophe in the Meuse Valley, Belgium, where SO_2 and particulate matter, combined with a high relative humidity, caused 63 excess deaths in five days. In the year 1948 similar conditions in Donora, Pennsylvania, a small industrial city, caused 20 excess deaths in five days, and in the early 1950s in London, England, two episodes of "killer fogs" claimed the lives of more than 6,000 people.

Atmospheric pollution causes serious damage for human health and to all natural ecosystems. The composition of the air pollutants can be inorganic, organic, or a complex mixture of both. Air pollutants also are classified as primary or secondary. Primary pollutants are those that are emitted directly into the atmosphere, CO and SO_2, and secondary pollutants are those that are produced in the atmosphere by chemical and physical processes from primary pollutants and natural constituents. Nowadays, when the biggest provocative of atmospheric pollution is anthropogenic human activities and transport sector, main pollutants are toxic metals, Blagnytė and Paliulis (2010). Monitoring toxic air pollutants is needed for understanding their spatial and temporal distribution and ultimately to minimize their harmful effects. Among the many inorganic air pollutants originating from anthropogenic activities, heavy metals such as As, Cd, Cr, Hg, and Pb are of a major concern due to their toxic and potentially carcinogenic characteristics. One of the most important sources of Cd, Cr and Pb in the urban environment is road traffic, Melaku et al. (2008).

Environmental sources for pollutants could include construction and agricultural activities, mining and mineral processing, wind-blown dust, traffic activities on the road. Power plants are the largest source of several harmful pollutants. They are responsible for 50 % of Hg and about 25 % of toxic metal emissions in the United States. Coal-fired power plants are responsible for 99 % of Hg emissions and the bulk the other pollutants from the power sector. Thus, power plants are the largest source of Hg emissions to the air. The Hg concentration in air is in most areas close to the mean global background value, which is 1.5–2.0 ng/m^3, WHO (2007). The total concentration of Hg is 2 ng/m^3, which is valid for the northern hemisphere. In recent years, emissions of Hg into the air in Europe have been declining. The decrease in Hg emissions in 1990–1996 in 39 countries ranged from 0 % to 56 %, WHO (2007). According to EEA (2003), the global anthropogenic emissions of Hg may amount to 30,000 t/year.

According to Harmens et al. (2008) in some European countries during 1990–2005/6 the main sources of Pb emissions were from manufacturing industries and construction (41 %) and road traffic has become the second source of Pb emission (17 %). Pb levels in ambient air in Europe have decreased in recent decades: Pb levels in air in Europe between 1990 and 2003 fell by 50–70 %. Low concentrations of Pb in the air rural areas are between 0.05 and 0.1 μg/m^3. Natural Pb content is estimated to be 6.0 ng/m^3 in the atmosphere, WHO (2007).

Cd pollution levels exhibit about a twofold decrease for Europe as a whole in the period 1990–2003, WHO (2007). The WHO 2000 air quality guideline has been recommended 5 ng/m^3 for Cd in order to prevent any further increases in Cd levels in agricultural soils, WHO (2007).

In addition to direct physical and chemical methods of air pollution monitoring, bioindicators have also been used to evaluate air pollution risk, Dmuchowski and Bytnerowicz (2009). Biomonitoring based on plant as heavy metal air pollution indicators has been widespread all over the world since the 1970s. Biomonitoring of environmental pollution with toxic metals using vascular plants, mosses, and lichens have been used for many years in Europe and North America. Foliage and bark of trees has also been widely used in this regard, Dmuchowski and Bytnerowicz (2009).

5.3.2 Water

Metals have many sources from which they can flow into the water. Heavy metals in aquatic system can be naturally produced by the slow leaching from soil/rock to water, which are usually at low levels, causing no serious effects on human health, Zhou et al. (2008). Nowadays, the industry and agriculture development promotes the rapid increase of water metal pollution. Industry, mining, advanced agriculture, household waste, and motor traffic are all among the activities considered to be major sources of aquatic metal pollution. Metals can accumulate in aquatic organisms and persist in water and sediments, Sevcikova et al. (2011).

Aquatic heavy metal pollution usually represents high levels of Hg, Cr, Pb, Cd, Cu, Zn, and Ni in water system, Zhou et al. (2008). Cu, Ni, Cr, and Zn are essential trace metals to living organisms, but become toxic at higher concentrations. Toxic elements Pb, Hg, As, and Cd have no known biological function. The aquatic environment with its water quality is considered the main factor controlling the state of health and disease in man and biota, Markovic et al. (2012).

Some heavy metals including Hg, Cr, Cd, Ni, Cu, As, and Pb introduced into environmental water system may pose high toxicities on the aquatic organisms, Zhou et al. (2008), Stankovic et al. (2011a), Stankovic and Jovic (2012). Concentrations of inorganic Hg in surface and groundwaters are generally below concentrations known to cause adverse health effects. MeHg in freshwater fish originates from inorganic Hg in the soil and direct atmospheric deposition, Chen et al. (2008). Although, the anthropogenic Hg emission in Europe decreased approximately by 50 % after 1990, but MeHg concentration decreasing has not been observed in freshwater fish WHO (2007). In fishes as a predatory animals Hg levels can be biomagnified. In general, fish as food has important beneficial effects on human health, Chen et al. (2008).

The physicochemical properties of water play an important role in metal solubility. The amount of dissolved metal strongly depends on water pH and metal can accumulate in aquatic organisms. Organic substances found in water can bind metal ions influencing their lower bioavailability to aquatic organisms and thereby somewhat reduce metal toxicity. Some metals can rapidly be bonded to organic substances and cannot be detected in water, but they can later become accessible to water organisms and their food, Sevcikova et al. (2011).

5.3.3 *Soil*

Soil is one of the principal regulatory compartments of all terrestrial and benthic ecosystems. Soil is a fundamental natural resource for agriculture. Preservation of healthy, high-quality soil is essential to maintain the health of plants, animals and human beings.

Soil is not only subjected to physical degradation due to deforestation, flooding, and wind erosion, it is recently intensively polluted by various toxic metals and other pollutants as well. The contamination of soil occurs through air pollution, wastewater intake and fertilizers in agriculture. A special type of soil contamination is related to the accumulation of industrial and municipal wastes, especially those that are hard to break. Soil pollution by heavy metals which are released into the environment through various anthropogenic activities such as mining, energy and fuel production, electroplating, wastewater sludge treatment and agriculture is one of the world's major environmental problem. An extensive literature review of trace metal contamination of the urban terrestrial environment, in particular of urban soils, in Europe, America, Africa, and Asia region is given by Wong et al. (2006).

Heavy metals refer to a large group of elements which are both industrially and biologically important. Initially, heavy metals are naturally present in soils as natural components but as of now, the presence of heavy metals in the environment has accelerated due to human activities. This is a widespread problem around the world where excessive concentration of heavy metals such as Ni, Pb, Zn, Cr, Cu, Cd, Hg, and As can be found in soils, Wong et al. (2006).

The soil environment is a major sink for multitude of chemicals and heavy metals, which inevitably leads to environmental contamination. Millions of tonnes of trace elements are produced every year from the mines in demands for newer materials. On being discharged into soil, the heavy metals get accumulated and may disturb the soil ecosystem, plant productivity, and also pose threat to human health and environment, Musarrat et al. (2011).

In areas of agricultural and industrial activity, higher concentrations of heavy metals, in comparison with background levels, can be detected. Especially, soils near heavy metals mines are exposed to metalloids pollution by Zn, Pb, Cr, Mn, Fe, Tl, In or As, Babula et al. (2008). Landfilling is still the main waste disposal method in Europe, Paoli et al. (2012). Consequence of the waste disposal in landfills is the emission of pollutants via the leachate and gas generation. Recent EU legislation prescribes strict rules for waste disposal in landfills, EU Directive 2008/98/EC, Paoli et al. (2012).

Cadmium is accumulating in soils and catchments under certain environmental conditions, thus increasing the risk of future exposure through food. The transfer of Cd from soil to the food-chain depends on a number of factors, such as the type of plant, the type and pH of the soil and the zinc and organic matter content in the soil, WHO (2007). Soil pH is the principal factor governing the concentration of Cd in the soil solution. As soil pH consequence, Cd plant uptake decreases as the soil pH increases.

Cd normally occurs in low concentrations in soils and is a nonessential element for plants and potentially toxic pollutant all over the world. The toxicity of Cd to plant growth, phytosynthesis, carbonhydrate metabolism, and enzyme activities is well documented, Javed and Greger (2011). The cropping land with total soil Cd concentration exceeding 5 mg/kg would be announced as Soil Pollution Control Site (SPCS) and all farming activities were not allowed, since elevated levels of Cd in soil may potentially enter a food chain, Meharg and Hartley-Whitaker (2002) and increase the risk of cancer development in humans, Stankovic et al. (2011a).

Trace metal assessments of urban soils frequently examined trace metals that were traditionally significant for the environment and health, particularly Cd, Cu, Pb, and Zn, Wong et al. (2006). In the EU the limit value for Cu in soil is 50–140 mg/kg of dry weight (Council Directive, 86/278) and the mean levels for Cu in soils vary from 13 to 24 mg/kg; the Zn soil levels usually fall in the range of 10–300 mg/kg and in the EU the limit value for Zn in soils from 150 to 300 mg/kg of dry weight (Council Directive, 86/278); the limit values in the EU are: for Pb from 50 to 300 mg/kg, for Cd from 1 to 3 mg/kg and for Hg from 1 to 1.5 mg/kg dry weight (Council Directive, 86/278), Serbula et al. (2012).

The mean Pb concentration for surface soils on the world scale is estimated at 25 mg/kg; the average contents of Cd in soils are between 0.06 and 1.1 mg/kg; for various soils on a world level mean concentrations of Hg in surface soils do not exceed 0.4 mg/kg, Serbula et al. (2012). The terrestrial abundance of As is around 1.5–3.0 mg/kg. The world-average value for As in soils is 5 mg/kg, Reimann et al. (2009) and EU limit for As in agricultural soil is 20 mg/kg, Bhattacharya et al. (2010). In well-aerated soils, arsenate (As(V)) is the predominate form. Lai et al. (2010) showed that As(V) in aerated soils will be reduced to more mobile and toxic As(III) in paddy soils and transferred to plants. As(III) is more soluble and more mobile, and is much more toxic than As(V), Lai et al. (2010), Tangahu et al. (2011).

Traffic-related activities continue to exert a widespread influence on the urban environment contaminating urban land. Nowadays, platinum group elements, and manganese (Mn) have been recommended as environmental indicators of traffic and other "urban" activities, especially where leaded gasoline is no longer in use, Wong et al. (2006). Also, soil pollution problems arise from agrochemicals being used and dispersed into the environment, Reijnders and Huijbregts (2009). Currently about half of the world's land and 70 % of total water are used for agriculture. In recent years, agriculture materials come out as high priority from an environmental impact point of view, Reijnders and Huijbregts (2009).

Since metals are not degradable they accumulate in soils and sediments. This affects human and ecosystem health. Their hazardous potential can be reduced by transformations via geological routes or via soil processes to a less harmful state.

Heavy metals as soil contaminants typically occur in low concentrations: 10,000 mg kg^{-1} dry soil, Schwartz et al. (2011). The contaminants from soil are able to infiltrate deep into the layer of underground waters and pollute the groundwater as well as the surface water. Also, toxic metals in the soil subsequently enter the human food web through plants and animals and they constitute risk to the ecosystem as they tend to bioaccumulate and can be transferred from one food chain to another.

5.3.4 Biota

Flora, fauna and other forms of life such as fungi are collectively referred to as biota. Fauna is all of the animal life of any particular region or time. The corresponding term for plants is flora. Biota is the total collection of organisms of a geographic region or a time period, from local geographic scales and all the way up to whole-planet and whole-timescale spatiotemporal scales. The biota or biotic components of the Earth makes up the biosphere.

Biota growth and development are essential processes of life and propagation of the species. They are continuous and mainly depend on external resources present in water, soil and air. Growth is chiefly expressed as a function of genotype and environment, which consists of external and internal growth factors, Shanker et al.

(2005). Presence of toxic metals in the external environment leads to changes in the growth and development pattern of the biota. Metals are commonly considered as simulators or inhibiting factors of life processes, due to which they may appear toxic for living organisms, Szyczewski et al. (2009).

The first reports started to appear in 1950–1970s on the toxic effects of Hg and Cd contamination on wild animals, including foxes, seafood and cultivated rice in Japan, Basu and Head (2010), when countries with highly industrialized agriculture were widely using organic Hg in chemicals for plant protection and seafood and rice as food. Some of the toxic metals such as Cd, Pb, Hg and As are strongly poisonous to metal-sensitive enzymes, resulting in growth inhibition and death of organisms, Nagajyoti et al. (2010), Stankovic et al. (2011a, c).

Various metals play a key role to maintain and control various vital functions of living organisms. The essential metals include, among others, Cu, Zn, Fe, Mn, Cr, Co, Mo, and Ni and play biochemical and physiological functions in living organisms. Two major functions of essential metals are the following: (a) participation in redox reaction, and (b) direct participation, being an integral part of several enzymes, Szyczewski et al. (2009), Nagajyoti et al. (2010). However, in the literature there are several references regarding the essential characters of Ni, Nagajyoti et al. (2010), but at the some Ni is classified as 'probably essential', Szyczewski et al. (2009). Too large doses of essential metals can damage living organisms or in some cases the organism can be injured or die, Szyczewski et al. (2009). The availability of metals in medium varies, and metals such as Cu, Zn, Fe, Mn, Mo, Cr, Ni and Co are essential micronutrients whose uptake in excess results in toxic effects, Nagajyoti et al. (2010).

Fauna

Metals uptake by fauna occurs through diffusion as well as ingestion of food. Metals can be stored in the skeletal structure and intracellular matrices of an organism, and excreted in feces and eggs. Fauna organisms have additional defense mechanisms: metallothioneins and other heavy metal-binding proteins bind metals within the organism helping them to control toxic metal levels, Garcia et al. (2008). In comparison to plants, animals have generally developed a greater arsenal of defense mechanisms; in addition, non-sessile animals can avoid a certain number of environmental or anthropogenic stressors by their mobility.

Prey, predatory birds and mammals, are accumulators of significant amounts of organic Hg by eating beans and small warm-blooded vertebrate herbivores contaminated with Hg, Basu and Head (2010). Toxicity of methylmercury (MeHg) and Cd in mammals has been well documented, Basu and Head (2010). Metals are toxic to mammalian renal cells, causing damage that leads to leakage of phosphates, Ca, glycogen and proteins from the kidney, Lavery et al. (2009).

Terrestrial vertebrates, including wild animals, absorb Hg via food, water, air and through the skin. Organic Hg is strongly absorbed from the gastrointestinal tract (>90 %) and accumulates in internal organs, mainly in the liver, kidney and also in the brain, Kalisinska et al. (2012). In contrast, elemental Hg and its inorganic compounds are very poorly absorbed in the gastrointestinal tract (<10 %), and additionally their harmful effects on warm-blooded vertebrates, including man, is much lower than MeHg, which is widespread in the environment, especially in water ecosystem, Kalisinska et al. (2012).

Birds are primary and secondary consumers, feeding on plants, invertebrates and vertebrates alike. Colonial waterbirds (herons, storks, ibises, pelicans, cormorants, gulls, terns, and marine birds) often have been used or proposed as indicators of Pb, Cd, As, Se, and Hg. Many toxic metals (Cr, Mn, Cu, Zn, As, Se, Rb, Sr, Mo, Ag, Cd, Hg) have been studied in several terrestrial, aquatic and seabirds, as birds are ordinarily at the top of the food web, they are valuable for environmental monitoring, Horai et al. (2007), Hargreaves et al. (2011).

Heavy metals such as Hg, Cd, As, and Pb are toxic metals that have no known vital or beneficial effect on organisms and their accumulation over time in the bodies of animals can cause serious illness, Stankovic et al. (2011a), Stankovic and Jovic (2012). Certain elements that are normally toxic are, for certain organisms or under certain conditions, beneficial: examples include V, W, and even Cd, Hogan (2010). Kidneys and liver are important sites for metal accumulation and toxicity in vertebrates, Adham et al. (2011). Metals accumulate initially in the liver where they form a complex with metal-binding proteins, such as metallothionein, until an accumulation threshold is reached. Then the metal protein complex is transported via the plasma to the kidney, where they accumulate over time, Adham et al. (2011).

Flora

Uptake of elements into plants can happen via different ways. The elements can be taken up via roots from soil and transported to the leaves. Also, they may be taken up from the air or by precipitation directly via the leaves. Trace elements may be also taken up via both mentioned ways. Many factors influence the plants metals uptake and include the growing environment, such as temperature, soil pH, soil aeration, Eh condition (particularly of aquatic environment) and fertilization, competition between the plant species, the type of plant size, the root system, the availability of the elements in the soil or foliar deposits, the type of leaves, soil moisture and plant energy supply to roots and leaves, Nagajyoti et al. (2010).

Plants dig their roots in soils, sediments and water, and roots can take up organic compounds and inorganic substances. Roots can stabilize and bind substances on their external surfaces. Uptaken substances may be transported, stored, converted, and accumulated in the different cells and tissues of the plant. Finally, aerial parts of the plant may exchange gases with the atmosphere allowing uptake or release of molecules, Marmiroli et al. (2006). Accumulation and distribution of heavy metals in the plant depend on the plant species, the levels of the metals in the soil, water and air, the element species and bioavailiability, pH, cation exchange capacity, climacteric condition, vegetation period and multiple other factors, Filipovic-Trajkovic et al. (2012).

Metal toxicity can inhibit protein activity or disrupt their structures in plant. Metal detoxification mechanism in plants involves the activation of antioxidative enzymes systems that are responsible for protecting the plant cells from the toxic effects during metal contamination, Garcia et al. (2008). Plants' response to the presence of heavy metal ions includes synthesis of plant thiol compounds, namely phytochelatins, Babula et al. (2008). Some plants, called metallophytes, can demonstrate tolerance–hypertolerance to heavy metals, Babula et al. (2008) and hyperaccumule one or more metals/metalloids, Stojanovic et al. (2012).

Plants might react to environmental stress on various levels: on the cellular biochemical, or morphological scale, and at species or population level. The phototoxic effects typically cause physiological, morphological and ecological changes, expressed in many different ways, from reduced photosynthetic activity to changes in plant species distribution, the establishment of adapted species, and the dying-off of whole plants or communities, Johnson et al. (2011). However, stress can be induced in vegetation by a large number of other factors, including water deficiency, poor soil drainage, poor soil aeration, soil salinity, weed competition, nutrient deficiency, or nutrient poisoning, Schwartz et al. (2011).

For their development, plants require not only macronutrients (N, P, K, S, Ca, and Mg), but also essential micronutrients such as Fe, Zn, Mn, Ni, Cu, and Mo. Most of these micronutrients accumulate in the plant tissues for their metabolic needs, but they never exceed 10 mg, except Fe, Mn and Zn, Nagajyoti et al. (2010). Some metals which are considered contaminants are also essential micronutrients for all vegetation species; in particular Cu and Zn. Higher levels of these metals, and other heavy metals, in plant tissue might have phytotoxic effects, sometimes resulting in death, Szyczewski et al. (2009). The levels of heavy metals in plants, both terrestrial and aquatic, vary widely because of the influence of environmental factors and the type of plant itself. The metals range observed in plants was presented in the review paper of Nagajyoti et al. (2010).

The main sources of toxic metals to plant may be atmospheric or terrestrial, including mining and metal working industries, combustion of fossil fuels, disposal of ash residues from coal combustion, vehicular traffic and use of pesticides and fertilizers in agriculture. A few heavy metals such as Co, Mn, Fe, Cu, Se, Mo and Ni are essential, whereas As, Cd, Cr, Hg, Pb are nonessential elements and toxic to plants. Besides their toxicity, heavy metals are also quite persistent in the environment, Szyczewski et al. (2009).

Babula et al. (2008) claim that essential elements are important constituents of pigments and enzymes, mainly Cu, Ni, Zn, Co, Fe and Mo for algae and higher plants, but all metals/metalloids, especially Cd, Pb, Hg and Cu are toxic in higher concentrations because of disrupting enzyme functions, replacing essential metals in pigments or producing reactive oxygen species (ROS). The similarity of certain heavy metals to essential heavy metals, for example couples Cd–Zn, Se–S or As–P, predestinates their high toxicity due to the possibility to replace essential metals in enzymatic systems. The toxicity of less common heavy metals and metalloids (Tl, As, Cr, Sb, Se, and Bi) are still under investigation. The important factor of bioavailability of metals/metalloids is their presence in soil and water, although there are not many plants that are able to uptake them from an air, Babula et al. (2008).

Plants contribute to the circulation of heavy metals in the food chain through their active and passive absorption, accumulation in tissues as well as subsequent grazing by animals or consumption by humans. All plants have the ability to accumulate all metals, especially essential for their growth and development: Mg, Fe, Mn, Zn, Cu, Mo and Ni. Plants absorb metals through the root from the soil, water and through over ground vegetative organs from the atmosphere, Filipovic-Trajkovic et al. (2012). Levels of heavy metals in various species of wild plants growing in the same habitats may vary considerably, Wisłocka et al. (2006).

5.3.5 *Human-Health Implications*

Humans can be exposed to metals through inhalation of dust or gaseous particles, or ingestion through food and drink, Hu (2002). Once a metal is incorporated, it is distributed to tissues and organs. Although excretion typically occurs through the kidneys and digestive tract, metals tend to persist in sites like the liver, bones, and kidneys for years or decades. Low-level metals exposure likely contributes to chronic disease and impaired functioning, Hu (2002).

Living organisms, and as humans as well, require varying amounts of "heavy metals": Fe, Ni, Cu, Zn, Co, Mn, and Mo, Wong et al. (2006), Joksimovic et al. (2011a, 2012). Their excessive levels can be damaging to the organism. Other heavy metals such as Hg, Cd, As, and Pb are toxic metals that have no known vital or beneficial effect on humans, Stankovic et al. (2011a), Stankovic and Jovic (2012), and their accumulation over time in the bodies can cause serious illness.

Within the European community the eleven elements of highest concern are: As, Cd, Co, Cr, Cu, Hg, Mn, Fe, Zn, Ni, and Pb. Some of these elements are actually necessary for humans in trace amounts (Fe, Zn, Co, Cu, Cr, Mn, Ni), while more than necessary amounts of trace elements and some others are carcinogenic or toxic, affecting, among others, the central nervous system (Hg, Pb, As, Mn), the kidneys or liver (Hg, Pb, Cd, Cu) or skin, bones, or teeth (Ni, Cd, Cu, Cr), Rai and Pal (2002), Chen et al. (2008), Lavery et al. (2009), Zevenhoven and Kilpinen (2001), Stankovic et al. (2011a), Jovic et al. (2012), Stankovic and Jovic (2012), Markovic et al. (2012).

Metals are significant to human because some of them are most important trace elements in various metabolic enzymes and constituents of cells. Zn, Cu and Fe form important component of cell and co-factor in several metalloenzymes, while organically chelated Cr^{3+}-ion act as a co-factor in insulin hormone response controlling carbohydrate metabolism in humans, Rai and Pal (2002). Some non-essential elements are present in all tissue and organs of man. Under normal conditions, more than 90 % of Pb is retained in the skeleton, Stankovic et al. (2011a).

The importance of Cr as trace nutrient in humans and animals has been known for about 40 years. Cr is necessary for an optimal metabolism in humans and some animals. The daily need for Cr in humans should be around 50 μg. Most frequently occurring damages at increased uptake of Cr are damages to skin and mucus membranes. International Agency for Research on Cancer (IARC) has assessed that Cr (VI) compounds are carcinogenic to humans, Rai and Pal (2002). The most common effect of exposure to Ni is allergic contact eczema. Acute poisonings are very rare, as for Cr. Observed effects of Ni compounds are various damages to the respiratory system. IARC has assessed that Ni compounds are carcinogenic to humans, Jovic et al. (2012).

In the 1956 an unusual and painful disease of rheumatic nature was recorded in the humans from a village in Japan. During subsequent years it became known as 'itai-itai' disease, chronic Cd poisoning, Stankovic et al. (2011a). Cd induced disturbances in Ca metabolism accompanied by softening of bones, fractures and skeletal deformations. The main source of Cd exposure in the general population is food, representing >90 % of the total intake in non-smokers, Stankovic et al. (2011a).

Hg is considered to be highly toxic metal for living organisms. Even at very low concentration, Hg and its compounds present potential hazards due to enrichment in food chain. Poisoning by MeHg presents a bizarre neurological picture as observed in large-scale outbreaks in Japan, China, Iraq and various parts of the world, Rai and Pal (2002). In 1959 the humans who had consumed fish and shellfish around Minamata bay in Japan progressively suffered from the disease known as Minamata disease. The humans' deaths were caused by the consumption of the fish and other foodstuffs contaminated with MeHg, Stankovic et al. (2011a), Stankovic and Jovic (2012). Human exposure to MeHg occurs mainly through the diet, more specifically from the consumption of fish. Humans can also be exposed to elemental Hg vapors and other inorganic Hg species, but the contributions from the air and consuming drinking-water are insignificant, Rai and Pal (2002).

In 1960 fatal incidents of lung cancer were reported from the Kiryama factory of Nippon-Denki concern. The Nippon chemical industry has deposited approximately 530,000 t of wastes containing Cr(VI) around Tokyo in Japan. Due to its hardening properties the material has found extensive use for construction purposes. Medical warnings were issued that inhalation of dust containing Cr(VI) was associated with malignant growth in the respiratory tract, Rai and Pal (2002).

The importance of As as a health hazard, which is also known as 'slow killer' is now well recognised. As is the most common cause of acute heavy metal poisoning in humans, and does not leave the body once it enters. The most obvious signs are the blisters on the palms of the hands and soles of the feet, which can eventually turn gangrenous and cancerous, Rai and Pal (2002). Meanwhile, the poison also attacks internal organs, notably the lungs and kidneys, which can result in illnesses including cancer, Stankovic et al. (2011a), Stankovic and Jovic (2012), Markovic et al. (2012), Jovic et al. (2012).

From several places As poisoning has been reported: in the vicinity of a coal power plant in Czechoslovakia where coal that was used contained about 1,000–

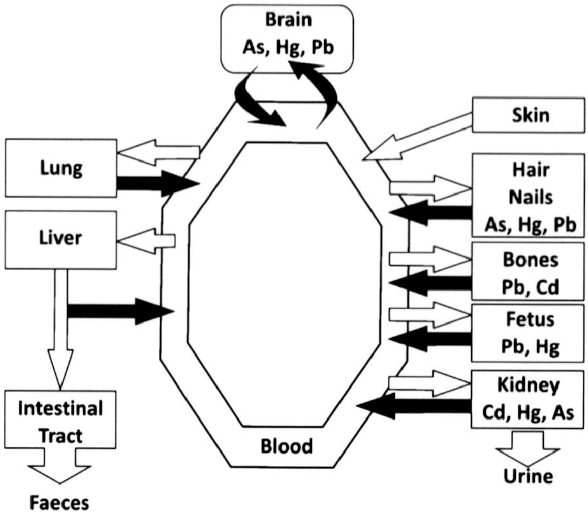

Fig. 5.3 Deposition of metals in human, Kakkar and Jaffery (2005)

1,500 g of As per tonne, children showed respiratory symptoms and hearing loss, Rai and Pal (2002). Similar cases of As poisoning have been reported from West Bengal (India) and Bangladesh in recent years. There is no medicine available for chronic As toxicity. The only treatment is safe water, nutritious food and time, Rai and Pal (2002). Long term ingestion of As contaminated drinking water produced gastrointestinal, skin, liver and nerve tissue injuries, Rai and Pal (2002) and cancer Stankovic et al. (2011a), Stankovic and Jovic (2012), Markovic et al. (2012), Jovic et al. (2012).

Even the Romans were aware that Pb could cause serious health problems. Pb is the number one environmental poison amongst the toxic heavy metals all over the world, Rai and Pal (2002), causing serious health hazards to humans, especially to young children, affecting the membrane permeability of kidney, liver and brain cells, resulting in either reduced functioning or a complete breakdown of these tissues since Pb is a cumulative poison, Stankovic et al. (2011a), Stankovic and Jovic (2012). The full impact of Pb poisoning on the health of children and adults is becoming clearer to most countries, and many governments have begun to take action.

Figure 5.3 depicts the deposition of metals in humans, Kakkar and Jaffery (2005). Different levels of exposure to heavy metals can lead to three different effects: acute effects in which symptoms appear immediately after exposure of heavy metals during a short exposure period; chronic effects are a result of low-level exposure over a long period. Lastly, lethal effects can be defined as responses that occur when physical or chemical agents interfere with cellular and subcellular processes in the organism at the high level thus causing death, Kakkar and Jaffery (2005).

Problems related to pollutants include the emission of certain substances, even in small quantities, which disturb natural processes and threaten the health of humans and ecosystems due to their toxicity, carcinogenity, mutagenicity or hormone mimicking properties, der Voet et al. (2000). Metal toxicity causes irregular metallothioneins (MT) protein synthesis, renal damage, and disruption of bone structure in humans. Metal-binding proteins have special functions in the detoxification of toxic metals and also play a role in the metabolism of essential metals. MTs are involved in the regulation of the essential metals Cu and Zn and in the detoxification of Cd and Hg, Zhou et al. (2008), Freisinger (2010), Sevcikova et al. (2011).

Pulmonary absorption of inhaled Cd ranges from 10 % to 50 % and the average normal gastrointestinal absorption of ingested Cd in humans' ranges from 3 % to 7 %, while in the tissues is mainly bound to metallothionein. The synthesis of this protein probably represents the body's defence mechanism against the toxic Cd-ion. Liver and kidney tissues are the two main sites of Cd storage and these organs accumulate considerable amounts of Cd, about 40–80 % of the body burden, WHO (2007). Important health endpoints include kidney and bone damage and cancer, Lavery et al. (2009), Stankovic et al. (2011a), Stankovic and Jovic (2012).

Heavy metals have been utilized by humans for thousands of years. Exposure to heavy metals continues although several adverse health effects of heavy metals have been known for a long time, Rai and Pal (2002), Schwartz et al. (2011), Stankovic et al. (2011a), Stankovic and Jovic (2012), Markovic et al. (2012). The main threats to human health from heavy metals are related with exposure to Pb, Cd, Hg and As. Pb is a toxic metal and can accumulate in the blood, bones, and soft tissues, Markovic et al. (2012); even low exposure to Pb can cause mental retardation in children. More toxic is Hg. Once Hg from the air reaches water, microorganisms can change it into MeHg, a highly toxic form that builds up in fishes. People are primarily exposed to Hg by eating contaminated fish and seafood, Stankovic et al. (2011a), Stankovic and Jovic (2012), Markovic et al. (2012), Jovic et al. (2012). MeHg exposure is a particular concern for women of childbearing age, unborn babies, and young children, because studies have linked that high level of MeHg damaging a developing nervous system. Other toxic metals As, Cr and Ni emitted from power plants and burning coal, can cause cancer in humans, Stankovic et al. (2011a), Stankovic and Jovic (2012). For example, Hg is still used in gold mining in many parts of Latin America. Pb remains a common additive to petrol, although this use has decreased in the developed countries, and As is still common in wood preservatives, fungicides and pesticides. Coal fuels and waste especially contain heavy metals which should be a central concern in the consideration of their use. Since the mid-nineteenth century the production of heavy metals has risen abruptly and their emissions into the environment, especially in less developed countries.

5.4 Regulations

Experience gained in the1960s and early the 1970s has led to the conclusion that the geographical marine toxic metal contamination and the associated biological impact were largely unknown and undocumented. The first monitoring program of that kind was established by Goldberg (1975), 'Mussel Watch' monitoring program, and proposed to assess the spatial and temporal trends in chemical contamination in estuarine and coastal areas. Nowadays Agenda 21 is an action plan of the United Nations (UN) related to world sustainable development and their environment protection.

On the basis of the Agenda 21 the action should be taken globally, nationally, and locally by organizations of the UN, governments, and major groups in every area in which humans directly affect the environment. This meeting, known in UN circles as simply "Rio," was the first time that the international community agreed that the environment was something worth protecting and that long term economic development cannot be achieved without attention to principles of sustainability.

Today, compelled by the growing environmental and health awareness of the public, assessments of an array of trace metals in soils, sediments, water, air, as well as foodstuffs are demanded by regulatory guidelines, Wong et al. (2006), Smolders et al. (2009), Stankovic et al. (2011a), Joksimovic et al. (2011a), Jovic et al. (2011), Joksimovic and Stankovic (2012), Jovic et al. (2012). Those routinely regulated trace metals include As, Ba, Cd, Co, Cr, Cu, Hg, Pb, Mo, Ni, V, and Zn.

The significance of heavy metals for the environment quality has been reported in several European Union Directives 2004/107/EC (As, Cd, Hg, Pb, Ni, for instance) and Directive 2008/50/EC; Directive 2004/35/EC; Directive 2008/1/EC (for bioindicators), where it was stressed on the serious threat for human health evoked by the presence of these metals in the food chain, as well as in the environment. The aim of a directive is to protect the environment by monitoring water, air, sediment, soil, biota, and to establish common quality rules for their chemical analysis.

Within the European Union the first directive in the field of nature protection was the Bird Directive in 1978 (79/409/EEC), EEC (1979). Within the EU countries free access to environmental information is guaranteed according to the information directive 90/313/EEC, EEC (1990). Intensive Monitoring of Forest ecosystems were established in 1994 within the framework of the UN-ECE ICP Forest, UN/ECE and EC (2000). At gaining a better understanding of the effects of air pollution and other stress factors on forest ecosystems in 2001 was done by implementation of the multi-disciplinary Integrated Monitoring Programme ICP IM, UN/ECE Convention on Long-range Transboundary Air Pollution (2001).

Air pollution and the problems it causes are not confined by any geopolitical boundaries. In the last decade there has been significant improvement in air quality in urban areas as a result of strict legislation, the introduction of "cleaner" technologies and the migration industry in uninhabited areas. Monitoring of heavy metals (Pb, As, Cd, Hg, and Ni) and other air toxicants are planned according to the directives of the CAFE (Clean Air for Europe) program and those of the European Union (Directive 2004/107/EC 2004). EuroBionet – "European Network for Assessment of Air Quality by the Use of Plants Bioindicator" established and financially supported by the European Commission Environment Programme and EU obliges its member states to establish national networks for monitoring air quality.

The increasing importance of bioindicators is also encouraged within the European Union's Water Framework Directive (WFD), EC (2000). The directive aims to achieve a good ecological status in all European water bodies, such as rivers, lakes and coastal waters, and requires that the assessment of the ecological status of a system be accomplished primarily utilizing biological indicators. Among the wide range of bioindicators, five biological elements are listed within the WFD: phytoplankton, macroalgae, angiosperms, benthic invertebrates and fish, Frontalini and Coccioni (2011). One major goal of the WFD is to achieve the "good ecological status" and "good chemical status" in all surface water bodies of the EU. The term ecological status corresponds to the general philosophy of an integrated approach for evaluating the ecological integrity of water bodies.

The Marine Strategy Framework Directive 2008/56/EC (MSFD) are formulated as "concentrations of contaminants are at levels not giving rise to pollution effects". Contaminants are defined as chemical elements and compounds or groups of substances that are toxic, persistent and liable to bioaccumulate, and other substances or groups of substances which give rise to an equivalent level of concern. This definition is in line with the definition of hazardous substances used in the Water Framework Directive 2000/60/EC (WFD), EC 2000. The Water Framework Directive (WFD) establishes requirements for good surface water chemical status.

The recent adoption of human biomonitoring (HBM) as Action 3 in the Environment and Health Action Plan 2004–2010 of the European Commission COM (2004) has motivated the implementation and application of HBM in European environment and health research.

5.4.1 *Toxic Metals Regulations for Human's Intake*

Humans can be exposed to metals through inhalation or ingestion through food and drink. Once a metal is incorporated in a human, it tends to distributed to human's tissues and organs and persists in sites like the liver, bones, and kidneys for years or decades, Hu (2002), although some its excretion can occur through the kidneys and digestive tract. With respect to essential and non-essential elements, the Joint Food and Agricultural Organization and World Health Organization (FAO/WHO) have established the Provisional Tolerable Weekly Intake (PTWI) level, which is defined as an upper intake limit above which adverse effects might be expected, FAO/WHO (2004, 2007, 2010). The term 'provisional' is used to emphasize the lack of safety data on contaminants; consequently, the levels of PTWIs are continually re-evaluated.

According to WHO (2007) in most foodstuffs the Hg concentration is below 0.02 mg/kg. To safeguard public health, limits on total Hg levels in seafood have been established in various countries. Fish and fish products are the predominant sources of human exposure to MeHg. The United States Food and Drug Administration have set a maximum total mercury limit of 1 mg/kg wet weight for fish. In

Table 5.2 Tolerable intake levels for trace elements appointed by the FAO/WHO (From http://www.inchem.org/ 6 November 2009)

Element	Provisional tolerable weekly intake (PTWI) (mg kg^{-1} bodyweight)	Provisional maximum tolerable daily intake (PMTDI) (mg kg^{-1} bodyweight)
Arsenic	0.015	–
Cadmium	0.007	–
Copper	–	0.50
Iron	–	0.80
Lead	0.025	–
Mercury	0.005	–
Methylmercury	0.0016	–
Zinc	–	1.00

Europe, a value of 0.5 mg/kg was set by the European Commission for seafood, EC (2006). In Japan, fish with total Hg concentrations higher than 0.4 mg/kg are considered unfit for human consumption, WHO (2007).

The provisional tolerable weekly intake (PTWI) set by WHO and the Food and Agriculture Organization of the United Nations (FAO) for Cd is 500 μg (a weekly intake of 7 μg/kg body weight (kgbw)), corresponding to a daily intake of 70 μg or 1 μg/kgbw/day, FAO/WHO (2010). The USEPA Reference Dose amounts to 1 μg/kgbw/day in food and 0.5 g/kgbw/day in drinking-water, USEPA (1999). These values are based on the chronic effects of Cd on kidney function. However, it has been suggested that the PTWI should be lowered, WHO (2007). Data on the toxicology of tolerable intake of the most important elements are given in Table 5.2.

5.5 Bioindicators

There is a continual influx of toxic metals into the biosphere from both natural and anthropogenic sources. The concentrations of contaminants may, however, be too small to be detected by using chemical methods. So, biological indicators, also called bioindicators, can be used to provide an indication of environmental conditions including the presence or absence of contaminants in trace. The term 'bioindicators' was first used by Clements in 1920 in order to signify the organisms, which by their presence in a particular habitat, clearly indicated the ecological conditions of that habitat.

The loss of ecosystem services (e.g., clean air, drinking water, plant pollinators) has increasingly focused world's attention on the health of natural ecosystems. All species tolerate a limited range of chemical, physical, and biological conditions, which we can use to evaluate environmental quality. Various species of biota can be potentially used or were used as bioindicator organisms for measuring diversified types of pollutants at different situations.

Considering the 1.7 million species currently documented on Earth, how do we chose just one as a bioindicator? The answer is simple: no single species can adequately indicate every type of disturbance or stress in all environments. Depending upon the specific environment, the present species and their local disturbances, an appropriate bioindicator species or groups of species need to be selected. Ecologists have established a broad set of criteria that species must exhibit to be considered as good bioindicators, McBride et al. (2011). An organism, to be employed as a bioindicator, has to have the next characteristics: to be easily identified and sampled; to be widely spread in the studied area; to have a low mobility and a long-life cycle; to have a good genetic uniformity on all the considered area, and to be present all-year long, Zhou et al. (2008).

In practice, a bioindicator can be any animal, plant or microbial system that can be used to formulate conclusions about the environmental conditions they are continuously exposed to. Bioindicator is an organism that can be used to establish geographical and temporal variations in the bioavailability of contaminants by measuring the accumulated concentrations of chemicals in the whole body or in specific tissues, Hall et al. (2008). Previously used or suggested indicator organisms include plants, beetles, benthic invertebrates, butterflies, amphibians, fishes, birds, and mammals. As it was shown in the review of Burger (2006) over 40 % of the bioindicator papers were about metal pollution, wherein fish, plants, invertebrates, mammals were the dominant used bioindicator species. For aquatic metal pollution, the common used bioindicators mainly contained organisms including plankton, insects, mollusks, fishes, plants, and birds, Zhou et al. (2008).

Methods to measure and evaluate the effects of metals in the environment are complicated not only by bioavailability and susceptibility, but by measurement tools and detection limits, environmental conditions, and the interrelatedness of the metals themselves. The numerous benefits of bioindicators have spurred legislative mandates for their use in countries around the world. Examples of environmental, ecological, and biodiversity indicators can be found in many different organisms inhabiting many different environments. For example, lichens (a symbiosis among fungi, algae, and/or cyanobacteria) and bryophytes (mosses and liverworts) are often used to assess air pollution. Lichens and mosses serve as effective bioindicators of air quality because they have no roots, no cuticle, and acquire all their nutrients from direct exposure to the atmosphere.

A bioindicator is an organism, or part of an organism or a community of organisms, that contains information on the quantitative aspects of the quality of the environment. Furthermore, exposure can be measured by either levels or effects. The importance of metals to ecosystems and their component parts was initially evaluated by measuring metal levels in air, water, and soil, and inferring potential effects on organisms. Specific physiological and behavioural changes in bioindicators are used to detect changes in environmental health.

According to Hodkinson and Jackson (2005) bioindicator is a species or group of species that reflects biotic and abiotic levels of contamination of an environment. However, to the organism used as a bioindicator of pollution it must meet certain criterions. First of all, the body must constantly accumulate and tolerate large amounts of pollutants, must be tied to one place to make it real "representative" for the environmental area: soil, air, water; must be readily available for the collection, identification and handling; must be sufficiently long lifetime to ensure sampling over a longer period of time; to have enough tissue for chemical analysis and must tolerate the physical and chemical changes of the environmental parameters. Bioindicators are species that are particularly sensitive to the environment, and provide information about ecosystem health, respond well both to the presence or absence of other species, as well as the presence or absence of contaminants.

5.5.1 Animals as Bioindicators

Ideally, metal concentrations in the animal's body reflect environmental pollution levels quantitatively. Such biological accumulators have often been used as accumulation indicators of environmental metal pollution. In reality, however, many factors like the nutritional, physiological and reproductive status, the sex and age of the animals influence such quantitative relationships. Generally, metal accumulation by animals is favoured by their limited ability to excrete these contaminants directly after their uptake because of metal inactivation by binding to metallothioneins, Zhou et al. (2008). The animal's kinds mostly used as bioindicators are zooplankton, invertebrate and vertebrate. Animal species have been commonly used as indicators of aquatic ecosystems.

Zooplankton

Plankton is composed of phyto- and zooplankton microscopic organisms that float freely with oceanic currents and in other bodies of water. As primary source of food in the aquatic food chain, phytoplankton use chlorophyll to convert energy (from sunlight), inorganic chemicals (like nitrogen), and dissolved carbon dioxide gas into carbohydrates. A phyto- and other plankton are zooplankton food.

The freshwater cladoceran *Daphnia magna* is one of the oldest and most widely used zooplankton as test organism in aquatic toxicology, Ratte et al. (2003). The genus *Daphnia* is an important link in freshwater trophic chain, as it is both a phytoplankton consumer, as well as a food source to invertebrate and vertebrate predators.

The species of the genus *Daphnia* differ in their world distribution. *Daphnia magna* is a pond species with limited geographical range (Eurasia), whereas *Daphnia pulex* is a more widespread species (North and South America, Greenland and Europe), Ratte et al. (2003).

The choice of *Daphnia magna* as standard test species was influenced by several advantageous characteristics. It is of small size (compared to mollusks, fish, or macrocrustaceans) and is relatively sensitive to chemicals, compared to other freshwater organisms. The ecology of *Daphnia* has long been studied so there are a lot of background information on its biology and ecology. Thus, *Daphnia magna* is the most commonly tested freshwater species, in acute and also in chronic tests, on toxic metals such as Cu, Cd, Zn and Se, Lam and Wang (2008).

Invertebrate

The earthworm *Eisenia fetida* is one of the mostly used test organism for the terrestrial toxic metals bioindicators. This terrestrial species representing the soil fauna is used in acute tests as well as in reproduction tests. The earthworm *Eisenia fetida/E. andrei* is recommended in various guidelines as standard test organism for the terrestrial environment, Ratte et al. (2003). The use of *Enchytraeids* as standard test organisms is discussed in the recent years. Soil-inhabiting enchytraeids are more ecologically relevant than *E. fetida/E.andrei* as important members of the soil biocoenosis in many different habitats, especially where earthworms are rare occurring, Ratte et al. (2003).

Due to their ubiquitous distribution and enormous species number, mollusks play important ecological roles in the different aquatic and terrestrial ecosystems of the world. Although mollusks are basically a marine group of animals, gastropods and bivalves have also expanded their distribution to various freshwater systems, Moloukhia and Sleem (2011). Gastropods have additionally penetrated into a huge variety of terrestrial habitats so that mollusks can be found today from the abysses of the sea to mudflats, from lakes and rivers and their banks to forests, alpine mountains, but also in steppes and desserts; they occur on nearly all latitudes of the planet from polar to tropical temperatures. Thus, mollusks have been successfully used to obtain information on the quality of terrestrial, marine and freshwater ecosystems and to quantify contaminants in their environment, Oehlmann and Schulte-Oehlmann (2002). This is particularly the case for the two most diverse classes of mollusks, gastropods and bivalves, Moloukhia and Sleem (2011).

Oehlmann and Schulte-Oehlmann (2002) listed the following reasons why mollusks are well suited as bioindicators:

- Gastropods and bivalves are widespread and abundant in all marine and freshwater ecosystems worldwide; additionally, gastropods can also be found in almost all terrestrial environments;
- Most mollusks, especially those living in the aquatic environment, exhibit a broad distribution within and even between continents, facilitating their use in geographical large scale surveys;
- A number of species and genera are even cosmopolitans: mussels of the genus *Mytilus* with the two species *M. edulis* and *M. galloprovincialis* being the most widespread, Stankovic et al. (2011a).

The majority of gastropod and bivalve species exhibits an extremely limited mobility or is completely sessile as adults. Mollusks are in direct contact with the ambient medium, water or soil/sediment. Therefore, these mollusks represent the contamination of their habitat ideally. Toxic metals can be taken up not only from the diet, via the gastro-intestinal tract, but also additionally from ambient water or soil/sediment, resulting in greater accumulation potency for contaminants.

Compared with vertebrates, mollusks exhibit only a limited ability to excrete pollutants directly via their kidneys or other excretory organs and tissues. As a consequence, mollusks attain higher bioaccumulation or bioconcentration factors for many toxicants than other groups. Most gastropod and bivalve species used for biomonitoring and bioindication purposes are relatively large and therefore easy to handle. Consequently, they can be used both under laboratory and field conditions, for active and passive biomonitoring.

Gastropods represent the only molluskan class in terrestrial ecosystems and consequently, snails are the only mollusks which can be used for bioindication and biomonitoring purposes in these environments. Biomonitoring attempts with snails have found an increasing interest during the last decade and surrounding pollution effects. Particularly terrestrial snails can be utilised as accumulation indicator for metal pollution. There are numerous reports on the use of snails as accumulation bioindicators so that only a limited number of examples can be presented here.

Fritsch et al. (2011) compared the concentrations of Cd, Cu, Pb and Zn in the grove snail *Cepaea sp* and the glass snail *Oxychilus draparnaudi*, an herbivorous and carnivorous species, respectively. Grove snails were abundant in both agricultural and woodland landscapes but rare in urban areas. The glass snail was often present in urban areas but less often in woodlands, and it was rarely found in agricultural areas. Toxic metal concentrations increased with total soil concentrations at the sampling points in the glass snail. Based on their results, toxic metals accumulation in snails and small mammals is governed by ecological (diet, habitat preferences and mobility) and physiological (assimilation and excretion of toxic metals) characteristics of animals, Fritsch et al. (2011).

The land snail *Eobania vermiculata* was used to monitor the Hg exposure in Italy in an area affected by a chlor-alkali complex. Additionally, Hg residues were also assessed in plants, soils and air from the same region. The results indicate a low bioavailability of Hg for the snails. The concentrations were in the same range as measured in plants at the same sampling sites indicating a low bioaccumulation of Hg in this snail species, Oehlmann and Schulte-Oehlmann (2002).

The so-called mussel watch, initiated in the United States of America in 1976, was one of the first environmental surveillance programmes, which made use of living organisms in an extended geographical area, Stankovic et al. (2011a), Stankovic and Jovic (2012). The mussel watch comprised a coordinated, standardised sampling and measurement of heavy metals and radionuclides in four marine bivalve species (*Mytilus edulis*, *Mytilus californicus*, *Crassostrea virginica* and *Ostrea equestris*) at more than 100 sample sites on the coast of North America. At the same time, the potential of freshwater bivalves was also analysed in first investigations of heavy metals accumulation in the shell and soft tissues of six clam species, Oehlmann and Schulte-Oehlmann (2002).

Since 1976, bivalves have been used to assess the levels of contamination in marine ecosystems, and certain systematic groups, notably mussels and oysters, have been extensively studied worldwide, Stankovic et al. (2011a). Bioavailability and thus uptake of metals and other pollutants are highly dependent on chemical and biological factors. Among biological factors, there are major differences in bioaccumulation between bivalve species. Within a single species, accumulation can be a function of age, size, sex, and genotype, nutritional and reproductive status. Factors that influence metal bioaccumulation are physico-chemical water properties: temperature, pH, dissolved oxygen, salinity, sediment grain size, and hydrologic features of the system, Stankovic et al. (2011a).

Mussels *Mytilus* spp. (*M. edulis, M. galloprovincialis,* and *M. californicus*), oysters (*Crassostrea virginica* and *Ostrea edulis*) and clams (*Mercenaria mercenaria, Venerupis* sp., and *Macoma balthica*) are the most commonly used bivalve groups for toxic metals bioindicators, Oehlmann and Schulte-Oehlmann (2002). Bellotto and Miekeley (2007) confirmed that the mussel *Perna perna* is efficient toxic metals bioaccumulator in Asia-Pacific. The same species was also employed as an indicator of heavy metal pollution in Brazil and in tropical areas.

The two bivalve species *Crassostrea cucullata* and *Anadara granosa* as suitable bioindicators for heavy metal pollution in India and other Asian countries and the mussel *Mytella strigata* for the same purpose proposed in central and south America. In many areas, marine gastropods have been considered as additional accumulation monitors and the highest concentrations of Cu, Zn, Cd, Pb and Cr were measured in gastropod species, while bivalves were better accumulators of Ni, As and Sn, Oehlmann and Schulte-Oehlmann (2002).

Freshwater mollusks have also been frequently used as bioaccumulator organisms, Moloukhia and Sleem (2011). Biomonitoring studies with freshwater mollusks have covered a wide diversity of species, contaminants, and water ecosystems. In a freshwater ecosystem, *Filopaludina (Siamopaludina) martensi* may be used as bioindicator of heavy metal contaminants. The results of Piyatiratitivorakul and Boonchamoi (2008) indicate that the chosen snail species may be used as a bioindicator for acute and subacute exposures to Hg and Cd.

Metal accumulation by the bivalve *Caelatura (Caelatura) companyoi* and the snail *Cleopatra bulimoides* was determined and results showed that both mollusks could accumulate Cr and Cd to a large extent and could tolerate their toxicity to high limits, Moloukhia and Sleem (2011). Two important advantages of snails and bivalves, over most other freshwater organisms for biomonitoring research, are their large size and limited mobility.

One of the most frequently employed freshwater bivalve accumulator species is the zebra mussel *Dreissena polymorpha*. The zebra mussel *Dreissena polymorpha*, a widespread invasive species in Europe and North America, is an important bioindicator of freshwater environments reflecting the site specific pollution, Minguez et al. (2012). Zebra mussel accumulates high amounts of potentially toxic metals and is therefore widely considered as a bioindicator, Rahnama et al. (2011).

Mollusks have been successfully used as bioindicators in monitoring waters programmes in the past. Terrestrial ecosystems were much less considered than the aquatic environment. It can be stated that terrestrial bivalves, gastropods and especially the other molluskan classes have not yet received the attention they probably deserve, speaking in terms of their ecological importance. A number of invertebrate species are known to be efficient accumulators of trace elements.

Vertebrate

The diet is appear to be the most important factor affecting the levels of metals in the tissues of a given animal. Usually, the higher levels of metals in plants and in animals from lower trophic levels, the greater the concentrations of metals in the tissues of higher-level of animals, such as fish, birds and mammals.

Fish

Heavy metals play an important role as substances affecting aquatic organisms. Their impact, particularly on fish, is receiving considerable attention. Fish are well recognized bioindicators of environmental changes, including metals pollution, and are adequate for waters monitoring programs, Hauser-Davis et al. (2012). In fishes some trace metals, such as Mn, Co, Fe, V, Cu, Zn and Se, are necessary in small amounts for metabolic processes being naturally assimilated by fish, but Ni, Pb, Cr, Cd, Hg as non-essential elements perform no biological role for fish and become toxic above certain concentrations, Hauser-Davis et al. (2012).

Fish are used as test organisms in aquatic toxicology because of their top-position in the trophic chain and their role as food for humans. The acute and chronic fish tests are used to assess effects of chemicals. Different test species are used as test organisms representing the aquatic vertebrates. Commonly used test species are the rainbow trout *Oncorrhynchus mykiss*, the Zebrafish *Danio rerio*, the Common carp *Cyprinus carpio*, the fathead minnow *Pimephales promelas*, and the guppy *Poecilia reticulate,* Chovanec et al. (2003). Up to now the main database of fish toxicity data exist for freshwater species. Bioindication using fish represents a good monitoring tool especially with regard to river engineering, e.g. river restoration and management. The mobility of many fish species makes difficult to identify not only the exact source of pollution, but also the time and duration of pollution exposure.

Comparative studies of toxicity data of freshwater fish and marine species show that marine species were more sensitive than freshwater species for the majority of substances tested, but for 91 % of all substances the sensitivity ratios between freshwater and saltwater fish were within a factor of 10, Chovanec et al. (2003). Gills, liver, kidney and muscle fish are the organs mostly used for toxic metals investigations.

Metallothioneins (MTs) and other selectively metal binding proteins have found a comparable attention and application for aquatic studies like in terrestrial surveys. It appears that in fish MTs should be considered as a kind of general stress protein, which is particularly responsive to heavy metals, Hauser-Davis et al. (2012). In aquatic mollusks and especially snails and mussels, Dallinger et al. (2000), Chovanec et al. (2003), Stankovic et al. (2011a), MTs seem to be more specifically involved in responses to heavy metals and can thus be considered as a biomarker of exposure to metal contamination. Increased expression of MT in response to harmful levels of toxic metals has been demonstrated for several aquatic species such as gastropods, insect, crustaceans, mussels, fishes, Zhou et al. (2008), Hauser-Davis et al. (2012).

MTs are low molecular weight proteins with many sulfhydryl groups binding a variety of metals, showing a strong affinity toward certain essential and nonessential trace elements, such as Cd, Cu, and Zn. So far, MTs have been identified in a large number of tissues and species throughout the animal kingdom, including a number of terrestrial snails, Dallinger et al. (2000), and mussels, Stankovic et al. (2011a). Although a variety of biochemical data prove MTs to be structurally well defined, but their biological function is still under discussion, Freisinger (2010).

A number of studies demonstrated that the synthesis of MTs can be induced by certain trace elements, but also by organic chemicals and other non-chemical stress factors, like infections, starvation and injuries. Nevertheless, it has been shown that Zn, Cu, Cd, Hg, and other trace elements, are the most potent inducers of MT synthesis and it has therefore been speculated that the detoxification of metals is the primary biological function of these proteins. The involvement of MTs in Cd detoxification of terrestrial gastropods has been proven in detail for a number of species, for example the Roman snail *Helix pomatia*, Dallinger et al. (2000). Compared to other invertebrate taxa such as earthworms, land snails have the advantage that their MTs are by far less unstable, facilitating their use in biomonitoring surveys, Dallinger et al. (2000).

MTs are found in all tissues of fish, particularly in the liver and kidney, and play an important role in the intracellular regulation of the essential metals Zn and Cu. MT concentrations in tissues increase by exposure of fish to Cu, Zn, Cd, or Hg, and the affinity of Cd and Hg to MTs is even higher, so that they may displace essential metals, Zhou et al. (2008). Increased levels of stress proteins reflect the gradual impact on the fish metabolism ranging from adaptive to degenerative responses with severe consequences for fish survival.

The number of quantitative studies of heavy metals in both fish and their parasites has increased recently, Sures (2008). Most investigations focus on either comparative studies of parasite community changes subsequent to environmental stress, or on single investigations of heavy metal concentrations in fish and their parasites where especially acanthocephalans accumulate high burdens of Pb and Cd and therefore may be used as sensitive indicators for monitoring heavy metals contaminations in aquatic ecosystems, Sures (2008). The use of parasites as indicators of heavy metals contamination is of increasing in environmental control.

Birds

A few groups of birds dominate the contaminant literature, particularly raptors, waterfowl, and seabirds, Burger and Gochfeld (2004). Birds can play an important role as bioindicators. The general biology and ecology of birds are well known and birds are easy to identify. However, despite their undoubted advantages as bioindicators, birds are not being used as often or as effectively as they could be. One disadvantage is that many species are migratory, making it difficult to determine where exposure occurred, Burger and Gochfeld (2004).

Birds have been successfully used to indicate temporal and spatial trends in toxic metal pollutions in terrestrial and aquatic ecosystems, Burger and Gochfeld (2004),

Deng et al. (2007), Hargreaves et al. (2011), Berglund et al. (2011), Zhang and Ma (2011), Kitowski et al. (2012). Top predators, like raptors, water birds and seabirds accumulate toxic metals, which affect physiology, reproduction and even cause birds death, all of which lead to population declines and which have often in the past been early warning of environmental pollution. If birds are chronically exposed to heavy metals it could result in detrimental effects on their growth, behavior, reproduction, and resistance to diseases, Deng et al. (2007).

Birds have been proposed as useful biomonitoring species of pollutants from the year 1993. Until than focus have been on waterfowl or raptors, terrestrial passerines, such as great tit (*Parus major*), coal tit (*Periparus ater*), blue tit (*Cyanistes caeruleus*), pied flycatcher (*Ficedula hypoleuca*) and collared flycatcher (*Ficedula albicollis*), Berglund et al. (2011). They have successfully been used to monitor the environment close to a variety of different metal industries.

Birds have the advantage of being large, wide-ranging, abundant, long-lived, easily observed, and important to people. Many species are at the top of the food chain where they bioaccumulate contaminants with age, Burger and Gochfeld (2004). The metal concentration in birds can be assessed in a variety of samples, though reports of metal content in egg, blood, feathers, kidney and liver dominate the literature. Samples such as feathers or eggs are easy to collect which reduce the costs of sampling and is the non-destructive and noninvasive sampling technique, although they might not represent the body burden of metals, Berglund et al. (2011).

Birds' contaminations with Cd, Hg, and Se have been shown to adversely affect the condition of birds by reducing their growth or body weight, Zhang and Ma (2011). Since birds usually head the food web, they can provide valuable monitoring of environmental Hg. In many European areas, and also on other continents, birds have proven to be very useful bioindicators of Hg, especially sea and waterbirds, Burger and Gochfeld (2004), Kitowski et al. (2012).

Burger and Gochfeld (2004) analyzed methyl and inorganic Hg separately, MeHg makes up more than 90 % of the total mercury in liver, kidney, muscle, and feathers of birds. Feathers are good indicators of metal pollution, Zhang and Ma (2011), especially for organic bound metals like Hg or Sn, which accumulate in the plumage, and for which feathers are the main elimination routes. There is usually a significant correlation between concentrations of Pb in feathers and those in internal tissues, including blood; therefore the concentration of Pb in feathers is a good predictor of internal bird dose, Burger and Gochfeld (2004). The quantities of a metal incorporated into the feather represent the body level at the time of feather growth, during the development of the young or during moult.

As feathers reflect a bird's body burden during feather growth, body feathers from adult birds or chicks can be used to show long term and spatial trends in Hg contamination. The most frequently used matrix is the bird egg too. Waterbirds' eggs were used a lot as biomonitor to detect heavy metals' concentration or its temporal-spatial trends, as they are easier to obtain and can be saved for a long period compared with soft tissues and nestlings, Zhang and Ma (2011).

Analysis of Hg concentration in bird kidneys has been recognized for years as a valuable tool in monitoring environmental contamination levels. After Hg gains entrance to a bird, it can be either stored in internal tissues such as kidneys and liver or it can be excreted in feathers and eggs, Kitowski et al. (2012). The level of Hg concentration was found to be significantly lower in young birds than in adults. Increased Hg concentration was also registered in the kidneys of birds, Kitowski et al. (2012).

Deng et al. (2007) analyzed the concentrations of 11 trace metals in tissues from 10 body parts of Great Tits and Greenfinches collected in the Western Mountains, China. The highest concentrations of Hg, Ni, Zn, and Mn were found in the feather; Pb and Co in the bone; Cd, Cr, and Se in the kidney, and Cu in the liver and heart of birds. In general, the concentrations of the non-essential elements (Hg, As, Pb, and Cd) were lower than that of essential elements (Zn, Se, Cu, Cr, and Mn). In both species, Hg had the highest concentration in the feather.

Kitowski et al. (2012) were determined Hg concentration levels in the kidneys of 131 specimens of 42 species of birds with different food preferences. The highest concentrations were found in piscivorous species such as the goosander, great white egret, white-tailed eagle and black stork. The following sequence was identified for concentration levels in kidneys of birds in different trophic guilds: piscivorous > small mammal eating > bird eating > aquatic invertebrate eating > terrestrial invertebrate eating > herbivorous > omnivorous.

Hargreaves et al. (2011) investigated As, Be, Cd, Co, Cr, Cu, Fe, Hg, Mn, Mo, Ni, Pb, Sb, Se, Tl, V, Zn, and Hg concentrations in the tissues, food and abiotic environment of Arctic shorebirds: Hg, Se, Cd, Cu, and Zn bioconcentrated from soil to invertebrates, and Hg, Se, and Fe biomagnified from invertebrates to shorebird blood; As, Ni, Pb, Co, and Mn showed significant biodilution from soil to invertebrates to shorebirds. They tested elements in blood, feathers and eggs of six shorebird species (*Pluvialis squatarola, Calidris alpina, C. fuscicollis, Phalaropus fulicarius, Charadrius semipalmatus,* and *Arenaria interpres*). Levels of most elements in shorebird tissues were well below those associated with known pollution sources or toxicological thresholds. The exception was Hg being in blood and eggs, Hargreaves et al. (2011). The highest levels of metals were recorded in tissues of carnivore seabirds, Jakimska et al. (2011).

Berglund et al. (2011) measured the concentration of several elements, As, Ca, Cd, Cu, Ni, Pb, Se, and Zn, in adult and nestling pied flycatchers (*Ficedula hypoleuca)* and great tits (*Parus major*) at different distances to a Cu-Ni smelter. The uptake of Cu and Ni were regulated, but As, Cd, Pb, and Se accumulated in liver tissue. Pied flycatchers had generally higher element concentrations than great tits. The higher accumulation of As and Pb in pied flycatcher livers was explained by a more efficient absorption, whereas the higher Cd concentration was primarily due to different intake of food items. Age-related differences occurred between the two species, though both Cd and Se accumulated with age.

Birds are indeed useful indicators of the state of the environment in which they live, and have proved effective bioindicators as well as biomonitors of various kinds of environmental change. Consequently, avian bioindicators are already important components in current monitoring programmes in many countries and in a variety of ecosystems. Over the last decade the International Council for the Exploration of the Sea (ICES) has established a working group on "Seabird Ecology", dealing with the question of the usefulness of seabirds as indicators of the state of marine environments.

Birds are abundant, widely distributed and, in some cases, long-lived. Birds occupy various positions in the food chain, especially the higher trophic levels, indicating toxic metals contamination in various compartments of ecosystems by biomagnification of persistent metals. Many birds are widespread species allowing comparisons between different ecosystems, countries or even continents. The widespread distribution of some bird species allows comparisons of contamination on large scales, even between oceans or continents. Not only can birds monitor local food webs, but also, if they are migratory they can be used to compare exposure in different regions. The mobility of birds implies an integrative value in bioindication over broad spatial scales.

Mammals

Mammals represent useful organisms for biomonitoring purposes and can be used when both temporal and spatial information is required. Among the numerous members of the class of mammals, free ranging animals or "wildlife" fit best the requirements for a biomonitor. This is because they depend exclusively on the quality of food, water and air in their habitat. They consume flora or fauna that reflect the local soil, water and air contamination. Any contamination present will influence the animal and can have an effect on its health. The metal toxic levels in mammals depend on their diet composition and often influenced by food chain effects: do they eat either other animals or/and plants.

An omnivore is a kind of animal that eats either other animals or plants (generally only the fruits and vegetables). Some omnivores will hunt and eat carnivores, eating herbivores and other omnivores. A carnivore is an animal that gets food from killing and eating other animals. The most widespread predator in the order Carnivora is the red fox *Vulpes vulpes*. It lives in almost the entire northern hemisphere, from the Arctic Circle to North Africa, Central America and the Asian steppes, with the exception of Iceland, the Arctic islands and parts of Siberia, Kalisinska et al. (2012). A herbivore is an animal that gets its energy from eating only plants, including grasses. Some others are scavengers and will eat dead matter. Many will eat eggs from other animals. Therefore a great number of environmental contaminants accumulate in mammals, potentially toxic trace of Pb, Cd, Hg, As and radionuclides, like the Cs and Sr isotopes, Pokorny (2006), Rudy (2010), and Kalisinska et al. (2012).

Domestic animals, like cattle, pigs or sheep, can also be used for assessing environmental quality, providing they feed only on foodstuffs originating from their immediate environment. Although grazing sheep still fall in this category, due to modern agricultural practice pigs and most cattle no longer do. Their feed is produced not only outside of the original living area, but possibly even in other continents, Rudy (2010). Studies on domestic animals have shown how diet can influence Pb absorption. Low Ca and vitamin D levels as well as an insufficient Fe and protein supply increase the absorption rate of Pb, Markovic et al. (2012).

Although many wildlife species could be used for toxic metals biomonitoring, the available literature concentrates on only a few species: wild boar, red deer, brown hare, and red fox, Pokorny (2006), Rudy (2010), and Kalisinska et al. (2012). The suitability of a species for study is influenced by different factors: herbivores accumulate metals to a higher extent than carnivores, which in contrast are better monitors for organohalogen compounds. Additionally, the size of the home range is important. Home range size should be relatively constant for all seasons and over the animal's life span. This factor would therefore exclude wild boar (*Sus scrofa*) as a monitoring species since it shows a sex-dependent difference of home range size. Groups of female boar with their juveniles have a relatively constant home range, whereas males live solitary with no stable territory and migrate over many kilometers, Rudy (2010).

The populations of the European brown hare (*Lepus europaeus*) in Central Europe represents one of the best biomonitors for agriculturally used land, Rudy (2010). A very good bioindicator is the roe deer (*Capreolus capreolus*), the species abundant in nearly all parts of Europe in agricultural, as well as in forest areas, which is one of the most suitable species for bioindication of toxic metals pollution in terrestrial ecosystems (Pokorny 2006). Consequently, Pokorny (2006) suggested the roe deer as a bioindicator of habitat quality/toxic metals pollution for the European region.

Nowadays, especially in Europe, anthropogenic barriers like roads and human settlements inhibit animals' migration and they have to stay in the same habitat during the entire year. This makes them in fact more useful for monitoring studies. However, if the food shortage in winter is compensated for by additional feeding provided by hunters, this can alter the results, since the supplementary foodstuff is often imported from other areas.

Several taxa of small mammals, like shrews (*Soricidae*) and rodents, can also be used for toxic metals biomonitoring of terrestrial, Sanchez-Chardi et al. (2007) and mink (*Mustela vison*) for aquatic ecosystems, Basu et al. (2007). Shrews accumulate higher concentrations of more toxic elements, especially non-essential metals, while wood mice (*Apodemus sylvaticus*) are less tolerant to toxic metals. Wood mice showed lower bioaccumulation of Pb, Hg, Cd, Fe, Mg, Zn, Cu, Mn, Mo, and Cr, than shrews, Sanchez-Chardi et al. (2007).

Mink are monogastric animals with a digestive tract more similar to that of humans than that of rodents, Basu et al. (2007). They can be found in wooded areas and fields near streams and lakes, it is carniverous – a meat- eater that consume a range of prey items available in their local habitat, including small mammals, frogs, snakes, fish and birds. Mink can also provide data on spatial and temporal trends in environmental pollution because they are found over a wide geographical area. For example, temporal analyses of the data have shown that Hg content in mink tissues from Ontario have not changed over the past 20 years, Basu et al. (2007).

Forty percent of mammal species are rodents, and they are found in vast numbers on all continents other than Antarctica. Common rodents include mice, rats, squirrels, porcupines, beavers, and hamsters. Different species of rodents in a small mammal community may feed on different food items, leading to differences in exposure potential among species. The advantages of these species are that they are very abundant, do not migrate over long distances, have a widespread distribution and have generalised food habits. The disadvantages of these animals to be bioindicators are their small body size, so that very often the whole carcasses have to be analysed and they are not easy to trap.

In the past decade marine mammals, seals and dolphins have been accepted as bioindicators due to their long lifespan and their position at the very top of the marine food chain, in particular for medium and long-term ecosystem changes, Agusa et al. (2011), Bellante et al. (2012), Kakuschke et al. (2012). Marine mammals are extremely susceptible to Hg contamination from both natural and anthropogenic inputs, Wintle et al. (2011), Kakuschke et al. (2012). They accumulate Hg throughout their lifespan, and other factors such as growth features, gender, reproductive and feeding preference are assumed to control Hg levels in these animals, Pompe-Gotal et al. (2009). Considering the high mobility of dolphins, the levels of Hg concentrations in their tissues reflect the general contamination of the broad and poorly defined area in which they live, Wintle et al. (2011).

Concentrations of Cd and As were determined in organs and tissues of different dolphin species during the period 2000–2009 in the Italian coastal area. Significant differences were found between Cd concentrations in the different analysed tissues, but there were no differences in As concentrations, Bellante et al. (2012). Considerable Cd and As concentration differences in samples from different geographical areas were found. Dolphins from the Mediterranean Sea generally show higher levels of toxic metals when compared to others from open oceans, possibly because of the anthropogenic pressure on the basin, Bellante et al. (2012).

Marine mammals such as harbor seals (*Phoca vitulina*) are also accepted indicators. Metals concentrations were measured in their liver, kidney and muscle tissue, as well as in blood and plasma in samples of harbor seals (*Phoca vitulina*) from the North Sea, Kakuschke et al. (2012). The concentrations of Cr, Fe, Mn, and Zn were significantly lower in the whole blood samples of free-living harbor seals of the Elbe estuary than the range measured in free-living seals of the German and Danish Wadden Sea, Kakuschke et al. (2012). In the harp seal (*Phoca groenlandica*) from the coastal waters of Canada, concentrations of V, Mn, Fe, Cu, Mo, Ag, and Hg in the liver, Co, Cd, and Tl in the kidney, and Ba and Pb in the hair were significantly higher than those in other tissues, Agusa et al. (2011).

Absorption and Distribution of Toxic Metals Among Mammalian Organs

In addition to uptake by feeding and drinking, animals can also absorb and accumulate Pb by breathing. Absorption via the gastro-intestinal tract is relatively low (in the range of 5–10 %) in comparison to absorption via the lungs which can reach up to 50 %. However, the amount of resorption depends on the size of the

Pb containing particles. In general, the concentrations of Pb in soils, vegetation, invertebrates and selected tissues of mammals correlated with traffic densities. There was also a gradient effect on tissue levels with distance from the road. In all species, specimens from the direct vicinity of heavily used highways exhibited the highest Pb concentrations.

Usually liver and kidney samples are analysed in order to monitor the Pb, As, Hg and Cd exposure of an animal, Pokorny (2000), Jakimska et al. (2011), Kakuschke et al. (2012), Bellante et al. (2012). The distribution pattern of Pb and As within these organs is not as uniform as it is found with Cd for example, Bellante et al. (2012). This in turn also varies depending on the species. For red and roe deer, many authors report variable ratios of Pb between the liver and kidneys, whereas for brown hares, rabbits and wild boar higher levels are always found in the liver, Pokorny (2000), Kolesarova et al. (2008), Rudy (2010). The Pb uptake of these animals with their diet is higher in autumn than in spring. One of the most suitable mammalian indicator species for Pb is the brown hare, which accumulates Pb very strongly. The Pb contamination of the strictly herbivorous brown hare is not only influenced by the Pb content of its feeding plants, but also by the fact that this animal is much more exposed to soil particles and dust, Kolesarova et al. (2008).

Cd is a non-essential element for mammals. Under normal conditions, absorption is minimal at around 5–7 % of the total oral uptake. The retention rate of Cd increases in mammals when proteins, Cu, Zn, Fe, Cu or vitamin D are low in their diet. In addition to dietary uptake, another exposure pathway for Cd is the inhalation of contaminated air. Because Cd is better absorbed from inhaled air (10–40 %, depending on particle size) than from ingested food, inhalation of Cd containing particles is likely to be the most important exposure pathway in areas with elevated atmospheric Cd levels, Tataruch and Kierdorf (2003).

After absorption in the lung and gut, Cd is transported via the bloodstream to body stores, particularly the liver. In blood, more than 95 % of Cd is bound to protein in the blood cells. In the liver Cd is bound to metallothionein, Zhou et al. (2008) and the formed complex is transported to the kidneys, where it is almost completely absorbed. Interestingly, approximately 50 % of the total Cd burden of the body is found in the liver and kidneys, Jakimska et al. (2011). In muscles Cd concentrations are low.

As with Pb, Cd concentrations in the blood are not useful as an indicator of the general body burden. In the kidneys, Cd has a very long retention time. For cattle, it was estimated a Cd half life time of up to 12 years in the kidneys compared to only about 2 years in the liver, Tataruch and Kierdorf (2003). These long retention times in the kidneys result in a highly significant, positive correlation between the kidney concentrations of Cd and the age of the animal. Since the kidneys are the main target organ for Cd, they are the best tissue samples for analysis of Cd in mammals, Tataruch and Kierdorf (2003), Bellante et al. (2012).

A significant influence on Cd burden of animals is exerted by the geological properties of the habitat, in particular the soil. Several authors found higher Cd concentrations in roe deer than in other species from the same habitat, Tataruch and Kierdorf (2003). Chemical analysis of the preferred feeding plants of different ruminant species showed that the difference is due to the selective feeding habit of the roe deer. Its diet consists of a high percentage of Cd accumulating plants like herbs, weeds, fungi and ferns, Tataruch and Kierdorf (2003). Together with its narrow home range; its feeding habits make the roe deer a very sensitive bioindicator, especially in forest areas, Pokorny (2006).

For small mammals, a positive correlation between the Cd content in soils of their habitat and the Cd concentrations in soft tissues was found. As for other mammals, the highest values were found in the kidneys with average Cd levels in this organ being tenfold higher than in the liver. Other tissues displayed much lower concentrations, Tataruch and Kierdorf (2003).

The level of gastrointestinal absorption of Hg in animals depends on its chemical form. Hg in its inorganic form is absorbed up to about 7 % from food, which is similar to other heavy metals. On the other hand, the absorption of MeHg can be as high as 95 %. Elemental Hg cannot be absorbed orally, however, if its vapours are inhaled, around 80 % will be absorbed. The organ distribution of Hg in mammals follows the sequence of kidneys > liver > spleen > brain, in descending order. Concentrations in blood and muscle are low, Tataruch and Kierdorf (2003).

Hg concentrations in the organs of terrestrial mammals have declined considerably in recent years. Only those species that feed on fungi, such as roe deer and wild boar, showed higher levels. Fungi are known to accumulate metals, like Hg and Cd. Roe deer feed intensively on growing fungi, Tataruch and Kierdorf (2003). Other species do not show such an extreme preference for fungi, resulting in a lower Hg level.

The literature on Hg in small mammals indicates that concentration levels are influenced by feeding patterns, Basu et al. (2007). Small mammals, which feed on grasses, always showed comparably low Hg level in their organs than mammals whose food includes earthworms, insects, snails, fish and showed a remarkably higher Hg uptake compared to herbivorous species, Rimmer et al. (2009). Among terrestrial mammals, Hg concentrations increased from herbivores to omnivores and carnivores. For example, herbivores such as mule deer and various species of rabbits usually contained less than 1.0 mg Hg/kg fresh weight in liver and kidney, but carnivores such as red fox (*Vulpes vulpes*) contained 10 mg/kg ww, Kalisinska et al. (2012).

Animals as accumulative monitors of pollution by heavy metals have some advantages over plants such as area related results and comparability to man, Pokorny (2006). Many animal groups have been hailed as valuable sentinels, including birds, marine mammals, aquatic organisms, and domestic animals. Mammalian wildlife has physiological systems that are similar to those of humans in mediating the uptake, distribution, metabolism, and elimination of toxicants, Basu et al. (2007). Humans and many mammalian wildlife species inhabit similar ecosystems and are exposed to common climates, food sources, and pollutants.

5.5.2 Plants as Bioindicators

Accumulation and distribution of heavy metals in algae and plant depend on the species, the levels of the metals in the soil, water and air, the element species and bioavailiability, pH, cation exchange capacity, vegetation period and multiple other factors. Algae and plants absorb heavy metals through the water and root from the soil/sediment and through over ground vegetative organs from the water and atmosphere. Which plant species will be used as bioindicators also depends on how widely they are distributed throughout the environment.

The main sources of trace elements in algae and plants are their growth media. Biomaterials such as micro and macro algae, fungi, lichens, mosses, tree bark and leaves of higher plants have been used to detect the deposition, accumulation and distribution of metal pollution in water and air. Mosses and lichens, due to lack of root, are probably the most frequently used organisms for monitoring metal pollution in an air, Serbula et al. (2012).

Algae, fungi (micro and macro), and other plant communities play a fundamental role for nutrition and life on earth. As non-mobile organisms they are always exposed to the environmental conditions: to air, soil and water pollutants at their sites of growth. Algae and plants growing in metal-polluted sites exhibit altered metabolism, growth reduction, lower biomass production and metal accumulation. Various physiological and biochemical processes in them are affected by metals.

Chemical analyses of plant materials have been used for biomonitoring and biogeochemical mapping for many years. Trace element input to plants is governed by a wide variety of factors, Suchara et al. (2011). Accumulation and distribution of heavy metals in the plant depend on the plant species, the levels of the metals in the soil/sediment, water and air, the element species and bioavailiability, pH, vegetation period and multiple other factors, Nagajyoti et al. (2010). The quantity or level of heavy metals absorption in plant depends not only on the concentration levels of the metals in the physical and chemical composition of the soil/sediment and water, but also varies in different parts of the plant. Plants also have the ability to accumulate heavy metals which have no known biological function, Filipovic-Trajkovic et al. (2012). Plants are able to minimize the adverse effects of excess heavy metals by regulating the distribution and translocation of them within their organs or cells, Hossain et al. (2012).

Elements are often classified as being macronutrients or micronutrients, and as either essential or non-essential for the plant, whereas their concentrations are generally indicated as deficient, sufficient or toxic. Plant nutrients yet identified and best known as essential are C, H, O, N, P, K, S, Ca, Mg (as macronutrients) and B, Cl, Co, Cu, Fe, Mn, Mo, Ni, Si, Na and Zn (as micronutrients). When these metals are present in bioavailable forms and at excessive levels, they have the potential to become toxic to plants, Nagajyoti et al. (2010).

5.5.3 Lower Plants

Algae, Fungi, Lichine and Moss

Many toxic and bioaccumulative pollutants are found in only trace amounts in water, and often at elevated levels in sediments. Data from sediments may not be representative of pollutant concentrations in the overlying water column and cannot give information on patterns of contamination at the higher levels of the food chain. For example, the uptake of toxic metals by phytoplankton is the first step in the bioaccumulation in aquatic food webs. Macro- and microalgae also play an important role in the removal of toxic metals, Torres et al. (2008), Azizi et al. (2012), Rybak et al. (2012).

In terrestrial environments, bacteria, fungi, algae and other lower plants play the predominant role in the biochemical cycling of heavy metals. In aquatic environments algae play a key role in biogeochemical cycling of metals and their accumulation in sediments. Metals sequestered by microalgae are a major contributor to the metal load of the water column as well as to the metal content of sediments, Torres et al. (2008).

Some green algae or phytoplankton like *Scenedesmus subspicatus*, *Chlorella vulgaris* or *Pseudokirchneriella subcapitata* are in use as standard bioindicators representing primary producers, Ratte et al. (2003). For example, dominated algal mats of the green algae genera *Klebsormidium* are good indicators of high Fe concentration in water, whereas the presence of *Fucus vesiculosus* suggesting heavy metal pollution in marine environment, Das et al. (2009). At low metal concentration, algae, accumulating metals, pass them to other trophic levels.

The adsorption of heavy metals by algae is highly variable, depending on the metal, the taxon, and other conditions. Algal and fungal biomasses are reported to show efficient metal removal from wastewater, Das et al. (2009). Many heavy metals are necessary micronutrients of algae and fungi in low concentrations, however at high concentrations those can be fatal. Metal toxicity to both, algae and fungi, occurs by affecting their essential metabolic cell processes. Significantly, many nonessential metals compete with essential for uptake into cells. Several of those nonessential metals are toxic at very low concentrations (0.1–1.0 ppm).

Mosses and lichines are particularly used as bioindicators of aerial heavy metals contamination because of their bioaccumulative properties, Backor and Loppi (2009), Blagnytė and Paliulis (2010). Analysis of indigenous mosses is currently used in international and national monitoring programmes, particularly in Europe, Harmens et al. (2008). Mosses have been applied to measure heavy metal levels and trends within and around urban and industrial areas, Suchara et al. (2011). The use of fungi in the monitoring of heavy metal pollution is limited but some fungal groups are better bioaccumulators than others. Little published literature exists with regard to fungi distribution patterns in response to aerial metal contamination.

The use of marine organisms as bioindicators for trace metal pollution is currently very common. Macroalgae and mollusks are among the organisms most used for this purpose, Jakimska et al. (2011), Markovic et al. (2012), Joksimovic and Stankovic (2012). Freshwater and sea macroalgae are able to accumulate trace metals, reaching concentration values that are thousands of times higher than the corresponding concentrations in water, Vardanyan et al. (2008), Akcali and Kucuksezgin (2011), Joksimovic (2011b), Wolff et al. (2012).

Macroalgea is an important component of aquatic communities due to their role in oxygen production, nutrient cycling, sediment stabilization and providing habitat for aquatic life, Vardanyan et al. (2008). Fresh and sea water macrophites are probably the main source of metals for many animals like invertebrate and fish feeding on them. Invertebrate and fish are commonly consumed as human food. Therefore, the investigation of heavy metal concentrations in the macroalgae species may provide useful information on the transfer of potentially toxic elements from abiotic compartments, water and sediment, to higher consumers including man.

Higher Plants

Plants have been used as bioindicators in areas with significant air pollution in the absence of mosses. Some responses of higher plants to heavy metal contamination in the air have potential. The use of plants in the bioindication of heavy metal contamination is not routinely practised.

The most often used parts of higher plants are leaves/needles of: spruce, birch, pine, oak, olive, poplar; also a bark of: oak, pine, ash tree, Serbula et al. (2012). Different plant organs (leaves, flowers, bark or roots) from naturally occurring wild plants and trees, and cultivated plants (vegetables and fruits) were evaluated as possible bioindicators of heavy-metal pollution, Filipovic-Trajkovic et al. (2012). The bioavailability of trace elements from aerial sources through leaves also has a significant impact on plant contamination. Trace elements taken up by leaves can be translocated to other plant organs according to Filipovic-Trajkovic et al. (2012) and Serbula et al. (2012). Unlike Pb, Cd contamination cannot be removed from plants by washing them; it is distributed throughout the organism. It is often difficult to be certain of the cause of a Cd content found in fruit or vegetables, as the substance in its natural form exists everywhere in the soil and is absorbed by the roots.

However, excessive accumulation of metals can be toxic to most plant species. Generally, it is accepted that the normal Cd concentrations in plants are between 0.2 and 0.8 mg kg^{-1} and toxic Cd concentrations are defined as 5–30 mg kg^{-1} and Zn is not considered to be highly phytotoxic and the toxicity limit for Zn (300–400 mg kg^{-1}) depends on the plant species as well as on the growth stage. According to Kabata-Pendias and Pendias (1992) normal concentrations of Pb in plants are 0.1–10 mg kg–^{1} d.w. and also expressed that uptake and release of elements depend on plant species, growth stage and composition of the soil/sediment, especially Ca. Generally, toxic Pb concentrations are defined as 30–300 mg kg^{-1}, Filipovic-Trajkovic et al. (2012).

The results of the Filipovic-Trajkovic et al. (2012) indicate the existence of differences in accumulation of Pb, Zn and Cd in different plant organs, tissues and different plant species. Distribution of heavy metals is unequal and the largest is in the tree's bark, which is explained by the time of its exposure to the environmental conditions and structure (roughness), enabling a larger absorption of heavy metals from the air. After the tree bark, heavy metals are mostly accumulated in the roots, then in the leaves and as well as in the plant fruits. In over ground organs the highest amounts of Pb were found in the leaves and then in the fruits and vegetables, Filipovic-Trajkovic et al. (2012).

Hg present in soil has a very low availability to plants as the roots function as a barrier. Therefore, in general Hg concentrations in the environment are comparably low and are limited mostly to specific and/or localised aerogenous emissions. Significant accumulation of Hg has been observed in fungi, carrots and potatoes. When organomercuric compounds were still in use as fungicides it was shown that in wheat, barley, oats and corn, Hg was transferred from the seed dressing into the new seedling. During growth, dilution of the mercury occurred and the concentration in the grain was reduced, Tataruch and Kierdorf (2003).

The terrestrial plants are promising indicators for metal pollution in the soil too. It is important during metal biomonitoring programmes that the background of examined elements concentrations in the soil should be established. The design of a monitoring programme should involve the selection of appropriate species, sampling locations, sample collection, sampling frequency, metals to be analysed, chemical technique and data analysis.

It was well known that chemicals themselves can interact significantly with each other, showing either antagonistic (less than additive effects) or synergistic (more than additive effects) behaviour, Joksimovic et al. (2011a), Joksimovic and Stankovic (2012), but combined effects of natural environmental conditions and toxic metals are more important on the biota than interactions among chemicals. It has been shown in a number of studies that natural environmental conditions can significantly modify responses of organisms to toxicants, Laskowski et al. (2010), Holmstrup et al. (2010), Avgın and Luff (2010), Dondero et al. (2011).

The knowledge about these interactions and how they affect organisms is scarce. For example, already over 20 years ago it was noticed that salinity and temperature may significantly affect results of aquatic toxicity tests, Laskowski et al. (2010). Most studies on interactions between toxicants and natural factors showed that effects of toxic chemicals on organisms can differ vastly depending purely on external conditions: in 62.3 % cases Laskowski et al. (2010) were found significant interactions between natural factors and chemicals.

5.5.4 *Biomarkers of Metal Toxicity in Living Organisms*

A conventional monitoring system of environmental metal pollution includes measuring the level of selected metals in the whole organism or in respective organs. However, measuring only the metal content in particular organs does not give information about its effect at the subcellular level. Therefore, the evaluation of biochemical biomarkers metallothioneins (MTs), phytochelatins (PCs) and antioxidant enzymes (catalase – CAT, superoxide dismutase – SOD, glutathione S-transferases – GST and glutathione peroxidases – GPXs, lipid peroxidation – LP) may be useful in assessing metal exposure and the prediction of potential detrimental effects induced by environmental metal contaminations, Valavanidis et al. (2006), Vlahogianni and Valavanidis (2007), Vlahogianni et al. (2007), Torres et al. (2008), Nordberg and Nordberg (2009), Shariati and Shariati (2011), Semedo et al. (2012).

After having crossed the plasma membrane, the incoming metal is bound immediately by ligands and distributed between sites of storage, efflux, or toxic action. Toxicity does not depend on total accumulated metal concentration. Accumulated metal concentrations should be interpreted in terms of different trace metal accumulation patterns dividing accumulated metals into two components: metabolically available metal and stored detoxified metal. The relationship between metal accumulation and toxicity is influenced by physiological activity of living organisms, Shariati and Shariati (2011).

A number of trace metals are used by living organisms to stabilize protein structures, facilitate electron transfer reactions and catalyze enzymatic reactions, Nagajyoti et al. (2010). For example, Cu, Zn and Fe are essential as constituents of the catalytic sites of several enzymes. Although some heavy metals are essential micronutrients for animals, plants and many microorganisms, depending on the route and dose, all heavy metals demonstrate toxic effects on living organisms via metabolic interference and mutagenesis. The mechanisms by which metals exert their toxicity in living organisms is very diverse, especially their involvement in oxidative biochemical reactions through the formation of reactive oxygen species (ROS), Torres et al. (2008). Heavy metals are involved in toxic redox mechanisms through the generation reactive oxygen species, associated with oxidative damage to important biomolecules, and molecular mechanisms of metal toxicity and carcinogenicity, Vlahogianni and Valavanidis (2007).

Metal ions can penetrate inside the cell, interrupting cellular metabolism and in some cases can enter the nucleus. The entrance of the metal in the cell can mobilize several metabolic pathways and genetic processes to neutralize the source of toxicity, Azevedo and Rodriguez (2012). The entrance of certain metals into the nucleus can enhance the synthesis of RNA that codes from metallothioneins. Metallothieneins (MTs) are peptides found mainly in the cytosol, lysosomes and in the nucleus, low molecular weight peptides, high in the amino acid cysteine which contains a thiol group (-SH). The thiol group enables MTs to bind heavy metals, Nordberg and Nordberg (2009).

MT in the physiological system has several roles, especially in the metabolism and kinetics of metals: transport and detoxification of metal ions and protection from metal toxicity, free radical scavenging, storage of metal ions, metabolism of essential metal ions, immune response, genotoxicity and carcinogenicity, chemoresistance, and radiotherapy resistance, Shariati and Shariati (2011). Considering the heavy metal detoxification significance of MTs, these proteins can serve as biomarkers of heavy metal pollution of the environment, Valavanidis et al. (2006), Nordberg and Nordberg (2009), Shariati and Shariati (2011).

MT can be used as an indicator in both environmental and biological monitoring reflecting exposure to metals, and as a good biomarker of renal dysfunction. MT in urine can be used as a sensitive biomarker for metal induced nephrotoxicity. MT is an established biomarker in biomontoring of human Cd exposure and may also be useful in the risk assessment of other metal exposures. MT mRNA in lymphocytes in humans has been suggested as an indicator of susceptible groups in relation to metal exposure, Nordberg and Nordberg (2009).

Oxygen and nitrogen free radicals are essential in the physiological control of cell function in biological systems and are continuously produced in living cells. Basic cellular metabolism in aerobic organisms involves the production of oxygen free radicals and nonradical reactive species, referred to as reactive oxygen species (ROS). The potential of oxygen free radicals and other reactive oxygen species, able to damage tissues and cellular components in biological systems, called oxidative stress, have become a topic of significant interest for environmental toxicology studies, Valavanidis et al. (2006).

Both animal and plant cells are capable of generating – via multiple sources – a number of different reactive oxygen species (ROS), including the superoxide anion ($O_2^{\cdot -}$), hydrogen peroxide (H_2O_2) and the hydroxyl radical (·OH), Torres et al. (2008). These species occur transiently and are regular products of oxidative metabolism. Although some ROS may function as important signaling molecules that alter gene expression and modulate the activity of specific defense proteins, but all ROS are harmful to organisms at high concentrations, Torres et al. (2008).

Metal ions possess the ability to produce reactive radicals, resulting in DNA damage, lipid peroxidation and depletion of protein sulphydryls. When ROS levels exceed antioxidant defences, the cells go into oxidative stress which causes membrane lipid peroxidation and changes in the activity of the antioxidant defence enzymes, Vlahogianni et al. (2007). This system includes antioxidant enzymes like superoxide dismutase (SOD), catalase (CAT) and glutathione peroxidase (GPX), and nonenzymatic antioxidants, such as glutathione, vitamin E, ascorbate, bcarotene, and urate, whose activities are modified in response to cellular oxidative stress, Valavanidis et al. (2006).

Antioxidant defence enzymes play an important role in cellular antioxidant defence systems and protection from oxidative damage by ROS. Heavy metals are involved in toxic redox mechanisms through the generation of ROS, associated with oxidative damage to important biomolecules and molecular mechanisms of metal toxicity and carcinogenicity. Although there are considerable gaps in the knowledge of cellular damage, response mechanisms, repair processes, and disease etiology in biological systems, but free radical reactions and the production of toxic ROS are known to be responsible for a variety of oxidative damages leading to adverse health effects and diseases, Valavanidis et al. (2006).

Many enzymes need cofactors to work properly, such as Fe^{2+}, Mg^{2+}, Cu^{2+} Ca^{2+}. The substitution of one heavy metal ion by another leads to the inhibition or loss of enzymatic activity. Antioxidant enzyme activities, oxidative damages (lipid peroxidation – LP) and metal content in the marine species have been studied, Vlahogianni et al. (2007), Duarte et al. (2011), Giarratano et al. (2011), Semedo et al. (2012). Lipid peroxidation (LP) by ROS is considered to be a major mechanism by which heavy metals can cause tissue damage. The negative correlations found for LP with SOD and mainly GST activity highlighted the importance of these enzymes in preventing oxidative damage in the digestive mussel gland. CAT activity was also positively correlated with SOD and GST activities, which emphasizes that the three enzymes respond in a coordinated way to metal induced oxidative stress, Semedo et al. (2012). Correlations between metal accumulation and biomarkers of oxidative stress (LP, CAT and SOD) in marine mollusks were found, Vlahogianni and Valavanidis (2007), Duarte et al. (2011), Giarratano et al. (2011). The study of Vlahogianni et al. (2007) showed that seasonal variations of the activity of antioxidant defence enzymes and LP concentrations in mussels can be used as potential biomarkers of toxicity for long-term monitoring in marine coastal ecosystems.

Heavy metal toxicity is one of the major abiotic stresses also leading to hazardous effects in plants. A common consequence of heavy metal toxicity is the excessive accumulation of reactive oxygen species (ROS) and methylglyoxal (MG). Both can cause peroxidation of lipids, oxidation of protein, inactivation of enzymes, DNA damage and/or interact with other vital constituents of plant cells. Higher plants have evolved a sophisticated antioxidant defense system to scavenge ROS and MG, Hossain et al. (2012).

Heavy metal toxicity results in the accumulation of excessive ROS inside the plant cell. For example, Cu can directly generate ROS, whereas Cd is a redox-inactive heavy metal and can only generate ROS indirectly by enzyme inactivation. Potentially a very important mechanism of heavy metal detoxification and tolerance in plants under heavy metal stress is chelation of heavy metals in the cytosol or intracellular fluid by high affinity ligands. Plants make two types of peptide metal binding ligands: phytochelatins (PCs) and metallothioneins (MTs), Hossain et al. (2012).

MTs and PCs have been identified in a wide variety of plant species and in some microorganisms, Hegelund et al. (2012), Hossain et al. (2012). PCs and MTs are different classes of cysteine-rich heavy metal-binding protein molecules. MTs are cysteine-rich polypeptides encoded by a family of genes. In contrast, PCs are a family of enzymatically synthesized cysteine-rich peptides, Hossain et al. (2012). PCs are a family of Cys-rich polypeptides, although the most common PC forms have 2–4 peptides, Hossain et al. (2012).

PCs are structurally related to glutathione (GSH) and numerous physiological, biochemical, and genetic studies have confirmed that GSH is the substrate for PC biosynthesis. GSH is a tripeptide with thiol group (-SH) of cysteine. It can be synthesized in the human body from the amino acids. While all cells in the human body are capable of synthesizing glutathione, liver glutathione synthesis has been shown to be essential. In animal cells GSH is catalyzed by glutathione S-transferase enzymes. It is an antioxidant, preventing damage to important cellular components caused by reactive oxygen species such as free radicals and peroxides, Pompella et al. (2003). GHS has a vital function in Fe metabolism, Kumar et al. (2011) and it is the major free radical scavenger in the brain, Gawryluk et al. (2011).

Recent plant molecular studies have shown that GSH by itself and its metabolizing enzymes act additively and coordinately for efficient protection against heavy metals damage in plants. PC synthase is primarily regulated by activation of the enzyme in the presence of heavy metals, Hossain et al. (2012). The biosynthesis of PCs is induced by many heavy metals, including Cd, Hg, Ag, Cu, Ni, Au, Pb, As, and Zn; however, Cd is by far the strongest inducer.

Plants are not able to metabolize or eliminate Cd, Hossain et al. (2012). They adopt the strategy of making Cd-GSH and Cd-PCs complexes. Zhang and Ge (2008) found a close relationship between Cd level and GSH content as well as enzyme glutathione *S*-transferase (GST) activity in rice, suggesting that these two parameters of antioxidant defense system may be used as biomarkers of Cd-induced stress in plants. Glutathione peroxidases (GPXs) are key enzymes of the antioxidant network in plants. Cuypers et al. (2002) suggested that peroxidase activity can be used as a potential biomarker for heavy metal toxicity in plants. Similarly, a significant increase in GPX activity was also observed in red onion exposed to a variety of Hg, Pb, Cr, Cu, Zn, or Cd concentrations suggesting that the elevated activity of GPX was a result of heavy metal-induced free radical generation, Fatima and Ahmad (2005).

Research on plant MTs lags behind what is known about the vertebrate forms, Freisinger (2010). The large diversity in the metalbinding regions of plant MTs suggests that they have the ability to bind a greater range of metals than their animal counterparts and, consequently, a greater range of function, Cobbett and Goldsbrough (2002). In plants MTs are extremely diverse, Hossain et al. (2012) and their role in the detoxification process has not been conclusively shown, Freisinger (2010). The high metal ion binding capacity of MTs suggests a role in metal ion storage, metabolism and trafficking of essential Cu^{+} and Zn^{2+} ions, as well as the detoxification of nonessential metal ions such as Cd^{2+} and Hg^{2+} in living organisms. The precise function of the MT role in living organisms remains elusive, Freisinger (2010).

PCs have been shown to play an important role in the detoxification of certain heavy metals in both plants and animals. PCs play a wider role in heavy metal detoxification in biology than previously expected, but it appears that some organisms probably do not express a PCs synthesis. There is, for example, no evidence for PC synthase in mouse and human genomes. Organisms with an aquatic or soil habitat are more likely to express PCs, Cobbett and Goldsbrough (2002).

5.6 Bioindicators of Toxic Metals

Different species are used worldwide in scientific studies evaluating the risk of toxic chemicals, but only a few of this species have been established as standard organisms as bioindicators of worldwide use. The standardized bioindicators of

Table 5.3 The most analyzed heavy metals in the biota of environmental compartments from reviewed literature

Environmental compartment	Taxa studied	Heavy metals
Air	Mosses, lichens, high plants	Pb, Zn, Cd, Cu
Water	Bacteria, plankton, macro algae	Pb, Cd, Cu, Zn, Hg, Cr, As
	Mollusks, crustaceans, fish, birds, mammals	
Soil/sediment	Bacteria, protozoa, nematodes, oligochaeta (earthworms), amphibians, birds, mammals	Pb, Cd, Cu, Zn, Hg, Cr, As

the different trophic levels for the most commonly studied toxic metals in the environment compartments (air, soil and water) is given in Table 5.3. This part will mainly focus on organisms as a standard and some potential bioindicators of toxic metals in air, soil and aquatic ecosystems.

The oldest and widely used species for toxic metal bioindications are fish, bird and mosses species, while species representing soil organisms are currently becoming more important in toxic metals bioindications. Representative species traditionally used for risk assessment of chemicals are various freshwater species representing bacteria, microalgae, invertebrates and fishes. Up until now, among the terrestrial species mainly used in risk assessment of chemicals are the earthworm *Eisenia fetida* and some higher plants. Generally speaking, toxic metals distributions were considered more homogeneously in aquatic systems as compared to terrestrial ones.

5.6.1 Bioindicators for Air Metal Pollution

Biological monitors and vegetations are used to measure the levels of atmospheric trace metal concentration as the cheapest and simplest indicators for monitoring the trace metal concentrations in the air, Kord et al. (2010). Numerous different bio-indicators are used in monitoring air pollution: bacteria, fungi, mosses, lichens, grasses, agricultural crops, ornamental plants, vascular plants, and woody plants, Joshi (2008), Baslar et al. (2009), Kord et al. (2010), Suchara et al. (2011), Filipovic-Trajkovic et al. (2012). Samples were collected from industry, roadside, suburban and rural areas.

Which plant species will be used as bioindicators also depends on how widely they are distributed throughout the region. The most often used parts of higher plants are leaves/needles of: spruce, birch, pine, oak, olive, and poplar; also a bark of oak, pine, ash tree, Serbula et al. (2012). The usage of air pollution biondicators usually covers metal toxicity in traffics, smelters, mining industries, industrial pollution, coal-burning power plants and agriculture, Conti and Cecchetti (2001), Wolterbeek (2002), Harmens et al. (2008), Blagnytė and Paliulis (2010), Suchara et al. (2011), Stankovic et al. (2011c), Paoli et al. (2012).

Lichens and mosses may be considered as the most commonly applied organisms as bioindicators. Typical examples of a biological indicator of air pollution are lichens and mosses, Conti and Cecchetti (2001), Wolterbeek (2002), Blagnytė and Paliulis (2010). Today, lichens and mosses are mostly used as lower plants for atmospheric trace elements bioindicators due to their capacity to accumulate and store heavy metals and other toxins and, because they do not have roots, they are thus able to provide a direct indication of the air pollution, Suchara et al. (2011).

The idea of using mosses to measure atmospheric heavy metal deposition was developed already in the late 1960s, well established in the Scandinavian countries in the early 1980s, Sabovljevic et al. (2009) and in Europe a moss monitoring programme has been repeated every 5 years since then, Suchara et al. (2011). Naturally growing mosses have been widely used as effective bioindicators of heavy-metal air pollution, Harmens et al. (2008). Mosses properties as good bioindicator are: large surface to weight ratio, their slow growth rate and a habit of growing in groups, minimal morphological changes during the mosses lifetime, ease sampling, an ability to survive in highly polluted environment and the possibility to determine concentrations in the annual growth segments, Dragovič and Mihailovič (2009). Moss appears to be an excellent deposition bioindicator for monitoring elements such as Pb, Cd, Cu, V, and partially Zn. Secondary processes seriously affect monitoring of Ni, Cr, Mn, Fe, such as rain washing down of entrapped particles from mosses, which could result in loses reaching up to 20 %.

A high proportion of the pollutant load accumulates in mosses through wet deposition. The amount, duration and intensity of precipitation affect accumulation and leaching by mosses, depending on the meteorological conditions and seasonal growth. The metal uptake by mosses decrease in the order: $Cu > Pb > Ni > Co > Cd > Zn$, Mn. The most commonly used lichen species for toxic metals biomonitoring were *Parmelia sulcata*, *Hypnum cupressiforme*, *Hylocomium splendens*, and *Pleurozium schreberi,* Blagnytė and Paliulis (2010).

Many European countries have used mosses in national and multinational surveys of atmospheric-metal deposition. According to Harmens et al. (2008) metals deposition trends in some European countries during 1990–2005/6 measured on the basis of mosses were next: the lowest toxic metals concentrations in air were generally found in (north) Scandinavia, the Baltic States and northern parts of the United Kingdom in 2005/6. The concentration in European mosses since 1990 has declined the most for As, Cd, Fe, Pb and V (45–72 %) followed by Cu, Ni and Zn (20–30 %) with no significant reduction being observed for Cr (2 %) and Hg (12 % since 1995).

Moss (*Pleurozium schreberi*), grass (*Avenella flexuosa*), and 1- and 2-year old spruce (*Picea abies*) needles were collected over the territory of the Czech Republic. The samples were analysed for 36 elements: Ag, Al, As, Ba, Be, Bi, Cd, Ce, Co, Cr, Cs, Cu, Fe, Ga, Hg, La, Li, Mn, Mo, Nd, Ni, Pb, Pr, Rb, S, Sb, Se, Sn, Sr, Th, Tl, U, V, Y, and Zn, Suchara et al. (2011). Most elements showed the highest concentrations in moss, considerably lower concentrations in grass, even lower concentrations in the 1-year old spruce needles and again somewhat higher concentrations in the 2-year old spruce needles: As, Bi, Ce, Cr, Fe, La, Nd, Pb, Pr,

Sb, Se, Sn, Th, U, V, and Y. Mo is the element that was especially enriched in grass. Grass appeared to be a good model crop and allowed estimation of the level of Hg contamination in leafy vegetables.

Different moss types are currently widely used as bioindicators since they obtain most of their nutrients directly from the air and by dry deposition. The procedures for the moss as a bioindicator, the methods of collection, processing and analyses of moss samples, are outlined in the international moss monitoring manual, Markert et al. (2011). But it is important to note that a unique species that can be a suitable bioindicator for biomonitoring of toxic metal pollution all over the world has not been found yet. For this reason, different species of mosses are useful as bioindicators in different parts of the world, Blagnytė and Paliulis (2010).

Moss still appears to be one of the best air pollution bioindicator, but on the basis of literature search of Aboal et al. (2010) it could be concluded that mosses are possible to be bioindicators for certain metals such as Pb or Cd, probably because these elements are of almost exclusively atmospheric origin. According to Suchara et al. (2011) mosses can be identified as the most effective collectors of fine atmospheric dust than of wet depositions and can be used as bioindicator of atmospheric pollution in megacities, because this plant is able to absorb water and nutrients directly from the air, Markert et al. (2011).

Lichens have been widely used as trace elements air biomonitors as they are widespread and capable of absorbing elements directly from the atmosphere and accumulating them in their tissues. Many countries have performed biomonitoring research on metal levels in industrial areas and in clean areas based on lichens as air toxic metal bioindicators. But, lichens of similar composition are not easy to find, because of the differences caused by the tree on which lichens are growing. As different lichen species in the same location contain varying amounts of metal, it is obvious that the amount of metal contained by lichen is species-dependent.

Lichens are symbiotic organisms of fungus and algae, and have been widely used in biomonitoring of air pollution by trace elements. It is well known that the chemical composition of lichens reflects the availability of trace elements in the air environment, Backor and Loppi (2009) and can be used to monitor air quality changes in urban areas over intervals of a few years, Paoli et al. (2012).

Paoli et al. (2012) used the lichen *Flavoparmelia caperata* as indicators of pollution around a landfill in central Italy along 14 years of waste management. Lichens revealed an increased deposition for some elements (Cd, Cr, Fe, and Ni) and a decrease of the lichen diversity was subjected to high airborne dust concentrations, Paoli et al. (2012).

Numbers of phylloplane fungi and bacteria were used as a bioindicator for toxic metals air pollution, Joshi (2008). Analysis of phylloplane microorganisms, Pb, Zn, Cu, and Cd from plant leaves indicated that the bacterial and fungal population was higher at the unpolluted site. Numbers of phylloplane fungi and bacteria were significantly reduced in the toxic metals polluted sites. Bacterial population showed a significant negative correlation with Pb, Zn, Cu, and Cd. Microfungal species composition and their diversity in the phylloplane were directly correlated to the roadside toxic metals pollution, as Joshi (2008) suggested for bacteria.

In the last decades of the twentieth century a rapid increase in bioindication studies for pollutant loads in higher plants can be observed. Both the broadleaves and coniferous trees are used in toxic metals air pollution studies. Air pollutants can be taken up via stomata, cuticle or indirectly by uptake via roots after deposition of the air pollutants to the soil. The surface wax layer of leaves and needles works as an accumulator of particle bound air pollutants, Weiss et al. (2003). This observation clearly demonstrates the effective accumulation of particle bound elements on plant surfaces and supports the principal suitability of higher plants to monitor the corresponding atmospheric load.

Particularly for the following elements a significant contribution of the plant cuticle layer to the overall needle or leaf concentration was shown: Al, As, Br, Co, Cr, Cs, Cu, Fe, Hg, La, Mo, Na, Ni, Pb, Sc, Sb, Sn, and Th, Weiss et al. (2003). On the contrary, cuticular deposits usually do not contribute significantly to the overall needle or leaf concentration of Mg, K, Ca, Ba, Cd, Mn, Tl, and Zn, Weiss et al. (2003). These general statements, however, may be not valid for all plants.

Other plant materials are occasionally used as bioindicators of air toxic metal deposition includes pine and spruce bark, and grass. Tree bark is appropriate in indicating longer term of air pollution. In literature, there are many studies focusing on using barks of different tree species as bioindicator. Results of such studies identified that the barks of the pine trees are good adsorbents of airborne pollutants, including anthropogenic heavy metals. Among the studied barks of pine tree are Turkish red pine (*Pinus brutia Ten.*), Italian stone pine (*Pinus pinea L.*), Austrian pine (*Pinus nigra Arnold.*) Masson pine (*Pinus massoniana Lamb.*) and Scots pine (*Pinus sylvestris L.*), Baslar et al. (2009). Coniferous trees indicate pollution over a longer time period, and since bark is exposed to air pollutants directly from the atmosphere, tree bark is appropriate in indicating longer term air pollution.

Depend upon the type of the tree, the deployment of heavy metal content and ways of accumulation show a great variety: in some trees heavy metals are filtered out by the leaves from the air, while in others they are taken up by their crown or by their roots. According to Baslar et al. (2009) Mn or Zn are mainly taken up by the root system and are then transported from there to the crowns, since they are highly mobile elements. Therefore, it is better to investigate its deposition in the crowns.

Different plant organs (leaves, flowers, stems or roots) from naturally occurring wild plants and trees, and cultivated plants (vegetables and fruits) were evaluated as possible bioindicators of heavy-metal pollution too, Filipovic-Trajkovic et al. (2012). Filipovic-Trajkovic et al. (2012) were found the highest amounts of heavy metals in the leaves, especially Pb, and then in the fruits and vegetables. They concluded that fruits and vegetables were metal avoiders. Naveed et al. (2010) used leaves of roadside plants (*Dalbergia sissoo* Roxb., *Prosopis juliflora* L. and *Eucalyptus* spp.) as bioindicator of traffic related to Pb pollution. They conclude that all the three selected plants proved to be good indicators of Pb air pollution and, due to their diverse distribution in different parts of the world, their leaves can be used as bioindicators of Pb pollution.

General problems of air bioindicators that measure toxic metals concentrations do not only depend on anthropogenic elements inputs via the atmosphere, but also depend on different uptake mechanisms and element preference/rejection by certain plant species, Suchara et al. (2011). Lower plants, especially mosses and lichens, are probably the organisms most frequently used for monitoring metal pollutions in urban environments. They have an important advantage because they do not need soil substrate, but there are limitations in using them as in situ bioindicators, because of difficulties to find a single species spanning a large area, Dmuchowski and Bytnerowicz (2009), Backor and Loppi (2009). Tree bark and higher plants leaves started to be used to detect the deposition, accumulation and distribution of air metal pollution in large-scale of air pollution, Baslar et al. (2009), Kord et al. (2010), Stankovic et al. (2011c), Serbula et al. (2012).

5.6.2 *Bioindicators for Aquatic Metal Pollution*

In this part, principles of algal bio-indication and bio-monitoring in the environment are outlined for streams and rivers, lakes and reservoirs, as well as for marine ecosystems. Aspects of sediment testing are included. Since environmental contamination and pollution has severely expanded in the recent years, using bioindicators became increasingly important.

Monitoring of trace toxic metals in the aquatic environment using biological indicators has been well established since chemical analysis of the environment matrix such as water and sediment is the most direct approach to reveal the heavy metal pollution status in the environment, while it cannot afford the powerful evidence on the integrated influence and possible toxicity of such pollution on the water organisms. Speciation of metals, their solubility and complexation, are important factors that influence the toxicity of metals in the aquatic environment. The amount of dissolved metal strongly depends on water pH. The interaction of metals can alter their toxic effects on aquatic organisms both positively and negatively, Sevcikova et al. (2011).

Aquatic heavy metal pollution usually represents high levels of Hg, Cr, Pb, Cd, Cu, Zn, Ni etc. and introduced into environmental water system may pose high toxicities on the aquatic organisms, Zhou et al. (2008). The most important application of biomonitoring is for the evaluation of metal pollution in aquatic ecosystem including harbors and continental waters. Many pollutants are associated with sediments in aquatic systems. Hazard assessment to establish sediment quality criteria require rapid, inexpensive examination too. Direct bioassay with algae has proven to be very sensitive indicators of contaminant stress.

Since the 1950s, numerous studies have demonstrated the value of benthic foraminifera in detecting ecosystem contamination. They are among the most common marine plankton species. Foraminifera have typically been included in the group *Protozoa* as unicellular protists with animal-like behavior and movement. Living benthic foraminiferal taxa were studied in surface sediment samples collected from the Italian sea, marsh and likes sediments over the last three decades, Frontalini and Coccioni (2011) and foraminifera have been proven to be successful candidates as part of an integrated monitoring program of pollution. The statistical analysis reveals a strong relationship between trace elements, in particular Hg, Mn, Ni, Pb, and Zn, and the occurrence of abnormalities in foraminiferal taxa, Frontalini and Coccioni (2011).

The value of algae as biomonitors for fresh waters has already been recognised in the mid nineteenth century. Species which might be used as indicators are *Aulacoseira granulata, Actinocyclus normanii, Stephanodiscus neoastrea, Cyclotella meneghiniana* among many others, the usefulness of river plankton for bio-indication the toxicity of chemicals, mixtures of chemicals or polluted waters is measured as their inhibitory effect on the photosynthesis of natural algal assemblages. Several algal species accumulate considerable amounts of metals and can thus be used as monitors for elements such as Cd, Cu, or Pb. Most metals are slightly to highly toxic to algae, As, Cu, Hg, and Zn having the greatest effect. Impacts on algae in natural waters are highly variable. Cyanobacteria and green algae, both species concentrate various metals proportionally to ambient concentrations, Ratte et al. (2003), Azizi et al. (2012).

The most effective way of studying algal populations with respect to lake water quality is to analyse the microfossils present in surface sediments from lakes with known water chemistry, since, lake water quality changes in the last 5–10 years and several toxic elements can be monitored using sedimentary algal remains, Ratte et al. (2003).

Most what has been used for freshwater indicators, equally applies to bio-indicators in the marine environment, such as bacteria and phytoplankton, but macroalgae in coastal marine waters are far more important as bioindicators than macrophytic algae in fresh-waters. Macroalgae have several intrinsic advantages as organisms for monitoring environmental impacts. Because of their sessile nature they can be easily collected in abundance and used to characterize locations over time. Accumulation of heavy metals from the surrounding water makes them ideal as bioindicators.

Red algae (*Glaciralia* sp.), often dominating macroalgae of sea communities are frequently employed for Cd, Cu, and Zn coastal monitoring, Jakimska et al. (2011). Members of the Fucales (*Fucus, Ascophyllum*) and Laminariales (*Laminaria, Macrocystis*) received most attention. Pollution assessment studies have also used red algae (*Rodophyceae*). Among the green macroalgae, the genera *Ulva, Enteromorpha,* and *Posidonia,* have attracted considerable attention as toxic metal bioindicators, Richir et al. (2010), Akcali and Kucuksezgin (2011), Wolff et al. (2012), Rybak et al. (2012), Luy et al. (2012).

Specific programs for monitoring toxic metals in aquatic systems were undertaken as early as in 1960s using various animals like invertebrate, mollusks, and vertebrate, like mammals and birds. Various aquatic organisms occur in rivers, lakes, seas and marines, and they are potentially useful as metal pollution bioindicators of sediments and waters, Zhou et al. (2008), Markovic et al. (2012), Joksimovic and Stankovic (2012), including fish, Yilmaz (2009), shellfish, oyster, mussels, Özden et al. (2010), Tsangaris et al. (2010), Valavanidis and Vlachogianni (2007),

Jovic et al. (2011), Jakimska et al. (2011), Stankovic et al. (2011a), Hauser-Davis et al. (2012), clams, Singh et al. (2012), aquatic plankton, Das et al. (2009) and macroalgae, Joksimovic et al. (2011b), Rybak et al. (2012).

Microorganisms as Bioindicators of Water Quality

Plankton (phytoplankton, zooplankton and bacteriaplankton) are any organisms that live in the water column, on the surface and on the bottom. Phytoplankton live near the water surface with sufficient light to support photosynthesis. Bacterioplankton play an important role in the remineralising organic material down the water column. This scheme divides the plankton community into broad producer, consumer and recycler groups. They provide a crucial source of food to many large aquatic organisms, such as fish and marine mammals.

Many bacteria live in surface water, ground water, and other natural environmental water, which offers the possibility for water quality assessment especially for hygiene using bacteria test, Zhou et al. (2008). Trace elements in aquatic ecosystem accumulate on organic, inorganic sediment particles and their associated microorganism community. Fungi, bacteria, protists, and exudates of microorganisms form the biofilm on sediments. Microorganisms of the biofilm like bacteria and fungi are known to accumulate high amounts of metals and metalloids. Many species associate with metals and metalloids polluted sediments in aquatic systems, Schaller et al. (2011).

Various zooplankton species show their special advantages in the actual biomonitoring researches. Zooplankton containing protozoa are the important component of aquatic ecosystem. Many zooplankton species can accumulate and metabolize heavy metals which offer the possibility for its use in biomonitoring of water quality, Zhou et al. (2008). For example protozoa communities can be used for the monitoring of Hg and Cu levels in the water, Zhou et al. (2008). Most of protozoa species are worldwide distributed and not restricted by seasonal variations and regional discrepancies making them good bioindicators.

Zooplankton species *Daphniidae* as the representative of crustacean in biomonitoring of aquatic metal can sensitive response to heavy metals and pesticides. Thus, *Daphnia magna* is the most commonly tested freshwater species in acute, as well as in chronic tests, Zhou et al. (2008). Daphniidae toxicity test is the essential assay for worldwide water quality assessment.

Phytoplankton as the important elementary producers in marine and inland water plays key role to the whole ecosystem. Some phytoplanktons are bacteria, some are protists, and most are single-celled plants. Among the common kinds are cyanobacteria, green algae and fungi, Das et al. (2009). The aquatic algae species and amounts can directly reflect the water quality, Zhou et al. (2008). As an example, the green algae *Chlorella ellipsoidea* was reported to exhibit growth inhibition due to Cu, Zn, Ni, and Cd exposure. *C. ellipsoidea* also showed different toxicity response to different metal speciation exposure. The accumulation rate of some metals like Zn, Cu, and Ni in different plant organs shows significant accumulation difference, like roots than others, Zhou et al. (2008).

Microorganisms, such as protozoa, green algae or bacteria, reflect the water quality of only 1 or 2 weeks prior to their sampling and analysis, whereas insect larvae, worms, snails, and other macroinvertebrate organisms reflect more than a month, and possibly several years. Microorganisms have proven capability to take up heavy metals from aqueous solutions especially, when the metal concentrations range from less than 1 to about 20 mg/L, Ahmad (2006). Microbes such as bacteria also exist at the lowest trophic level, so bacteria have the ability to detect toxic compound before others do. Thus, bioindicators using bacteria have been commercialized such as the *Lux-Fluoro*, the Polytox™, the Deltatox™ and the Microtox™, Ahmad (2006).

The environmental monitoring of toxic metals by bioindicators like bioluminescent bacteria (BLB) in marine environment is reported by Ahmad (2006). The test method is design as a rapid evaluation of the toxicity of wastewaters or aqueous extracts from contaminated soils and sediments using luminescent marine bacteria. The *Photobacteria phosoreum* was used to the test wastewater or aqueous soil/sediment extract and the inhibition of light output measured over a specified time, Ahmad (2006).

Microorganisms also have evolved various measures to respond to heavy metal stress in aquatic systems via processes such as transport across the cell membrane, biosorption to cell walls and entrapment in extracellular capsules, precipitation, complexation and oxidation reduction reactions, Ahmad (2006), Burnat et al. (2010), Schaller et al. (2011), Jackson et al. (2012), and Murthy et al. (2012).

Animals as Bioindicators of Water Quality

Since the mid-1970s, scientists of several countries have used bivalve-filter feeding mollusks to monitor selected contaminants in coastal marine waters. This then led to the establishment of similar local or regional 'Mussel Watch' programs in many countries of the world, OSPAR (1992). The mussel watch program was initially used for analyzing trace metals in marine coastal waters. Mollusks (mainly mussels, *Mytilus* sp.) and fish (*Mullus* sp., *Platichthys flesus* L., *Zoarces viviparus*, *Perca* sp.) have both been employed as sentinel organisms in routine biomonitoring programs, at a national and an international level (Med Pol, UNEP Mediterranean Biomonitoring Program and OSPAR Convention).

Mussels are recognized as pollution bioindicator organisms because they accumulate pollutants in their tissues at elevated levels in relation to pollutant availability in the marine environment. Moreover, this ability has led to the adoption of the international "Mussel Watch Project" and several national programs on Mussel Watch in the marine environment have been carried out. As a result, much work on invertebrates has focussed in the aquatic environment, where mollusks have been used as effective bioindicators of toxic Hg, As, Cd, Pb, and trace elements: Fe, Mn, Ni, Co, Zn, Cu, Özden et al. (2010), Tsangaris et al. (2010), Jovic et al. (2011), Stankovic et al. (2011a), Joksimovic et al. (2011a), Jakimska et al. (2011).

Freshwater mussels are used as bioindicator of Pb, Cd, Hg, and Zn too, Mosher (2008), Faria et al. (2010), Rahnama et al. (2011), Minguez et al. (2012). It appears that freshwater gastropods and freshwater species in general, may be less sensitive than their marine relatives, although there are insufficient freshwater data to be sure if there is a real difference in sensitivity. Freshwater invertebrate toxicity tests are in current use and after many years of experimenting have reached a satisfactory level of precision. Most of these tests are described in the book Standard Method for the Examination of Water and Wastewater, APHA (1989).

Reptiles, particularly water snakes (*Nerodia* spp.), could serve as bioindicators of toxic metals contamination too, Burger et al. (2005), because some are comparatively long-lived, exhibit different trophic levels, and are at the top of their food chains. Water snakes can be useful bioindicators of As, Se, Cd, Cr, Pb, and Hg accumulation for multiple spatial and temporal scales, Burger et al. (2005).

Studies with fish species were introduced in the 1930s for the toxic effects determination in the field of toxicology investigation, especially for the screening of toxic chemicals, agricultural and industrial effluents and for pollution studies in rivers, lakes and the marine environment, Valavanidis and Vlachogianni (2010). Fish are used as test organisms in aquatic toxicology because of their top-position in the trophic chain and their role as food for humans. During their life-cycle, fish feed on algae, rotifers, microcrustaceans, macroinvertebrates, higher plants and other small fish. The use of fish in ecotoxicological studies has been advanced by various international organizations, Valavanidis and Vlachogianni (2010).

Fish are one of the most frequently used groups of bioindicators in ecotoxicological field studies. Various metal ions are involved in an oxidative stress in fishes. In the review Sevcikova et al. (2011) concentrated on the most important and most studied metals (Fe, Cu, Cr, Hg, and Pb) and metalloids (As, Se) in a fish. Apart from the mentioned ones, other metals are also connected with oxidative stress in fish (Ni, V, and Co), and can be detected in aquatic environments, Kandemir et al. (2010).

Due to quality assurance issues and to achieve adequate geographical and temporal representation, mussels (blue mussel – *Mytilus edulis*, Mediterranean mussel – *Mytilus galloprovincialis*) and fish (Atlantic cod – *Gadus morhua*, herring – *Clupea harengus* and flounder – *Platichthys flesus*) were selected as state indicators, EEA (2003). They are widely monitored in a comparable way. Mussels are attached to shallow-water surfaces, thus reflecting exposure at a fixed point, i.e., local pollution. Mussels are, however, restricted to the coastal zone. Atlantic cod is a widely distributed and commercially important fish species in the North Atlantic. It is a predator and as such will also to some extent reflect contamination levels in its prey. Herring is a commercially and ecologically important species both in the Baltic and the North-East Atlantic. Flounder is found in some of the most highly contaminated estuaries and is a common species along most European coasts, EEA (2003).

Other marine species, like marine mammals (seal, sea lion) and seabirds can also be used for the marine pollution biomonitoring due to their strong toxic metals bioaccumulation ability, Jakimska et al. (2011). Levels of accumulated toxic metals like Hg and Cd in marine mammals can be high, but without showing any signs of having been poisoned, Jakimska et al. (2011). Many studies have reported that Hg, Cd, and Se are accumulated in high levels in several seabirds, Horai et al. (2007), Hargreaves et al. (2011), and seabirds also have higher Hg levels than land birds, Jakimska et al. (2011). The high migration, however, limits their wide use in pollution biomonitoring in the investigated areas.

Plants as Bioindicators of Water Quality

Aquatic macrophytes have been used for monitoring the contamination level by various pollutants in aquatic environments, including heavy metals, such as Zn, Pb, Cd, and Hg and more others, as these plants have the ability to accumulate metallic ions, Vardanyan et al. (2008), Joksimovic et al. (2011b), Richir et al. (2010), Akcali and Kucuksezgin (2011), Rybak et al. (2011), Copat et al. (2012), Luy et al. (2012).

The results suggest that freshwater macroalgae *S. auriculata* and *Ulva* show good potential for use as a bioindicator and it can be used in the biomonitoring of aquatic ecosystems contaminated by Cd, Ni, and Pb, Rybak et al. (2011), Wolff et al. (2012). The studies of Richir et al. (2010), Joksimovic et al. (2011b), Copat et al. (2012), and Luy et al. (2012) confirm the relevance of the use of sea macrophyte *P. oceanica* as a biological indicator of metal contamination in coastal ecosystems and as a good tool for seawater quality evaluation. By biomonitoring study of heavy metals in macroalgae (the brown algae *Cystoseira* sp., the green algae *Ulva* sp. and *Enteromorpha* sp.) from eastern Mediterranean coastal areas, Akcali and Kucuksezgin (2011) concluded that *Cystoseira* and *Entromorpha* were the strongest accumulators of Cd, Cr, Fe, and Hg, Pb, Zn, respectively. *Ulva* turned out to be the highest Cu accumulator whereas this species was seen to accumulate Pb, Zn, and Fe with the minimum aptitude, Akcali and Kucuksezgin (2011). Freshwater populations of *Ulva* exhibited a greater efficiency to bioaccumulate Ni as compared to species derived from marine ecosystems, Rybak et al. (2012).

The use of these plants as bioindicators of contamination by heavy metals has advantages: (a) high tolerance to the pollutant, (b) easy sampling, (c) selective absorption ability, and (d) individuals are relatively big and easy to handle in laboratories. Data for better quantification of contaminants fluxes and inputs into water bodies and their water/air and water/sediments interfaces exchanges is lacking. These data are also essential for predictive and mass balance modelling of contaminants fates in the aquatic systems, Law et al. (2010).

5.6.3 Bioindicators for Soil Metal Pollution

Population growth combined with the increasing industrialization is responsible for generating tonnes of waste per day, which, many times, are accumulated in the environment without any previous treatment. Soil becomes a cheaper and practice alternative for the final disposal of these residues, but not without consequences: contaminated soil is a broad problem, since the contaminants can be leached into groundwater, rivers, lakes, and seas and adsorbed into plants and animals, and finally, into a soil food web.

Various abiotic and biotic soil characteristics can be used as indicators for evaluating soil health. With increasing heavy metal concentrations, the activities of microbes, soil enzymes, and nitrogen fixation are inhibited, and growth of microfloral communities such as fungi, algae, and photosynthetic bacteria is reduced. In this context, some taxonomic groups of soil invertebrates and higher plants have been proposed as bioindicator organisms. Plants and soil-inhabiting organisms such as soil microflora, and microfauna (protozoa), fungi, nematodes, earthworms, mites, and insects have been used as biotic indicators of soil toxicity, Park et al. (2011).

The microbial community concept is based on the interactions among in all of the various species of bacteria, fungi, protozoa, and microfauna that carry out the various broad level functions of the soil, ranging from nutrient cycling to organic matter formation, and plant disease protection, Crowley (2008). Acute responses of microbial populations to metal toxicities include reduction in species diversity and in biomass of microorganisms, and can result in long term, in reduction of soil quality and primary productivity. Some microbial communities can adapt to metal pollution, but loss of species diversity results in potential loss of biological functions and reduced soil resilience. The extent to which this occurs depends on soil physicochemical properties that affect the bioavailability of metals, Crowley (2008).

There are many ways in which bacteria and microfungi and algae can take up toxic metal ions from soil, Ahmad (2006). Heavy metal ions can be entrapped in the cellular structure and subsequently biosorbed onto the binding sites present in the cellular structure. There is also reported that a cell may develop metal resistance systems in an attempt to protect sensitive cellular components, Fontanetti et al. (2011).

The environmental monitoring of toxic pollutants by bioluminescent bacteria (BLB) in terrestrial environment is reported, Ahmad (2006). The presence of toxicants in the sample reduced light emission of the bioluminescent microorganisms. General reductions in microfungal number have often been noted in soils polluted with Cu, Cd, Pb, As, and Zn, Gadd (1993). The soil protozoa *Tetrahymena pyriformanis* is used for Cu and Zn determination in soil, Ahmad (2006). Other soil protozoa that were used as bioindicators for toxic metals were naked amoebae, testate amoebae, flagellates and sporozoans.

Mosses and also higher fungi have developed accumulation mechanisms with regard to heavy metals from the soil. Their heavy metal uptake is distinctly higher than from atmospheric deposition. Some higher fungi are known to have the ability to accumulate toxic elements such As, Cd or Pb from the environment. According to Kalac (2010), content of many trace elements, especially of Cd and Hg, increases in wild growing edible fungi from polluted areas as compared with those from unpolluted rural sites, but the relationship between substrate contamination with a trace element and its content in wild edible fungi is not tight enough to enable usage of a mushroom species as a reliable bioindicator, Kalac (2010).

Fauna Soil Bioindicators

Much work on invertebrates has focussed on the aquatic environment, where mollusks have been used as effective bioindicators of toxic metals and sea water quality, Markovic et al. (2012), Jovic et al. (2011), Joksimovic and Stankovic (2012). The types of invertebrate soil fauna used in monitoring pollutant effects include nematodes, oligochaetes (ertworms), gastropods, springtails, isopods, arachnids, Park et al. (2011), Sanchez-Hernandez (2006), Nahmani et al. (2006), Hirano and Tamae (2010), Olayinka et al. (2011).

Isopods are more abundant in grasslands and woodlands and accumulate tolerate high levels of Cu, Zn, Pb, and Cd, Nahmani et al. (2006). They can also potentially be used as bioindicators in assessing levels of heavy metals in soil. The soil nematode community has been suggested to be a useful indicator of the status of soils pollutants and ecological disturbance, because of their influence in soil food-webs and plant–soil interactions, Sochova et al. (2006).

The concentrations of heavy metals such as Cr, Cd, Pb, Zn, and Ni influence on the soil nematode community structure, Park et al. (2011). Single nematode candidates have been investigated far more widely, Boyle and Kakouli-Duarte (2008). The nematode *Caenorhabditis elegans* is a widely investigated in many studies demonstrating its usefulness as a test species for toxicity testing. In the United States a guide was accepted describing the use of *C. elegans* in soil toxicity tests (American Society for Testing and Materials 2002) and updated and also published in 2008, Boyle and Kakouli-Duarte (2008).

By far, the most common invertebrate soil bioindicator used to assess soil and contaminant toxicity are members of the Family *Lubricidae* and *Eisenia* spp. earthworms (Anellida, Oligochaeta), Sanchez-Hernandez (2006), Hirano and Tamae (2010). They are important members of the soil fauna. In soils, earthworms play a critical ecological role. For example, it is estimated that under favourable conditions, earthworms can move up to 18 t of soil per acre per year, Olayinka et al. (2011). Earthworms (*E. foetida*) are capable of accumulating Hg, Cd, Cu, Pb, and Zn and a positive correlation exists between Hg levels in worm tissues, the substrate they consume and the length of exposure. Earthworms have been mainly studied in response to Cd, Cu, Pb, and Zn exposure because significant positive correlations have been found between Cd, Cu, Pb, and Zn concentrations in the earthworm and the metals concentrations in soil, Hirano and Tamae (2010), Olayinka et al. (2011).

Earthworms may be available alternative to traditionally applied organisms in aquatic ecosystems (e.g. fish), as they are simple and they can provide indications of bioavailability in a short time at relatively low costs. Earthworms accumulate heavy metals from polluted soil; they ingest large quantities of soil and are in full contact with the substrate they consume. As well, they constitute up to 92 % of the invertebrate biomass of soils and participate in many food chains, acting as a food source for a wide variety of organisms including birds, fish, insects, amphibians, various mammals, and reptiles, and they can be used to assess bioavailability and to estimate food web transfer and impacts. In addition, they are easily bred, have been extensively studied, and are approved for use in toxicity testing by the US EPA and the European Economic Community and the Organization for Economic Cooperation and Development.

The effectiveness of invertebrates as bioindicators has been demonstrated using a range of soil invertebrates, Nahmani et al. (2006). These include families and ecological groups of soil macrofauna chilopoda, diplopoda, dermaptera, isopods, earthworms, gastropoda, but most work has focussed on earthworms or insects, Hall et al. (2008). In the terrestrial trophic systems, invertebrates are of great ecological importance in terms of pollution transfer. Beside aquatic and marine gastropods, terrestrial gastropods (snails) are recognized as adequate bioindicators, because of their ability to accumulate Pb, Zn, Cu, and Cd, Madoz-Escande and Simon (2006). They also exhibit a very wide distribution of a limited number of species, which would allow the accurate comparison of data from a variety of habitats. For example *Deroceras reticulatum* species is known to be found across much of North America, Europe, North Africa and Atlantic islands, Hall et al. (2008). Snails are known to play an important role in the diet of many species, including snakes, toads, beetles and birds.

In the case of vertebrates, in general only a few vertebrate species spend the majority of their time throughout the year in close contact with the soil ecosystem. Examples would include some small mammals, Hubbart (2012), such as ground squirrels, and some larger mammals, such as fox, and several species of snakes and amphibians, Alleva et al. (2006), Kalisinska et al. (2012). California ground squirrels (*Spermophilus beecheyi*) are ideal mammals for such studies, Hubbart (2012). Red fox is a species with a wide geographical range, occurring for the entire year in a small home range. It has a high position in the trophic pyramid and accumulates various toxic metals, including Hg. Taking into account Kalisinska et al. (2012) results and findings of other authors, Alleva et al. (2006), it may be argued that the fox exhibits a measurable response to environmental Hg pollution and meets the requirements for bioindicators not only between European countries but entire continents, e.g. Europe and North America.

Exposure among terrestrial vertebrates occurs through ingestion of contaminated biotic or abiotic matter, contaminant absorption through skin, or via inhalation of volatile, aerosolized, or particle bound contaminants. Terrestrial organisms consume a wide array of vegetative and animal matter, and in many cases free-standing water. In most cases, exposure of wild mammals to contaminants is likely through oral consumption, either by ingesting contaminants incorporated into dietary food and water. Dietary exposure in mammals is a function of age, sex, and season, with mammalian dietary range from pure herbivory to an exclusive carnivorous diet, with virtually all gradients in between, also influence contaminant exposure, Smith et al. (2007).

Nahmani et al. (2006) were investigated grasslands, forests, and cultivated lands using soil macrofauna at different taxonomic levels: species, families and ecological groups. They come to conclusion that the species level is the most accurate taxonomic level in bioindication studies. But it is difficult to use animals in environmental studies, because of their mobile lifestyle, especially when analyzing the effects of chronic exposure.

Flora Soil Bioindicators

As a result of their sedentary nature, plants have evolved diverse abilities for dealing with toxic compounds in their environment. Plants act as solar-driven pumping and filtering systems as they take up contaminants mainly water soluble, through their roots and transport/translocate them through various plant tissues where they can be metabolized, sequestered, or volatilized, Abhilash et al. (2009). Plants utilize different types of mechanisms for dealing with environmental pollutants in soils. Variety of contaminant-degrading enzymes can be found in plants, Singer et al. (2004). Long-term exposure to soil contaminants has a strong impact on the physiological composition of plant populations with many evidences of toxic metals pollutions.

Investigations of the content of heavy metals in soil are mainly focused on heavily urbanized areas such as industrial regions and city agglomerations, as well as on the areas of constant and linear emitters, which include industrial plants, waste landfills and roads. The investigations also concern the soils used for agricultural purposes, Szyczewski et al. (2009) and the results confirm degradation of urban and industrial areas where the concentrations of the studied elements are usually high, Szyczewski et al. (2009).

Most heavy metals accumulate in the top soil and in the long term their contaminations increase in the soil as a result of an increased absorption and accumulation in plants. There is a potential of using plants in the monitoring of heavy metals in the environment on the basis of a significant correlation between heavy metal content of foliage and soil, Wisłocka et al. (2006). Chemical composition of foliage varies with season and rainfall.

Plants require for their development not only macronutrients (N, P, K, S, Ca, and Mg), but also essential micronutrients such as Fe, Zn, Mn, Ni, Cu, and Mo. Most of these micronutrients accumulate in the plant tissues for their metabolic needs, but they never exceed 10 mg kg^{-1}. Yildiz et al. (2010) have reported the normal natural concentration intervals for toxic metals in terrestrial plants as Cd: 0.2–2.4 μg/g, Ni: 1–5 μg/g, Zn: 20–400 μg/g, Fe: 70–700 μg/g, Pb: 1–13 μg/g, Mn: 20–700 μg/g. They found that the level of accumulation in the high plant sample was soil-oriented. Higher levels of these metals and other heavy metals in plant tissue might have phytotoxic effects, sometimes resulting in death, Winkelmann (2005).

Grass has been used as a bioindicator of both: atmospheric metal deposition and metal levels in soils, Baslar et al. (2009), Suchara et al. (2011). They assume that grass will for Pb predominantly reflect atmospheric Pb input because the uptake of other toxic metals is strongly regulate via the root system. According to Brun et al. (2010) the relative importance of atmospheric input versus root uptake differs widely.

Depend upon the type of the tree, the deployment of heavy metal content and ways of accumulation show a great variety. For example, in some trees heavy metals are filtered out by the leaves from the air, while in others they are taken up by their crown or by their roots. According to Baslar et al. (2009) Mn or Zn are taken mainly up by the root system and are then transported from there to the crowns, since they are highly mobile elements. Therefore, it is better to investigate its deposition in the crowns.

The observed elevated levels of heavy metals in the soils are reflected by the high content of Fe, Zn, Pb, and Cd in the leaves of *B. pendula*, Zn, Pb, Cd, and Ni in *S. caprea* and Pb and Cd in *R. idaeus*, Wisłocka et al. (2006). The highest amounts of heavy metals were found in the root of the sensitive *Plantago major* and less in the resistant *Rumex acetosella*, Filipovic-Trajkovic et al. (2012). Thus, high levels of heavy metals in the soil do not always indicate similar high concentrations in plants. Brej (1998) showed that in a polluted soil most seeds are not sensitive to heavy metals occurring in the study area (Cu, Zn, Pb, Cd, and Ni). Also the root/shoot ratio is determinant in the plant tolerance to soil pollutants.

Metallophytes (chalcophytes) are characteristic of sites with heavy metal concentrations in soil above the normal background values of the micronutrients Fe, Mn, Zn, Cu, Co, Mo, Ni, V, and the trace metals Cd, As, U, Pb, Tl, Cr, and Hg. A classic Zn indicator is *Viola calaminaria* which occurs in association with the heavy metal ecotypes of *M. verna* and *S. cucubalus*. *B. pendula* and *S. caprea* have proved to be good bioindicators of Pb and Cd in contaminated soil, Wisłocka et al. (2006) and *Lolium* ssp. and *Sinapis* ssp. are considered to be suitable Pt bioindicators, especially because of their ability to grow along roads, Babula et al. (2008).

In an investigation of Cd, Cu, Ni, and Pb uptake from air and soil by *Achillea millefolium* (milfoil) and *Hordeum vulgare* (barley) in Denmark, Cu and Pb plant concentrations correlated with aerial deposition but not with soil concentrations; in contrast, Ni and Cd content in the plants correlated with aerial deposition and soil content. Contrary to Pb, Zn is a micro plant nutrient and Zn concentrations will thus be much more influenced by soil uptake into the higher plants than Pb values, Suchara et al. (2011).

Vascular plants have many species-specific mechanisms to restrict the cellular uptake of heavy metals and to detoxify them internally. The symptoms of reduced root growth, reduced seed sprouting and seedling stunting, necrosis, and chlorosis appear in susceptible plants grown in soils contaminated with heavy metals, Park et al. (2011). The use of physiological and biochemical plant parameters in the bioindication of heavy metal soil contamination is not routinely practiced yet. Some responses of higher plants to soil heavy metal contamination have potential.

5.6.4 Metal Bioindicators for Humans

For human contaminants, the approach involve first examining contaminants in all the media which form pathways for human exposure – food, drinking water, air and soil, and then setting out data on contaminants in different human tissues: adipose, blood, urine, breast milk, hair, nails, and so on. There has been a huge interest in developing alternative analytical methods to monitor the presence and levels of trace elements in children, apart from invasive blood method, especially in developing countries, Cole et al. (1998). Low cost and easy sample collection methods can be done through hair, nails, exhaled air, serum, placenta, breast milk, urine and saliva, Mehra and Juneja (2004), Khuder et al. (2007), Hussein et al. (2008), Dastych et al.

(2008), Esteban and Castano (2009), Priya and Geetha (2010), Ayodele and Bayero (2010), Ogboko (2011), Hubbart (2012), Gundacker and Hengstschläger (2012).

The use of hair as an indicator is not new. Over 200 years ago, hair was analyzed to measure As levels in the body, Hubbart (2012). Hair is a complex matrix, formed over a period of time, consisting of high levels of both organic and inorganic materials. Hair can provide a more permanent record of trace element associated with normal and abnormal metabolism, as well as trace element assimilated from the environment, Hashem and Abed (2007). Further, hair is easily collected, and may better reflect the total body pool of some elements, than either blood or urine as short-term indicators, Hubbart (2012). Hair retains an incorporated substance for long periods, since there is no active metabolism/excretion to remove the substance once deposited, Hubbart (2012). For example, hair is a long-term exposure bioidicator to MeHg: once Hg is incorporated into the hair, it remains unchanged, WHO (2007). One approach to assessing the status of environmental health is to compare the levels of trace element concentrations of hair in populations living in areas with toxic metal pollution of soil, water or air, Hubbart (2012).

The placenta can be used to detect the presence of As, Hg or Pb, and also urine, finger/toenails, and human milk have been repeatedly used as indicators of toxic metals exposure, Smolders et al. (2009). The heavy metals Hg, Pb, and Cd are toxicants that are well-known to cross the placenta and to accumulate in fetal tissues, Gundacker and Hengstschläger (2012). Prenatal exposure to Hg and Pb poses a health threat particularly to the developing brain. Fetal exposures to Pb and Cd correlate with reduced birth weight and birth size. The placental passage of Cd is limited suggesting a partial barrier for this metal. It is very likely that metallothionein is responsible for placental storage of the metals, especially of Cd, Gundacker and Hengstschläger (2012).

Unlike the urine, blood, placenta, and human milk, hair and fingernails can record the level and changes of elements in the body over a long period of time, Ayodele and Bayero (2010). Mehra and Juneja (2004, 2005) suggested that hair and nail samples provide reliable indications of exposure to many toxic metals over periods of a year or more, and can be used in epidemiological studies related to environmental pollution.

Humans can be directly affected by air, water and soil metal pollutants, as well as indirectly, through contaminated food supplies. In recent past, a large number of silent epidemics have been reported due to metal contamination, which is escalating day by day. Even metals that are biologically essential have the potential to be harmful to humans and other living organisms at high levels of exposure, Valavanidis et al. (2006), Freisinger (2010), Jovic et al. (2012). An increased concentration of heavy metals in biotic and abiotic environment can affect mineral and enzyme status of human beings, Nordberg and Nordberg (2009). The metals irreversibly bind to active sites of enzymes, thereby destroying normal metabolism by producing high-level toxicity, Valavanidis et al. (2006). The excessive content of heavy metals in humans' body may affect the body and psychophysical development, Szyczewski et al. (2009).

5.7 Conclusion

Toxic metals contamination is a great concern at global, regional and local level and influencing the functional and structural integrity of an ecosystem. Toxic metals or "heavy metals" are individual metals and metal compounds that negatively affect living organisms. In very small amounts, many of metals are necessary to support life. However, in larger amounts, they become toxic. They may build up in biological systems and become a significant health hazard.

There are hundreds of sources of metal pollution, including natural and much more anthropogenic sources: industries, mines, agricultural, domestic, and urban pollution. They are emitted to atmosphere, soil, fresh and marine waters via a number of pathways and can have detrimental effects on biota and man. Humans can be exposed to heavy metals through the ingestion of contaminated drinking water, of contaminated food and throughout a food web. Therefore, the content of heavy metals needs to be known not only in water, air, soil/sediment, but also in biological samples: plants, animals, and on the end, in the human population.

Toxic metals introduced into the environment can penetrate ecosystems and can be found in the whole biosphere causing changes in the functions of organisms. Adverse effects of toxic metals on living organisms can be observed. In the last few decades investigations have focused on searching for bioindicators like microorganisms, plants and animals that accumulate toxic metals, even a man.

Biondicators represent organisms that may indicate the levels of environmental pollution. Bioindication is the use of an organism, a part of an organism or a society of organisms, to obtain information on the quality of its environment. It is important to highlight that the choice of the bioindicator organism is essential to the success of environmental monitoring. Thus, the use of bioindicators should help to describe the natural environment, to detect and assess human impacts. Considering all the results presented here, it can be concluded that Zn, Cu, Pb, Cd, and Hg are the most intensively investigated elements in presented bioidicators of air, water and soil. Their concentrations were depending on the investigated species and environmental compartment.

The largest number of bacteria, microrganisms, plant and animal species are used as bioindicators of heavy metals in the aquatic ecosystems. The lowest number is used for the soil contamination. Mosses are particularly effective bioindicators of aerial heavy metal contamination. Analysis of mosses is currently used in international and national monitoring programmes, particularly in Europe. Mosses mainly applied to measure heavy metal levels and trends within and around urban and industrial areas. Higher plants have appeal as indicators in air pollution monitoring in highly polluted areas where mosses are often absent. The use of plants in the bioindication of heavy metal contamination is not routinely practiced. Some responses of higher plants as bioindicators of soil contaminations to heavy metals have potential.

Without question, economic growth and social development are critical for improving environment, human health and well-being. The toxic metals are widely used and consequently distributed in the environment and their early identification is fundamental to prevent or to control possible damages to the environment and humans. Because of that, a wide range of legislations exists in Europe and other countries related to the metal release and impact on the environment, including air, water and soil, and legislations related to toxic metals levels in a food.

Acknowledgement This research was financed by the Ministry of Science and Technological Development of the Republic of Serbia, Contract No. III43009.

References

Abhilash PC, Jamil S, Singh N (2009) Transgenic plants for enhanced biodegradation and phytoremediation of organic xenobiotics. Biotechnol Adv 27:474–488

Aboal RJ, Fernández AJ, Boquete T, Carballeira A (2010) Is it possible to estimate atmospheric deposition of heavy metals by analysis of terrestrial mosses? Sci Total Environ 408:6291–6297

Adham GK, Al-Eisa AN, Farhood HM (2011) Risk assessment of heavy metal contamination in soil and wild Libyan jird *Meriones libycus* in Riyadh, Saudi Arabia. J Environ Biol 32:813–819

Agusa T, Nomura K, Kunito T, Anan Y, Iwata H, Tanabe S (2011) Accumulation of trace elements in harp seals (*Phoca groenlandica*) from Pangnirtung in the Baffin Island, Canada. Mar Pollut Bull 63:489–499

Ahmad BF (2006) Use of microorganisms as bioindicators for detection of heavy metals. M.Sc. thesis, Universiti Putra, Malaysia

Akcali I, Kucuksezgin F (2011) A biomonitoring study: heavy metals in macroalgae from eastern Aegean coastal areas. Mar Pollut Bull 62:637–645

Alleva E, Francia N, Pandolfi M, De Marinis AM, Chiarotti F, Santucci D (2006) Organochlorine and heavy-metal contaminants in wild mammals and birds of Urbino-Pesaro province, Italy: an analytic overview for potential bioindicators. Arch Environ Contam Toxicol 51:123–134

APHA (1989) American Public Health Association. Standard methods for the examination of water and wastewater. In: Clesceri LS, Greenberg AE, Trussell RR (eds) American Water Works Association and Water Pollution Control Federation, 17th edn. Washington, DC

Avgın SS, Luff ML (2010) Ground beetles (coleoptera: carabidae) as bioindicators of human impact. Mun Entomol Zool 5(1):209–215

Ayodele JT, Bayero AS (2010) Manganese concentrations in hair and fingernail of some Kano inhabitants. J Appl Sci Environ Manage 14(1):17–21

Azevedo R, Rodriguez E (2012) Phytotoxicity of mercury in plants: a review. J Botany: 6 Article ID 848614. doi:10.1155/2012/848614

Azizi NS, Colagar HA, Hafeziyan MS (2012) Removal of Cd (II) from aquatic system using *Oscillatoria* sp. biosorbent. The Sci World J, Article ID 347053, 7 pp. doi:10.1100/2012/347053

Babula P, Adam V, Opatrilova R, Zehnalek J, Havel L, Kizek R (2008) Uncommon heavy metals, metalloids and their plant toxicity: a review. Environ Chem Lett 6:189–213

Backor M, Loppi S (2009) Interactions of lichens with heavy metals – a review. Biol Plant 53:214–222

Bardi U (2010) Extracting minerals from seawater: an energy analysis. Sustainability 2:980–992. doi:10.3390/su2040980

Baslar S, Dogan Y, Durkan N, Bag H (2009) Biomonitoring of zinc and manganese in bark of Turkish red pine of western Anatolia. J Environ Biol 30(5):831–834

Basu N, Head J (2010) Mammalian wildlife as complementary models in environmental neurotoxicology. Neurotoxicol Teratol 32:114–119

Basu N, Scheuhammer MA, Bursian JS, Elliott J, Rouvinen-Watt K, Chan MH (2007) Mink as a sentinel species in environmental health. Environ Res 103:130–144

Bellante A, Sprovieri M, Buscaino G, Buffa G, Di Stefano V, Manta SD, Barra M, Filiciotto F, Bonanno A, Mazzola S (2012) Distribution of Cd and As in organs and tissues of four marine mammal species stranded along the Italian coasts. J Environ Monit 14:2382–2391

Bellotto VR, Miekeley N (2007) Trace metals in mussel shells and corresponding soft tissue samples: a validation experiment for the use of Perna perna shells in pollution monitoring. Anal Bioanal Chem 389(3):769–776

Bencko V (1997) Health aspects of burning coal with a high arsenic content: the central Slovakia experience. In: Abernathy CO, Calderon RL, Chapell WR (eds) Arsenic exposure and health effects. Chapman & Hall, London, pp 84–92

Berglund MMA, Koivula JM, Eeva T (2011) Species- and age-related variation in metal exposure and accumulation of two passerine bird species. Environ Pollut 159:2368–2374

Bhattacharya P, Samal CA, Majumdar J, Santra CS (2010) Arsenic contamination in rice, wheat, pulses, and vegetables: a study in an arsenic affected area of West Bengal, India. Water Air Soil Pollut 213:3–13

Blagnytė R, Paliulis D (2010) Research into heavy metals pollution of atmosphere applying moss as bioindicator: a literature review. Environ Res Eng Manage 4(54):26–33, 2029–2139

Boyle S, Kakouli-Duarte T (2008) Soils and bioindicators – the development of the nematode steinernema feltiae as a bioindicator for chromium VI soil pollution (2008-FS-28-M1). EPA STRIVE Programme 2007–2013

Brej T (1998) Heavy metal tolerance in *Agropyron repens* (L.) P. Bauv. Populations from the Legnica copper smelter area, Lower Silesia. Acta Soc Bot Pol 67:325–333

Brun CB, Peltola P, Aaström ME, Johansson M-B (2010) Spatial distribution of major, trace and ultra trace elements in three Norway spruce (*Picea abies*) stands in boreal forests Forsmark. Sweden Geoderma 159:252–261

Burger J (2006) Bioindicators: a review of their use in the environmental literature 1970–2005. Environ Bioind 1:136–144

Burger J, Gochfeld M (2004) Marine birds as sentinels of environmental pollution. Ecohealth 1:263–274. doi:10.1007/s10393-004-0096-4

Burger J, Campbell KR, Campbell TS, Shukla T, Jeitner C, Gochfeld M (2005) Use of skin and blood as nonlethal indicators of heavy metal contamination in northern water snakes (*Nerodia sipedon*). Environ Monit Assess 49:232–238

Burnat M, Diestra E, Esteve I, Solé A (2010) Confocal laser scanning microscopy coupled to a spectrofluorometric detector as a rapid tool for determining the in vivo effect of metals on phototrophic bacteria. Bull Environ Contam Toxicol 84(1):55–60

CAOBISCO (1996) Heavy metals: 1–11. http://www.caobisco.com/doc_uploads/nutritional_factsheets/metals.pdf. Accessed 18 June 2012

Chen YC, Serrell N, Evers CD, Fleishman JB, Lambert FK, Weiss J, Robert P, Mason PR, Bank SM (2008) Methylmercury in marine ecosystems – from sources to seafood consumers. Environ Health Perspect 116(12):1706–1712

Chovanec A, Rudolf Hofer R, Schiemer F (2003) Fish as bioindicators. In: Markert BA, Breure AM, Zechmeister HG (eds) Bioindicators and biomonitors. Elsevier Science Ltd, Bostan, pp 639–676

Cobbett C, Goldsbrough P (2002) Phytochelatins and metallothioneins: roles in heavy metal detoxification and homeostasis. Annu Rev Plant Biol 53:159–182

Cole CD, Eyles J, Gibson LB (1998) Indicators of human health in ecosystems: what do we measure? Sci Total Environ 224:201–213

COM (2004) Commission of the European Communities: communication from the Commission to the Council, the European Parliament, the European Economic and Social Committee. "The European environment & health action plan 2004–2010". Brussels 9 June 2004. COM (2004) Volume I and II, Brussels 2004:8; 2004:22

Conti ME, Cecchetti G (2001) Biological monitoring: lichens as bioindicators of air pollution assessment – a review. Environ Pollut 114:471–492

Copat C, Maggiore R, Arena G, Lanzafame S, Fallico R, Sciacca S, Ferrante M (2012) Evaluation of a temporal trend heavy metals contamination in *Posidonia oceanica* (L.) Delile, (1813) along the western coastline of Sicily (Italy). J Environ Monit 14:187–192

Crowley D (2008) Impacts of metals and metalloids on soil microbial diversity and ecosystem function. In: The 5th international symposium ISMOM, Pucón, 24–28th Nov 2008, Plenary paper, pp 1–7

Cuypers A, Vangronsveld J, Clijsters H (2002) Peroxidases in roots and primary leaves of *Phaseolus vulgaris* copper and zinc phytotoxicity: a comparison. J Plant Physiol 159:869–876

Dallinger R, Berger B, Gruber C, Hunziker P, Sturzenbaum S (2000) Metallothioneins in terrestrial invertebrates: structural aspects, biological significance and implications for their use as biomarkers. Cell Mol Biol 46:331–346

Das KB, Roy A, Koschorreck M, Mandal MS, Wendt-Potthoff K, Bhattacharya J (2009) Occurrence and role of algae and fungi in acid mine drainage environment with special reference to metals and sulfate immobilization. Water Res 43:883–894

Dastych M, Cienciala J, Krbec M (2008) Changes of selenium, copper, and zinc content in hair and serum of patients with idiopathic scoliosis. J Orthop Res 26(9):1279–1282

Deng H, Zhang Z, Chang C, Wang Y (2007) Trace metal concentration in Great Tit (Parus major) and Greenfinch (Carduelis sinica) at the Western Mountains of Beijing, China. Environ Pollut 148:620–626

Dmuchowski W, Bytnerowicz A (2009) Long-term (1992–2004) record of lead, cadmium, and zinc air contamination in Warsaw, Poland: determination by chemical analysis of moss bags and leaves of Crimean linden. Environ Pollut 157:3413–3421

Dondero F, Banni M, Negri A, Boatti L, Dagnino A, Viarengo A (2011) Interactions of a pesticide/heavy metal mixture in marine bivalves: a transcriptomic assessment. BMC Genomics 12:195–212

Dragovič S, Mihailovič N (2009) Analysis of mosses and topsoils for detecting sources of heavy metal pollution: multivariate and enrichment factor analysis. Environ Monit Assess 157:383–390

Duarte AC, Giarratano E, Amin AO, Comoglio IL (2011) Heavy metal concentrations and biomarkers of oxidative stress in native mussels (*Mytilus edulis chilensis*) from Beagle Channel coast (Tierra del Fuego, Argentina). Mar Pollut Bull 62:1895–1904

Duffus HJ (2002) "Heavy metals"a meaningless term? (IUPAC Technical Report). Pure Appl Chem 74:793–807

EC (2006) Commission regulation (EC) No. 188/2006. Off J Eur Union 31(11)

EEA (2003) Hazardous substances in the European marine environment: trends in metals and persistent organic pollutants. Topic report 2/2003

EEC (1979) Council Directive 79/409/EEC of 2 April 1979 on the conservation of wild birds. http://eur-lex.europa.eu/LexUriServ/site/en/consleg/1979/L/01979L0409-20070101-en.pdf

EEC (1990) Council Directive 90/313/EEC of 7 June 1990 on the freedom of access to information on the environment. Off J L 158, 23 June 1990

EEC (2000) Concil Directive 2000/14/EC of the European parliament and of the Council of 8 May 2000. Off J Eur Commun L 162(1), 3 July 2000

EEC (2004) Concil Directive 2004/107/ec of the European parliament and of the Council of 15 December 2004 relating to arsenic, cadmium, mercury, nickel and polycyclic aromatic hydrocarbons in ambient air. Off J Eur Union L 23(3), 26 Jan 2005

EPA TRI (2002–2005) USEPA toxic release inventory, U.S. Environmental Protection Agency, Washington, DC, 2000, 2001, 2003, 2003, 2004, 2005. www.epa.gov/tri

Esteban M, Castano A (2009) Non-invasive matrices in human biomonitoring: a review. Environ Int 35:438–449

FAO/WHO (2004) Summary of evaluations performed by the joint FAO/WHO expert committee on food additives (JECFA 1956–2003). ILSI Press International Life Sciences Institute, Washington, DC

FAO/WHO (2007) Summary of evaluations performed by the joint FAO/WHO expert committee on food additives (JECFA 1956–2007) (first through 68th meetings). Food and agriculture organization of the United Nations and the World Health Organization. ILSI Press International Life Sciences Institute, Washington, DC

FAO/WHO (2010) Joint FAO/WHO expert committee on food additives. Seventy-third meeting, Geneva, 8–17 June 2010

Faria M, López AM, Díez S, Barata C (2010) Are native naiads more tolerant to pollution than exotic freshwater bivalve species? An hypothesis tested using physiological responses of three species transplanted to mercury contaminated sites in the Ebro River (NE, Spain). Chemosphere 81:1218–1226

Fatima AR, Ahmad M (2005) Certain antioxidant enzymes of *Allium cepa* as biomarkers for the detection of toxic heavy metals in wastewater. Sci Total Environ 346:256–273

Filipovic-Trajkovic R, Ilic SZ, Lj S, Andjelkovic S (2012) The potential of different plant species for heavy metals accumulation and distribution. J Food Agric Environ 10(1):959–964

Fontanetti SC, Nogarol RL, de Souza BR, Perez GD, Maziviero TG (2011) Bioindicators and biomarkers in the assessment of soil toxicity. In: Pascucci S (ed) Soil contamination, InTech Rijeka, Croatia, pp 143–168, ISBN 978-953-307-647-8

Freisinger E (2010) The metal-thiolate clusters of plant metallothioneins. Chimia 64(4):216–224. doi:10.2533/chimia.2010.217

Fritsch C, Coeurdassier M, Giraudoux P, Raoul F, Douay F et al (2011) Spatially explicit analysis of metal transfer to biota: influence of soil contamination and landscape. PLoS One 6(5):e20682. doi:10.1371/journal.pone.0020682

Frontalini F, Coccioni R (2011) Benthic foraminifera as bioindicators of pollution: a review of Italian research over the last three decades. Rev Micropaléontol 54:115–127

Gadd MG (1993) Interactions of fungy with toxic metals. New Phytol 124:25–60

Garcia SJ, Souza FMHG, Eberlin NM, Arruda ZAM (2008) Evaluation of metal-ion stress in sunflower (*Helianthus annuus L.*) leaves through proteomic changes. Metallomics 1:107–113

Gawryluk JW, Wang JF, Andreazza AC, Shao L, Young LT (2011) Decreased levels of glutathione, the major brain antioxidant, in post-mortem prefrontal cortex from patients with psychiatric disorders. Int J Neuropsychopharmacol (CINP) 14(1):123–130

Giarratano E, Gil MN, Malanga G (2011) Seasonal and pollution-induced variations in biomarkers of transplantedmussels within the Beagle Channel. Mar Pollut Bull 62:1337–1344

Goldberg ED (1975) The mussel watch. A first step in global marine monitoring. Mar Pollut Bull 6(7):111–111. doi:10.1016/0025-326X(75)90271-4

Gundacker C, Hengstschläger M (2012) The role of the placenta in fetal exposure to heavy metals. Wien Med Wochenschr 162(9–10):201–206

Hall MC, Rhind MS, Wilson JM (2008) The potential for use of gastropod mollusks as bioindicators of endocrine disrupting compounds in the terrestrial environment. J Environ Monit 11:491–497

Hargreaves LA, Whiteside PD, Grant Gilchrist G (2011) Concentrations of 17 elements, including mercury, in the tissues, food and abiotic environment of Arctic shorebirds. Sci Total Environ 409:3757–3770

Harmens H, Norris DA, Koerber GR, Buse A, Steinnes E, Ruhling A (2008) Temporal trends (1990–2000) in the concentration of cadmium, lead and mercury in mosses across Europe. Environ Pollut 151:368–376

Hashem RA, Abed FK (2007) Aluminum, cadmium and microorganisms in female hair and nails from Riyadh. Saudi Arabia J Med Sci 7(2):263–266

Hauser-Davis AR, de Campos CR, Ziolli LR (2012) Fish metalloproteins as biomarkers of environmental contamination. In: Whitacre DM (ed) Reviews of environmental contamination and toxicology, vol 218, Reviews of environmental contamination and toxicology. Springer, New York. doi:10.1007/978-1-4614-3137-4_2

Hegelund NJ, Schiller M, Kichey T, Hansen HT, Pedas P, Husted S, Schjoerring KJ (2012) Barley metallothioneins. Plant Physiol Preview. doi:10.1104/pp.112.197798

Hirano T, Tamae K (2010) Heavy metal-induced oxidative DNA damage in earthworms: a review. Appl Environ Soil Sci. doi:10.1155/2010/726946

Hodkinson ID, Jackson JK (2005) Terrestrial and aquatic invertebrates as bioindicators for environmental monitoring, with particular reference to mountain ecosystems. Environ Manage 35(5):649–666

Hogan MC (2010) Heavy metal. In: Monosson E, Cleveland C (eds) Encyclopedia of earth. National Council for Science and the Environment. Washington, DC

Holmstrup M, Bindesbøl A-M, Oostingh JG, Duschl A, Scheil V, Köhler R-H, Loureiro S, Soares MVMA, Ferreira GLA, Kienle C, Gerhardt A, Laskowski R, Kramarz EP, Bayley M, Svendsen C, Spurgeon JD (2010) Interactions between effects of environmental chemicals and natural stressors: a review. Sci Total Environ 408:3746–3762

Holt EA, Miller SW (2011) Bioindicators: using organisms to measure environmental impacts. Nat Educ Knowl 2(2):8

Horai S, Watanabe I, Takada H, Iwamizu Y, Hayashi T, Tanabe S, Kuno K (2007) Trace element accumulations in 13 avian species collected from the Kanto area, Japan. Sci Total Environ 373:512–525

Hossain AM, Piyatida P, Teixeira da Silva AJ, Fujita M (2012) Molecular mechanism of heavy metal toxicity and tolerance in plants: central role of glutathione in detoxification of reactive oxygen species and methylglyoxal and in heavy metal chelation. J Bot, Article ID 872875. doi:10.1155/2012/872875

Hu H (2002) Human health and heavy metals exposure. In: McCally M (ed) Life support: the environment and human health. MIT Press, Cambridge, pp 65–82

Hubbart AJ (2012) Hair analysis as an environmental health bioindicator: a case-study using pelage of the California ground squirrel (*Spermophilus beecheyi*). Int J Appl Sci Technol 2:227–294

Hussein WF, Njue W, Murungi J, Wanjau R (2008) Use of human nails as bio-indicators of heavy metals environmental exposure among school age children in Kenya. Sci Total Environ 393(2–3):376–384

Jackson AV, Paulse NA, Odendaal PJ, Khan S, Khan W (2012) Identification of metal-tolerant organisms isolated from the Plankenburg River, Western Cape, South Africa. Water SA 38(1):29–38, http://dx.doi.org/10.4314/wsa.v38i1.5

Jakimska A, Konieczka P, Skora K, Namiesnik J (2011) Bioakumulation of metals in tissues of marine animals, part II: metal concentrations in animal tissues. Pol J Environ Stud 20(5):1127–1146

Javed MT, Greger M (2011) Cadmium triggers Elodea Canadensis to change the surrounding water pH and thereby Cd uptake. Int J Phytoremediation 13:95–106

Johnson A, Singhal N, Hashmatt M (2011) Metal–plant interactions: toxicity and tolerance. Environ Pollut 20:29–63

Joksimović D, Stanković S (2012) The trace metal accumulation in marine organisms of the southeastern Adriatic coast, Montenegro. J Serb Chem Soc 77(1):105–117

Joksimović D, Tomić I, Stanković RA, Jović M, Stanković S (2011a) Trace metal concentrations in Mediterranean blue mussel and surface sediments and evaluation of the mussels quality and possible risks of high human consumption. Food Chem 127:632–637

Joksimović D, Stanković RA, Stanković S (2011b) Metal accumulation in the biological indicator *(Posidonia oceanica)* from the Montenegrin coast. Stud Mar 25(1):37–58

Joshi RS (2008) Influence of roadside pollution on the phylloplane microbial community of *Alnus nepalensis* (Betulaceae). Rev Biol Trop 56(3):1521–1529

Jovic M, Stankovic RA, Slavković Beskoski L, Tomic I, Degetto S, Stankovic S (2011) The environmental quality of the coastal water of the Boka Kotorska Bay (Montenegro) using mussels as a bioindicator. J Serb Chem Soc 76(6):933–946

Jovic M, Onjia A, Stankovic S (2012) Toxic metal health risk by mussel consumption. Environ Chem Lett 1:69–77

Kabata-Pendias A, Pendias H (1992) Trace elements in soils and plants, 2nd edn. CRC Press, Boca Ratón

Kakkar P, Jaffery FN (2005) Biological markers for metal toxicity. J Environ Toxicol Pharmacol 19:335–349
Kakuschke A, Gandrass J, Luzardo PO, Boada DL, Zaccaroni A, Griesel S, Grebe M, Profrock D, Erbsloeh H-B, Valentine-Thon E, Prange A, Kramer K (2012). Postmortem health and pollution investigations on harbor seals (*Phoca vitulina*) of the islands Helgoland and Sylt. Int Sch Res Netw ISRN Zool Article ID 106259: 1–8. doi:10.5402/2012/106259
Kalac P (2010) Trace element contents in European species of wild growing edible mushrooms: a review for the period 2000–2009. Food Chem 122:2–15
Kalisinska E, Lisowski P, Kosik-Bogacka ID (2012) Red fox *Vulpes vulpes* (L., 1758) as a bioindicator of mercury contamination in terrestrial ecosystems of North-Western Poland. Biol Trace Elem Res 145:172–180
Kandemir S, Dogru MI, Orun I, Dogru A, Altas L, Erdogan K, Orun G, Polat N (2010) Determination of heavy metal levels, oxidative status, biochemical and hematological parameters in *Cyprinus carpio* L., 1785 from Bafra (Samsun) fish lakes. J Anim Vet Adv 9:617–622
Khuder A, Bakir MA, Hasan R, Mohammad A (2007) Determination of nickel, copper, zinc and lead in human scalp hair in Syrian occupationally exposed workers by total reflection X-ray fluorescence. Environ Monit Assess 143(1–3):67–74
Kitowski L, Kowalski R, Komosa A, Lechowski J, Grzywaczewski G, Scibior R, Pilucha G, Chrapowicki M (2012) Diversity of total mercury concentrations in kidneys of birds from Eastern Poland. Ekológia (Bratislava) 31(1):12–21
Kolesarova A, Slamecka J, Jurcik R, Tataruch F, Lukac N, Kovacik J, Capcarova M, Valent M, Massanyi P (2008) Environmental levels of cadmium, lead and mercury in brown hares and their relation to blood metabolic parameters. J Environ Sci Health A Tox Hazard Subst Environ Eng 43(6):646–650
Kord B, Mataji A, Babaie S (2010) Pine (*Pinus Eldarica* Medw.) needles as indicator for heavy metals pollution. Int J Environ Sci Technol 7(1):79–84
Krystofova O, Shestivska V, Galiova M, Novotny K, Kaiser J, Zehnalek J, Babula P, Opatrilova R, Adam V, Kizek R (2009) Sunflower plants as bioindicators of environmental pollution with lead (II) ions. Sensors 9:5040–5058
Kumar C, Igbaria A, D'Autreaux B, Planson A-G, Junot C, Godat E, Bachhawat KA, Delaunay-Moisan A, Toledano BM (2011) Glutathione revisited: a vital function in iron metabolism and ancillary role in thiol-redox control. EMBO J 30(10):2044–2056
Lai HY, Juang KW, Chen ZS (2010) Large-area experiment on uptake of metals by twelve plant species growing in soil contaminated with multiple metals. Int J Phytoremediation 12:785–797
Lam SKI, Wang XW (2008) Trace element deficiency in freshwater cladoceran *Daphnia magna*. Aquat Biol 1:217–224
Laskowski R, Bednarska JA, Kramarz EP, Loureiro S, Scheil V, Kudłek J, Holmstrup M (2010) Interactions between toxic chemicals and natural environmental factors – a meta-analysis and case studies. Sci Total Environ 408:3763–3774
Lavery JT, Kemper MC, Sanderson K, Schultz GC, Coyle P, James G, Mitchell GJ, Seuront L (2009) Heavy metal toxicity of kidney and bone tissues in South Australian adult bottlenose dolphins (*Tursiops aduncus*). Mar Environ Res 67:1–7
Law R, Hanke G, Angelidis M, Batty J, Bignert A, Dachs J, Davies I, Denga Y, Duffek A, Herut B, Hylland K, Lepom P, Leonards P, Mehtonen J, Piha H, Roose P, Tronczynski J, Velikova V, Vethaak D (2010) Contaminants and pollution effects. Joint Report EUR 24335 EN – 2010
Luy N, Gobert S, Sartoretto S, Biondo R, Bouquegneau J-M, Richir J (2012) Chemical contamination along the Mediterranean French coast using *Posidonia oceanica* (L.) Delile above-ground tissues: a multiple trace element study. Ecol Indic 18:269–277
Madoz-Escande C, Simon O (2006) Contamination of terrestrial gastropods, Helix aspersa maxima, with ^{137}Cs, ^{85}Sr, ^{133}Ba and ^{123m}Te by direct, trophic and combined pathways. J Environ Radioact 89:30–47
Markert B, Wuenschmann S, Fraenzle S, Figueiredo GMA, Ribeiro PA, Wang M (2011) Bioindication of atmospheric trace metals – with special references to megacities. Environ Pollut 159:1991–1995

Markovic J, Joksimovic D, Stankovic S (2012) Trace elements concentrations determined in collected wild mussels in the coastal area of southeastern Adriatic, Montenegro. Arch Biol Sci Belgrade 64(1):265–275

Marmiroli N, Marmiroli M, Maestri E (2006) Phytoremediation and phytotechnologies: a review for the present and the future. In: Twardowska I, Allen HE, Haggblom MH (eds) Soil and water pollution monitoring, protection and remediation. Springer, Dordrecht

McBride CA, Dale HV, Baskaran ML, Downing EM, Eaton ML, Efroymson AR, Garten TC Jr, Kline LK, Jager IH, Mulholland JP, Parish SE, Schweizer PE, Storey MJ (2011) Indicators to support environmental sustainability of bioenergy systems. Ecol Indic 11:1277–1289

Meharg AA, Hartley-Whitaker J (2002) Arsenic uptake and metabolism in arsenic resistant and nonresistant plant species. New Phytol 154:29–43

Mehra R, Juneja M (2004) Biological monitoring of lead and cadmium in human hair and nail and their correlations with biopsy materials, age and exposure. Indian J Biochem Biophys 41:53–56

Mehra R, Juneja M (2005) Fingernails as biological indices of metal exposure. J Biosci 30(2):253–257

Melaku S, Morris V, Raghavan D, Hosten C (2008) Seasonal variation of heavy metals in ambient air and precipitation at a single site in Washington, DC. Environ Pollut 155:88–98

Minguez L, Boiche A, Sroda S, Mastitsky S, Brule N, Bouquerel J, Giamberini L (2012) Cross-effects of nickel contamination and parasitism on zebra mussel physiology. Ecotoxicology 21:538–547. doi:10.1007/s10646-011-0814-y

Moloukhia H, Sleem S (2011) Bioaccumulation, fate and toxicity of two heavy metals common in industrial wastes in two aquatic mollusks. J Am Sci 7(8):459–464

Morillo J, Usero J, Gracia I (2005) Biomonitoring of trace metals in a minepolluted estuarine system (Spain). Chemosphere 58:1421–1430

Mosher S (2008) Ph.D. theses, North Carolina State University, Raleigh

Murthy S, Bali G, Sarangi SK (2012) Lead biosorption by a bacterium isolated from industrial effluents. Int J Microbiol Res 4(3):196–200, 0975-9174

Musarrat J, Zaidi A, Khan SM, Siddiqui AM, Al-Khedhairy AA (2011) Genotoxicity assessment of heavy metal–contaminated soils. Environ Pollut 20:323–342

Nagajyoti CP, Lee DK, Sreekanth MVT (2010) Heavy metals, occurrence and toxicity for plants: a review. Environ Chem Lett 8:199–216

Nahmani J, Lavelle P, Rossi P-J (2006) Does changing the taxonomical resolution alter the value of soil macroinvertebrates as bioindicators of metal pollution? Soil Biol Biochem 38:385–396

Naveed HN, Batool IA, Ur Rehman F, Hameed U (2010) Leaves of roadside plants as bioindicator of traffic related lead pollution during different seasons in Sargodha, Pakistan. Afr J Environ Sci Technol 4(11):770–774

NCM (2003) Cadmium review, Nordic Council of Ministers

Nordberg M, Nordberg FG (2009) Metallothioneins: historical development and overview. Met Ions Life Sci 5:1–29

Norgate T, Jahanshahi S, Rankin JW (2007) Assessing the environmental impact of metal production processes. J Cleaner Prod 15:838–848

Nriagu OA (1996) A history of global metal pollution. Science 127

OECD (1994) Data requirements for pesticide registration in OECD member countries: survey results. Report. Series on pesticides no. 1, Organisation for Economic Co-operation and Development, Paris

Oehlmann J, Schulte-Oehlmann U (2002) Mollusks as bioindicators. In: Markert BA, Breure AM, Zechmeister HG (eds) bioindicators and biomonitors. Elsevier Science B.V

Ogboko B (2011) Copper and zinc concentration in hair of healthy children in Ceres District of South Africa. J Basic Appl Sci Res 1(8):818–824

Olayinka OT, Idowu AB, Dedeke GA, Akinloye OA, Ademolu KO, Bamgbola AA (2011) Earthworm as bio-indicator of heavy metal pollution around Lafarge, Wapco Cement Factory, Ewekoro, Nigeria. In: Proceedings of the environmental management conference, Federal University of Agriculture, Abeokuta, 2011

OSPAR (1992) The convention for the protection of the marine environment of the North-East Atlantic, OSPAR Convention

OSPAR (2002) Cadmium. Hazardous substances series. OSPAR Commission 2002

Östensson O (2006) Mineral and metals production: an overview. Caromb Consulting, Stockholm

Özden O, Ulusoy S, Erkan N (2010) Study on the behavior of the trace metal and macro minerals in *Mytilus galloprovincialis* as a bioindicator species: the case of Marmara Sea, Turkey. J Verbraucherschutz Lebensmittelsicherheit 5:407–412

Paoli L, Corsini A, Bigagli V, Vannini J, Bruscoli C, Loppi S (2012) Long-term biological monitoring of environmental quality around a solid waste landfill assessed with lichens. Environ Pollut 161:70–75

Park Y-B, Lee K-J, Ro M-H, Kim HJ (2011) Effects of heavy metal contamination from an abandoned mine on nematode community structure as an indicator of soil ecosystem health. Appl Soil Ecol 5:17–24

Peralta-Videa RJ, Lopez LM, Narayan M, Saupe G, Gardea-Torresdey J (2009) The biochemistry of environmental heavy metal uptake by plants: implications for the food chain. IntJ Biochem Cell Biol 41:1665–1677

Peternel R (2007) Biomonitoring of air pollution. Zrak i zdravlje 3(9):7–12, ISSN 1845-3082

Piyatiratitivorakul P, Boonchamoi P (2008) Comparative toxicity of mercury and cadmium to the juvenile freshwater snail, *Filopaludina martensi martensi*. Science Asia 34:367–370

Pokorny B (2000) Roe deer (*Capreolus capreolus*) as an accumulative bioindicator of heavy metals in Slovenia. Web Ecol 1:54–62

Pokorny B (2006) Roe deer (*Capreolus capreolus* L.) antlers as an accumulative and reactive bioindicator of lead pollution near the largest Slovene thermal power plant. Vet Arh 76(Suppl):S131–S142

Pompe-Gotal J, Srebocan E, Gomercic H, Prevendar Crnic A (2009) Mercury concentrations in the tissues of bottlenose dolphins (*Tursiops truncatus*) and striped dolphins (*Stenella coeruloalba*) stranded on the Croatian Adriatic coast. Vet Med 54(12):598–604

Pompella A, Visvikis A, Paolicchi A, De Tata V, Casini AF (2003) The changing faces of glutathione, a cellular protagonist. Biochem Pharmacol 66(8):1499–1503

Priya MDL, Geetha A (2010) Level of trace elements (copper, zinc, magnesium and selenium) and toxic elements (lead and mercury) in the hair and nail of children with autism. Biol Trace Elem Res 142:148–158

Rahnama R, Javanshir A, Mashinchian A (2011) The effect of lead bioaccumulation on condition indices of zebra mussel (*Dreissena polymorpha*) from Anzali wetland-Caspian Sea. Turk J Fish Aquat Sci 11:561–568

Rai N, Pal A (2002) Health hazards of heavy metals. Environ News Newslett ISEB India 8(1)

Ratte TH, Hammers-Wirtz M, Cleuvers M (2003) Ecotoxicity testing. In: Markert AB, Breure MA, Zechmeister GH (eds) Bioindicators and biomonitors, vol 6. Elsevier Science Ltd., Oxford, pp 221–256

Reijnders L, Huijbregts MAJ (2009) Biofuels for road transport: a seed to wheel perspective, Green energy and technology. Springer, London. doi:dx.org/10.1007/978-1-84882-138-5. ISBN 978-1-84882-137-8

Reimann C, Matschullat J, Birke M, Salminen R (2009) Arsenic distribution in the environment: the effects of scale. Appl Geochem 24:1147–1167

Richir J, Gobert S, Sartoretto S, Biondo B, Bouquegneau MJ, Luy N (2010) *Posidonia oceanica* (l.) Delile, a useful tool for the biomonitoring of chemical contamination along the Mediterranean coast: a multiple trace element study. In: Proceedings of the 4th Mediterranean symposium on marine vegetation, Yasmine-Hammamet, December 2010. http://www.rac-spa.org/sites/default/files/doc_vegetation/actes_4ieme_sympos_veg_2010.pdf

Rimmer CC, Miller KE, McFarland PK, Steven D. Faccio DS, Strong BA, Taylor JR (2009) Mercury bioaccumulation in a terrestrial food web of a montane forest. Atmospheric mercury in Vermont and New England. Final Project Report – 1/16/2009 – Terrestrial Food Web – 1–18

Roméo M, Frasila C, Barelli GM, Damiens G, Micu D, Mustata G (2005) Biomonitoring of trace metals in the Black Sea (Romania) using mussels *Mytilus Galloprovincialis*. Water Res 39:596–604

Rudy M (2010) Dependences between the age and the level of bioaccumulation of heavy metals in tissues and the chemical composition of wild boars' meat. Food Addit Contam 27:464–472

Rybak A, Messyasz B, Łeska B (2012) Freshwater *Ulva* (Chlorophyta) as a bioaccumulator of selected heavy metals (Cd, Ni and Pb) and alkaline earth metals (Ca and Mg). Chemosphere 89:1066–1076

Sabovljevic M, Vukojevic V, Sabovljevic A, Vujicic M (2009) Deposition of heavy metals (Pb, Sr and Zn) in the county of Obrenovac (Serbia) using mosses as bioindicators. J Ecol Nat Environ 1(6):147–155

Sanchez-Chardi A, Penarroja-Matutano C, Ribeiro OAC, Nadal J (2007) Bioaccumulation of metals and effects of a landfill in small mammals. Part II. The wood mouse, *Apodemus sylvaticus*. Chemosphere 70:101–109

Sanchez-Hernandez CJ (2006) Earthworm biomarkers in ecological risk assessment. Rev Environ Contam Toxicol 188:85–126

Schaller J, Brackhage C, Mkandawire M, Gert Dudel E (2011) Metal/metalloid accumulation/remobilization during aquatic litter decomposition in freshwater: a review. Sci Total Environ 409:4891–4898

Schwartz G, Eshel G, Ben-Dor E (2011) Reflectance spectroscopy as a tool for monitoring contaminated soils. In: Simone Pascucci (ed) Soil contamination. InTech, Rijeka, ISBN 978-953-307-647-8

Semedo M, Reis-Henriques AM, Rey-Salgueiro L, Oliveira M, Delerue-Matos C, Morais S, Ferreira M (2012) Metal accumulation and oxidative stress biomarkers in octopus (*Octopus vulgaris*) from Northwest Atlantic. Sci Total Environ 433:230–237

Serbula MS, Miljkovic DD, Kovacevic MR, Ilic AA (2012) Assessment of airborne heavy metal pollution using plant parts and topsoil. Ecotoxicol Environ Saf 76:209–214

Sevcikova M, Modra H, Slaninova A, Svobodova Z (2011) Metals as a cause of oxidative stress in fish: a review. Vet Med 56(11):537–546

Shanker KA, Cervantes C, Loza-Tavera H, Avudainayagam S (2005) Chromium toxicity in plants. Environ Int 31:739–753

Shariati F, Shariati S (2011) Review on methods for determination of metallothioneins in aquatic organisms. Biol Trace Elem Res 141:340–366

Siegel FR (2002) Environmental geochemistry of potentially toxic metals. Springer, New York

Singer AC, Thompson IP, Bailey MJ (2004) The tritrophic trinity: a source of pollutant degrading enzymes and its implication for phytoremediation. Curr Opin Microbiol 7:239–244

Singh TY, Krishnamoorthy M, Thippeswamy S (2012) Status of heavy metals in tissues of wedge clam, Donax faba (Bivalvia: Donacidae) collected from the Panambur beach near industrial areas. Recent Res Sci Technol 4(5):30–35, ISSN: 2076-5061 http://recent-science.com/

Smith NP, Cobb PG, Godard-Codding C, Hoff D, McMurry TS, Rainwater RT, Reynolds DK (2007) Contaminant exposure in terrestrial vertebrates. Environ Pollut 150:41–64

Smolders R, Schramm K-W, Nickmilder M, Schoeters G (2009) Applicability of non-invasively collected matrices for human biomonitoring. Environ Health 8:8. doi:10.1186/1476-069X-8-8

Sochova I, Hofman J, Holoubek I (2006) Using nematodes in soil ecotoxicology. Environ Int 32:374–383

Stankovic S, Jovic M (2012) Health risks of heavy metals in the Mediterranean mussels as seafood. Environ Chem Lett 2:119–130

Stankovic S, Jovic M, Stankovic RA, Katsikas L (2011a) Heavy metals in seafood mussels. In: Lichtfouse E et al (eds) Risks for human health in: environmental chemistry for a sustainable world: Volume 1. , Chapter 9, pages 64, pp 311–375, ISBN-10: 9400724411, ISBN-13: 978-94007244119. doi 10.1007/978-94-007-2442-6_9

Stankovic S, Jovic M, Milanov R, Joksimovic D (2011b) Trace elements concentrations (Zn, Cu, Pb, Cd, As and Hg) in the Mediterranean mussel (*Mytilus galloprovincialis*) and evaluation of mussel quality and possible human health risk from cultivated and wild sites of the southeastern Adriatic Sea, Montenegro. J Serb Chem Soc 76(12):1725–1737

Stankovic D, Krstic B, Orlovic S, Trivan G, Poljak PL, Nikolic SM (2011c) Woody plants and herbs as bioindicators of the current condition of the natural environment in Serbia. J Med Plant Res 5(15):3507–3512

Stojanovic DM, Mihajlovic LM, Milojkovic VM, Lopicic RZ, Adamovic M, Stankovic S (2012) Efficient phytoremediation of uranium mine tailings by tobacco. Environ Chem Lett. doi:10.1007/s10311-012-0362-6

Suchara I, Sucharova J, Hola M, Reimann C, Boyd R, Filzmoser P, Englmaier P (2011) The performance of moss, grass, and 1- and 2-year old spruce needles as bioindicators of contamination: a comparative study at the scale of the Czech Republic. Sci Total Environ 409:2281–2297

Sures B (2008) Host–parasite interactions in polluted environments. J Fish Biol 73:2133–2142

Szyczewski P, Siepak J, Niedzielski P, Sobczyński T (2009) Research on heavy metals in Poland. Pol J Environ Stud 5:755–768

Tangahu VB, Abdullah SRS, Basri H, Idris M, Anuar N, Mukhlisin M (2011) A review on heavymetals (As, Pb, and Hg) uptake by plants through phytoremediation. Hindawi Publishing Corporation Inter J Chem Eng Article ID 939161, 31 pp. doi:10.1155/2011/939161

Tataruch F, Kierdorf H (2003) Mammals as biomonitors. In: Markert AB, Breure MA, Zechmeister GH (eds) Bioindicators and biomonitors, vol 6. Elsevier Science Ltd., Oxford, pp 737–772

Torres AM, Barros PM, Campos GCS, Pinto E, Rajamani S, Sayre TR, Colepicolo P (2008) Biochemical biomarkers in algae and marine pollution: a review. Ecotoxicol Environ Saf 71:1–15

Tsangaris C, Cotou E, Papathanassiou E, Nicolaidou A (2010) Assessment of contaminant impacts in a semi-enclosed estuary (Amvrakikos Gulf, NW Greece): bioenergetics and biochemical biomarkers in mussels. Environ Monit Assess 161(1–4):259–269

UNEP (2008) www.chem.unep.ch/mercury/

USEPA (1999) National recommended water quality criteria – correction. EPA 822-Z-99-001. Office of water. Washington, DC.

Valavanidis A, Vlachogianni T (2010) Metal pollution in ecosystems. Ecotoxicology studies and risk assessment in the marine environment. www.chem-tox-ecotox-org

Valavanidis A, Vlahogianni T, Dassenakis M, Scoullos M (2006) Molecular biomarkers of oxidative stress in aquatic organisms in relation to toxic environmental pollutants. Ecotoxicol Environ Saf 64:178–189

Vardanyan L, Schmider K, Sayadyan H, Heege T, Heblinski J, Agyemang T, De J, Breuer J (2008) Heavy metal accumulation by certain aquatic macrophytes from lake Sevan (Armenia). In: Sengupta M, Dalwani R (eds) Proceedings of Taal2007: the world lake conference, pp 1028–1038

Vlahogianni HT, Valavanidis A (2007) Heavy-metal effects on lipid peroxidation and antioxidant defence enzymes in mussels *Mytilus galloprovincialis*. Chem Ecol 23:361–371

Vlahogianni T, Dassenakis M, Scoullos JM, Valavanidis A (2007) Integrated use of biomarkers (superoxide dismutase, catalase and lipid peroxidation) in mussels *Mytilus galloprovincialis* for assessing heavy metals' pollution in coastal areas from the Saronikos Gulf of Greece. Mar Pollut Bull 54:1361–1371

van der Voet E, Guinée JB, Udo de Haes HA (2000) Heavy metals: a problem solved? Methods and models to evaluate policy strategies for heavy metals, Series environment policy. Kluwer Academic, Boston

Weiss P, Offenthaler I, Ohlinger R, Wimmer J (2003) Higher plants as accumulative bioindicators. In: Markert AB, Breure MA, Zechmeister GH (eds) Bioindicators and biomonitors, vol 6. Elsevier Science Ltd., Oxford, pp 465–500

WHO (2007) Health risks of heavy metals from long-range transboundary air pollution. World Health Organization, WHO Regional Office for Europe, Copenhagen. ISBN 978 92 890 7179 6
Winkelmann KH (2005) On the applicability of imaging spectrometry for the detection and investigation of contaminated sites with particular consideration given to the detection of fuel hydrocarbon contaminants in soil. Ph.D. thesis, Brandenburg University of Technology
Wintle JN, Duffield AD, Barros BN, Jones DR, Rice MJ (2011) Total mercury in stranded marine mammals from the Oregon and southern Washington coasts. Mar Mammal Sci 27(4):E268–E278. doi:10.1111/j.1748-7692.2010.00461.x
Wisłocka M, Krawczyk J, Klink A, Morrison L (2006) Bioaccumulation of heavy metals by selected plant species from uranium mining dumps in the Sudety Mts., Poland. Pol J Environ Stud 15(5):811–818
Wolff G, Pereira GC, Castro EM, Louzada J, Coelho FF (2012) The use of *Salvinia auriculata* as a bioindicator in aquatic ecosystems: biomass and structure dependent on the cadmium concentration. Braz J Biol 72(1):71–77
Wolterbeek B (2002) Biomonitoring of trace element air pollution: principles, possibilities and perspectives. Environ Pollut 120:11–21
Wong CSC, Li X, Thornton I (2006) Urban environmental geochemistry of trace metals: a review. Environ Pollut 142:1–16
Wood JM (1974) Biological cycles for toxic elements in the environment. Science 183:1046–1052
Yildiz D, Kula I, Ay G, Baslar S, Dogan Y (2010) Determination of trace elements in the plants of Mt. Bozdag, Izmir, Turkey. Arch Biol Sci Belgrade 62(3):731–738
Yilmaz F (2009) The Comparison of Heavy Metal Concentrations (Cd, Cu, Mn, Pb, and Zn) in tissues of three economically important fish (*Anguilla anguilla, Mugil cephalus and Oreochromis niloticus*) inhabiting Köycegiz Lake-Mugla (Turkey). Turk J Sci Technol 4(1):7–15
Zevenhoven R, Kilpinen P (2001) Control of pollutants in flue gases and fuel gases. TKK Espoo 2001
Zhang HC, Ge Y (2008) Response of glutathione and glutathione *S*-transferase in rice seedlings exposed to cadmium stress. Rice Sci 15:73–76
Zhang WW, Ma ZJ (2011) Waterbirds as bioindicators of wetland heavy metal pollution. Procedia Environ Sci 10:2769–2774
Zhou Q, Zhang J, Fu J, Shi J, Jiang G (2008) Biomonitoring: an appealing tool for assessment of metal pollution in the aquatic ecosystem. Anal Chim Acta 606:135–150

Chapter 6
Natural Dyes and Antimicrobials for Textiles

Masoud B. Kasiri and Siyamak Safapour

Abstract Many studies on extraction, purification, and modification processes of natural dyes and antimicrobials, and their subsequent application on textiles in recent years, demonstrates the revival of natural dyeing and finishing. Natural dyes have been widely used in textile coloration since ancient times. However, with advent of man-made synthetic dyes in the mid-nineteenth century, the dye market has been captured due to variety of competitive properties of synthetic dyes. Such properties are lower cost, color variety, ability to dye synthetic fibers, and availability in large industrial scale. Subsequently the use of natural dyes fainted. However, most synthetic dyes raise serious problems for human health and cause environmental risks. As a consequence there is now a worldwide interest for the production of dyes from natural sources such as plants and microorganisms.

The use of natural dyes in textile processing is increasing because of higher dyestuff quality, environmental compatibility and lower costs. In addition many natural dyes have inherently antimicrobial properties. Natural dyes are extracted from microorganisms and plant organs such as bark, leaf, root, fruit, seed, and flowers. Natural dyes are not only a rich and varied source of dyestuff, but also could be considered as safe, environmentally friendly and low cost treatments with additional benefit of coloring in a single stage. In this article, we review extraction and application of natural dyes on textiles as effective coloring and antibacterial agents. Extraction and treatment methods are discussed

Keywords Textile dyes • Antimicrobials • Dye extraction • Bark • Leaf • Flower • Root • Chitosan • Berberine

---

M.B. Kasiri (✉) • S. Safapour
Tabriz Islamic Art University, Tabriz, Iran
e-mail: m_kasiri@tabriziau.ac.ir; masud.kasiri@gmail.com; s_safapour@tabriziau.ac.ir; s.safapour@gmail.com

E. Lichtfouse et al. (eds.), *Green Materials for Energy, Products and Depollution*, Environmental Chemistry for a Sustainable World 3, DOI 10.1007/978-94-007-6836-9_6,

Contents

6.1 Introduction

For centuries, natural colorants have been used for numerous purposes varying from tribal or funeral ceremonies, body painting, decorative art and clothing or domestic decoration. At first, colorants were mainly extracted from minerals, insects and plants (Guinot et al. 2006). But since 1856, the manufacture of synthetic colorants became more important as a result of: (I) moderate washing and light fastness of the natural colorants; (II) high cost of the production of natural dye; (III) the advancements in chemical synthesis and characterization techniques and; (IV) the greater production efficiency of synthetic dyes in terms of quality, quantity and the potential to produce low-cost raw materials (Samanta and Agarwal 2009). Consequently, natural colorants were increasingly replaced by synthetic ones, produced in cheaper ways, with moderate to excellent fastness. It is estimated that some 700,000 t of dyes are consumed annually and world production of pigments is also about five million tons, with inorganic pigments accounting for approximately 96 % (McCarthy 1997). Almost all these synthetic colorants being synthesized from petrochemical sources through hazardous chemical processes pose threat towards the environment and human body health (Kasiri et al. 2008; Safapour et al. 2010; Kasiri and Khataee 2011; Samanta and Konar 2011).

Global concern for environmental pollution has put stringent rules forward for high level pollutant industries. Textile industry with huge amount of hazardous wastewater production is one of such industries. Numerous researches in recent years have focused on development of totally new technologies and/or modification of conventional technologies for cleaner and environmentally friendlier production processes.

Hence, in the last few years the use of non toxic, antimicrobial and eco-friendly natural dyes on textiles, preferably natural fiber products, has become a matter of increasing importance due to the increased awareness of synthetic dyes environmental hazards (Hill 1997; Dawson 2009). Some of the advantages of natural dyes are

Fig. 6.1 Natural dyes are extracted from different parts of plants such as (**a**) bark (Purple bark; http://www.wickedsmileys.com/GothicGreetingCards/index2.html), (**b**) leaf (Eucalyptus; http://herbs.ygoy.com/what-is-eucalyptus/), (**c**) flower (Marigold; http://flowerwallpapers.co/marigold/download-yellow-marigolds-flower-images-inpretty-orange-marigolds-wallpaper.html), (**d**) fruits (Pomegranate rind; http://www.anniesremedy.com/herb_detail518.php?gc=518b) and (**e**) root (Madder; http://microsites.merton.gov.uk/riverandcloth/image/madder-roots.html)

eco-friendliness, i.e., they do not create any environmental problems at the stage of production or use and maintains the ecological balance (Sivakumar et al. 2011).

On the other hands, textiles, by virtue of their characteristics and proximity to human body, provide an excellent medium for the growth and multiplication of microorganisms, where they could find the required nutrition including; water, carbon, nitrogen and some inorganic salts. This could help microorganisms to transfer, propagation of infection and causing microbial diseases (Gupta and Bhaumik 2007).

Most of natural dyes have inherently antimicrobial properties and could consequently possess high medicinal activity. Natural dyes are extracted from different parts of plants including bark (e.g. Purple bark, Sappan wood, Shillicorai, Khair, Red and Sandalwood), leaf (e.g. Indigo, Henna, Eucalyptus, Tea, Cardamon, Coral Jasmine, Lemon Grass), root (e.g. Turmeric, Madder, Onions, Beet-root), fruits or seeds (e.g. Latkan, Pomegranate rind, Beetle nut, Myrobolan) and flower (e.g. Marigold, Dahlia, Tesu, Kusum) that contain coloring materials like tannin, flavonoids, quinonoids, etc (Fig. 6.1).

The natural dyes also come from some types of microorganisms such as fungi, algae and bacteria. These dyes can offer not only a rich and varied source of dyestuff, but also could be considered as safe, environmentally friendly and low cost treatments with additional benefit of coloring in a single stage (Samanta and Agarwal 2009; Samanta and Konar 2011).

Application of natural dyes has, of course, some limitations. It is difficult to reproduce shades by using natural dyes. Natural dyeing requires skilled workmanship and is therefore expensive. They have poor fastness and the colors produced are of poor quality. Moreover, these dyes require time-consuming extractions, are difficult to use and the dyed textile may change color when exposed to the sun, sweat and air (Hill 1997).

For using natural dyes in an industrial scale, the appropriate and standardized extraction methods and dyeing techniques must be developed without sacrificing the required quality and quantity of dyed textiles. Therefore, to obtain stable shades with acceptable color fastness behavior and reproducible color yield, appropriate extraction, identification, dyeing and finishing techniques need to be derived from the current scientific studies.

In this review, an attempt has been made to give the latest scientific overview on extraction and application of natural dyes/antimicrobials on textiles as effective coloring and antibacterial agents. Different methods of extraction of natural dyes/antimicrobials will be discussed and examples of early applications of these dyes on textile processing, properties, advantages and disadvantages will be reviewed.

6.2 Natural Dyes/Antimicrobials

Natural dyes/antimicrobials are dyes, colorants, and agents derived from natural resources. The resurgence of natural dyes in the early twenty-first century is due to some their peculiar benefits. However, there are some drawbacks restricting the application of these beneficial natural materials (Dawson 2008). In this section, some main benefits and restrictions of natural dyes/antimicrobials have been summarized.

6.2.1 Benefits

According to the results of previous works, growing global interest toward dyes/antimicrobials from natural resources is due to some main following benefits:

- Some natural dyes have intrinsic additional properties such as antibacterial, moth proof, anti-allergy, anti-ultra violet, etc.

- Natural dyes produce different uncommon, eye-catching, and soothing shades on textiles.
- Wide range of shades can be produced with an individual natural dye either in mixture with mordants or by change in dyeing condition.
- Different natural constituents along with natural dye may enhance dyeing/finishing process efficiency as auxiliary, mordant, fixing component.
- Natural dyes/antimicrobials are obtained from inexpensive renewable resources with huge potential.
- From two aspects, natural dyes/antimicrobials are environmentally friendlier than synthetic ones: first, in natural materials all synthesis processes are accomplished by nature with no pollution of environment. Second, these materials are readily biodegradable and do not produce hazardous wastewater upon degradation in environment, and so, there is no need for treatment of wastewater before discharging into the environment.
- Majority of natural dyes/antimicrobials are extracted from wild and self growing plants with no additional cost for their cultivation.
- Color characteristics of some natural dyes mellow with time. This characteristic imparts unique properties and appearance to the stuff dyed, e.g., old carpet piles dyed with natural dyes.
- Most of natural antimicrobial components do not have negative effects on beneficial non-target microorganisms.
- Most of natural antimicrobials do not have side effects on human body.
- Extraction, purification, and application of natural dyes/antimicrobials have become easier and more possible with different modern characterization techniques developed.

6.2.2 Restrictions

Despite various benefits of natural dyes/antimicrobials, they have some drawbacks limiting their application on textiles. Some important restrictions can be summarized as follows:

- Generally, natural dyes/antimicrobials have low wash, rub, light, sweat, and gas fastness on textiles probably due to weak bonding, weak interaction, etc. with textiles.
- To improve fastness of natural dyes/antimicrobials, some additional chemicals are used such as metallic mordants, crosslinking chemicals, etc. Discharge of wastewater of dyeing and finishing processes containing these hazardous materials poses some serious environmental risks.
- Only little information is available about extraction, purification, chemical structure, and application methods of natural dyes/antimicrobials onto textiles.

- Color matching and reproduction of color is another disadvantage of natural dyes since the quantity and quality of natural materials steadily change with climate, plant genus, etc.
- There is no standard method of natural dyeing.
- The efficiency of the natural dyes/antimicrobials extraction process is generally low, only few grams of natural component per kg of raw materials are normally obtained.

6.3 Extraction of Natural Dyes

Extraction of dyes/antimicrobials from the natural sources could be one of the most important steps for treatment of textiles to achieve the desired dyeing properties and/or antimicrobial activities. Moreover, obtaining a standard extraction process and optimizing the extraction variables for a particular natural source, are economically important and consequently, affect the end-products price.

Natural dyes/antimicrobials could be extracted from different types of microorganisms as well as various parts of the plants such as bark, leaf, root, fruit, seed, flower, etc. There are different methods to extract these materials from above-mentioned natural sources including; aqueous extraction i.e. using water for the extraction with or without addition of salt/acid/alkali/alcohol in the extraction bath, chelating with alum and precipitating with acid, enzyme assisted extraction, alcoholic/organic solvent extraction by using relevant extracting equipment, such as soxhlet apparatus, with use of alcohol, hexane or benzene solvents, adsorption of the pigments on silica gel and finally, ultrasonic extraction method. In the following paragraphs, four main important techniques, i.e. aqueous extraction methods, alcoholic/organic solvent extraction, ultrasonic extraction and enzyme assisted extraction will be reviewed.

6.3.1 Aqueous Extraction Methods

Due to the chemical structure of the most natural dyes/antimicrobials, simple set-up and cost-effective aqueous extraction method is the most widely used extraction technique. Routinely, dried and finely cut materials of natural source are grinded in powdered form and then the dye/antimicrobial is extracted in water employing different standard processes.

For instance, the aqueous solution could be filtered and concentrated under reduced pressure (rotary evaporator) to give a crude extract (Farizadeh et al. 2009).

Extraction time, pH of the solution, and temperature are the most important parameters influencing the yield of extraction. In an interesting work, the natural dye was extracted from petals under different operating conditions such as extraction

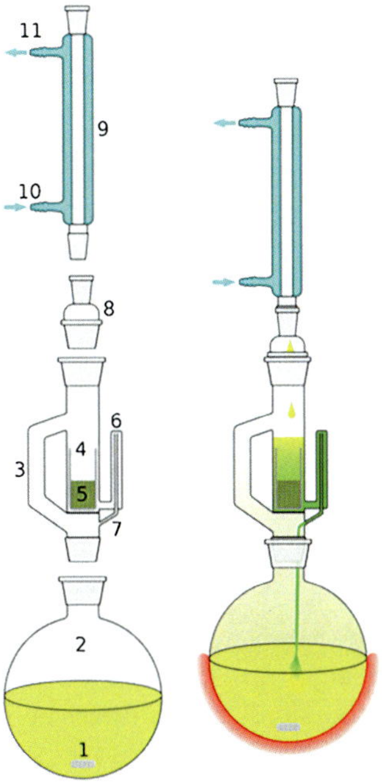

Fig. 6.2 A schematic representation of a soxhlet extractor; (**1**) stirrer bar (**2**) still pot (**3**) distillation path (**4**) thimble (**5**) solid (**6**) siphon top (**7**) siphon exit (**8**) expansion adapter (**9**) condenser (**10**) cooling water in (**11**) cooling water out (http://en.wikipedia.org/wiki/Soxhlet_extractor)

time (45–120 min), temperature (60–90 °C) and mass of the petals (0.5–2 g) by conventional extraction technique, where response surface methodology (RSM) was used for optimization of the extraction process (Sinha et al. 2012).

6.3.2 Alcoholic/Organic Solvent Extraction Methods

The organic nature of some dyes/antimicrobials from natural sources implies the application of non-aqueous extraction methods. During this method, a mixture of organic solvents, usually alcohols with hexane, is used for extraction of natural agents.

As an example, the roots of the authentic sample of *Ratanjot* were air-dried and ground to coarse powder. The powdered roots were extracted with *n*-hexane in soxhlet apparatus (Fig. 6.2) till the color of the decoction became very light. Solvent was removed and the percentage yield of the hexane extraction from roots was estimated (Arora et al. 2009).

Similarly, for extraction of carminic acid, a natural colorant that can be obtained from the dried bodies of females of the *Dactylopius coccus* Costa insect species, Borges et al. (2012) have tried the following method. Three different mixtures of solvents i.e. methanol:water, ethanol:water and pure ethanol were employed to achieve extracts with different compositions. Moreover, extractions were performed at three different extraction temperatures (100 °C, 150 °C, and 200 °C) and 30 min as extraction time. All extractions were performed in 11 ml extraction cells, containing 2.0 g of sample.

Another example has been explained at flavonoid dye extraction from marigold flower (Guinot et al. 2008). Dried ground flowers were soaked by decoction for 10 min in distilled water (1:100 w/v). Extracts of the dried ground flowers were obtained by maceration (1:100 w/v) at 60 °C for 30 min and 60 min under mechanical stirring (180 rpm). The solvents used were ethanol:water mixtures of various compositions, i.e. 2:8, 3:7, 5:5 v/v. The resulting suspensions were filtered through gauze and the filtrates were made up to their initial volume with the appropriate solvent. Finally, 2 ml of filtrate were centrifuged for 10 min (7,176 g) before analysis.

6.3.3 Ultrasonic Extraction Method

State of arts process like ultrasonic, enzymatic and enzyme mediated ultrasound extraction techniques are nowadays widely being studied for their high efficiency over the conventional methods. In this method, the energy of the ultrasonic waves enhances the yield of extraction.

An example of application of this extraction technique has been explained by Mishra et al. (2012). The petals of the flowers were picked and dried at 50 °C in tray drier. Methanol (15 ml/g of petals) was added in dried petals and then pH was adjusted to the desired value. The beakers was then placed in ultrasonic bath and sonicated for 5–30 min at 27–30 MHz and 160 V. After sonication, the contents of beakers were filtered through standard test sieve to remove solid materials. The filtrated dark red liquid was vacuum evaporated in a rotary vacuum evaporator to about half of the original volume and then, the concentrated red liquid was spray-dried.

In this method, extraction bath pH, salt concentration, ultrasonic power, extraction time and temperature are the most influencing parameters. Kamel et al. (2005) have studied the extracting of lac as a natural dye with ultrasonic assisted method varying amounts of the dye materials (2–12 %) at different temperatures (50–80 °C) using different ultrasonic powers (100–500 W) and for different time intervals (20–120 min), yielding different amount of dyeing agent.

Similarly, plant extracts (1:100 w/v) were obtained from dried leaves and stems in ethanol/water (3:7 v/v) with ultrasonication (15 min, 24 kHz, R.E.U.S-GEX 180). After passing through filter paper, the filtrate was centrifuged for 10 min (7,176 g) (Guinot et al. 2009).

As another example of ultrasonic extraction described by Sivakumar et al. (2011), 1 g of the plant samples was taken and added to 50 ml water taken in the beaker. The beaker was covered using aluminum foil to prevent loss of solvent by evaporation. The ultrasonic probe was positioned in the beaker and the ultrasound power set at 80 W, while the temperature was maintained at around 45 °C.

6.3.4 *Enzyme Assisted Extraction Method*

As a recipe for enzymatic extraction technique, a 2 % solution of pectinase:cellulase (2:1) was sprayed on pomegranate rind (25 g) for better soaking and contact and then, left overnight. The enzyme treated material was washed with little amount of distilled water and pH of the solution was adjusted to 10. The sample was shaken at 150 rpm for 20–80 min at desired temperature. The contents of beakers were filtered through standard test sieve to remove solid materials and the concentrated dye extract was vacuum evaporated in a rotary vacuum evaporator to about half of the original volume and finally, the concentrated liquid was spray-dried (Tiwari et al. 2010).

6.4 Application of Natural Dyes/Antimicrobials in Textiles

6.4.1 *Treatment Methods*

In application of materials from natural sources onto textiles, fastness properties such as wash, light, perspiration, etc. are often in low and undesired level. This can be due to heterogeneity, diversity and complexity of natural matters as well as their weak or negligible interactions with textiles. Low fastness properties limit the application area of such useful, eco-safe, low cost natural materials in the treatments of textiles in spite of their various beneficial properties. Hence, to surpass this drawback, additional treatments are often needed to enhance the durability and value of textile products. Treatment processes can be performed on textiles (Das et al. 2007; Pruthi et al. 2008; Ammayappan and Moses 2009; Çalıs et al. 2009; Giri Dev et al. 2009; Kiumarsi et al. 2009; Mongkholrattanasit et al. 2011) or on natural matters (Iqbal et al. 2008; Syamili et al. 2012) or on both (Iqbal et al. 2008), where the first one is more common and frequently used. In literature survey to prepare the present review, we have found some interesting treatment methods to enhance the properties and values of final products. Although addressing all available treatment methods is not possible here, some important examples will be briefly reviewed in the following paragraphs.

Structure Modification of Textiles

In the dyeing process with indicaxanthin natural dye, Guesmi et al. (2013) used modified acrylic fibers with amino and hydroxyl functional groups. They reported that the affinity of the dye and dyeability of fibers were strongly dependent on the pH value of dyeing bath with the highest amount of dye up-take at pH 3. It was attributed to the action of functional groups available in both fiber and dye structures controlling ionic interaction of carboxyl groups of dye with the protonated amino groups of the modified acrylic fibers.

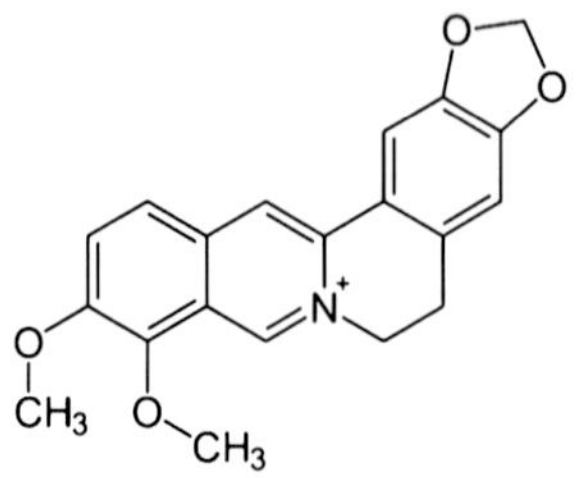

Fig. 6.3 Chemical structure of berberine extracted from a Chinese herb, *Rhizoma coptidis*

Treatment with Chemicals

In a series of studies, cotton was dyed with Berberine as natural colorant (Kim et al. 2004; Kim and Son 2005; Son et al. 2008). Berberine is a quaternary ammonium salt and strong yellow colored natural dye with both dyeing and antimicrobial properties (Fig. 6.3). Berberine has high affinity for protein fibers such as wool and silk whereas it shows low affinity for cellulosic fibers because of lack of ionic charges on cotton surface compared to protein fibers. Accordingly, cotton was reacted with an anionic bridging agent to produce negative charges on the surface, being responsible for electrostatic interactions with cationic natural dye, in order to enhance the affinity of Berberine to cotton. Dyeing tests demonstrated significantly enhanced properties in both Berberine exhaustion and fastness properties.

In another work, Glyoxal as crosslinking chemical were employed to enhance the application of *Azadirachta indica*, a natural antibacterial agent, on polyester/cotton blend fabrics (Joshi et al. 2007). Antibacterial properties retained after different wash cycles demonstrated the efficiency of chemicals in crosslinking yield of bioactive matter on the substrate. Similarly, citric acid was used as crosslinking agent in antibacterial finishing of cotton with green tea leaves extracts to increase the durability against wash (Syamili et al. 2012).

Another example of treatment with chemicals is the work of Ammayappan and Moses (2009). They reported that pre-treatment of cotton with hydrogen peroxide and wool fibers with formic acid increase antibacterial activity of fibers treated with aloevera, chitosan, and curcumin as natural antimicrobials. It was attributed to the increase in the accessible areas in the case of cotton, and disulphide linkage modification and change in wettability in the case of wool/hair fibers. Moreover, additional reasons stated were the increase in the quantity of antimicrobials absorbed and decrease of impurities for all fibers after treatment with selective chemicals.

Treatment with Chitosan

Chitosan is a linear polysaccharide produced commercially by deacetylation of chitin (Fig. 6.4), the structural element in the crustaceans such as crabs and shrimp

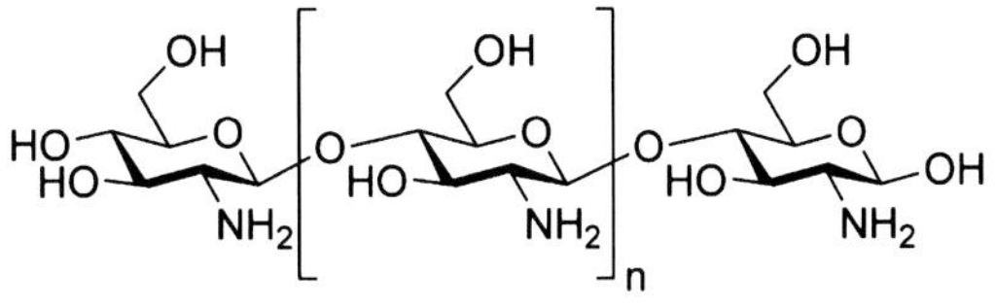

Fig. 6.4 Chemical structure of chitosan, a linear polysaccharide

and cell walls of fungi. It has been widely used in various industries because of its multifunctional properties. Chitosan is also extensively used in treatment of textiles such as dyeing, finishing, printing, etc.

For instance, chitosan has been used in pre-treatment of cotton fibers prior to dyeing with lac dye (Rattanaphani et al. 2007). Due to lack of cationic sites in cotton fibers, lac dye has low affinity for cotton. It was shown that treatment with chitosan produced cationic sites on cotton and subsequently the affinity of lac dye toward fiber, dye up-take, and fatness properties have been enhanced. It was discussed that chitosan could be a suitable eco-friend alternative in treatment of cotton compared to use other toxic chemicals.

In another work, cotton has been pre-treated with chitosan and then dyed with turmeric (Kavitha et al. 2007). Treatment with chitosan enhances antibacterial activity and improves some mechanical properties of fibers. Similar attempt has been made to pre-treat wool fibers with this biopolymer prior dyeing with henna dye (Giri Dev et al. 2009). Increasing in henna dye up-take and antimicrobial finish were properties imparted to wool fibers after treatment with chitosan.

Treatment with Enzyme

Protease is an enzyme, biological molecules that catalyze chemical reactions, that conducts proteolysis. In the study of antimicrobial characteristics of some natural dyes on wool fibers, effects of protease enzyme treatment on dyeing and antimicrobial yield were studied (Raja and Thilagavathi 2011). It was demonstrated that alkaline protease enzyme process enhances the quantity of natural dye exhausted. It was ascribed to partial or complete damage of cuticle facilitating easier penetration of natural dye molecules into fibers.

Treatment with Ultraviolet (UV) Radiation

Iqbal et al. (2008) have studied the effect of UV irradiation on the henna dyed cotton fabrics. Henna and cotton were separately UV irradiated and then after coloration, the results were compared with non-irradiated samples. At first step, untreated fabrics were dyed with both virgin and irradiated henna. It was observed that both color strength and shade of fabrics dyed with irradiated henna are higher compared to non-irradiated henna. It was concluded that lawsone, the coloring matter in henna, upon irradiation is hydrolytically degraded and thereby considerable increase in the sorption of degraded dye occurs. In the next step, effect of radiation treatment of cotton substrate on its dyeability with henna was investigated and similarly higher color strength values were obtained for treated samples. It was inferred that increase in the interaction between dye molecules and cellulose macromolecules, resulted from developed spaces within fibers after irradiation, was responsible for higher dye up-take values. Finally, Iqbal et al. (2008) concluded that to obtain the highest color strength both henna dye and cotton must be UV irradiated.

Treatment with Ultrasound

Nowadays, ultrasound energy is exploited in dyeing of textiles so-called sonicator dyeing. Sonicator dyeing has some significant advantages over the conventional dyeing, e.g., decrease in temperature and time of dyeing, and decrease in quantity of dye and electrolyte required. Wool fibers were dyed with lac as a natural dye using both conventional and ultrasonic techniques (Kamel et al. 2005). About 47 % increase in lac dye up-take was reported using ultrasonic dyeing. Also, the amount of dye extracted by ultrasonic technique was 41 % higher than conventional extraction methods. Similar study was conducted on the dyeing of cationized cotton fabric with same dye using both conventional and ultrasonic methods, exhibiting 66.5 % improvement in lac dye up-take using ultrasonic dyeing technique (Kamel et al. 2007).

In another study, cotton was dyed with natural dye Eclipta using both conventional and sonicator dyeing. The efficiency of sonicator dyeing was reported to be 7–9 % higher than conventional method (Vankar et al. 2007).

Guesmi et al. (2013) also compared dyeing properties of modified acrylic fibers with indicaxanthin natural dye using both conventional and sonicator dyeing and reported 49.62 % higher natural dye up-take, better wash and light fastness with less process time and energy consumption for sonicator dyeing technique.

Similarly, improved natural dye exhaustion, fastness properties as well as waste reduction of dyeing were reported in the ultrasonic dyeing study of cotton with natural dye from plant species *Symplocos spicata* (Vankar et al. 2008).

Treatment with Plasma

Plasma treatment is a modern method of textiles treatment. Plasma is the fourth state of the matter in the form of partly ionized gases. When a gas under controlled pressure interacts with electromagnetic field, plasma is produced. Plasma technology is gradually being replaced by chemical methods in treatment of textiles. Low-temperature plasma is widely used in non-destructive surface modification of textiles where a wide range of properties can be obtained. Ghoranneviss et al. (2010) have studied dyeing properties of plasma treated wool fibers with two natural dyes, namely, madder and weld. It was proved that the treatment of wool fibers with plasma enhances natural coloration process and properties. Moreover, plasma treatment was introduced as a suitable replacement for metallic mordant treatment with some additional benefits. Chen and Chang (2007) used low temperature microwave plasma to treat cotton fabric prior to grafting with onion skin and onion pulp extracts in order to impart antimicrobial properties to cotton fabrics. It was discussed that functional groups produced on the fiber surface after low-temperature oxygen plasma treatment significantly increase the hydrophilic nature of cotton and thereby the grafting yield of both onion skin and pulp extracts enhances.

6.4.2 Examples of Early Applications

Numerous reports have been published on natural dyes/antimicrobials application, including variability in the quantity and quality in the natural source, their use in combinations, their dyeing and antimicrobial properties, the effect of mordants and auxiliaries, their fastness, improvements in extraction methods and the development of alternative sources. Natural dyes/antimicrobials are extracted mainly from two different resources: plants and microorganisms. In the following parts, examples of early applications of dyes/antimicrobials from these two resources will be reviewed.

Dyes/Antimicrobials from Plants

Plants are the major source of natural dyes/antimicrobials in the nature. These materials are extracted from different parts of the plants such as bark, leaf, root and flower containing commonly coloring materials like tannin, flavonoids, quinonoids, etc. During the following paragraphs, the examples of applications are classified according to the different parts of plants used as the natural source.

Root

Root is one of the common parts of the plants that has been extensively used for dyes/antimicrobials extraction (Table 6.1). **Madder**, the ground root of the *Rubia* species with anthraquinone substituted chemical structure, has traditionally been an important natural source of red and brown color on wool and cotton along with different mordants (Gupta et al. 2001a, 2004a). Chemical structure of some colorants presented in madder root is shown in Fig. 6.5.

Gupta et al. (2001a) have tried the dyeing of nylon with the pigment extracted from indian madder. Nordamncanthal, a bright red pigment with a hydroxyanthraquinone chemical structure, was extracted from madder by chelating with alum and precipitating with acid. This pigment was used for dyeing of nylon and polyester, and the results showed good affinity for the fibers studied.

The authors have also studied the dyeing process of nylon with another extract of madder (Gupta et al. 2001b). Purpurin with 1,2,4-trihydroxyanthraquinone structure

Table 6.1 Characteristics of dyes/antimicrobials extracted from roots of the plants and their application properties

No.	Source (used part)	Botanic name	Chemical base	Shade on substrate	Substrate applied	Properties achieved	Dye fastness[a]	Extraction method	Reference
1	Indian madder	*Rubia*	*Hydroxyanthra-quinones* (nordamncanthal)	Bright red	Nylon and polyester	Good affinity for the fibers	–	Chelating with alum and precipitating with acid	Gupta et al. (2001a)
2	Indian madder	*Rubia*	*1,2,4-trihydroxy-anthraquinone (purpurin)*	Red	Nylon and polyester	Good affinity for the fibers	–	Chelating with alum and precipitating with acid	Gupta et al. (2001b)
3	Indian madder	*Rubia*	*1,2,4-trihydroxy-anthraquinone* (purpurin), *1,3-dihydroxy-2-carboxy anthraquinone* (munjistin)	–	Nylon with mordant	Color	Purpurin*(G)* Munjistin*(P)*	Extraction with alum	Gupta et al. (2004a)
4	Madder	–	Alizarin, purpurin and 1,4-dihydroxy-anthraquinone	–	Wool	Adsorption dependent on the pH and temperature	–	Extraction with distilled water	*Farizadeh et al.* (2009)
5	Madder	–	Alizarin, rubierythric acid, rubiadin, munjistin purpuroxanthin or xanthopurpurin	–	Wool	Accelerate dye diffusion	–	Extraction with distilled water	*Montazer et al.* (2007)
6	Madder	–	*Hydroxyanthra-quinones*	Red	Wool with mordant	Fading of the madder-dyed yarns	–	–	*Montazer and Parvinzadeh* (2004)

7	Madder, cochineal, walnut, weld, white onion, red onion, pomegranate	*Rubia tinctoria, Dactylopius coccus, Juglens regia, Reseda luteola, Allium cepa, Allium cepa, Punica granatum*	*Anthraquinone, naphthaquinone, flavone and tannin*	–	Wool with mordant	–	–	–	Montazer et al. (2004)
8	Ratanjot	*Arnebia nobilis* Rech.f	Naphthoquinonoid, hydroxynaphthoquinone, isohexenylnaphthazarin	Pink, blue, purple	Polyester, nylon, cotton, wool, silk, acrylic	Color	Wash(*E*) Light(*P*) Rub(*E*)	Soxhlet extraction with n-hexane	Arora et al. (2009); Arora et al. (2012a); Arora et al. (2012b)
9	Turmeric	*Curcuma longa L.*	Natural phenols	Yellow	Wool	Color, anti-microbial	Wash(*G*) Light(*P*)	Aqueous ethanol extraction	Han and Yang (2005)
10	Turmeric	*Curcuma longa L.*	Natural phenols	Yellow	Cotton coated with chitosan	Color, anti-bacterial	–	Aqueous ethanol extraction	Kavitha et al. (2007)
11	Aloevera, chitosan, curcumin	–	–	–	Cotton, wool, rabbit hair	Antimicrobial	Wash(*G*)	–	Ammayappan and Moses (2009)

(continued)

Table 6.1 (continued)

No.	Source (used part)	Botanic name	Chemical base	Shade on substrate	Substrate applied	Properties achieved	Dye fastness[a]	Extraction method	Reference
12	Turmeric, sapanwood, phellodendron, chinese yellow berries, gardenia, safflower, madder, gromwell, indigo	*Curcuma longa* L., *Caesalpinia sappon* L., *Phellodendron amurense Ruprecht, Sophora japonica* L., *Gardenia jasmoides Ellis, Carthamus tinctorius* L., *Rubia cordifolia,* L., *Lithospermum erythrorhizon, Indigo*	Curcumin, berberine, rutin, crocin, safflower yellow, safflower red, purpurin, munjistin, brazilin, shikonin, indigo	–	–	–	–	–	Choo and Lee (2002)
13	Cochineal, madder, alkanna, henna, brazilwood, red sandalwood, safflower, indigo, logwood	*Porphyrophora polonica* L., *Rubia tinctorum* L., *Alkanna tinctoria* Tausch., *Lawsonia inermis* L., *Caesalpinia* trees, *Pterocarpus santalinus* L., *Carthamus tincorius* L., *Indigofera tinctoria, Hematoxylon compechianum* L.	–	Red, red, red, orange, red, red, red, blue, black	Cotton and wool with mordants	Significant improvement of the dye adsorption on the cotton fabrics	Wool: Wash*(G)*	Extraction with boiling water	Zarkogianni et al. (2010)

14	Acacia, indigo, lac, kamala, pomegranate, gall nuts, cutch, myrobolan, himalayan rhubarb, indian madder, golden dock	*Acacia nilotica, Indigofera tinctoria, Kerria lacca, Mallotus philippinensis, Punica granatum, Quercus infectoria, Acacia catechu, Terminalia chebula, Rheum emodi, Rubia cordifolia, Rumex maritimus*	*Bark, indigo, laccaic acid, flavonoid, tanin, tanin, tanin, tanin, hydroxyanthraquinones, hydroxyanthraquinones, hydroxyanthraquinones*	–	Cotton	Antibacterial	Wash(*G*)	–	Gupta et al. (2004b)
15	Chinese herb	*Rhizoma coptidis*	Berberine	Brownish yellow	Wool with mordant	Deeper shades and darker colors, good antibacterial property	Wash(*P*)	Extraction with boiling water	Ke et al. (2006)
16	Amur cork tree, goldenseal coptis, oregon grape, barberry, tree turmeric, yerba mansa	*Coptis chinensis, Berberis aquifolium, Berberis vulgaris, Berberis aristata, Anemopsis californica*	Isoquinoline alkaloid	Yellow	Nylon 66	Color, antimicrobial	–	–	Son et al. (2007)
17	Dolu	*Rheum emodi*	Anthraquinone	Yellow, olive green	Wool, silk with metallic, mordants	Color	Wash(*G*) Light(*M*)	–	Das et al. (2008)

[a]Excellent (E), Very good (VG), Good (G), Medium (M), Poor (P)

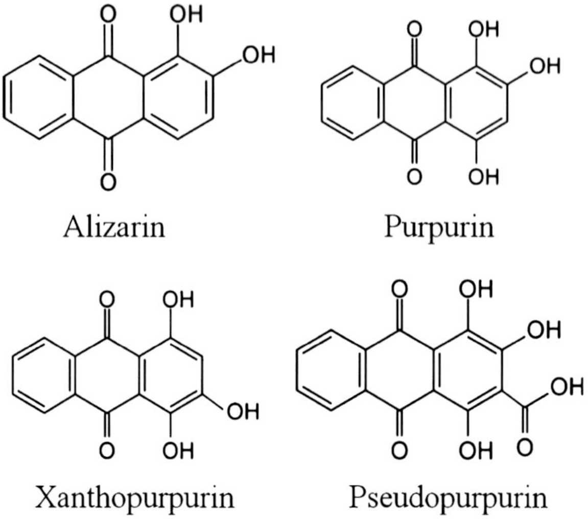

Fig. 6.5 Chemical structure of alizarin, purpurin, xanthopurpurin and pseudopurpurin extracted from madder root

was extracted from madder in a similar way and was used for dyeing of nylon fabrics. The authors have reported that this red pigment has also good affinity, like nordamncanthal, for the nylon fibers.

The light fastness of madder extracts applied on nylon has been reported in a related work (Gupta et al. 2004a). The authors have extracted two pigments, purpurin and munjistin, from the indian madder by soaking and extraction with alum and applied it for dyeing of nylon with ferrous sulphate, copper sulphate, alum and stannous chloride as mordants. The authors have reported a good light fastness for the fabrics while the dye fastness of fabrics with munjistin was poor.

Extraction of dyes from madder at different conditions has been studied by Farizadeh and his colleagues (2009), too. The extracted pigments with distilled water were analyzed by thin layer chromatography (TLC) and high performance liquid chromatography (HPLC) techniques and identified as alizarin, purpurin and 1,4-dihydroxyanthraquinone. Extracted pigments were used for dyeing of wool fabrics at different temperatures and dyebath pHs and the results proved that the affinity of the dye on the fabrics dependent mainly on the pH and temperature of the bath, where increasing of temperature led to decrease the values of partition ratio and affinity.

In an attempt, dyeing process of wool with madder and liposomes as an auxiliary in dyeing has also been optimized (Montazer et al. 2007). The pigment were extracted with steeping the madder in water, heated and passed through a filter. The dyeing process was optimized by response surface methodology, where the bath temperature, time, and liposome percentage were chosen as input variables and the fabrics color strength as the response. The dyeing process was optimized, but it was also found that wash, light, wet and dry rub fastness properties of samples dyed with madder including liposomes have not changed significantly.

Fig. 6.6 Chemical structure of alkannin extracted from Ratanjot

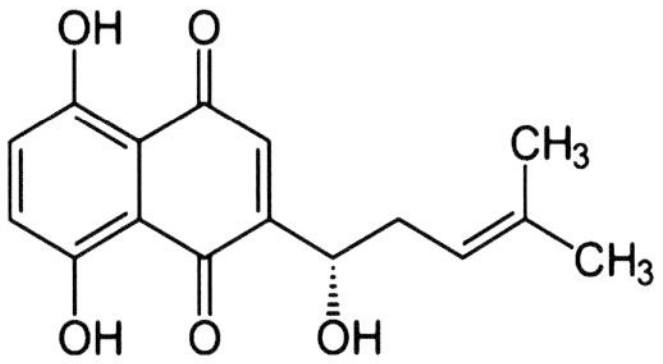

On the other hand, wool yarns were dyed with madder along with aluminum potassium sulfate as the mordant and then treated with different ammonia solutions by Montazer and Parvinzadeh (2004). It has been reported that a change of color is observed on the dyed samples when wash-fastness tests were carried out, while the light-fastness tests showed more fading of the madder-dyed yarns after ammonia treatment.

They have also studied the effect of ammonia on the coloristic properties of wool dyed with madder and other natural dyes containing anthraquinone, naphthaquinone, flavone and tannin structures (Montazer et al. 2004). Wool was first mordanted with aluminium potassium sulphate and then dyed using madder, cochineal, walnut husks, weld, red and white onion skins, and pomegranate shells. The results showed that the color changes could be due to the extension of dye molecule conjugated system by one lone pair electron in ammonia solution.

Ratanjot, *Arnebia nobilis* Rech.f is the root of herbal plant that has been widely used as a potent resource for obtaining both herbal drugs and color matter (Arora et al. 2009). It contains a variety of naturally occurring chemicals, mainly a naphthoquinone derivative called alkannin (Fig. 6.6). Alkannin has a deep red color in a greasy or oily environment and a violet color in an alkaline environment (Papageorgiou et al. 1999).

Arora et al. (2009) have tried to extract coloring matters from Ratanjot. Due to variety of plant species, they used different identification methods such as macroscopic, microscopic, chemical and TLC analysis of the roots. The extraction was performed using soxhlet system with n-hexane and the TLC analysis showed the presence of five naphthoquinone derivatives with red and pink colors.

In another work, Arora et al. (2012a) have used ratanjot extract for coloration of variety of textiles, i.e., cotton, wool, silk, nylon, polyester, and acrylic. Interestingly, the strong sensitivity of extracted natural dye solution to pH and temperature changes was observed. With gradual increasing of pH in the range 3–12, continuous changes in color from red to purple and then to dark blue was observed. It was attributed to the conversion of dye structure from quinonoid in acidic solution to benzenoid in alkaline solution. Dyeing results showed that as the dyeing pH increases, the dye up-take decreases which was ascribed to decomposition of dye structure in alkaline solution.

Wool, cotton, silk, nylon, polyester, and acrylic were dyed with ratanjot extract in similar condition and different shades were obtained such as pink on polyester, blue on nylon, and purple on the rest fibers. Different results for affinity and dye up-take were observed for fibers. Very low dye up-take on cotton fabric was attributed to non-polar character of dye and low affinity to cotton fibers. Wool had higher

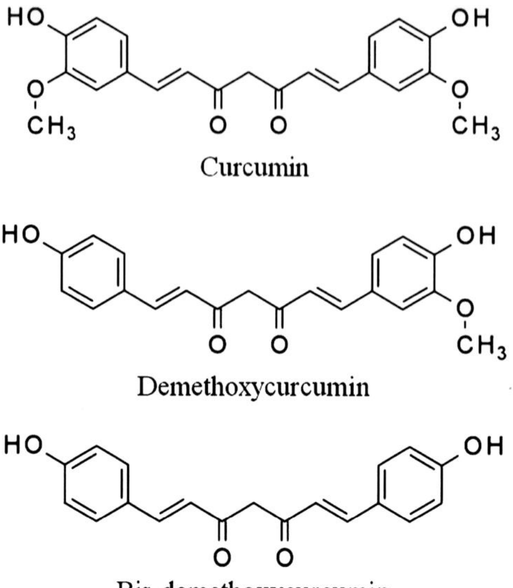

Fig. 6.7 Chemical structure of curcuminoids extracted from turmeric

dye up-take compared to silk possibly because of lower crystallinity of wool fibers. Polyester showed lower dye up-take at 130 °C compared to 100 °C, not due to lower affinity of the dye to polyester but owing to more decomposition of dye at enhanced temperatures decreasing effective dye concentration in the solution (Arora et al. 2012a).

Interestingly, by means of oxidation/reduction test, the reason of completely different observed shades was addressed to be the structure transformation of dye from normal state (quinonoid) to the reduced reversible state (benzenoid) on all substrates. Fastness tests data demonstrated poor light fastness whereas wash, rub and perspiration fastnesses were excellent. Finally, the findings of the study showed that *Arnebia nobilis* Rech.f can be used as a good natural dye resource for dyeing and production of variety of eco-friend textiles where light fastness is not key factor in dye selection (Arora et al. 2012a).

Using *Arnebia nobilis* Rech.f dye, kinetics and thermodynamics of dyeing on wool has also been studied by Aora et al. (2012b) and the results were compared with those of juglone, lawsone, and Rheum emodi from literature. Lower diffusion coefficient than other three natural dyes with exothermic dyeing process was reported for this dye.

Turmeric, *Curcuma longa*, is a rhizomatous herbaceous perennial plant of the ginger family native to tropical south Asia. The dried rhizomes are ground into a deep orange-yellow powder commonly used as spice in cuisine and for dyeing. Turmeric's active ingredient is curcumin which is the principal curcuminoids, natural phenols that are responsible for the yellow color of turmeric. Other two curcuminoids present in turmeric are desmethoxycurcumin and bis-desmethoxycurcumin, known as curcumin I-III (Chan et al. 2009). Chemical structures of curcuminoids are shown in Fig. 6.7.

Turmeric has been used to impart either color or antibacterial property on textiles (Popoola 2000; Han and Yang 2005; Kavitha et al. 2007; Ammayappan and Moses 2009). Han and Yang (2005) used curcumin as antimicrobial finish for wool against two bacteria *Staphylococcus aureus* (*S. aureus*) and *Escherichia coli* (*E. coli*) with no additional treatment through conventional dyeing process. Very low concentration of curcumin was effective in bacterial growth inhibition on wool, e.g., only 0.05 % and 0.2 % curcumin were able to impart inhibition rate of 70 % and 95 % against *E. coli*, respectively. An equation was also established with good accuracy to predict the inhibition rate of bacteria growth using color strength values (K/S). The advantage of the developed equation was in the measuring of antibacterial property indirectly with no need to perform antimicrobial tests.

Using chitosan and turmeric, an attempt has also been made to develop green textiles for medical application. Kavitha et al. (2007) have studied the effects of chitosan pretreatment and subsequent dyeing with turmeric on the properties of cotton yarns. Improvements in mechanical properties such as tensile strength, flexural rigidity, and shear strength as well as natural dye up-take were achieved through chitosan treatment. Excellent antibacterial activity against two pathogenic bacteria, *E. coli* and *S. Aureus*, was another beneficial result of the treatment of cotton yarns with chitosan and turmeric.

In an interesting attempt to develop an eco-friend antibacterial finish using materials from green resources, curcumin along with two other natural bioactive agents, aloevera and chitosan, have been applied lonely or in binary or in ternary combinations on cotton, wool and rabbit hair (Ammayappan and Moses 2009). Then, their bactericidal properties in both pretreated and untreated substrates were examined. It was found that antibacterial activity of substrates finished after pretreatment with selective chemicals is higher than directly finished fabrics. Such behaviors were explained through enhanced wettability, critical surface tension change, and available area in the fiber responsible for better diffusion, adhesion and up-take of antibacterial agent.

When natural antibacterials applied individually on substrates, the antibacterial activity was in descending order: aloevera > curcumin > chitosan (Ammayappan and Moses 2009). The higher antibacterial activity of aloevera was concluded to be related to diverse ingredients, such as acemannon, anthraquinone, salicylic acid, amino acids, zinc, saponins, and many others compared to curcumin with individual benzenoid structure (only phenolic groups) and chitosan with polysaccharide structure (only hydroxyl and amino groups). Antimicrobial performance of binary or ternary combined agents was much better than corresponding individuals. It was explained through some determining factors, e.g., (I) the number and concentration of components present in natural material; (II) the number and diversity of functional groups and their determining role on affinity/binding to fiber; (III) Their mutual interaction with other chemical components and/or fibers and; (IV) synergistic, antagonistic or neutral effects of ingredients on each other in combined applications. Wash fastness tests proved that antibacterial finish durability on substrates are quite high where up to 25 wash cycles completely retained.

Choo and Lee (2002) have also tried to recreate traditional korean dyeing methods using colorants extracts from nine plants, used either singly or in combination, to produce a wide range of colors. In addition to madder, the extracts of indigo, turmeric, sapanwood, phellodendron, chinese yellow berries, gardenia, safflower and gromwell were used as dyeing agents and the fibers dyed were studied by microchemical tests, visible spectroscopy, thin layer chromatography and high performance liquid chromatography.

In a similar study, reconstruction of traditional textile dyeing techniques was the subject of the scientific research done by Zarkogianni et al. (2010). In this work, extractions of cochineal, madder, alkanna, henna, brazilwood, red sandalwood, safflower, indigo and logwood were obtained by aqueous extraction of the powdered mixture of plant/insect. These nine pigments were used for dyeing of wool and cotton fabrics after mordanting of the textile material with a variety of metallic salts. The results demonstrated significant improvement of the dye adsorption on the cotton fabrics and very good washing fastness of the wool samples.

The antimicrobial properties of 11 natural dyes from madder and other plants against three types of gram-negative bacteria were studied experimentally in an interesting work (Gupta et al. 2004b). The pigments were extracted from different parts of acacia, indigo, lac, kamala, pomegranate, gall nuts, cutch, myrobolan, himalayan rhubarb, indian madder and golden dock and were used for dyeing of wool fibers. The authors have reported that certain dyes are able to reduce microbial growth almost completely in the case of *Escherichia coli* and *Proteus vulgaris*, and since the selected dyes showed reasonably good wash fastness it could be assumed that the antimicrobial effect will be durable in practice.

The roots of a chinese herb, *Rhizoma coptidis,* is a source of natural cationic yellow colorant called berberine (Fig. 6.7). In a research work by Ke et al. (2006), the plant root was minced and dipped in distilled water for 30 min, then boiled in distilled water to extract the brownish yellow pigment and used for dyeing of wool fibers with mordanting.

Wool fabrics dyed with extraction of *R. coptidis* under different conditions showed various yellow colors with deep shades and good antibacterial properties, but the washing fastness of dyed fibers was poor.

Berberine has been similarly used as colorant and antimicrobial finish on nylon 66 fibers (Son et al. 2007). It imparted excellent bactericidal property to nylon fibers. It was concluded that finishing with berberine could be practical, eco-friendly, very simple and effective finish.

Das et al. (2008) carried out an interesting research on the dyeing of protein fibers using a natural dye. Wool and silk fibers were dyed with *Rheum emodi* extracted dye with/without Mn, Al, Fe metallic mordants and various shades from yellow to olive green were obtained. Chrysophanic acid, an anthraquinone derivative, is the main coloring matter of *Rheum emodi* whose structure is shown in Fig. 6.8. Interestingly, dyeing results exhibited no changes in the dye up-take of this dye in the pH range 4–8. Another finding was the unusual higher dye up-take, dye affinity, and dyeing rate of silk than wool fibers. The dyeing mechanism was shown corresponded to partition mechanism similar to those of disperse dyes.

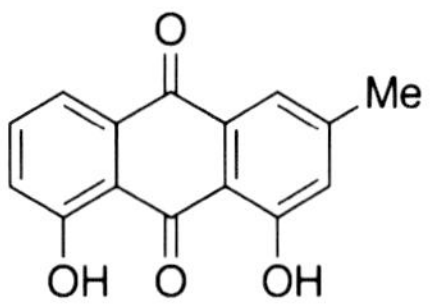

Fig. 6.8 Chemical structure of Chrysophanic acid the main color matter of *Rheum emodi*

It was concluded that those observations were related to typical structure of dye component behaving like disperse dyes in the dyeing process. Wash and light fastness of dyed fibers were moderate where wash fastness was further enhanced with iron mordant.

Seed

Seed of the plants could be a rich source of natural dyes/antimicrobuals (Table 6.2). **Bixin** is a natural dye which has been exploited in yellow to orange coloration of textiles. It is an apocarotenoid found in annatto, a natural food coloring obtained from the seeds of the achiote tree, *Bixa orellana*. Annatto seeds contain about 5 % pigments, which consist of 70–80 % bixin. The chemical structure of bixin is shown in Fig. 6.9. Bixin is chemically unstable when isolated and converts through isomerization into trans-bixin, double-bond isomer. This extract is soluble in fats but insoluble in water. Upon exposure to alkali, the methyl ester is hydrolyzed to produce the dicarboxylic acid norbixin, a water-soluble derivative.

Wool and silk have been dyed using bixin natural dye with/without metallic mordants (Das et al. 2007). Bixin colorant was extracted from the seeds of annato in alkali solution. Wool and silk in the fiber and fabric form, respectively, were first pre-treated, then pre-, post- and simultaneous mordanted, and finally dyed with (10 % on weight of fiber) annatto solution at 90 °C at wide range of pH 2–10. The shades obtained were from yellow to brown depending on the mordants employed.

Annato dye up-take was increased when dyeing bath pH decreased and maximum dye up-take and color strength was obtained at pH 4.5. Bixin with carboxylate groups can be considered as an acid dye (Fig. 6.9). It was discussed that with decreasing pH, the amino groups in protein fibers were protonated, ionic interaction of bixin carboxylate anions with amino cations increased, and as a result the affinity and dye up-take increased (Das et al. 2007). Interestingly, the color imparting ability of mordants for both wool and silk fibers was found to be in the descending order with bixin dye: ferrous sulphate > aluminium sulphate > magnesium sulphate > no salt. It was attributed to the ease and rapid salt formation of iron and aluminium with bixin compared to magnesium.

Generally low light fastness was reported for bixin on both wool and silk fibers. After mordants incorporation through pre- and post mordanting, light fastness was enhanced by one point scale. It was ascribed to the good complex formation ability of iron with tannin, associated with bixin, and subsequent deposition on the fibers. However, such improvement was not that much marked in simultaneous mordanting due to early combination of iron with dye in the solution before sorption with fibers.

Table 6.2 Characteristics of dyes/antimicrobials extracted from seed of the plants and their application properties

No.	Source (used part)	Botanic name	Chemical base	Shade on substrate	Substrate applied	Properties achieved	Dye fastness[a]	Extraction method	Reference
1	Annato	*Bixa orellana*	Carotenoids (bixin, norbixin)	Yellow, orange, brown	Wool, silk with metallic mordants	Color	Wash*(M)* Light*(P)*	–	Das et al. (2007)
2	Bixin	*Bixa orellana*	Carotenoids	Yellow, orange, brown	Cotton	Color	Wash*(P)* Light*(M)*	Soxhlet extraction with n-hexane	Popoola (2000)
	Curcumin	*Curcuma longa*	Natural phenols	Yellow	Cotton	Wash*(M)* Light*(G)*	Soxhlet extraction with methanol		
3	Ban-ajwain	*Thymus serpyllum*	–	–	Cotton	Antibacterial	Wash*(VG)*	Soxhlet extraction with methanol	Sathianarayanan et al. (2011)

[a]Excellent (E), Very good (VG), Good (G), Medium (M), Poor (P)

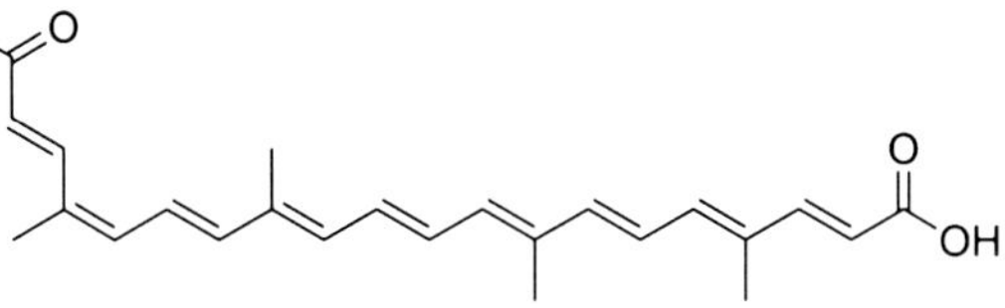

Fig. 6.9 Chemical structure of bixin found in annatto seed

Magnesium did not improve the light fastness. Wash fastness data were in the range 2–3 without mordants. Wash fastness was improved by one point scale only in pre-mordanting condition (Das et al. 2007).

Popoola (2000) have studied bixin and curcumin in the coloration of cotton and compared their fastness properties. Bixin seeds and turmeric roots were used for extraction of coloring matter using soxhlet apparatus reflux system with n-hexane and methanol as extracting solvents, respectively. Afterwards, dyeing was performed without any mordant. The produced shades on cotton were yellow, orange and brown for bixin and yellow for curcumin. Both natural dyes had moderate light and alkaline wash fastness on cotton but, overall curcumin had higher light and alkaline wash fastness than bixin in identical conditions. It was attributed to the difference in chemical structure of coloring components, bixin with extended open-chain conjugated structure making it less stable against light and related photo-reactions compared to curcumin with two benzenoid rings, giving rise resonance stability, in its structure. Alkaline wash fatness data of both dyes were slightly higher than medium. But, similar to light fastness, curcumin dyed cotton had slightly higher wash fastness probably due to less solubility in equal alkaline condition.

In another report, Sathianarayanan et al. (2011) have tried to impart antibacterial finish on cotton fabrics using the bio-active ingredients extracted from ban-ajwain seeds, *Thymus serpyllum*. The extract was applied using different methods such as direct application, micro-encapsulation with acacia gum, crosslinking with resin, and the mixture of them. In direct application method, durability of finish against wash was very low. Wash fastness was enhanced with crosslinking and micro-encapsulation, however, some physical/mechanical properties were lost. Combined method gave the best fastness results so that the finish was durable up to 15 wash cycles.

Leaf

In some researches, leaves of different plants have been used as color resource (Iqbal et al. 2008; Ali et al. 2009; Giri Dev et al. 2009; Mongkholrattanasit et al. 2010, 2011) (Table 6.3). **Henna**, *Lawsonia inermis*, is a shrub or small tree frequently cultivated in India, Pakistan, Egypt, Yemen, Iran and Afghanistan. Henna has been used since ancient times to dye skin, hair, fingernails, leather, silk and wool. Henna's coloring and antimicrobial properties are due to lawsone (2-hydroxy-1,4-naphthoquinone), an organic compound also known as hennotannic

Table 6.3 Characteristics of dyes/antimicrobials extracted from leaves of the plants and their application properties

No.	Source (used part)	Botanic name	Chemical base	Shade on substrate	Substrate applied	Properties achieved	Dye fastness[a]	Extraction method	Reference
1	Henna	*Lawsonia inermis* L.	Lawsone: 2-hydroxy-1,4-naphthaquinone	–	Cotton with mordants	Much higher color strength with alkaline extract method	Light*(M to G)*	Alkaline extract method	Ali et al. (2009)
2	Henna	*Lawsonia inermis*	Naphthoquinone	–	Wool treated with chitosan	Color, antimicrobial	Wash*(VG)* Rub*(M)* Perspiration*(M)*	Aqueous ethanol extraction	Giri Dev et al. (2009)
3	Eucalyptus	*Eucalyptus camaldulensis*	Polyphenol (quercetin, rutin, ellagic acid, gallic acid)	Yellow to brown, grey	Wool, silk with metallic mordants	Color, UV protection	Wash*(G to VG)* Light*(M to G)* Rub*(M to G)*	Aqueous extraction	Mongkholrattanasit et al. (2011)
4	Tea	*Camellia sinensis*	Tannin	Black	Cotton, Jute with mordants	Deep shad on jute	Wash*(G to E)* Light*(G to E)*	Extracted with boiling water	Deo and Desai (1999)
5	Green tea	*Camellia sinensis*	–	–	Cotton	Antibacterial	Wash(M)	Methanolic extraction	Syamili et al. (2012)
6	Deciduous	*Silver oak, Flame of the forest, Tanner's senna, Wattle, Serviceberry*	Tannin	Yellow, brown, yellow, orange, dark grey	Wool with mordants	Color	Light*(M to G)*	Extraction with boiled water	Raja and Thilagavathi (2008)
7	Dom Sheng	*Symplocos spicata*	Flavonol	Yellowish brown	Cotton, wool, silk with metallic mordants	Color	Wash*(VG)* Light*(VG)* Rub*(VG)*	Aqueous extraction	Vankar et al. (2008)

[a]Excellent (E), Very good (VG), Good (G), Medium (M), Poor (P)

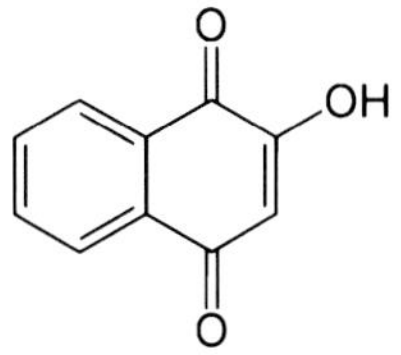

Fig. 6.10 Chemical structure of lawsone, a red orange dye extracted from henna leaves

acid (Fig. 6.10). Lawsone is a red-orange dye primarily concentrated in the leaves, especially in the petioles of the leaf. Lawsone reacts chemically with the protein known as keratin in skin and hair, in a process known as Michael addition, resulting in a strong permanent stain that lasts until the skin or hair is shed. Lawsone strongly absorbs UV light, and aqueous extracts can be effective sunless tanning and sunscreens.

As an example of application, the extract of henna leaves has been used for dyeing of cotton fibers (Ali et al. 2009). The results revealed that extracts produced with alkaline extraction of henna leaves have moderate to good fastness properties and better color strength than the dye extracts obtained in distilled water.

In a similar study to develop an eco-friend, nontoxic and multifunctional finish on wool, Giri Dev et al. (2009) studied the role of chitosan treatment, individually and in combination with natural henna dye, on the antimicrobial property and dyeability of wool fabrics. They found that chitosan simultaneously showed bi-functional effects, i.e., imparted antimicrobial property against two common pathogenic bacteria, *S. aureus* and *E. coli*, and enhanced dye up-take. Fastness tests demonstrated good wash, light, and perspiration durability of finish on fabrics.

In another attempt to extend eco-friendly textiles processing techniques, Iqbal et al. (2008) have tried to improve the color strength and color fastness properties of henna dye on cotton fabric through UV irradiation treatment. They irradiated cotton and henna dye separately, extracted coloring matter using water, alum, and methanol, dyed irradiated cotton with the dye extract and subsequently compared the results with non-irradiated samples. They found that, generally, irradiated dyed samples demonstrate more color strength and deeper shades, and color fastness enhanced from medium to good level.

The observed behaviors were explained from one hand via enhanced affinity and facile sorption of hydrolytically degraded lawsone, decrease of insoluble impurities in the case of irradiated henna, and on the other hand via an increase of number and size of irradiated fiber pores and carboxylic acid groups produced upon UV photo-oxidation of cotton fiber (Iqbal et al. 2008). The industrial aspects of the process were discussed as well and the optimum dyeing condition were introduced as following: 65 °C dye bath temperature, 120 min dyeing time, 3 % NaCl, and 4 % copper sulphate (pre-mordanting).

Eucalyptus, *Eucalyptus camaldulensis*, is another example of natural dye resource. It is a rich source of natural polyphenols and tannins. Leaves contain 11 % quercetin as major component. Other components are rutin, gallic acid, ellagic acid, etc. that are very useful in dyeing yield and durability of substrate (Mongkholrattanasit et al. 2010, 2011). Chemical structures of these coloring matters are shown in Fig. 6.11.

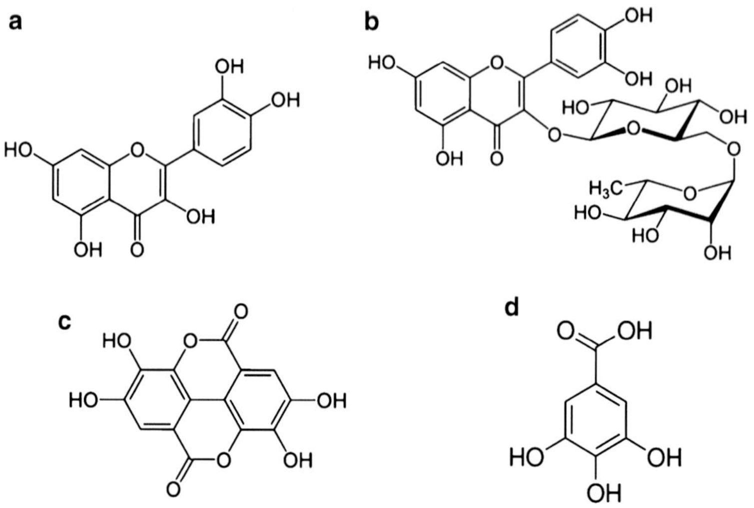

Fig. 6.11 Chemical structures of (**a**) quercetin, (**b**) rutin, (**c**) ellagic acid and (**d**) gallic acid, extracted from eucalyptus leaf

Mongkholrattanasit et al. (2011) demonstrated that water extract of eucalyptus leaves could be effectively used in exhaustion coloration process of wool and silk. They used aluminium, iron, copper, and tin as mordants. The optimum dyeing condition was determined to be at pH 4, time 40 min, and temperature 90 °C. Wide variety of shades was achieved where wool fibers had higher color strength than silk fibers. Shades obtained were yellowish brown with aluminium and copper, dark grayish brown with iron and bright yellow with tin on both wool and silk substrates. Protection against UV was additional property achieved for both substrates. Both fibers showed good to very good wash fastness and medium to good light and rub fastness.

Tea plant, *Camellia sinensis*, is the rich resource of different beneficial natural components such as antioxidants, flavanols, flavonoids, polyphenols, and catechins. It is processed using various methods to produce variety kinds of teas: green, yellow, dark, white, oolong, and black. Aqueous extract of tea could also be used for textiles. Deo and Desai (1999) have studied the dyeing of cotton and jute with an aqueous extraction of black tea where the tannins were the main colorant. The mixture of black tea was stirred, heated, held at the boil for 30 min, allowed to stand for 15 min and then filtered and used as black pigment for dyeing of fabrics, where ferrous sulphate heptahydrate, alum and copper sulphate pentahydrate were used as mordants. The results showed deep color shad on jute fabrics, while good to excellent washing and light fastnesses were observed.

In a related study and with the aim to production of eco-friendly medical textiles, Syamili et al. (2012) have tried to apply extract of green tea leaves onto cotton to impart antibacterial properties. Various tests were performed to identify the effective ingredients of green tea. The ingredients were extracted using ethanol, chelated with copper to exploit intrinsic bactericidal property of copper, applied on cotton with and without citric acid as crosslinker agent. Significant inhibition against bacteria, *S. aureus* and *E. coli,* growth was obtained for finished fabrics.

The wash durability of antibacterial finish was tested and the results showed that after 10 wash cycles the bacterial inhibition rate was less than 25 % for fabrics finished directly with no crosslinker agent and more than 50 % for crosslinked finished fabrics. It was concluded that to increase the durability of antibacterial finish, extracted components should be chemically fixed onto substrate surface using proper crosslinker. This finish process was also introduced as an eco-safe, facile and economic process (Syamili et al. 2012).

Similarly, the leaves of deciduous plant with high content of tannin have been used as a natural dye sources for dyeing of wool (Raja and Thilagavathi 2008). The colorant agent of five different deciduous plants, namely, silver oak, flame of the forest, tanner's senna, wattle and serviceberry, was extracted by boiling 250 g of the leaf material with 5 l of water for 1 h and, were used with mordants to dye wool. The authors have introduced the use of plant leaves more eco-friendly and economical than using their barks or other parts, showing a moderate to good fastness of the fibers dyed.

Symplocos spicata plant leaves extract have also been used in ultrasound coloration of cotton, wool, and silk with Al, Cu, Fe, Cr, Sn(*II*), Sn(*IV*) mordants (Vankar et al. 2008). An additional pretreatment with tannic acid was also done for cotton to provide carboxylic acid groups on the fiber surface for better mordanting. Dye up-take was markedly improved and became almost double in ultrasonic dyeing compared to common exhaustion dyeing process. Yellow to brown shades with high color fastness properties were obtained for all fibers. Lower required time, energy, and dye amount to obtain the same results of conventional dyeing process proved the superiority of ultrasonic natural dyeing technique.

Fruit

Fruits of the plants are another source of natural dyes and have been extensively used for textile dyeing (Table 6.4). **Pomegranate** is the fruit of deciduous shrub or small tree, *Punica granatun*. Fruit rind has been widely used in traditional coloration of textiles since ancient times. However, in open literature there are rare reports addressing the application of pomegranate rind on textiles scientifically except some recently published papers (Çalıs et al. 2009; Sathianarayanan et al. 2010; Kulkarni et al. 2011a; Rajendran et al. 2011). Rind of the pomegranate is a rich resource of tannins, phenolic contents, alkaloids, etc. Granatonine is the main coloring component found in the fruit rind in the form N-methyl granatonine (Fig. 6.12).

The extract of pomegranate rind has been used in environmentally friendly dyeing (Kulkarni et al. 2011a) and bactericidal treatment (Sathianarayanan et al. 2010; Rajendran et al. 2011) of cotton fabric. Wide range of shades such as yellow, brown, and black with good fastness properties were achieved using copper and iron mordants with different ratios (Kulkarni et al. 2011a).

Table 6.4 Characteristics of dyes/antimicrobials extracted from fruits of the plants and their application properties

No.	Source (used part)	Botanic name	Chemical base	Shade on substrate	Substrate applied	Properties achieved	Dye fastness[a]	Extraction method	Reference
1	Pomegranate *(peel)*	*Punica granatum*	Alkaloid (granatonine)	Yellow, brown, black	Cotton with metallic mordants	Color	Wash*(G)* Light*(G)* Rub*(G)*	Soxhlet extraction with aqueous ethanol	Kulkarni et al. (2011a)
2	Tulsi*(leaf)* Pomegranate *(rind)*	*Ocimum sanctum Punica granatum*	Eugenol, germacrene, phytol, tannin	–	Cotton	Antibacterial	Wash*(G)*	Soxhlet extraction with methanol	Sathianarayanan et al. (2010)
3	Pomegranate *(peel)*	*Punica granatum*	Alkaloid	–	Cotton with metallic mordants	Color, anti-microbial	Wash*(G)* Light*(G)* Rub*(G)* Per-spiration*(G)*	Aqueous extraction, Ethanolic extraction	Rajendran et al. (2011)
4	Pomegranate *(rind)*	*Punica granatum*	Tannin, alkaloid	Yellow, grey	Wool, silk with metallic mordants	Color	Wash*(G)* Light*(G)* Rub*(G)*	Aqueous extraction, Ethanolic extraction	Das et al. (2006)
5	Madder*(root)* Onion*(skin, pulp)* Pomegranate *(rind)* Pep-permint*(leaf)*	*Rubia tinctorum, Allium cepa, Punia granatum L., Mentha SP*	Anthraquinone, flavoniod, tannin, terpinene menthol	–	Wool with metallic mordants	Color, anti-microbial	–	Aqueous extraction	Çalıs et al. (2009)
6	Fruit	*Opuntia ficus-indica*	Indicaxanthin, betanin	–	Wool with mordant	Color	Wash*(G)*	Extraction with ethanol	Guesmi et al. (2012)

7	Fruit	*Opuntia ficus-indica*	Indicaxanthin	Orange-yellow	Modified acrylic fibers	Improvement of dye uptake by sonicator dyeing	Light(*M*)	Extraction with ethanol	Guesmi et al. (2013)
8	Prickly pear(*fruit*)	*Opuntia Lasiacantha* Pfeiffer	Betalain	Red	Wool with mordants	Showing high bacteria inhibition activity	Light(*G*)	Extraction with water	Ali and El-Mohamedy (2011)
9	Mangosteen (*hull*)	*Garcinia mangostana Linn*	Anthocyanin	Red	Cotton, Silk with metallic mordants	Color	Wash(*G*) Light(*G*) Rub(*G*)	With aqueous citric acid solution	Chairat et al. (2007)
10	Amla(*fruit*)	*Emblica officinalis G.*	Tannin	–	Cotton, Silk with natural and metallic mordants	Antibacterial	Wash(*G*)	Ethanolic extraction	Prabhu et al. (2011)
11	Green chili	*Capsicum annum*	Oleoresin	Yellow	Cotton with modrants	–	Wash(*M*) Light(*G*) Rub(*G*)	Solvent extraction method	Kulkarni et al. (2011b)

[a]Excellent (E), Very good (VG), Good (G), Medium (M), Poor (P)

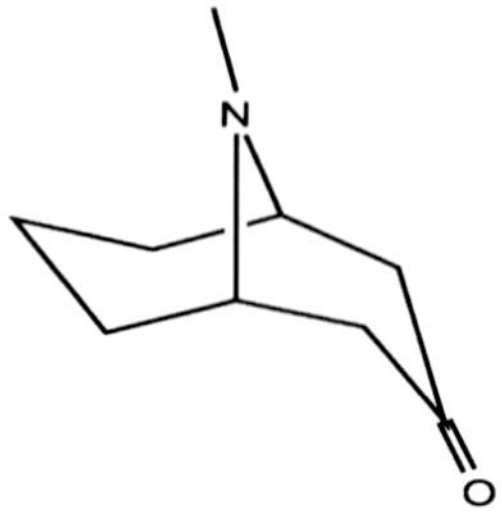

Fig. 6.12 Chemical structure of N-methyl granatonine found in the fruit rind of pomegranate

As an example, pomegranate rind extract was used for bi-functional treatment of cotton, i.e., color and bactericidal finish (Rajendran et al. 2011). The dye was extracted using water and ethanol, separately. The dye extract showed good affinity for cotton, and was applied with alum and copper. Good wash, light, rub, perspiration fastnesses and excellent antibacterial property against common bacteria, *S. Aureus* and *E. coli*, was resulted. However, antibacterial property steadily decreased by successive wash cycles.

Pomegranate rind has also been utilized for dyeing of wool and silk with/without mordants (Das et al. 2006; Çalıs et al. 2009). The best dyeing pH was found to be four. Iron and aluminium were resulted in grey and yellow shades on fibers, respectively. Dye up-take, and light fastness were improved with mordants, however, wash fastness was not markedly changed (Das et al. 2006).

In a similar study, four common natural dyes were applied on wool and resulting antimicrobial effect was examined (Çalıs et al. 2009). Bacterial growth reduction of 4–80 % for pomegranate rind, *Punica granatum*, 53–86 % for onion, *Allium cepa L.*, 32–52 % for madder, *Rubia tinctorium L.*, and 28–91 % for peppermint, *Mentha piperita L.*, was obtained against variety of common pathogenic bacteria used.

Guesmi et al. (2012) have extracted indicaxanthin from *Opuntia ficus-indica* fruit and used for dyeing of textile fabrics. Extraction method was to mix 50 g of juice from cactus pears with 100 ml of 80 % aqueous ethanol as solvent. This plant contains mainly two dyes, indicaxanthin and betanin (Fig. 6.13).

Extracted indicaxanthin was used for dyeing of wool fabrics with different mordants and the results showed that the fabrics dyed have good washing fastness, where $KAl(SO_4)_2$, $MnSO_4$ and $CoSO_4$ gave the best results.

In a similar work, indicaxanthin has been used for dyeing of modified acrylic fibers by both conventional and ultrasonic techniques (Guesmi et al. 2013). The authors have reported that sonicator dyeing with this orange-yellow natural dye showed marked improvement in dye uptake and fair to good fastness properties of the dyed fabrics.

Relevantly, Ali and El-Mohamedy (2011) have tried to extract the dyeing agent of red prickly pear plant, *Opuntia lasiacantha* Pfeiffer, and used it for dying of wool with different types of mordants. Betalain (Fig. 6.14), the red dye extracted from this plant, was used for dyeing of wool fabrics with ferrous sulfate, copper sulfate, potassium dichromate and tannic acid as mordants.

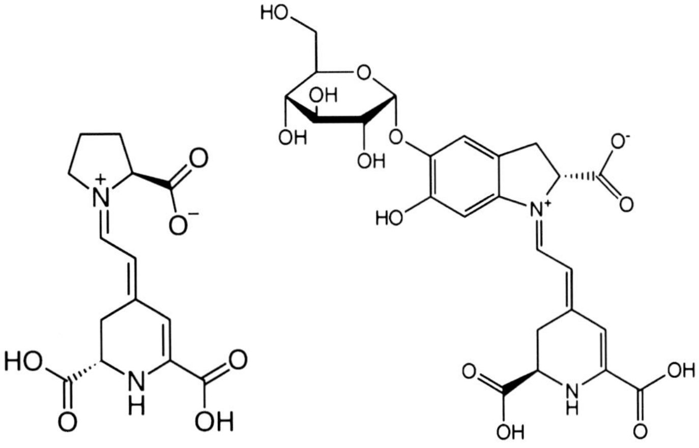

Indicaxanthin Betanin

Fig. 6.13 Chemical structures of indicaxanthin and betanin extracted from *Opuntia ficus-indica* fruit

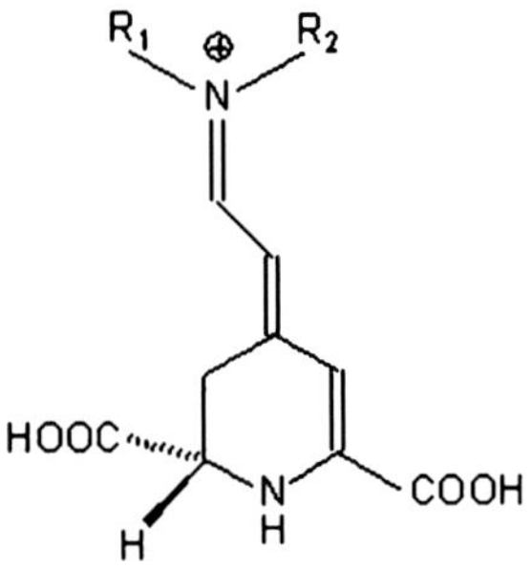

Fig. 6.14 Betalain pigment structure extracted from *Opuntia lasiacantha* Pfeiffer fruits

The fabrics dyed with betalain showed high bacteria inhibition activity and good fastness property.

Garcinia mangostana Linn, is a tropical evergreen tree which grows especially in Thailand, Indonesia, Malaysia, and Philippines. It is known for its best tasting fruits. Fruit shells contain various useful natural components such as mangosteen that is a red pigment. Chairat et al. (2007) have extracted mangosteen using an aqueous acetic acid solution where extract solution had dark red color. The extract solution was used in dyeing of silk and cotton with Al, Ca, Fe, Zn, and a natural mordant. Generally, color depth, wash and light fastnesses of dye were markedly improved using mordants and the best results obtained in post-mordanting method with Fe and Ca.

Emlica officialis G. fruit is a rich resource of natural tannins used as natural mordants. It contains mainly two hydrolysable tannins that on hydrolysis give gallic acid and ellagic acid. Prabhu et al. (2011) have used the ethanol isolated tannins

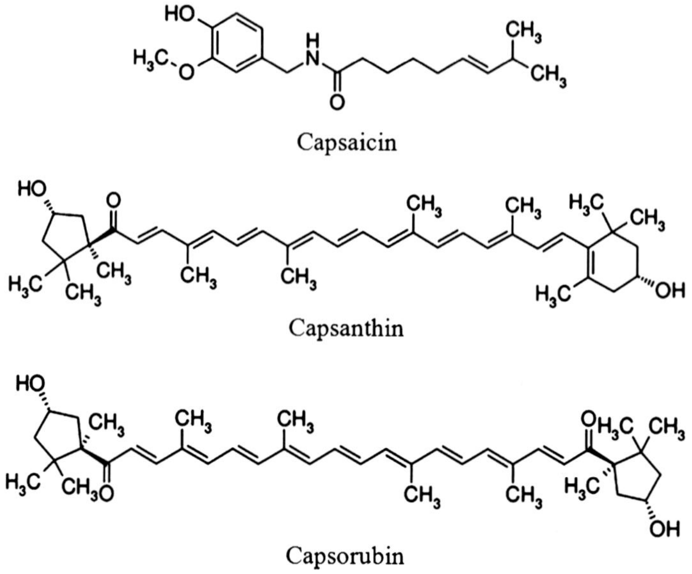

Fig. 6.15 Chemical structure of oleoresin ingredients; capsaicin, capsanthin and capsorubin

individually and/or in mixture with copper sulphate for enhancement of color yield and color fastness properties in coloration of cotton and silk using natural dyes. Four natural dyes used were turmeric, *Curcuma longa L.*, pomegranate rind, *Punica granatum L.*, henna, *Lawsonia inermis L.*, and indian madder, *Rubia cordifolia L.* Different shades were obtained using these dyes together with mordants showing good color fastness and color strength. Antibacterial test was also carried out to evaluate the bactericidal property of *Emlica officialis G.*on cotton and silk and it was found that cotton and silk treated with this natural mordant have good antibacterial activity which was durable up to 20 washes.

In a similar work, Kulkarni et al. (2011b) have tried to produce different shades of yellow color using the dye extracted from chili skin. The yellow pigment of *Capsicum annum*, that mainly contains oleoresin, was extracted by solvent extraction method and was used for dyeing of cotton fibers by mordants. Oleoresin is an oil soluble extract from the fruits of *Capsicum annum Linn*, and is primarily used as a coloring and/or flavouring in food products. It is composed of capsaicin, the main flavouring compound giving pungency in higher concentrations, and capsanthin and capsorubin, the main coloring compounds (Pérez-Gálvez et al. 2003) (Fig. 6.15).

Based on the results obtained, the authors have reported good light and rub fastness, and moderate wash fastness for the fabrics dyed with the green chili extract.

Bark

Some studies have reported the use of bark of various plants as natural dye resource (Table 6.5). Reuse of ash-tree bark for extraction of a natural dye for textile dyeing

Table 6.5 Characteristics of dyes/antimicrobials extracted from bark of the plants and their application properties

No.	Source (used part)	Botanic name	Chemical base	Shade on substrate	Substrate applied	Properties achieved	Dye fastness[a]	Extraction method	Reference
1	Ash-tree	*Fraxinus excelsior* L.	*Quercetin*	Beige olive	Wool with mordant	Good shade reproducibility and levelness of dyed fibers	Light(*G*)	Extracted with distilled water	Bechtold et al. (2007)
2	Chandada(*bark*)	*Macaranga Peltata*	–	Yellow, red	Silk	Color	–	Soxhlet extraction with methanol	Vinod et al. (2010)
3	Barberry(*bark*)	*Berberis aristata DC.*	–	Wide range of shades	Silk with metallic mordants	Color	Wash(*P*) Light(*VG*) Rub(*VG*)	Aqueous extraction	Pruthi et al. (2008)
4	–	*Sticta coronate lichen*	–	Lilac (pale purple)	Silk with metallic, natural mordants	Color	–	Ultrasonic extraction with aqueous acetone	Mansour and Heffernan (2011)
5	Ficus(*bark*)	*Ficus Religiosa L.*	–	Orange, red, brown	Silk with metallic, natural mordants	Color	Wash(*G*) Light(*G*) Rub(*G*)	Aqueous extraction	Saravanan and Chandramohan (2011)
6	Jackfruit(*wood*)	*Artocarpus heterophyllus Lam*	Flavanol (Hydroxyflavone) (Morol)	Yellow to golden brown	Cotton, jute with metallic, natural mordants	Color	Wash(*VG*) Light(*M*) Rub(*G*)	Aqueous extraction	Samanta et al. (2007)

(continued)

Table 6.5 (continued)

No.	Source (used part)	Botanic name	Chemical base	Shade on substrate	Substrate applied	Properties achieved	Dye fastness[a]	Extraction method	Reference
7	Jackfruit wood Manjistha/ Madder *(root)* Red sandalwood Marigold *(flower)* Babool*(bark, wood)*	*Artocarpus heterophyllus L., Rubia cordifolia, Pterocarpus santalinius, Tagetes patula/Tagetes erecta, Acacia arabica*	Morol putpurin, manjistin, pupuroxanthin, pseudo-purpurin, nordamncanthal, santalin A, santalin B, deoxyxantalin, flavonoid, cretonoid, polyphenol	Variety of shades	Jute with natural, metallic mordants	Color	Wash*(VG)* Light*(M)* Rub*(VG)*	Aqueous extraction	Samanta et al. (2008)

[a]Excellent (E), Very good (VG), Good (G), Medium (M), Poor (P)

was the subject of an interesting research (Bechtold et al. 2007). A quercetin-rich beige olive pigment extracted with distilled water from ash-tree bark, *Fraxinus excelsior* L., was used for dyeing wool fabrics with different mordants. Good shade reproducibility and levelness of the dyed material along with the acceptable fastness to light and wash were obtained.

Again, the colorant extracted from the tree bark has been used by Vinod et al. (2010) for dyeing of textile fabrics. In this work, pigment from the bark of *Macaranga peltata* was extracted by methanol and, using spectral techniques, ellagic acid was identified as the main coloring agent. Then, this yellow pigment was used for dyeing silk with tannic acid and potash alum [$Al(NH_4)(SO_4)_2.12H_2O$] as mordant, and the results demonstrated good fastness of dyed fabrics.

Barberry plant, *Berberis aristata DC.*, is another plant that its bark has been used in textile coloration (Pruthi et al. 2008). Silk was dyed with aqueous extract of barberry bark and post-mordanted with different combination of various mordants. Wide range of shades was obtained with high dye fastness properties. Based on the results several conclusions were drawn: best dyeing method for silk is simultaneous mordanting because of the highest dye up-take in this method; alum must be used in combination with other mordants due to its dye-up take improvement ability; the use of combination of mordants produce wide range of shades.

Similar attempt has been made to make use of *Sticta coronata* lichen as an abundant resource of natural dye in New Zealand in ultrasonic dyeing of silk (Mansour and Heffernan 2011). Catechu, *Acacia catechu*, a natural mordant and alum were used in dyeing process. Ultrasound energy improved dye up-take markedly. The dye showed good affinity and produced brilliant lilac color with good fastness on silk fibers.

In similar study, bark of Peepal tree, *Ficus religiosa* L., were used in natural dyeing of silk (Saravanan and Chandramohan 2011). The color component was extracted with water and then was used for dyeing of silk with metallic mordants such as Al, Cu, Cr, Ni, and Sn and natural mordants like myrobolan, *Terminalia chebula*, and cow dung. Myrobolan or harda is an important natural mordant with chebulinic acid as main ingredient whose chemical structure is shown in Fig. 6.16. The color fastness properties on fibers were overall acceptable.

Similarly, Samanta et al. (2007) have used the extraction of Jackfruit wood, *Artocarpus heterophyllus* Lam, in coloration of cotton and jut fibers with myrobolan, aluminium, iron, tin and EDTA mordants. The main coloring component of jackfruit is morol, a typical flavanol that imparted yellow to golden brown shade with various mordants on cotton and jut fibers. The optimum dyeing condition was, time 90 min, temperature 70–90 °C, 20–30 % mordant, 30–40 % dye with pre-mordanting. Good color fastness properties were reported for both dyed fibers.

After that, Samanta et al. (2008) in another study have used binary mixture of jackfruit extract with other natural dyes, manjistha, red sandal wood, marigold, sappan wood, and babool in dyeing of jute fibers with myrobolan and aluminium mordants. They also examined the compatibility of mixed dyes. After dyeing, additional treatment with various chemicals was performed to enhance wash and light fastness.

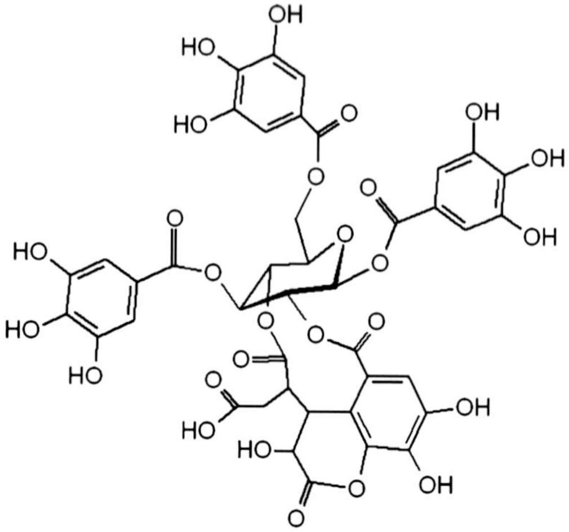

Fig. 6.16 Chemical structure of chebulinic acid, the main component of myrobolan natural mordant

Flower

Natural dyes/antimicrobials can also be extracted from flower part of various plants (Table 6.6). Pigments extracted from marigold flowers, *Tagetes,* with a quite dark and saturated yellow to orange red color on textiles is another example of natural dyes/antimicrobials. The aqueous extraction of african marigold, *Tagetes erecta* L., that contains mainly lutein has been used for dyeing of cotton and silk fabrics with mordants (Jothi 2008). The author has reported good light fastness and good to excellent wash fastness for the fabrics dyed with this natural yellow pigment.

In a similar work, Guinot and co-workers (2008) have evaluated the dyeing potential of pigments from marigold flowers, *Tagetes patula* L. Flavonoids were obtained by water-ethanol mixture extraction technique, giving the highest extraction efficiency, and were used for dyeing of woven wool with mordants. Based on the results observed, the authors have introduced marigold flower as a potential dyeing plant.

A greenish yellow natural dye called luteolin was also extracted by methanol from leaves, stems and flowers of weld (*Reseda luteola*) (Cerrato et al. 2002). Then, it was used for dyeing of cotton and wool fibers with alum as mordant. A composite design experiment allowed the effects of particle size and liquid/solid ratio to be determined and the results demonstrated that luteolin concentrated extract could be used for home dyeing of fabrics of kits specially designed textiles.

Also, Vankar and Shanker (2009) have tried the pretreatment of silk fiber in an eco-friendly dyeing with flower extract. A bright reddish brown pigment, containing quercetin, was extracted by soaking the dried and ground flower sample, *Delonix regia,* in distilled water and then, was used for fiber dyeing with mordants.

Wool fabrics have also been dyed with catechu (*Acacia catechu*) extract by Bhattacharya and Shah (2000). The light brown pigment, containing mainly tannin,

Table 6.6 Characteristics of dyes/antimicrobials extracted from flower of the plants and their application properties

No.	Source (used part)	Botanic name	Chemical base	Shade on substrate	Substrate applied	Properties achieved	Dye fastness[a]	Extraction method	Reference
1	African marigold	*Tagetes erecta* L.	Lutin	Yellow to orange red	Cotton and silk fabrics with mordants	–	Wash(*G to E*) Light(*G*)	Extraction with water	Jothi (2008)
2	Marigold flower	*Tagetes patula* L.	Flavonoid	–	Woven wool with mordant	Development of new natural coloring matters	–	Extraction with water-ethanol mixture	Guinot et al. (2008)
3	Weld	*Reseda luteola*	Luteolin	Greenish yellow	Cotton and wool with mordants	Color	–	Extraction with methanol	Cerrato et al. (2002)
4	Flower	*Delonix regia*	Quercetin	Bright reddish brown	Silk with mordant	Eco-friendly textile pretreatment that does not utilize metal mordanting	Light(*M*)	Extraction with water	Vankar and Shanker (2009)
5	Catechu	*Acacia catechu*	Tannin	Light brown	Wool with mordant	Production of different colors	Light(*G*)	–	Bhattacharya and Shah (2000)
6	African tulip tree	*Spathodea campanulata*	–	Various shades	Cotton with metallic natural mordants	Color	Wash(*E*) Light(*M*) Rub(*E*) Perspiration (*G-E*)	Aqueous extraction	Kumaresan et al. (2012)

(continued)

Table 6.6 (continued)

No.	Source (used part)	Botanic name	Chemical base	Shade on substrate	Substrate applied	Properties achieved	Dye fastness[a]	Extraction method	Reference
7	Plants	*Acacia catechu, Kerria lacca, Quercus infectoria, Rubia cordifolia, Rumex maritimus*	–	–	Wool	Antimicrobial activity	–	–	Singh et al. (2005)
8	Oak	*Quercus infectoria*	–	–	Cotton with mordants	Antimicrobial	–	–	Gupta and Laha (2007)
9	Herb leaves, stem and flower	*Ricinus communis, Senna auriculata, Euphorbia hirta*	–	–	Denim fabric	Antibacterial	–	Methanol extraction	Sumithra and Raaja (2012)
10	Herb	*Sappanwood*	–	–	Wool with mordant	Anti-ultraviolet effect	Rub*(G)*	Dissolution and extraction	Zhang and Cai (2011)

[a]Excellent (E), Very good (VG), Good (G), Medium (M), Poor (P)

was used for dyeing of fabrics with different mordants. Different colors produced on wool fabric with catechu using various metal mordants. Moreover, the fastness properties were sufficiently good for practical dyeing.

Spathodea campanulata is a flowering tree that is planted extensively in tropical countries as an ornamental tree. It has very showy reddish-orange or crimson (rarely yellow), campanulate flowers. Kumaresan et al. (2012) have investigated the dyeing properties of cotton with dye extracted from *Spathodea campanulata* flowers without and with various binary mixtures of natural/metallic mordants. Extraction of dyed was carried out in boiling water and ethanol separately. Ethanol extract was good to obtain brown shades. Generally, in almost all mordant-dye-fiber cases, color fastness to wash, rub, and perspiration was quite high. It was inferred that this dye can be used as an eco-friendly natural dye in coloration of textiles.

On the other hands, antibacterial activity of fibers dyed with extracts from plants has been widely investigated by the scientists. The antimicrobial activity of five natural dyes, *Acacia catechu, Kerria lacca, Quercus infectoria, Rubia cordifolia, Rumex maritimus*, on wool fabrics has been shown by Singh et al. (2005). Moreover, tannin-rich extract of oak has been used as an antibacterial agent for textiles by Gupta and Laha (2007). In this work, the extract of *Quercus infectoria* plant in combination with alum, copper and ferrous mordants have been used as antibacterial agents for treatment of cotton fabrics. The results showed that the cotton textiles can be successfully treated with this extract to show antibacterial activities against both Gram-positive and Gram-negative bacteria.

In a similar work, the antibacterial activity of extracts from herb leaves, stem and flower has been studied (Sumithra and Raaja 2012). The methanol extracts of *Ricinus communis*, *Senna auriculata* and *Euphorbia hirta* were examined for their antimicrobial efficiency on the four variant of denim fabrics directly by using dip method and the findings demonstrated that the fabrics had good antibacterial activities. Similarly, the natural pigment from sappan was used for coloration of wool fabrics after treatment with the protease and transglutaminase (Zhang and Cai 2011). The pigment of sappan was extracted by dissolving the herb in distilled water and allowed to boil in a beaker kept over water bath for quick extraction. Then, this pigment was used for dyeing of wool fabrics with ferrous sulphate, potassium alum, and tannic acid as mordants. This process had a good anti-ultraviolet effect on the fibers while modification with protease led to some decrease in wet rub fastness, whereas transglutaminase had almost no influence on rub fastness. But, the treated fibers had good dry rub fastness.

Other Parts of Plants

Natural extracts of other parts of plants have also important dyeing/antimicrobial properties and their application on textile has been widely studied (Table 6.7).

A natural dye was extracted by ultrasonic method from a vegetable with botanic name *Sargentodoxa cuneata,* and its application on wool fabric was studied (Xinsheng et al. 2008). The authors have reported that the color of fabrics dyed with ultrasonic extraction solution is deeper.

Table 6.7 Characteristics of dyes/antibacterial extracted from other parts of plants and their application properties

No.	Source (used part)	Botanic name	Chemical base	Shade on substrate	Substrate applied	Properties achieved	Dye fastness[a]	Extraction method	Reference
1	Vegetable	*Sargentodoxa cuneata*	–	–	Wool with mordant	Color	–	Ultrasonic extraction technique	Xinsheng et al. (2008)
2	Weld, buckthorn, bastard hemp	*Rhamnus petiolaris* Boiss, *Datisca cannabina* L., *Reseda luteola* L.	Flavonoid and anthra-quinone	Yellow	Wool with mordant	Color	–	Extraction of with hydrogen chloridemethanol-water mixture	Deveoglu et al. (2012)
3	Plant	*Dahlia variabilis*	Flavonoids – antho-cyanins	–	Wool with mordant	Good potential to replace phe-nolphthalein as an indicator	Light*(G)*	Ultra sound assisted extraction	Mishra et al. (2012)
4	Madder, weld and woad	*Reseda luteola* L., *Isatis tinctoria, Rubia tinctoria*	–	Red Yellow Blue	Cotton with mordant	Color	Light*(P)*	–	Cristea and Vilarem (2006)
5	Tesu	*Buteamonosperma frondosa*	Butein	Yellow, Orange, Brown	Jute with metallic, natural mordants	Color	Wash*(G)* Light*(G)* Rub*(G)*	Aqueous extraction	Samanta et al. (2011)
6	Silver oak	*Grevillea robusta*	Tannin	–	Wool with enzyme and mordants	Color, antimicrobial	–	Aqueous extraction	Raja and Thilaghavati (2011)
	Flame of forest	*Spathodea campanulata*	–						
	Tanner's senna	*Cassia auriculata*	–						
	Wattle	*Acacia decurrens*	Tannin						

[a]Excellent (E), Very good (VG), Good (G), Medium (M), Poor (P)

In a similar work, the mordanted wool fabrics with alum, were dyed in 50 % *Reseda luteola* L. (weld), 20 % *Rhamnus petiolaris* Boiss (buckthorn) and 50 % *Datisca cannabina* L. (bastard hemp) dye-baths (Deveoglu et al. 2012). Extraction of yellow pigments, containing mainly flavonoid and anthraquinone, was carried out with a hydrogen chloridemethanolwater mixture. The authors have reported as a conclusion that, it would be possible to analyze and to identify the natural dyes present in historical textiles using a reverse-phase high-performance liquid chromatography.

Natural dye from *Dahlia variabilis* have been similarly used for dyeing of textile (Mishra et al. 2012). The flavonoids – anthocyanins rich colorant was extracted by ultra sound assisted technique and was used for dyeing of wool fibers with alum and milk of tartar as mordants. The authors have reported good fastness properties of dyed fibers as well as intrinsic potential of this dye to replace phenolphthalein as a pH indicator.

Cristea and Vilarem (2006) have also studied the light fastness of selected natural dyes from madder, weld and woad. Red, yellow and blue extracts of *Rubia tinctoria*, *Reseda luteola L.* and *Isatis tinctoria*, respectively, were used for dyeing of cotton with mordants. The authors have concluded that, in spite of poor fastness of fiber dyed with these three natural dyes, using UV absorbers or antioxidants improves the light fastness of dyed fabrics.

Samanta et al. (2011) have used aqueous extract of flower petals of tesu flowering tree, *buteamonosperma frondosa*, in coloration of jute fabric in alkaline medium to obtain various shades like creamish orange, yellowish orange, ochre brown, and reddish brown. The main color component in the yellowish red/orange extract was butein (Fig. 6.17). Good color fastness and color yield was reported for proper selected dyeing condition (dye-fiber-mordant system) and further improvements in light fastness were obtained with proper post-chemical treatment.

In an interesting and recent research work, Raja and Thilaghavati (2011) have tried to make use the inherent antibacterial characteristic of some natural dyes in bi-functional environmentally friendly treatment of textiles. In their work, four natural dyes, with in vitro antimicrobial activity against both gram positive and gram negative bacteria, were chosen. The effect of mordant and protease enzyme treatment of wool on the bactericidal performance of natural dyes were examined and compared with untreated wool results. Various test results showed higher dye up-take and antibacterial property in enzyme treated wool that was ascribed to increased dye diffusion and affinity due to hydrolysis of outer layer of wool. Moreover, although the color fastness properties of dyed wool were increased upon mordant treatment, but none of dyed fibers mordant treated showed antibacterial activity. Such behavior was attributed to the blocking of bio-active ingredients in natural dye through possible complex formation of metal ion with those components (Raja and Thilaghavati 2011).

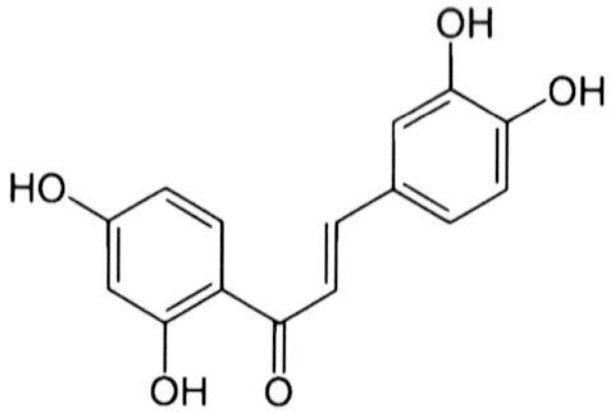

Fig. 6.17 Chemical structure of butein extracted from flower petals of tesu flowering tree

Dyes/Antimicrobials from Microorganisms

The main disadvantage of natural dyes/antimicrobials from plants is the low efficiency of the extraction process (a few grams of pigment per kg of dried raw material). This makes their application limited to high-value-added natural-colored garments only (De Santis et al. 2005). To overcome this drawback, it was suggested to exploit the potentiality of microorganisms such as fungi, bacteria and algae that are fast growing and have the potential of being standardized commercially (Sharma et al. 2012).

There are several microorganisms which can produce pigments, which are one of the important classes of secondary metabolites and are often referred to as biopigments. Microorganisms produce various pigments including carotenoids, melanins, flavins, quinones, prodigiosins and more specifically monascins, violacein or indigo (Khanafari et al. 2006; Venil and Lakshmanaperumalsamy 2009). So, researchers have recently focused on the production, identification and application of these pigments on textiles and have studied the dyeing and antimicrobial characteristics of the dyed fibers (Table 6.8).

Chitosan is a linear polysaccharide produced commercially by deacetylation of chitin, the structural element in the crustaceans such as crabs and shrimp and cell walls of fungi.

Chitosan, *N*-deacetylated derivative of chitin that is a linear polysaccharide, is also an important natural antibacterial agent. It has been used in wool finishing, but its weak binding to the wool constitutes the main problems in its application. For instance, Ranjbar-Mohammadi and her co-workers (2010) have studied the grafting of chitosan onto wool fabric using anhydride bridge and its antibacterial property. The results showed that the wool fabrics treated by chitosan had a considerable antimicrobial and antifelting activities as well as good wash fastness.

In a similar works, chitosan has been used to impart antibacterial properties to the cotton yarns (Karolia and Mendapara 2007; Kavitha et al. 2007). It was observed that the chitosan finish provides better functionality to the fabrics with increasing tensile strength, flexural rigidity and shear strength as well as showing excellent activity of the cotton fabrics against bacteria.

Prodigiosin is a bright red pigment with a pyrrolylpyrromethane chemical structure that possesses antibacterial activity, too. Venil and Lakshmanaperumalsamy (2009) have studied the prodigiosin potential of *Serratia* and the effect of different factors influencing the production of this pigment. They have reported the optimum

Table 6.8 Characteristics of dyes/antibacterial extracted from microorganisms and their application properties

No.	Source	Scientific name	Chemical base	Shade on substrate	Substrate applied	Properties achieved	Dye fastness[a]	Extraction method	Reference
1	–	*Chitosan*	Polysaccharide	–	Wool	Anti-microbial, antifelting	Wash(*G*)	–	Ranjbar-Mohammadi et al. (2010)
2	–	*Chitosan*	Polysaccharide	–	Cotton	Antimicrobial	Wash(*G*)	–	Karolia and Mendapara (2007)
3	–	*Chitosan*	Polysaccharide	–	Cotton	Increasing tensile strength, flexural rigidity and shear strength. Excellent activity against bacteria	–	–	Kavitha et al. (2007)
4	Bacteria	*Serratia*	Tri-pyrrylmethene	Bright red	–	Antibacterial	–	–	Venil and Lakshmanaperumalsamy (2009)
5	Bacteria	*Serratia marcescens*	Pyrrolylpyrromethane	Bright red	wool, silk, nylon, cotton and polyester	Antibacterial, antifungal, antioxidant	Resistant to acid, alkali and detergents	Centrifugation	Gulani et al. (2012)
6	Bacteria, marine sediments	*Vibrio gazogenes*	–	Bright red	wool, nylon, acrylics, and silk	Antibacterial	–	Fermentations	Alihosseini et al. (2008)

(continued)

Table 6.8 (continued)

No.	Source	Scientific name	Chemical base	Shade on substrate	Substrate applied	Properties achieved	Dye fastness[a]	Extraction method	Reference
7	Fungi	*Monascus purpureus, Isaria farinosa, Emericella nidulans, Fusarium verticillioides, Penicillium purpurogenum*	–	Red, pink, red, yellow, yellow	Cotton fabric and leather, with mordant	Anti-microbial	Light(*G*)	Absorption of the pigments on silica gel	Velmurugan et al. (2009)
8	Fungi	*Monascus purpureus*	–	Red or orange	Wool with mordant	Cost-effective production	Light(*G*)	Extraction by ethanol	De Santis et al. (2005)
9	Bacteria: bacillus	*Protease*	–	–	Wool and Silk	Increasing the degree of dye uptake	–	Dilution and incubation	Kim et al. (2005)
10	Bacteria	*Serratia sakuensis*	–	Red	Silk, wood and cotton	High color strength values	Color(*G*)	Solid-state cultivation	Vaidyanathan et al. (2012)
11	Fungi	*Trichoderma virens, Alternaria alternate, Curvularia lunata*	–	–	Wool and silk with mordants	Antifungal	Wash(*G*) Rub(*G*)	–	Sharma et al. (2012)
12	Fungi	*Monascus purpureus, Isaria farinosa, Emericella nidulans, Fusariumverti-cillioides, Penicillium purpurogenum*	–	Red pink, reddish brown, and yellow	Cotton with mordant	Considerable shade	Wash(*G*)	Extraction with ethanol	Velmurugan et al. (2010)

[a]Excellent (E), Very good (VG), Good (G), Medium (M), Poor (P)

conditions for prodigiosin biosynthesis as well as the different techniques used for its extraction. It has been concluded that prodigiosin is a remarkable species for its diverse range of biological and dyeing effects, although much deeper insight into the mode of action of these compounds is required.

Gulani et al. (2012) have studied the effect of different parameters on the production of prodigiosin form *Serratia marcescens*. The effects of various media components and process parameters like carbon and nitrogen sources, temperature, pH, incubation period were investigated. The pigment was extracted by centrifugation and applied on wool, silk, nylon, cotton and polyester fibers. They have reported that the dyed fibers show good antibacterial, antifungal, and antioxidant activity where a good resistant to acid, alkali and detergents was observed.

This natural pigment has also been isolated from marine sediments by Alihosseini and her co-workers (2008). They have used this bright red pigment for dyeing wool, nylon, acrylics, and silk fibers and have reported that fabrics dyed with the microbial prodiginines demonstrate antibacterial activity.

Velmurugan et al. (2009) have studied the anti-bacterial activity and dyeing property of the pigments obtained from five fungal species. The fungi species were *Monascus purpureus, Isaria farinosa, Emericella nidulans, Fusarium verticillioides,* and *Penicillium purpurogenum,* producing red, pink, red, yellow and yellow pigments, respectively. The pigments were extracted by absorption on silica gel and used for dyeing of cotton fabric and processed leather with mordanting. The results showed that the dyed fibers have anti-microbial activity and good color fastness property.

In a similar work, dyeing properties of pigment from fungi *Monascus purpureus* has been investigated by De Santis and co-workers (2005). Red to orange pigment has been extracted by ethanol and along with alum or stannic chloride as mordants applied for dyeing wool fibers. The authors have reported good dye fastness for the dyed fibers and introduced this process as a cost-effective production.

Kim et al. (2005) have used the protease extracted from bacteria "bacillus" for improvement of dyeing quality in wool and silk. The enzyme was extracted by dilution and incubation of the soil sample containing the bacteria and was used for dyeing the fibers. Their results showed that by increasing the protease treatment, the dyeing characteristics of the fibers are enhanced, and as a result of soil removal, the fibers become smoother.

Extraction, characterization and use of a pigment produced by bacteria, *Serratia sakuensis* has been tried by Vaidyanathan et al. (2012). This novel red pigment was produced by solid-state cultivation of a bacterial isolate obtained from garden soil. It was a prodigiosin-like pigment that shows properties similar to prodigiosin. Then, the pigment was used for dyeing of silk, wool and cotton fabrics and the results showed that the dyed fabrics have good color fastness, high color strength values and dye uptake as well as antibacterial activity.

Similarly an attempt has also been made to apply the dyes obtained from fungus, *Trichoderma virens, Alternaria alternate,* and *Curvularia lunata* to the fibers by Sharma et al. (2012). Wool and silk fibers were dyed unmordanted and the results obtained show very good wash and rub fastnesses as well as antifungal activity.

Velmurugan et al. (2010) have also studied the dyeing of fibers with soluble pigments of selected fungi. They have used water soluble fungal pigments obtained from *Monascus purpureus, Isaria farinose, Emericella nidulans, Fusariumverticillioides,* and *Penicillium purpurogenum* species. These red pink, reddish brown, and yellow pigments extracted by ethanol were used for dyeing of cotton fibers accompanying mordanting. The authors have reported that a strong variation in shade and color depth could be achieved by using alum and ferrous sulfate that gives good wash fastness properties to the fabrics.

Reusing and Recycling of Natural Dyes

Some researchers have focused on the economical aspects of the application of natural dyes for textile such as reusing or recycling (Table 6.9). Mirjalili et al. (2011) have studied the dyeing of wool fibers with the extract of weld and concluded that in the case of dyeing with this extract, and keeping the good light fastness, the exhaustion rate for the extracted dye increased by 49 % compared to the raw dye.

In a similar work, Meksi and co-workers (2012) have used the olive mill wastewater as a potential source of natural dyes for wool dyeing. This wastewater was filtered and the coloring agents such as pectines, mucilages and tannins were extracted with hexane. Along with the darker shades and with generally good fastnesses, the results proved that mordanting give deeper shades and enhanced fastness properties.

In another work, plant waste materials including *Terminalia catappa, Artocarpus heterophyllus, Tectona grandis* and *Morinda citrofolia* were extracted by methanol and used for dyeing of silk fibers (Prusty et al. 2010). The fibers dyed with the extracts showed excellent antimicrobial activity and good durable fastness properties.

Another example for recycling of useless by-products of natural materials in sustainable treatment of textiles is the study that has been accomplished by Ahmad et al. (2011). They extracted color component from *Gluta aptera* wood (Rengas) saw dust and used it in environmentally friendly dyeing of cotton and silk. Water at boil and methanol at room temperature were used as extracting solvents. Rice husk ash and Al were used as natural and metallic mordants, respectively. Different brownish shades with good color fastness properties were obtained for both substrates.

In another similar work, Lee (2007) has tried to use the coffee sludge extract for dyeing of textile. The natural colorant was extracted from *Coffea arabica* L., using water as extractant and was applied for dyeing of cotton, silk, and wool with a variety of metallic salts as mordants. The author has reported that the fabrics dyed show good deodorization performance as well as excellent light fastness with mordanting.

Interestingly, Shams-Nateri (2011) has tried to reuse the dyebath containing madder extract for dyeing of wool fibers. The dyebath contained mainly alizarin, purpurin, and water, was reconstructed by adding madder. The author has reported

Table 6.9 Characteristics of dyes/antimicrobial studied for economical aspects of their application

No.	Source (used part)	Botanic name	Chemical base	Shade on substrate	Substrate applied	Properties achieved	Dye fastness[a]	Extraction method	Reference
1	Weld *(flower)*	–	–	–	Wool	Good potential of weld extracts for commercial dyeing	Light*(M)*	Extraction with soxhlet apparatus	Mirjalili et al. (2011)
2	Olive mill *(wastewater)*	–	Pectines, mucilages and tannins	–	Wool with mordants	Deeper shades and enhanced fastness properties	Light*(M)*	Filtration of wastewater and extraction with hexane	Meksi et al. (2012)
3	Plant *(waste materials)*	*Terminalia catappa, Artocarpus heterophyllus, Tectona grandis, Morinda citrofolia*	–	–	Silk with mordants	Excellent antimicrobial activity	Light*(M)*	Solvent extraction	Prusty et al. (2010)
4	Rengas *(wood)*	*Gluta Aptera*	Flavonoid (phenolic compounds)	Brown	Cotton, Silk with metallic mordants	Color	Wash*(VG)* Light*(M)* Rub*(VG)*	Aqueous extraction, Methanolic extraction	Ahmad et al. (2011)
5	Coffee *(sludge)*	*Coffea arabica* L.	–	–	Cotton, silk, and wool with mordants	Good deodorization performance	Light*(E)* with mordanting	Extraction with water	Lee (2007)

(continued)

Table 6.9 (continued)

No.	Source (used part)	Botanic name	Chemical base	Shade on substrate	Substrate applied	Properties achieved	Dye fastness[a]	Extraction method	Reference
6	Madder dyebath *(wastewater)*	*Rubia tinctorum* L.	Alizarin: 1,2-dihydroxy-anthraquinone Purpurin: 1,2,4-trihydroxyanthra-quinone	–	Wool	Reusing successfully the dyebath, 19.91% cost saving in wool dyeing	Light*(VG)*	Reconstruction of dyebath by adding madder and water	Shams-Nateri (2011)
7	Banana *(peel)*	*Musa cavendish*	Flavonoid	–	Cotton with metallic mordants	Color, antibacterial, UV protection	–	Alkaline extraction	Salah (2011)
8	Neem*(seed)* Mexican daisy*(flower)*	–	Alkaloids	–	Cotton	Antimicrobial	Wash*(VG)*	Methanolic extraction	Thilagavathi and Krishna Bala (2007)
9	Tulsi*(leaf)* Pomegranate *(rind)*	*Ocimum sanctum, Punica granatum*	Eugenol, germacrene, phytol, tannin	–	Cotton	Antibacterial	Wash*(G)*	Soxhlet extraction with methanol	Sathianarayanan et al. (2010)
10	Neem*(seed)*	*Azadirachta indica*	Azadirachtin, nimbin, salannin	–	Polyester and cotton blend fabric	Antimicrobial	Wash*(M)*	Methanolic extraction	Joshi et al. (2007)
11	Yarrow*(flower)*	*Achillea millefolium*	Flavonoids, luteolin and apigenin	–	Wool with metallic mordants	Color	Wash*(G)* Light*(G)*	Aqueous extraction	Kiumarsi et al. (2009)
12	Saffron*(flower petal)*	*Crocus sativus L.*	Anthocyanidin	Yellow to brown	Wool with metallic mordants	Color	Wash*(G)* Light*(G)*	Aqueous extraction	Mortazavi et al. (In press)
13	Indeciduous	*Caesalpinia sappan* L.	–	–	Poly(ethylene terphthalate)	Higher fastness, shortening of the dyeing time	–	–	Park et al. (2008)

[a]Excellent (E), Very good (VG), Good (G), Medium (M), Poor (P)

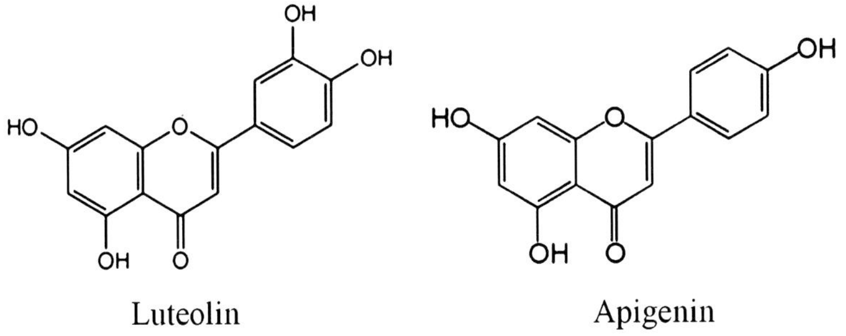

Fig. 6.18 Chemical structure of luteolin and apigenin extracted from *Achillea millefolium* plant

that dye fastness of the samples dyed in reconstructed dyebath is the same as initial wool dyeing while the economic analysis showed that the reusing wastewater causes 19.91 % cost saving in wool dyeing with madder.

In exploring to find an alternative to decrease environmental pollution from burning of useless by product of banana fields, the studies led to find out the coloring components present in banana peel (Salah 2011). Natural dye extracted from banana peel wastes, *Musa cavendish*, by alkaline extraction was used for coloration of cotton with iron mordant. In addition to good color imparted, dyed cotton showed excellent antibacterial activity and high UV protection properties.

To develop an eco-friendly and economical alternative for production of medical cotton and polyester/cotton, the extracts of pomegranate rind, *Punica granatum*, and leaves of tulsi, *Ocimum sanctum* (Sathianarayanan et al. 2010), neem, *Azadirachta indica*, and mexican daisy (Thilagavathi and Krishna Bala 2007) have been applied on cotton through variety of methods such as direct application, pad-dry-cure, micro-encapsulation, resin crosslinking and their combinations. Excellent antibacterial property was obtained in all methods with good wash durability except for direct application of extracts where wash fastness was poor (Joshi et al. 2007).

Interestingly, Kiumarsi et al. (2009) have recently optimized the coloration process of yarrow, *Achillea millefolium*, on wool using Taguchi statistical method. *Achillea millefolium* is a pharmaceutical plant known as a powerful healing herb containing flavoniods such as luteolin and apigenin as natural dyes (Fig. 6.18).

The dye aqueous extract was used for dyeing of wool with aluminium, tin, and copper mordants. Interesting variety of shades from pale yellow to dark green was obtained with very good wash and light fastness for all dyed samples. The process was claimed to be economical due to saving time and energy that could also be commercialized.

In a related and most recent work, Mortazavi et al. (2012) introduced saffron, *Crocus sativus L.*, as a potent natural dye resource. They used saffron flower petals and sepals extract, a rich resource of anthocyanidin natural dye previously discarded as useless wastes, in coloration of wool. They used different metallic mordants such as tin, copper, aluminium, iron, and chrome together with oxalic and tartaric acid to obtain a variety of shades as well as to enhance fastness properties of natural dye on wool. Although the color of fresh petals and sepals was light purple, but the shade range obtained on wool varied from yellow to brown with different mordants. Wash and light fastnesses of wool fibers were excellent with iron and chrome, moderate with copper and aluminium, and poor with tin mordants.

Dyeing of poly(ethylene terphthalate) fabrics with an aqueous extract of an indeciduous tree, *Caesalpinia sappan* L., was also investigated by Park and co-workers (2008). The results showed that the pretreatment with chitosan and/or plasma is better than a metal mordant in terms of the dye uptake and shortening of the dyeing time.

6.5 Conclusion

Due to the increasing awareness of the high risk impact associated with the synthetic colorants used for textiles, the application of natural or green extracts as colorants or antimicrobials has been these days increasingly considered. Review of the related published papers revealed that the non-reproducibility, time-consuming extraction methods and poor fastness properties are the main problems ahead the dyers and finishers to use these products in textile industry and consequently, must be the cores of scientific researches in this field.

The papers published in this field show that most of natural dyes have inherently antimicrobial properties and could consequently possess high medicinal activity. They are extracted from the different types of microorganisms as well as various parts of the plants such as bark, leaf, root, seed, fruit, and flower containing coloring materials like tannin, flavonoids, quinonoids, etc. Natural sources can offer not only a rich and varied source of dyestuffs, but also could be considered as safe, environmentally friendly and low cost treatments with additional benefit of coloring in a single stage. There have also been undeniable developments in extraction techniques of natural dyes/antimicrobials such as the invention of ultrasonic or enzyme assisted ultrasonic extraction methods where they prove high extraction efficiency over the conventional methods. But, there are still some important issues to solve. Most particularly, the application of metallic mordants for better fastness purpose is one of the problems concerning the environmental impact of natural dyes. The result of this survey showed that the shade and colorfastness of the fiber dyed depend not only on the natural colorant but also on the mordant and mordant assistants used. While dyes from natural sources are eco-friendlier than synthetic ones, the level of mordants used is normally higher than the permitted official concentration and poses environmental problem. It is expected that mordants will be replaced by modern environmentally friendlier technologies like plasma treatment. Moreover, studying and enlightening different processes involving natural dyes will be helpful in identification and analyzing color characteristics of historic/old textiles.

Real effort has been put to include all the relevant referred publications on the exploiting natural dyes/antimicrobials in green treatments of textiles in this review.

However, the limitation of our resources and the sheer number of publications in this field may have prevented the comprehensiveness of this report. Our sincere apologies are extended to any and all authors whose works are not included in this report.

Acknowledgments The authors are grateful to Tabriz Islamic Art University, Tabriz, Iran for the all supports.

References

Ahmad WYW, Rahim R, Ahmad MR, Abdul Kadir MI, Misnon MI (2011) The application of *Gluta Aptera* wood (Rengas) as natural dye on silk and cotton fabrics. Universal J Environ Res Technol 1(4):545–551

Ali NF, El-Mohamedy RSR (2011) Eco-friendly and protective natural dye from red prickly pear (*Opuntia Lasiacantha* Pfeiffer) plant. J Saudi Chem Soc 15:257–261

Ali S, Hussain T, Nawaz R (2009) Optimization of alkaline extraction of natural dye from Henna leaves and its dyeing on cotton by exhaust method. J Clean Prod 17:61–66

Alihosseini F, Ju K, Lango J, Hammock BD, Sun G (2008) Antibacterial colorants: characterization of prodiginines and their applications on textile materials. Biotechnol Prog 24(3):742–747

Ammayappan L, Moses JJ (2009) Study of antimicrobial activity of aloevera, chitosan, and curcumin on cotton, wool, and rabbit hair. Fiber Polym 10(2):161–166

Arora A, Gulrajani ML, Gupta D (2009) Identification and characterization of Ratanjot (*Arnebia nobilis* Reichb.f.). Nat Prod Radiance 8(2):142–145

Arora A, Gupta D, Rastogi D, Gulrajanii ML (2012a) Kinetics and thermodynamics of dye extracted from *Arnebia nobilis* Rech.f on wool. Indian J Fibre Text Res 37:178–182

Arora A, Rastogi D, Gupta D, Gulrajanii ML (2012b) Dyeing parameters of hydroxynaphthoquinones extracted from *Arnebia nobilis* Rech.f. Indian J Fibre Text Res 37:91–97

Bechtold T, Mahmud-Ali A, Mussak RAM (2007) Reuse of ash-tree (*Fraxinus excelsior* L.) bark as natural dyes for textile dyeing: process conditions and process stability. Color Technol 123:271–279

Bhattacharya SD, Shah AK (2000) Metal ion effect on dyeing of wool fabric with catechu. J Soc Dyers Colour 116:10–12

Borges ME, Tejera RL, Díaz L, Esparza P, Ibáñez E (2012) Natural dyes extraction from cochineal (*Dactylopius coccus*). New extraction methods. Food Chem 132:1855–1860

Çalıs A, Yuvalı Çelik G, Katırcıoglu H (2009) Antimicrobial effect of natural dyes on some pathogenic bacteria. Afr J Biotechnol 8(2):291–293

Cerrato A, De Santis D, Moresi M (2002) Production of luteolin extracts from *Reseda luteola* and assessment of their dyeing properties. J Sci Food Agric 82:1189–1199

Chairat M, Bremner JB, Chantrapromma K (2007) Dyeing of cotton and silk yarn with the extracted dye from the fruit hulls of mangosteen, *Garcinia mangostana* linn. Fiber Polym 8(6):613–619

Chan EWC, Lim Y, Wong S, Lim K, Tan S, Lianto F, Yong M (2009) Effects of different drying methods on the antioxidant properties of leaves and tea of ginger species. Food Chem 113(1):166–172

Chen C, Chang W (2007) Antimicrobial activity of cotton fabric pretreated by microwave plasma and dyed with onion skin and onion pulp. Indian J Fibre Text Res 32:122–125

Choo CKK, Lee YE (2002) Analysis of dyeings produced by traditional Korean methods using colorants from plant extracts. Color Technol 118:35–42

Cristea D, Vilarem G (2006) Improving light fastness of natural dyes on cotton yarn. Dyes Pigm 70:238–245

Das D, Bhattacharya SC, Maulik SR (2006) Dyeing of wool and silk with *Punica granatum*. Indian J Fibre Text Res 31:559–564
Das D, Maulik SR, Bhattacharya SC (2007) Dyeing of wool and silk with *Bixa orellana*. Indian J Fibre Text Res 32:366–372
Das D, Maulik SR, Bhattacharya SC (2008) Colouration of wool and silk with *Rheum emodi*. Indian J Fibre Text Res 33:163–170
Dawson TL (2008) It must be green: meeting society's environmental concerns. Color Technol 124:67–78
Dawson TL (2009) Biosynthesis and synthesis of natural colours. Color Technol 125:61–73
De Santis D, Moresi M, Gallo AM, Petruccioli M (2005) Assessment of the dyeing properties of pigments from *Monascus purpureus*. J Chem Technol Biotechnol 80:1072–1079
Deo HT, Desai BK (1999) Dyeing of cotton and jute with tea as a natural dye. J Soc Dyers Colour 115:224–227
Deveoglu O, Torgan E, Karadag R (2012) High-performance liquid chromatography of some natural dyes: analysis of plant extracts and dyed textiles. Color Technol 128:133–138
Farizadeh K, Montazer M, Yazdanshenas ME, Rashidi A, Mohammad Ali Malek R (2009) Extraction, identification and sorption studies of dyes from madder on wool. J Appl Polym Sci 113:3799–3808
Ghoranneviss M, Shahidi S, Anvari A, Motaghi Z, Wiener J, Šlamborová I (2010) Influence of plasma sputtering treatment on natural dyeing and antibacterial activity of wool fabrics. Prog Org Coat 70(4):388–393
Giri Dev VR, Venugopal J, Sudha S, Deepika G, Ramakrishna S (2009) Dyeing and antimicrobial characteristics of chitosan treated wool fabrics with henna dye. Carbohydr Polym 75:646–650
Guesmi A, Ben Hamadi N, Ladhari L, Sakli F (2012) Dyeing properties and colour fastness of wool dyed with indicaxanthin natural dye. Ind Crop Prod 37:493–499
Guesmi A, Ben Hamadi N, Ladhari N, Sakli F (2013) Sonicator dyeing of modified acrylic fabrics with indicaxanthin natural dye. Ind Crop Prod 42:63–69
Guinot P, Roge A, Gargadennec A, Garcia M, Dupont D, Lecoeur E, Candelier L, Andary C (2006) Dyeing plants screening: an approach to combine past heritage and present development. Color Technol 122:93–101
Guinot P, Gargadennec A, Valette G, Fruchier A, Andary C (2008) Primary flavonoids in marigold dye: extraction, structure and involvement in the dyeing process. Phytochem Anal 19:46–51
Guinot P, Gargadennec A, La Fisca P, Fruchier A, Andary C, Mondolot L (2009) *Serratula tinctoria*, a source of natural dye: flavonoid pattern and histolocalization. Ind Crop Prod 29:320–325
Gulani C, Bhattacharya S, Das A (2012) Assessment of process parameters influencing the enhanced production of prodigiosin from *Serratia marcescens* and evaluation of its antimicrobial, antioxidant and dyeing potentials. Malays J Microbiol 8(2):116–122
Gupta D, Bhaumik S (2007) Antimicrobial treatments for textiles. Indian J Fibre Text Res 32:254–263
Gupta D, Laha A (2007) Antimicrobial activity of cotton fabric treated with *Quercus infectoria* extract. Indian J Fibre Text Res 32:88–92
Gupta D, Kumari S, Gulrajani M (2001a) Dyeing studies with hydroxyanthraquinones extracted from Indian madder. Part 2: Dyeing of nylon and polyester with nordamncanthal. Color Technol 117:333–336
Gupta D, Kumari S, Gulrajani M (2001b) Dyeing studies with hydroxyanthraquinones extracted from Indian madder. Part 1: Dyeing of nylon with purpurin. Color Technol 117:328–332
Gupta D, Gulrajani M, Kumari S (2004a) Light fastness of naturally occurring anthraquinone dyes on nylon. Color Technol 120:205–212
Gupta D, Khareb SK, Lahaa A (2004b) Antimicrobial properties of natural dyes against Gram-negative bacteria. Color Technol 120:167–171
Han S, Yang Y (2005) Antimicrobial activity of wool fabric treated with curcumin. Dyes Pigm 64:157–161
Hill DJ (1997) Is there a future for natural dyes? Rev Prog Color 27:18–25

Iqbal J, Bhatti IA, Adeel S (2008) Effect of UV radiation on dyeing of cotton fabric with extract of henna leaves. Indian J Fibre Text Res 33:157–162
Joshi M, Wazed Ali S, Rajendran S (2007) Antibacterial finishing of polyester/cotton blend fabrics using neem (*Azadirachta indica*): a natural bioactive agent. J Appl Polym Sci 106:793–800
Jothi D (2008) Extraction of natural dyes from African marigold flower (*Tagetes ereectal*) for textile coloration. Autex Res J 8(2):49–53
Kamel MM, El-Shishtawy RM, Yussef BM, Mashaly H (2005) Ultrasonic assisted dyeing III. Dyeing of wool with lac as a natural dye. Dyes Pigm 65:103–110
Kamel MM, El-Shishtawy RM, Youssef BM, Mashaly H (2007) Ultrasonic assisted dyeing. IV. Dyeing of cationised cotton with lac natural dye. Dyes Pigm 73(3):279–284
Karolia A, Mendapara S (2007) Imparting antimicrobial and fragrance finish on cotton using chitosan with silicon softener. Indian J Fibre Text Res 32:99–104
Kasiri MB, Khataee AR (2011) Photooxidative decolorization of two organic dyes with different chemical structures by UV/H_2O_2 process: experimental design. Desalination 270:151–159
Kasiri MB, Aleboyeh H, Aleboyeh A (2008) Degradation of acid blue 74 using Fe-ZSM5 zeolite as a heterogeneous photo-Fenton catalyst. Appl Catal B 84:9–15
Kavitha T, Padmashwini R, Swarna A, Giri Dev VR, Neelakandan R (2007) Effect of chitosan treatment on the properties of turmeric dyed cotton yarn. Indian J Fibre Text Res 32:53–56
Ke G, Yu W, Xu W (2006) Color evaluation of wool fabric dyed with *Rhizoma coptidis* extract. J Appl Polym Sci 101:3376–3380
Khanafari A, Assadi MM, Fakhr FA (2006) Review of prodigiosin, pigmentation in *Serratia marcescens*. Online J Biol Sci 6(1):1–13
Kim T, Son Y (2005) Effect of reactive anionic agent on dyeing of cellulosic fibers with a Berberine colorant – part 2: anionic agent treatment and antimicrobial activity of a Berberine dyeing. Dyes Pigm 64(1):85–89
Kim T, Yoon S, Son Y (2004) Effect of reactive anionic agent on dyeing of cellulosic fibers with a Berberine colorant. Dyes Pigm 60(2):121–127
Kim S, Cha M, Oh ET, Kang S, So J, Kwon Y (2005) Use of protease produced by *Bacillus* sp. SJ-121 for improvement of dyeing quality in wool and silk. Biotechnol Bioprocess Eng 10:186–191
Kiumarsi A, Abomahboub R, Rashedi SM, Parvinzadeh M (2009) *Achillea millefolium*, a new source of natural dye for wool dyeing. Prog Color Colorants Coat 2:87–93
Kulkarni SS, Gokhale AV, Bodake UM, Pathade GR (2011a) Cotton dyeing with natural dye extracted from pomegranate (*Punica granatum*) peel. Universal J Environ Res Technol 1(2):135–139
Kulkarni SS, Bodake UM, Pathade GR (2011b) Extraction of natural dye from chili (*Capsicum annum*) for textile coloration. Universal J Environ Res Technol 1:58–63
Kumaresan M, Palanisamy PN, Kumar PE (2012) Application of eco-friendly natural dye on cotton using combination of mordants. Indian J Fibre Text Res 37:194–198
Lee YH (2007) Dyeing, fastness, and deodorizing properties of cotton, silk, and wool fabrics dyed with coffee sludge (*Coffea arabica* L.) extract. J Appl Polym Sci 103:251–257
Mansour HF, Heffernan S (2011) Environmental aspects on dyeing silk fabric with sticta coronata lichen using ultrasonic energy and mild mordants. Clean Technol Environ Policy 13:207–213
McCarthy BJ (1997) Biotechnology and coloration. Rev Prog Color 27:26–31
Meksi N, Haddar W, Hammami S, Mhenni MF (2012) Olive mill wastewater: a potential source of natural dyes for textile dyeing. Ind Crop Prod 40:103–109
Mirjalili M, Nazarpoor K, Karimi L (2011) Eco-friendly dyeing of wool using natural dye from weld as co-partner with synthetic dye. J Clean Prod 19:1045–1051
Mishra PK, Singh K, Gupta KK, Tiwari H, Srivastave P (2012) Extraction of natural dye from *Dahlia variabilis* using ultrasound. Indian J Fibre Text Res 37:83–86
Mongkholrattanasit R, Krystufek J, Wiener J (2010) Dyeing and fastness properties of natural dyes extracted from Ecalyptus leaves using padding techniques. Fiber Polym 11(3):346–350

Mongkholrattanasit R, Kryštůfek J, Wiener J, Viková M (2011) Dyeing, fastness, and UV protection properties of silk and wool fabrics dyed with eucalyptus leaf extract by the exhaustion process. Fibres Text East Eur 19(3):94–99

Montazer M, Parvinzadeh M (2004) Effect of ammonia on madder-dyed natural protein fiber. J Appl Polym Sci 93:2704–2710

Montazer M, Parvinzadeh M, Kiumarsi A (2004) Colorimetric properties of wool dyed with natural dyes after treatment with ammonia. Color Technol 120:161–166

Montazer M, Taghavi FA, Toliyat T, Moghadam MB (2007) Optimization of dyeing of wool with madder and liposomes by central composite design. J Appl Polym Sci 106:1614–1621

Mortazavi SM, Kamali Moghaddam M, Safi S, Salehi R (2012) Saffron petals, a by-product for dyeing of wool fibers. Prog Color Colorants Coat 5:75–84

Papageorgiou VP, Assimopoulou AN, Couladouros EA, Hepworth D, Nicolaou KC (1999) The chemistry and biology of alkannin, shikonin, and related naphthazarin natural products. Angew Chem Int Ed 38(3):270–300

Park Y, Koo K, Kim S, Choe J (2008) Improving the color fastness of poly (ethylene tetraphthalate) fabrics with the natural dye of *Caesalpinia sappan* L. wood extract and the effect of chitosan and low-temperature plasma. J Appl Polym Sci 109:160–166

Pérez-Gálvez A, Martin HD, Sies H, Stahl W (2003) Incorporation of carotenoids from paprika oleoresin into human chylomicrons. Br J Nutr 89(6):787–793

Popoola AV (2000) Comparative fastness assessment performance of cellulosic fibers dyed using natural colorants. J Appl Polym Sci 77:752–755

Prabhu KH, Teli MD, Waghmare NG (2011) Eco-friendly dyeing using natural mordant extracted from *Emblica officinalis* G. fruit on cotton and silk fabrics with antibacterial activity. Fiber Polym 12(6):753–759

Prusty AK, Das T, Nayak A, Das NB (2010) Colourimetric analysis and antimicrobial study of natural dyes and dyed silk. J Clean Prod 18:1750–1756

Pruthi N, Chawla GD, Yadav S (2008) Dyeing of silk with barberry bark dye using mordant combination. Nat Prod Radiance 7(1):40–44

Raja ASM, Thilagavathi G (2008) Dyes from the leaves of deciduous plants with a high tannin content for wool. Color Technol 124:285–289

Raja ASM, Thilagavathi G (2011) Influence of enzyme and mordant treatments on the antimicrobial efficiency of natural dyes on wool materials. Asian J Text 1(3):138–144

Rajendran R, Balakumar C, Kalaivani J, Sivakumar R (2011) Dyeing and antimicrobial properties of cotton fabrics finished with *Punica Granatum* extracts. J Text Apparel Technol Manage 7(2):1–12

Ranjbar-Mohammadi M, Arami M, Bahrami H, Mazaheri F, Mahmoodi NM (2010) Grafting of chitosan as a biopolymer onto wool fabric using anhydride bridge and its antibacterial property. Colloids Surf B 76:397–403

Rattanaphani S, Chairat M, Bremner JB, Rattanaphani V (2007) An adsorption and thermodynamic study of lac dyeing on cotton pretreated with chitosan. Dyes Pigm 72(1):88–96

Safapour S, Seyed-Esfahani M, Auriemma F, Ruiz de Ballesteros O, Vollaro P, Di Girolamo R, De Rosa C, Khosroshahi A (2010) Reactive blending as a tool for obtaining poly (ethylene terephthalate)-based engineering materials with tailored properties. Polymer 51(19):4340–4350

Salah SM (2011) Antibacterial activity and ultraviolet (UV) protection property of some Egyptian cotton fabrics treated with aqueous extract from banana peel. Afr J Agric Res 6(20):4746–4752

Samanta AK, Agarwal P (2009) Application of natural dyes on textiles. Indian J Fibre Text Res 34:384–399

Samanta AK, Konar A (2011) Dyeing of textiles with natural dyes, 3rd chapter. In: Kumbasar EPA (ed) Natural dyes. InTech, Rijeka

Samanta AK, Agarwal P, Datta S (2007) Dyeing of jute and cotton fabrics using jackfruit wood extract: Part I – effects of mordanting and dyeing process variables on colour yield and colour fastness properties. Indian J Fibre Text Res 32:466–476

Samanta AK, Agarwal P, Datta S (2008) Dyeing of jute with binary mixture of jackfruit wood and other natural dyes – study on colour performance and dye compatibility. Indian J Fibre Text Res 33:171–180

Samanta AK, Konar A, Chakraborti S (2011) Dyeing of jute fabric with tesu extract: Part 1 – effects of different mordants and dyeing process variables. Indian J Fibre Text Res 36:63–73

Saravanan P, Chandramohan G (2011) Dyeing of silk with ecofriendly natural dye obtained from barks of *Ficus Religiosa*.L. Universal J Environ Res Technol 1(3):268–273

Sathianarayanan MP, Bhat NV, Kokate SS, Walunj VE (2010) Antimicrobial finish for cotton fabrics from herbal products. Indian J Fibre Text Res 35:50–58

Sathianarayanan MP, Chaudhari BM, Bhat NV (2011) Development of durable antibacterial agent from ban-ajwain seed (*Thymus serpyllum*) for cotton fabric. Indian J Fibre Text Res 36:234–241

Shams-Nateri A (2011) Reusing wastewater of madder natural dye for wool dyeing. J Clean Prod 19:775–781

Sharma D, Gupta C, Aggarwal S, Nagpal N (2012) Pigment extraction from fungus for textile dyeing. Indian J Fibre Text Res 37:68–73

Singh R, Jain A, Panwar S, Gupta D, Khare SK (2005) Antimicrobial activity of some natural dyes. Dyes Pigm 66:99–102

Sinha K, Saha PD, Datta S (2012) Extraction of natural dye from petals of flame of forest (*Butea monosperma*) flower: process optimization using response surface methodology (RSM). Dyes Pigm 94:212–216

Sivakumar V, Vijaeeswarri J, Anna JL (2011) Effective natural dye extraction from different plant materials using ultrasound. Ind Crop Prod 33:116–122

Son Y, Kim B, Ravikumar K, Kim T (2007) Berberine finishing for developing antimicrobial nylon 66 fibers: % exhaustion, colorimetric analysis, antimicrobial study, and empirical modeling. J Appl Polym Sci 103:1175–1182

Son Y, Ravikumar K, Kim B (2008) Development of berberine attraction sites onto cellulosic substrates modified by reactive bridging agent: statistical optimization and analysis. Colloids Surf A 325(3):120–126

Sumithra M, Raaja NV (2012) Antibacterial efficacy analysis of *Ricinus communis*, *Senna auriculata* and *Euphorbia hirta* extract treated on the four variant of denim fabric against *Escherichia coli* and *Staphylococcus aureus*. J Text Sci Eng 2(3):1–4

Syamili E, Elayarajah B, Kulanthaivelu, Rajendran R, Venkatrajah B, Ajith Kumar P (2012) Antimicrobial cotton finish using green tea leaf extracts interacted with copper. Asian J Text 2(1):6–12

Thilagavathi G, Krishna Bala S (2007) Microencapsulation of herbal extracts for microbial resistance in healthcare textiles. Indian J Fibre Text Res 32:351–354

Tiwari HC, Singh P, Mishra PK, Srivastava P (2010) Evaluation of various techniques for extraction of natural colorants from pomegranate rind – ultrasound and enzyme assisted extraction. Indian J Fibre Text Res 35:272–276

Vaidyanathan J, Bhathena-Langdana Z, Adivarekar RV, Nerurkar M (2012) Production, partial characterization, and use of a red biochrome produced by *Serratia sakuensis* subsp. nov strain KRED for dyeing natural fibers. Appl Biochem Biotechnol 166:321–335

Vankar PS, Shanker R (2009) Eco-friendly pretreatment of silk fabric for dyeing with *Delonix regia* extract. Color Technol 125:155–160

Vankar PS, Shanker R, Srivastava J (2007) Ultrasonic dyeing of cotton fabric with aqueous extract of *Eclipta alba*. Dyes Pigm 72(1):33–37

Vankar PS, Shanker R, Dixit S, Mahanta D, Tiwari SC (2008) Sonicator dyeing of natural polymers with *Symplocos spicata* by metal chelation. Fiber Polym 9(2):121–127

Velmurugan P, Chae J, Lakshmanaperumalsamy P, Yun B, Lee K, Oh B (2009) Assessment of the dyeing properties of pigments from five fungi and anti-bacterial activity of dyed cotton fabric and leather. Color Technol 125:334–341

Velmurugan P, Kim M, Park J, Karthikeyan K, Lakshmanaperumalsamy P, Lee K, Park Y, Oh B (2010) Dyeing of cotton yarn with five water soluble fungal pigments obtained from five fungi. Fiber Polym 11(4):598–605

Venil CK, Lakshmanaperumalsamy P (2009) An insightful overview on microbial pigment, prodigiosin. Electron J Biol 5(3):49–61

Vinod KN, Puttaswamy, Gowda KNN, Sudhakar R (2010) Natural colorant from the bark of *Macaranga peltata*: kinetic and adsorption studies on silk. Color Technol 126:48–53

Xinsheng X, Lu W, Shunhua J, Qicheng Z (2008) Extraction of coloring matter from *Sargentodoxa cuneata* by ultrasonic technique and its application on wool fabric. Indian J Fibre Text Res 33:426–430

Zarkogianni M, Mikropoulou E, Varella E, Tsatsaroni E (2010) Colour and fastness of natural dyes: revival of traditional dyeing techniques. Color Technol 127:18–27

Zhang R, Cai Z (2011) Study on the natural dyeing of wool modified with enzyme. Fiber Polym 12(4):478–483

Chapter 7
Surfactants in Agriculture

Mariano J.L. Castro, Carlos Ojeda, and Alicia Fernández Cirelli

Abstract Adjuvants such as surfactants improve pesticide efficiency by multiple mechanisms. In particular surfactants increase the foliar uptake of herbicides, growth regulators, and defoliants. Therefore, the choice of the adjuvant in an agrochemical formulation is crucial. Surfactants include anionic, nonionic, amphoteric and cationic surfactants. This review describes the role and properties of new adjuvants for agriculture. In particular adjuvants such as glyphosate formulations are modified to decrease ecotoxicity.

Keywords Surfactant formulation • Pesticide • Agriculture • Adjuvant • Plant cell • Glyphosate • Cuticular • Alkyl polyglycosides • Saccharose esters • Chitosan

Contents

M.J.L. Castro • C. Ojeda • A. Fernández Cirelli (✉)
Centro de Estudios Transdisciplinarios del Agua (CETA-INBA-CONICET), Universidad de Buenos Aires, Chorroarin 280, C1427CWO Ciudad de Buenos Aires, Argentina
e-mail: ceta@fvet.uba.ar; ceta@fvet.uba.ar; AFCirelli@fvet.uba.ar; gemini21@chemist.com

E. Lichtfouse et al. (eds.), *Green Materials for Energy, Products and Depollution*, Environmental Chemistry for a Sustainable World 3, DOI 10.1007/978-94-007-6836-9_7,

7.1 Introduction

The world population is projected to continue increasing into the next century. Population growth is assumed to follow the United Nation medium projection leading to about 10 billion people by 2,050. A central question is how global food production may be increased to provide for the coming population expansion. It would be necessary to increase current levels of food production more than proportional to population growth so as to provide most humans with an adequate diet (Kindall and Pimentel 1994).

Agronomic sciences and their different disciplines should be able to supply solution in order to increase food production. Agricultural science includes knowledge of the land, conducted by the geology, geochemistry and soil science, environmental analysis by the ecology, biogeography, bioclimatology, biometeorology and crops sown, with its potential for improvement through the creation of new seed varieties designed for hybridization and genetic modification. Moreover, there are a lot of factors with influence in the crops yields, such as irrigation, drainage, tillage and fertilizer applied to soil. The type of treatment above described improves the physical qualities of soil, while the use of fertilizers improves their chemical qualities. Finally, the fight against pests and weeds through pesticides is of paramount importance within the modern agricultural activity. The agricultural breakthrough occurred during recent decades by the incorporation of new laboratory techniques, statistics, computer science, satellite information. On the other hand, crop modified by genetic engineering produced a huge increase in yields per hectare of different crops. All the disciplines above described are modifying year by year through the incorporation of new technologies.

Top-10 drivers for the changes of the agricultural industry are (Heinemann 2010):

1. Rising number and age of human beings and their nutritional behaviour
2. Increasing competition of food, feed and energy for agricultural outputs
3. Scarcity of water and arable land relative to the rising demand
4. Constantly increasing regulations & requirements for healthiness and sustainability
5. Genetically modified organism (GMO) and precision-farming enabled increase of productivity and quality
6. Higher volatility of demand vs. supply due to political and financials interventions
7. Ongoing commoditization of essential inputs like agrochemicals and fertilizer
8. Beyond consolidation: business model deconstruction and recomposition
9. More and quicker information due to novel media platforms
10. Personalization-driven convergence of health and nutrition

Agrochemical formulations cover a wide range of systems forms for crop protection that are prepared to suit a specific application. All these formulations require the use of a surface active agent or surfactant, which is not only essential for its preparation and maintenance of long-term physical stability, but also to enhance biological performance of the agrochemical. Most active ingredient used in agrochemical formulation are water-insoluble compounds. The obvious reason for adding surfactants is to enable the spray solution to adhere to the target surface, and spread over it to cover a large area. In addition, the surface active agent plays a very important role in the optimization of the biological efficacy. The surfactants commonly used in adjuvants illustrate the wide array of surfactant chemistry available which includes anionic, nonionic, amphoteric and cationic surfactants; all can be found in today's adjuvant market. Generally some solvent or glycol can also present to hold the composition in a homogeneous state or control viscosity. Increasingly, if a solvent is essential it will have a high flash point and low vapour pressure since the practice of using isopropanol, acetone, or other flammable materials in adjuvant formulations is quickly disappearing due to the dangers of low flash point products. Other components common to adjuvants depending on type include oils (based on natural raw materials and petrochemical), gums and/or polymers, and oleochemicals like fatty acids or glycerin.

Adjuvants usually have multiple functions in relation to pesticide efficacy (Green and Beestman 2007). Increasing the foliar uptake of active ingredients is of particular importance for herbicides, growth regulators, and defoliants (Fig. 7.1). Therefore, the choice of the adjuvant in an agrochemical formulation is crucial (Tadros 2005).

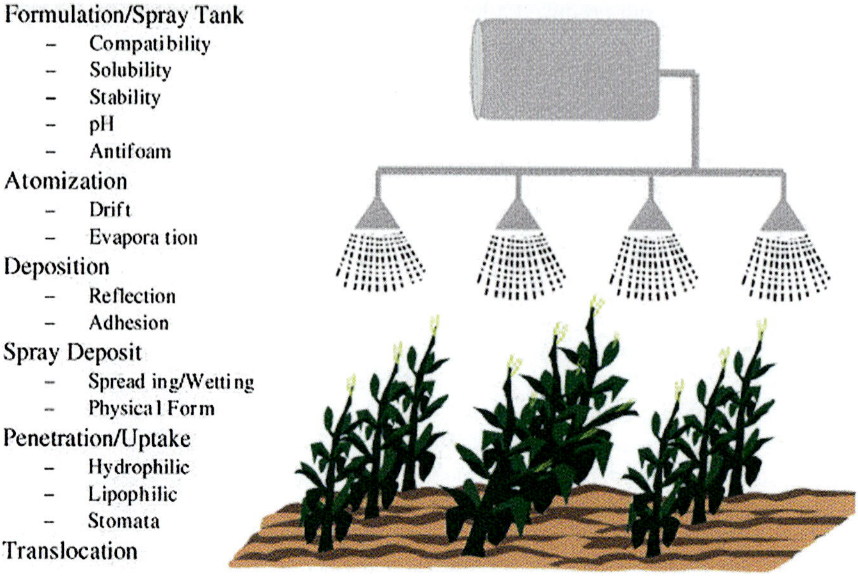

Fig. 7.1 Influence of formulation and adjuvants on herbicide performance (Reprint from Green and Beestman (2007))

It is actually difficult to ignore adjuvants when dealing with foliar uptake of pesticides. Indeed, the use of adjuvants dated back as early as about 200 years ago, although most of the synthetic adjuvant products started to emerge only a few decades ago.

Adjuvants are fascinating compounds. Typically showing no biological activity at all on their own, they help to exploit the full biological potential of many active ingredients. They can be absolutely necessary for some actives or to preserve activity, for example, under unfavorable environmental conditions or adverse water quality. Sometimes their mode of action appears obscure, since for a single active many different adjuvants with completely different modes of action, and used at differing rates per hectare, increase performance similarly. The other quite common extremes are situations where a particular adjuvant fits best to a particular active at a particular ratio in a particular formulation type, or conversely, where adjuvants are antagonistic to the product performance. In the case of adjuvants that somehow improve coverage, the visible mode of action correlates often with the achieved gain in performance.

However, most adjuvant effects are not visible at all since they occur at the molecular level, are time dependent and are often variable, since the adjuvant effects are based on interactions with the biological target and environmental factors. There are several good reasons to predict that adjuvant innovations will become even more important in the future.

The variety of products and technologies in the agricultural adjuvant industry has grown with the need to maximize the performance of pesticide products. These products have been shown to improve the performance of pesticides through aiding on-target delivery, maximizing efficacy of the active ingredient and improving spray water quality just to mention a few general areas of use. Development of pesticide formulations can be a rigorous science with the competing demands of maximizing performance of the active ingredient, maximizing the physical stability of the formulation, ensuring the human health and environmental effects of the formulation are minimized. Meanwhile, the costs of the product have to keep low enough so it makes business sense to bring it to market and for the customer to have value from its use. All of these factors must be balanced and can only be balanced by a team approach to developing the product.

Climate changes are likely to occur and possibly become more pronounced in the next decades. Weather is becoming more variable and drought spells will increase in many regions. Challenges for crop production are thus to increase crop productivity further considering that drought is only but one of the adverse conditions which crop production is already facing. Others include e.g. soil compaction, salinity, acidity or alkaline conditions, nutrient fixation or depletion. All those conditions interfere as well with the nutrient uptake of crops from soil. The development of new surfactant based-system as bio-activator for actives is a key factor to improve the cost-effective performance increasing process efficiency, energy and raw material savings. On the other hand, sustainability should take into account the use of renewable resources and the improvement of eco-friendly product profile.

7.2 Generic Pesticides of Growing Interest

Agrochemicals are an aging industry. While over 800 pesticide active ingredients remain on the market, the number of new substances being developed has fallen considerably over the last 10 years. In some industrialized country markets older products are losing their registrations. For example, the European Union review has removed over 360 active ingredients to date. But these and many other chemicals are still available elsewhere.

The overall global agrochemical pesticide market was valued at US$32 billion in 2004 with generic products making up an increasing percentage of this market. Latest estimates indicate that generic products account for US$18 billion, or around 66 % of overall sales and by volume, generic active ingredients may account for approximately 95 % of all product sales worldwide.

The types of generic most sold are herbicides with 57 % of the market, followed by insecticides with 34 % and then fungicides with 11 % of sales. The biggest selling generic is glyphosate with annual sales approximating US$5 billion followed by the insecticide imidacloprid with annual sales of around US$1 billion. The biggest selling fungicide is mancozeb, seventeenth in the list with sales of $220 million (see Table 7.1). On the other hand, generic active ingredients glyphosate, chlorpyrifos and cypermethrin, are produced by between 31 and 39 manufacturers.

Table 7.1 Market profile of the top 24 generic active ingredients (Adapted from Agrow's Complete guide to generic pesticides. Volume 2, Products and Markets (DS250), August 2005, pp 22–23)

Rank	Active ingredient	Activity	Value sales Sales ($m)	Prospects
1	Glyphosate	Herbicide	5000	Rising
2	Imidacloprid	Insecticide	1000	Rising
3	Malathion	Insecticide	400	Stable
4	Paraquat	Herbicide	400	Falling
5	Acephate	Insecticide	350	Rising
6	Pendimethalin	Herbicide	350	Stable
7	2,4-D	Herbicide	335	Stable
8	Acetochlor	Herbicide	300	Rising
9	Chlorpyrifos	Insecticide	300	Falling
10	Trifluralin	Herbicide	300	Stable
11	Atrazine	Herbicide	280	Stable
12	Imazethapyr	Herbicide	280	Falling
13	Lambda-cyhalothrin	Insecticide	275	Rising
14	Permethrin	Insecticide	270	Falling
15	Carbofuran	Insecticide	250	Falling
16	Deltamethrin	Insecticide	250	Stable
17	Mancozeb	Fungicide	220	Rising
18	Delta-cypermethrin	Insecticide	200	Rising
19	Carbendazim	Fungicide	200	Stable
20	Chlorothalonil	Fungicide	200	Stable
21	Cypermethrin	Insecticide	200	Rising
22	Dichlorvos	Insecticide	200	Falling
23	Linuron	Herbicide	200	Falling
24	Methomyl	Insecticide	200	Stable

7.3 Adjuvant Classification, Functionality and Chemistry

Although the first recorded use of adjuvants to increase herbicide performance was in the late 1890s, it was not until after the discovery of effective organic herbicides like 2,4-D in the 1940s that significant adjuvant research began (Foy 1989).

Adjuvants are defined as "an ingredient in the pesticide prescription, which aids or modifies the action of the principal ingredient" (Foy 1987). Adjuvants can be divided into two general types: (1) formulation adjuvants and (2) spray adjuvants. The first type consists of adjuvants, which are part of the formulation, while the second type of adjuvants is added along with the formulated product to the water in the tank of the spray equipment before application on the fields. Spray adjuvants are sometimes called tank mixing additives or just adjuvants, whereas the formulation adjuvants are called additives or inerts (Hochberg 1996).

The term adjuvant has developed several meanings in the agrochemical industry. In some parts of the world, adjuvant can refer to anything other than the pesticide formulation which is added to the spray tank. In other parts of the world, the term adjuvant is used interchangeably with the term formulant and refers to

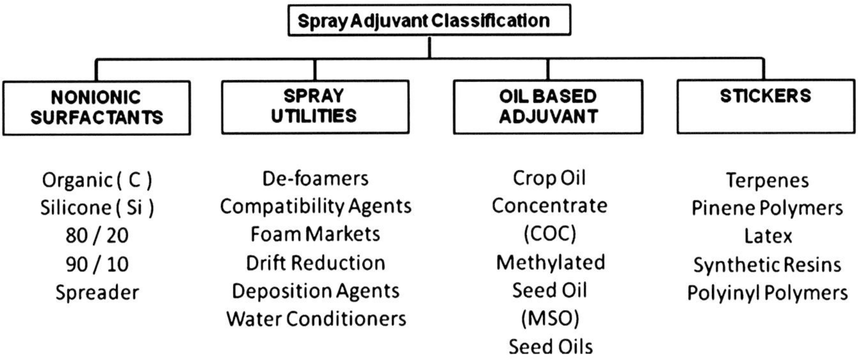

Fig. 7.2 Classification of adjuvants by tank mix function (Adapted from Roberts 2010)

any nonpesticidal ingredient in the formulated pest control product. Adjuvant terminology globally became such a large issue in the 1990s that several trade organizations decided to attempt standardizing both, the terminology and functionality of adjuvants.

Although several authors attempted to write materials for adjuvant terminology, only ASTM International (formerly American Society of Testing and Material) began to work on definitions for standard terms. ASTM E 1519 defines an adjuvant as a material added to a tank mix to aid or modify the action of an agrochemical, or the physical characteristics of the mixture (ASTM E 1519).

Although ASTM definitions are widely used, in some cases different terminology is adapted for unique circumstances. In 2009 the US Environmental Protection Agency (U.S. EPA) announced it was considering requiring the submission of residue data for active ingredients including an adjuvant as part of spray treatment when a spray tank adjuvant was included on the labelling. In order to facilitate EPA's investigation, industry proposed a set of classifications for tank mix adjuvants to minimize the testing necessary to provide the residue data EPA suggested it needed to establish a tolerance on food. The Spray Classification system presented to EPA utilized the ASTM definitions but aggregated specific adjuvants by broad type as a more general level of classification. Figure 7.2 illustrates the proposed classification of tank mix adjuvants (Roberts 2010).

The functions of adjuvants selected for use in tank mixtures vary widely depending on the needs of the active ingredient with which the product is applied. For example, certain active ingredients show enhanced performance from improved wetting while others may show enhanced performance from water conditioning or drift reduction. In some cases, the presence of an oil adjuvant may improve penetration and uptake of the active ingredient. A list of common crop protection products which respond to the use of a spray tank adjuvant along with the possible function provided by the adjuvant appear in Table 7.2 (Roberts 2010).

We have considered in the present review the term adjuvants to refer to any non-active ingredient in the formulated pest control product added to aid or modify the action of an agrochemical.

Table 7.2 Crop protection product combinations with spray tank adjuvants (Adapted from Roberts 2010)

Crop protection product	Desired adjuvant functionality
Glyphosate	Wetting, water conditioning
Sethoxydim	Cuticular penetration
Atrazine (post use)	Wetting, compatibility
Paraquat	Wetting, sticker
Glufosinate	Wetting, penetration, water conditioning
Imazapyr	Cuticular penetration
Acifluorfen	Wetting, cuticular penetration
Sulfonyl urea	Wetting, pH buffering
Fluazifop	Wetting, cuticular penetration
Pyraclostrobin	Cuticular penetration

7.4 Types of Surfactants

Surface-active agents have a characteristic molecular structure consisting on a structural group that has a very little attraction for water, known as a hydrophobic group, together with a group that has strong attraction for water, called hydrophilic group. This is known as an amphiphilic structure. The hydrophobic group is usually a long-chain hydrocarbon, and the hydrophilic group is an ionic or highly polar group. According to the nature of the hydrophilic group, surfactants are classified as: anionic, cationic, nonionic and amphoteric (Fernández Cirelli et al. 2008).

7.4.1 Anionic Surfactants

The hydrophilic groups of anionic surfactants consist in most cases of sulfonate, sulfate, or carboxylate groups with either a sodium or a calcium a counterion (Table 7.3). Amongst them, linear alkylbenzene sulfonate (LAS) are produced in the largest quantities worldwide. These are mainly used in powdery and liquid laundry detergents and household cleaners. It is important to point out that calcium linear alkylbenzenesulfonate is employed as adjuvant in many agrochemicals formulations.

7.4.2 Nonionic Surfactants

The hydrophilic behavior of nonionic surfactants is caused by polymerized glycol ether or glucose units (Table 7.3) (Fernández Cirelli et al. 2008). They are almost

Table 7.3 General classification and characteristic features of surfactants

	Alkyl tail	Polar head	Example
Anionic	C_8-C_{20} linear or branched-chain	–COOH	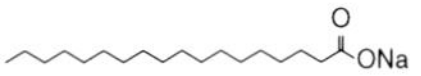 Soap
	C_8-C_{15} alkylbenzene residues	$–SO_3Na$	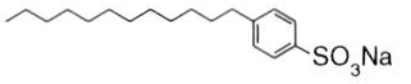 LAS
	C_8-C_{20} linear-chain ethoxylated	$–OSO_3Na$	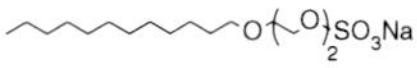 LES
Cationic	C_8-C_{18} linear-chain	$–N(CH_3)_3Cl$	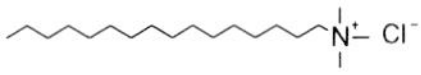 CTAC
	C_8-C_{18} linear-chain	$–N(CH_3)_2Cl$	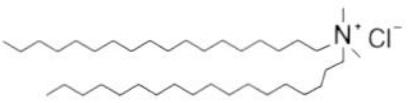 DODAC
Non-ionic	C_8-C_9 alkylphenol residues	$–(CH_2CH_2O)_n–OH$ n: 4–22	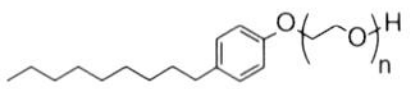 APEO
	C_8-C_{20} linear or branched-chain	$–COO(CH_2CH_2O)_n–OH$ n: 4–22	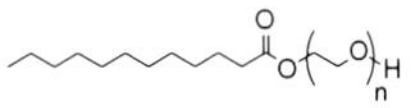 FAEO
	C_8-C_{20} linear or branched-chain	$–(CH_2CH_2O)_n–OH$ n: 2–22	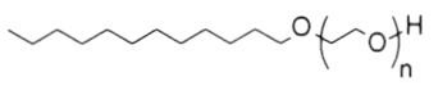 AEO
	C_8-C_{20} linear or branched-chain	$–NH(CH_2CH_2O)_n–OH$ n: 2–22	
	C_8-C_{20} linear or branched-chain	Glucose	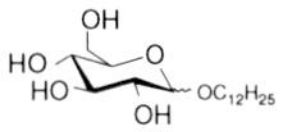 APG
Amphoteric	C_{10}-C_{16} amido-propylamine residue	$–N^+(CH_2)_2CH_2COO^–$	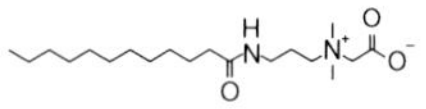 (CAPB)
	C_8-C_{18} linear-chain	$–N^+(CH_2)_2CH_2CH(OH)CH_2SO_3^–$	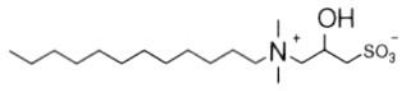 (CAHS)

exclusively synthesized by addition of ethylene oxide or propylene oxide to alkylphenols, fatty alcohols, fatty acids, fatty amines or fatty acid amides. Nonionic surfactants found major applications as detergents, emulsifiers, wetting agents, and dispersing agents. They are used in many sectors, including household, industrial and institutional cleaning products, textile processing, pulp and paper processing, emulsion polymerization, paints, coatings, and agrochemicals. A large amount of them are employed as adjuvant in many agrochemicals formulations.

7.4.3 Cationic Surfactants

Cationic surfactants contain quaternary ammonium ions as their hydrophilic parts (Table 7.3). This class of surfactants has gained importance because of its bacteriostatic properties. Therefore, cationic surfactants are applied as disinfectants and antiseptic components in personal care products and medicine. Because of their high adsorptivity to a wide variety of surfaces, they are used as antistatic agents, textile softeners, corrosion inhibitors, and flotation agents.

7.4.4 Amphoteric Surfactants

Amphoteric surfactants containing both cationic and anionic group in their structure and sometimes are referred to as zwitterionic molecules (Table 7.3). They are soluble in water and show excellent compatibility with other surfactants, forming mixed micelles. The change in charge with pH of amphoteric surfactants affects wetting, detergency, foaming, etc. The properties of amphoteric surfactants resemble those of non-ionics very closely. Zwitterionic surfactants have excellent dermatological properties, they also exhibit low eye irritation and are frequently used in shampoos and other personal care products. Amphoteric surfactants are now starting to be used in agrochemicals formulation.

7.5 Mode of Action of Adjuvants

Despite the negative perception of the public, pesticides are still going to be used for many decades to ensure the food supply for the ever growing world population. The simple reason for this is that products more costly for farmers mean more expensive foods for the population, alternative methods for plant protection are either inefficient or too costly for farmers.

The global pesticide production in 2000 amounted to over 3 million tons of active ingredients (Tilman et al. 2002). It is estimated that of the total amount of pesticides applied for weed and pest control, only a very small part (<0.1 %) actually reaches the sites of action, with the larger proportion being lost *via* spray drift, off-target deposition, run-off, photodegradation and so on (Pimentel 1995). Agrochemical industries are continuously seeking for new products and formulations of expanding area of use and market share. It typically takes at least 10 years of intense research and development from the discovery of an active ingredient until its market introduction as an agrochemical product (Cordiner 2004). In addition, it has become increasingly more and more difficult to find materials of high activity that are environmentally friendly, safe and cheaper to manufacture. Most government and regulatory authorities are now demanding formulations that are cleaner and safer for the users, have minimal impact on the environment, and can be applied at the lowest dose rate (Holden 1992).

The transporting-limiting barrier restricts the performance of foliar-applied agrochemicals molecules such as pesticides, fruit chemical thinning, and growth regulators (Schreiber 2006; Petracek et al. 1998). An agrochemical active ingredient is often ineffective and of little use to the end user if applied to the target surface alone (Tominack 2000). In most cases the active ingredient is formulated into complex, carefully designed multi-agent preparations that facilitate mixing, dilution, application and stability. To enhance the efficacy of foliar-applied agrochemicals, surfactants are widely used in spray solution to increase active ingredient solubility, to improve wetting of the plant cuticle, and to increase cuticular penetration.

Around 230,000 t of surfactants is used annually in agrochemical products, with a formulation typically containing 1–10 % of one or more surfactants (Edser 2007). Surfactant, as a plasticizer, softens the crystalline waxes in cuticle and thus increases the mobility of the agrochemicals across the cuticular membrane (Schönherr et al. 2000).

With the presence of surfactant, the sorption behavior of plant cuticle becomes very complex. Derived from similar studies, surfactant demonstrates two opposite effects on the cuticle sorption behavior: (i) surfactant solutions decrease the distribution of organic contaminant onto the cuticle by increasing the solute aqueous solubility, that is, a negative effect; (ii) surfactants increase the sorption capability by softening of cuticular wax (i.e., plasticizing effect) or by forming a new partition phase for the cuticle-sorbed surfactant, that is, a positive effect (Zhu et al. 2004; Lee et al. 2000, 2004). The apparent effects of surfactants on sorption depend on the balance of the two opposite effects, which are dominated by the compositional characteristics of plant cuticles, surfactant type and concentration, and the solute's properties (Tamura 2001).

On the other hand, it is well known that droplets of between 50 and 250 mm diameter will adhere to plant surfaces, but larger droplets may rebound or shatter (Hartley and Brunskill 1958). However, smaller droplets are more prone to drift, so we have the classic conundrum of how to achieve maximum on-target adhesion/retention, with minimum off-site drift. The exact mechanism by which these molecules elicit this acceleration of movement is not yet fully understood.

In the case of systemic pesticides, if there is little or no uptake, then regardless of the efficiency of deposition and retention, this will have been to no avail. Obviously environmental factors, such as rain (causing wash-off), wind and relative humidity (affecting droplet drying), will materially affect pesticide uptake. However, a major factor in all cases is the plant leaf itself, including its surface and cuticle. One of the main functions of spray adjuvants is to overcome or minimise the effect of leaf waxes and the cuticular barrier. Appropriate adjuvants will substantially aid droplet spread, redistribution and leaf coverage, but can also enhance uptake of the active ingredient. Penetration and absorption of a compound by leaves is determined by the ability of the compound to move through the cuticle, a process known as transcuticular penetration or movement (Solel and Edgington 1973).

Foliar uptake of pesticides is a diffusion process across the epicuticular wax, the cuticle, and the plasma membrane of epidermal cells. The diffusion rate of both lipophilic and hydrophilic compounds across isolated leaf cuticles was negatively correlated with their molecular weight (MW), or molecular size (Bauer and Schönherr 1992; Schönherr and Schreiber 2004). In fact, the plasmodesmata of plants can only allow free circulation of molecules smaller than 1 kDa (Oparka and Roberts 2001). It is thus not a coincidence that most foliar-applied herbicides have a molecular weight between 100 and 500, as any out-ranged molecules would be eliminated at the screening stage if they cannot enter plant leaves or move between cells (Tice 2001).

Lipophilicity is probably the single most important property of pesticides related to foliar uptake. Generally, an octanol/water partition coefficient (log Ko/w or log$P$)

is used to describe the lipophilicity of a compound. Chemicals with log $P < 0$ are considered as hydrophilic and those with log $P > 0$ lipophilic. Due to the lipoidal nature of the epicuticular wax and the cuticle, foliar uptake tends to increase with increasing lipophilicity of the chemicals (Baker et al. 1992; Stevens et al. 1988). A best example showing this is the faster uptake of 2,4-D ester (lipophilic) compared to 2,4-D salt (hydrophilic) (De Ruiter et al. 1993).

However, like molecular size, log$P$ is not the only property that determines the foliar penetration. Indeed, based on the uptake data obtained with 26 chemicals and four plant species, only a moderate correlation ($R2$: 24–45 %) was found between log$P$ and the foliar uptake of chemicals. In fact, some hydrophilic compounds, such as paraquat, are actually able to penetrate into plant leaves very fast.

7.6 Plants Cells Morphology

Plant cells, unlike animal cells, are surrounded by thick walls that form rigid tissues. They do not need the junctions found in animal tissues. But some higher plant cells are interconnected by plasmodesmata, tubelike structures that penetrate through cell walls and form fluid connections between adjacent cells. Like gap junctions in animal cells, plasmodesmata allow the free exchange of small molecules and help coordinate the activities of neighboring cells.

On the surface of the leaves of higher plants is a layer of waxy material known as the cuticle or cuticular membrane (Kirkwood 1999; Riederer and Schreiber 2001). It is theorized that after plants migrated from the sea onto land 400 million years ago, in the Silurian period of the Paleozoic era, cuticles evolved as a means of reducing water loss to the surrounding air (Stewart et al. 1993; Thomas and Spicer 1987). The plant cuticle is a thin continuous layer of predominantly lipid material synthesized by the epidermal cells and deposited on their outer. Cuticle composition and structure of modern higher plants vary greatly from species to species but are generally 0.1–10 μm thick, being composed of two major classes of lipids: (i) solid waxes, usually mixes of long chain aliphatic compounds and (ii) insoluble, high molecular weight polyester cutins. Plant cuticle contains too terpenoids and phenolic compounds, some of which serve as anti-feedants to discourage grazing by herbivores.

There are five basic types of cuticles: smooth, ridged, papillose, glaucous (having an additional covering of microcrystalline wax), and glandular where trichomes are present in high number and comprise the main surface of the leaf (Hess and Foy 2000).

Generally, the plant cuticle is composed of two regions. The lower region consists of cutin, long chain alkyl ketones, alcohols, and closest to the cells, polysaccharides; the upper region consists of epicuticular wax, mostly very long chain alkanes (Li and Chen 2009; Chen et al. 2008).

The epicuticular wax present in the outer surface of the membrane has two main forms, crystalline and amorphous (Jeffree 1996; Chen and Li 2007). Whereas there are some important exceptions, the dominant epicuticular wax on leaves of grass species is crystalline and on broadleaf species is amorphous. The crystalline wax is undoubtedly an obstacle for spray retention and an intimate contact between pesticide droplets and leaf surface, but it does not seem to be a formidable barrier for pesticide uptake. All aerial surfaces of plants are covered by the cuticle. It is widely known that the waxy sheet of cuticle not only prevents water loss, but also functions in defense by forming a barrier that resists physical damage and microbial invasion. The cuticle is incontestably the most important barrier for the penetration of pesticides. The microscopic structure and chemical composition of plant cuticles were reviewed by Holloway (Holloway 1993).

The cuticle permeability for pesticide molecules varies greatly with plant species and growth stage, but is not correlated to the thickness (Becker et al. 1986; Buchholz et al. 1998). It is clear that both lipophilic and hydrophilic compounds can diffuse through the cuticle. It was postulated that lipophilic and hydrophilic chemicals follow distinct pathways, i.e., polar (aqueous pores) and apolar (cutin matrix and wax) routes (Crafts 1960). Until now the existence of aqueous pores inside the cuticle is only supported by indirect evidence obtained through studying the diffusion characters of calcium salts and glyphosate across isolated leaf cuticles (Schönherr 2000, 2002).

Their high sorption capability may be seriously suppressed and even inhibited by the cuticular waxes deposited within and on the surface of the polyester matrix because of their partially cristaline structure. More importantly, it is the arrangement of the wax into amorphous and crystalline domains that is believed to be the transport limiting barrier (Reynhardt and Riederer 1994; Santier and Chamel 1998). Removal of waxes from bulk plant cuticles promotes their sorption capability and leads to an increase in the permeability by several orders of magnitude (Chefetz 2003; Popp et al. 2005).

7.7 Mechanism of Action of Adjuvants

It is generally recognized that not any surfactant can increase the uptake of any herbicide. At present, there is no practical theory or comprehensive model that can predict quantitatively the effect of a particular surfactant on the uptake of a specific herbicide. Applying the right concentration and type of surfactant is not an easy task for the user. Systematic studies will be needed to elucidate the mechanisms involved in the complex interactions between herbicides, surfactants and plants. In order to improve the uptake of post-emergence herbicides, numerous surfactants have been tested over the last 30 years. However, until the early 1990s, most experiments involved only a few randomly selected surfactants and it was not possible to know which parameter of surfactants was most important for enhancing herbicide uptake.

A first review of the mechanisms of action is listing various contributions like wetting and spreading, dissolution or disruption of epicuticular waxes, solubilisation of agrochemicals in deposits, prevention or delay of crystal formation in deposits, retention of moisture in deposits by humectant action, and promotion of uptake of solutions via stomatal infiltration (Stock and Holloway 1993).

A first mechanistic approach of the adjuvant/wax interactions was reported by Schreiber (Schreiber 1995). From the observation that the surfactant-induced acceleration of the diffusion of pentachlorophenol is completely reversible if the surfactant is removed, it was tentatively concluded that the effect of the surfactant on wax is an unspecific plasticising interaction. It was not before a decade later that again a plasticising effect was revealed from macro and micro-thermal analysis of plant/surfactant interactions (Perkins et al. 2005). Finally, a series of results from different experiments with reconstituted cuticular wax, like the already mentioned reversibility of diffusion rates when adjuvants are removed, like the decrease in size selectivity in the presence of adjuvants, and like the fluidity increase in the presence of adjuvants as determined by electron spin resonance (ESR) and by nuclear magnetic resonance (NMR), clearly allow the conclusion that the adjuvants act by a non-specific, plasticizing mechanism (Schreiber 2006; Petracek et al. 1998).

An even deeper insight into the uptake-enhancing mechanism of adjuvants is reported by Burghardt et al. (Burghardt et al. 1998). In this paper, a free volume model was applied to predict diffusion coefficients of active ingredients on the basis of the transport properties of the wax, the molar volume of the diffusing compound and the adjuvant concentration in the wax. Increase in the mean free volume by the plasticising action of adjuvants raises the probability for a diffusing compound to find a fitting void. Despite these experimental and theoretical advances, it was stated by Wang et al. that a more multidisciplinary approach is needed to elucidate the transcuticular diffusion behaviour of active ingredients and the mode of action of adjuvants (Wang and Liu 2007).

7.8 Efficiency Parameters to Increase the Agrochemicals Performance

7.8.1 Cuticular Uptake

One of the most important ways to improve the efficacy of pesticides and minimize their impact on off-target organisms is through increasing the penetration of active ingredients into plant foliage. If foliar uptake is important for the efficacy of systemic fungicides and insecticides, then diffusion into plant tissues is a prerequisite for the activity of foliar-applied herbicides, growth regulators, and defoliants. Foliar uptake of pesticides is a complex process, depending on leaf surface characters of plants, physicochemical properties of the chemicals, types and concentration of the additives, and environmental conditions. How each of these factors influences the uptake is only partially understood, despite an urgent demand for efficient use of pesticides.

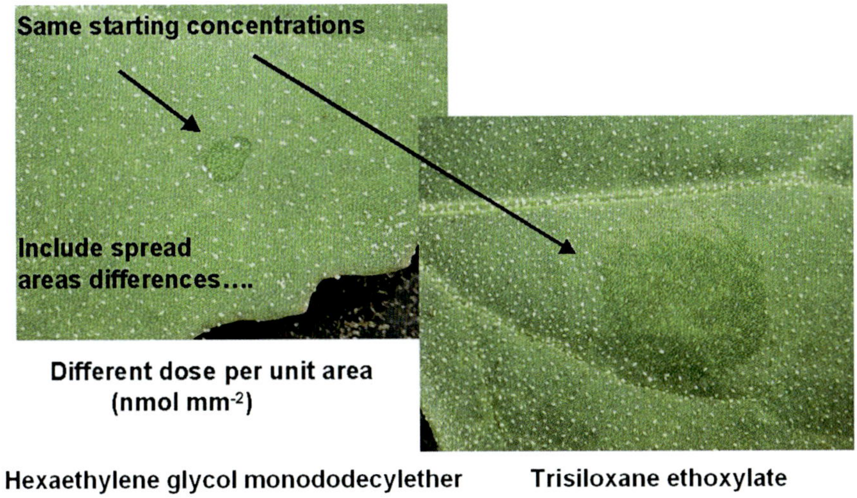

Fig. 7.3 Illustration of droplets spread effects on *Chenopodium album* with different spray formulation (Reprint from Zabkiewicz (2007))

Cuticular uptake (diffusion of the chemical directly through the cuticle) of pesticides has been studied for a longer time, but progress in understanding the factors controlling this mechanism has been much slower. Attention was focussed on relating the physical properties of the formulation to percentage uptake, with less focus on the interactions of adjuvant(s) in a spray formulation, or the importance of plant cuticle and epidermal layers (Stevens and Bukovac 1987; Stock et al. 1993). Just as it is essential to consider application and formulation together in the case of adhesion and retention by plants, so the interactions of plant, active and adjuvant must all be considered in the uptake process.

The fundamental mechanism of uptake has been considered, with most attention given to the epicuticular lipids and their role in modifying active ingredient diffusion through cuticles (Kirkwood 1999; Riederer and Markstädter 1996; Schönherr et al. 1999). However, there is a much simpler effect on the leaf surface that needs to be considered first. If a spray formulation contains adjuvants that cause droplet spread on a leaf surface (Fig. 7.3), this will in effect lower the mass of active per unit area, and without any change in concentration until the spray solution begins to evaporate. In any case, there will be a "solution residue", where the concentration of the active is many times more than in the starting spray solution (Zabkiewicz 2003).

7.8.2 *Translocation*

Absorption and translocation of herbicides can be significantly increased when adjuvants are added to foliar applied spray mixtures level of increased control from

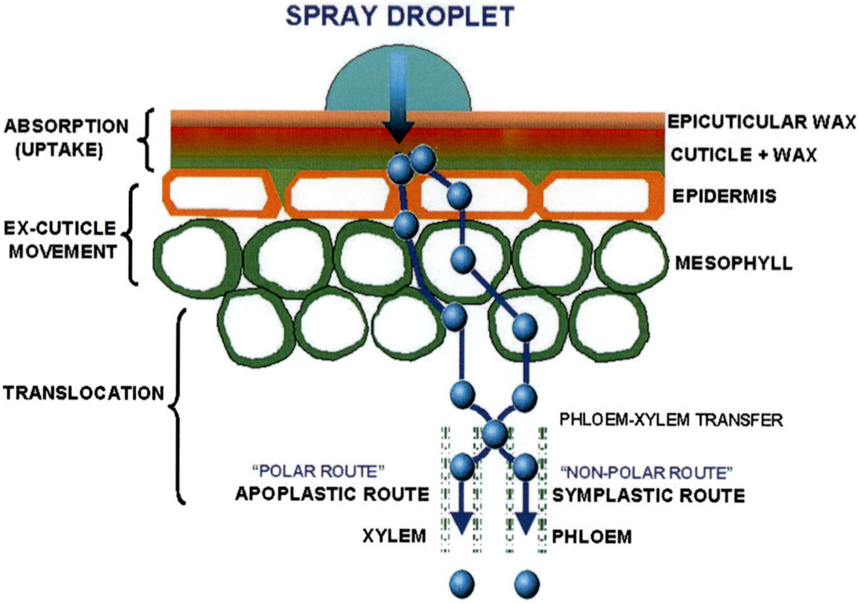

Fig. 7.4 Representation of different trans-cuticular pathways and subsequent apoplastic (*polar*) and symplastic (*non-polar*) pathways (Reprint from Zabkiewicz (2007))

adjuvants is usually affected by the environmental conditions at the time of herbicide application. In most cases, herbicide absorption and translocation are greatest under environmental conditions which are most favorable for growth of the treated plant. Adjuvants added to treatment solutions often increase absorption and translocation of foliar applied herbicides, especially when plants are under stress at time of treatment (Wills and McWhorter 1988).

Translocation has received the least attention, though long distance transport has been well studied and reviewed (Price 1976; Coupland 1988). In the case of foliar applied pesticides, it is known that lipophilicity is important, as these compounds have to cross hydrophobic membranes or structures other than the cuticle proper. Hydrophilic molecules are readily transported in either the phloem (downwards) or the xylem (upwards), though their initial movements through the cuticle, epidermal cells and into the mesophyll are not well understood (Devine and Hall 1990). The presence of separate "hydrophilic" and "lipophilic" pathways as part of the uptake process, may in turn determine the efficiency of the subsequent translocation pathway, but it is also difficult to define when uptake becomes translocation (Fig. 7.4).

Practically all herbicides and all systemic pesticides are weak electrolytes. Bromilow et al. (1990) represented polarity or lipophilicity and acid strength (pKa), as two main herbicide characteristics that help in understanding the transport pattern in plants. As the chemicals become more polar and more lipophilic, the permeation rate decreases. The pH in the exterior of the cell is approximately 5.5, herbicides

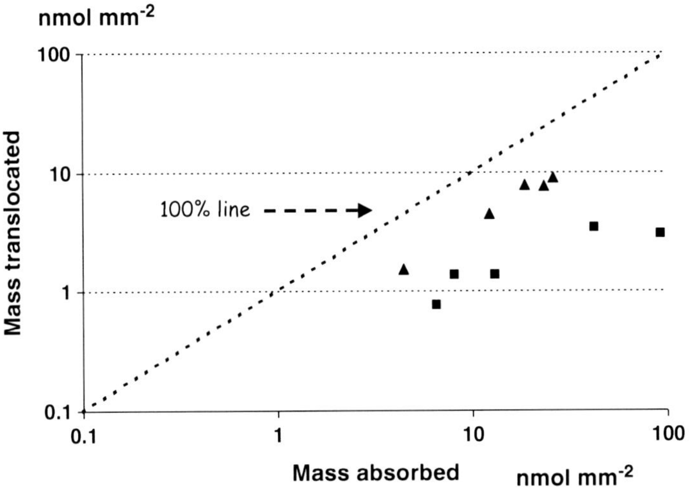

Fig. 7.5 Glyphosate mass uptake into *Chenopodium album* versus mass absorbed from Roundup® Ultra (▲) and Touchdown (■) commercial formulations; (- - -) maximum uptake line, representing 100 % uptake over the initial dose range (Reprint from Zabkiewicz (2007))

with pKa below 5.5 are in a non-dissociated form and are thus able to penetrate the cell more easily. Once at the cytoplasm, where the medium is more alkaline (pH approximately 7.5), these herbicides disassociate and turn into the most active, and less capable, translocating form of the molecule. Because the dissociated form is less capable of spreading to the exterior of the cell, the molecules of these herbicides get "stuck" in the cytoplasm; this behavior receives the name "ionic trap". In addition, plasmodesmata may have significant participation in translocation of molecules that have pKa (dissociation constant) below the pH of the xylem, which is approximately 5.5; it may also assist symplastic movement of these molecules by other routes besides phloem (Trapp 2004).

It is clear that "long-term" translocation will be affected by plant growth stage and environmental conditions, but "short-term" translocation (over hours rather than days) may have other rate limiting or modifying features.

Adjuvants are known to facilitate cuticular "transport" (foliar uptake) but are not thought to play any significant part in further short or long-distance translocation processes. However, in theory, if adjuvants could reach the cellular plasmalemma, then they could affect the initial stage of the sub-cuticular transport process (Fig. 7.4). The recent use of mass or molar relationships, instead of percentages, for xenobiotic uptake into plants from differing formulations, may be a means of elucidating some of the interactions among actives, adjuvants and plants (Forster et al. 2004).

It has been demonstrated that a relationship similar to that developed for mass uptake, for translocation vs. mass absorbed (which is related to the initial dose applied) can also apply to the translocation process in model systems but less well with complex formulations (Fig. 7.5) (Zabkiewicz 2007). For example, the commercial glyphosate formulations show considerable interaction, if the reduction in both mass uptake and translocation at higher initial dosages, is a correct indication. This manner of presenting mass uptake and translocation has the potential to identify anomalous behaviour or complex interactions.

The advantage of the mass relationship is that it can provide information similar to that used for drug delivery dose prescriptions. Knowing the mass uptake, an estimation can be made of the mass translocated in specific systems; with subsequent studies on the influence of physiological and environmental influences, appropriate "dosages" could be applied at specific growth stages or conditions. It can also be used to estimate the change in overall efficacy of spray formulations, as to date no generic quantitative relationship has been identified.

The principal factors controlling foliar uptake appear to be solute mobility in the cuticle, cuticle tortuosity and solute-driving force.

7.8.3 Environmental Conditions

It is well known that environmental conditions have an important influence on herbicide efficacy. In particular, the effect of humidity on herbicide uptake has been attributed to changes in cuticle hydration and droplet drying. As early as the 1950s, it was hypothesized that humectants such as glycerol would enhance herbicide uptake by not letting droplets dry, thus maintaining the herbicide in solution, and hence making it available for uptake. Shortly thereafter, evidence was found to support this hypothesis and humectants were used successfully in warm, dry areas to increase herbicide efficacy. Furthermore, current evidence suggests that highly water-soluble, ionic herbicides may be more sensitive to low humidity and rapid drop drying than lipophilic herbicides. The interaction of water-soluble herbicides with surfactants, the cuticle, and humidity, with particular emphasis on the impact of low humidity and humectants on herbicide uptake is presented by Ramsey et al. 2005. It was found that when one focuses on research performed at low humidity the importance of humectants emerges, which is not in keeping with what is now commonly accepted (Ramsey et al. 2005).

It is well known that the activity of foliar-herbicide sprays is influenced by the environmental conditions at the time of spraying. Current evidence suggests that water-soluble herbicides may be more sensitive to variations in spray conditions than lipophilic herbicides.

Among the many environmental factors that can affect herbicide uptake, two of the most important are temperature and humidity, with optimal uptake being favoured by warm, humid conditions (Muzik 1976).

Temperature can affect herbicide uptake by changing the viscosity of cuticle waxes, the rate of diffusion, and in conjunction with humidity, cuticle hydration (Price 1983). Increased diffusion of solutes into cuticles with increasing temperature has been explained as a result of lower partition coefficients (log P) as temperature increases (Schönherr et al. 1999).

By the mid-1980s, it appeared as though research on humectants gave way to investigations on the effect of ethylene oxide (EO) content on surfactant performance. Ethoxylated surfactants have been shown to be effective at increasing the uptake of both lipophilic and hydrophilic herbicides.

However, an important observation from research on ethoxylated surfactants is that for water-soluble herbicides, surfactants that increase droplet spreading often result in poor uptake compared to surfactants that are poor spreaders and have higher surface tension. The reason for this is unclear, but it may be related to the fact that droplets with high surface tension have smaller surface areas for evaporation than droplets with low surface tension, which disperse across the leaf surface (Zhu et al. 2008).

7.9 Agro-adjuvants Models

The traditional method of study and development of different formulations has been to conduct experimental trials (both field trials and controlled environment trials) which are time consuming, expensive, and product specific. More recently, models are being developed that take a more mechanistic approach to the different aspects of spray application with the hope of gaining insight into the underlying physical processes (Zabkiewicz 2000). Often these models consist of sub-models of individual aspects of interest. The challenge in the future is to bring these modeling approaches together in a coherent and systematic way so that an integrated formulation efficacy support system can be developed.

By their nature comprehensive detailed models have large numbers of parameters that tend to be plant and a.i. specific and may be difficult to determine. An example of such a comprehensive model is given by the series of papers by Satchivi et al. (2000a, b, 2001, 2006). The results from these types of models are usually excellent but care must be taken that they are not used outside of their limitations (Trapp 2004). By their very nature these comprehensive models include many competing mechanisms for active ingredient uptake. Due to this the effects of varying one parameter (say droplet size) is not always obvious. The benefit of simplified models, such as that presented here, is that results are not obscured by surfactant adjuvants (Tu et al. 2003). The precise nature of this increased foliar uptake is not well understood but hydrophilic compounds generally show larger increases in uptake than lipophilic ones (Stock et al. 1993). Initial dose per unit area has been shown to be a strong determinant of uptake for surfactants of various spread characteristics (Forster et al. 2004). The addition of a surfactant can lead to greater droplet spreading and so the dose per unit area is reduced but it is applied over a larger area. These phenomena were investigated to determine what effect an increased spread area has on uptake. To be able to model the effect of droplet size and spread area a model with both horizontal (along the leaf surface) and vertical (into the leaf) spatial dimensions are needed as well as time dependence. To simplify the analysis circular symmetry of the droplet is assumed so the along leaf dimension is simply

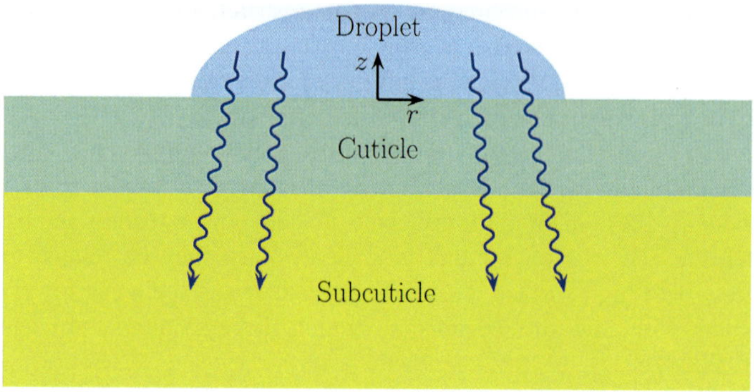

Fig. 7.6 The schematic of the layers used in the numerical solution (Reprint from Mercer (2007))

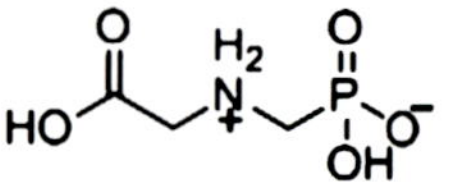

Fig. 7.7 Glyphosate structural formula

the radius from the centre of a circular droplet. A three layer model is used as shown in Fig. 7.6. The upper layer is the droplet, the middle layer the cuticle and the lower layer the subcuticle. The governing process is assumed to be a simple diffusion process with removal of active ingredient in the subcuticle region (Mercer 2007).

The effect spray droplet size and spread area have on the diffusion of a hydrophilic active ingredient through a leaf cuticle and therefore on its eventual uptake by the plant. The model presented here is not designed to be a comprehensive detailed model of all the physical processes involved but rather a simplified model so that specific size and area related factors can be investigated.

7.10 Toxicity of Adjuvants for Glyphosate

Toxicity of wetting agents employed in the formulations of glyphosate will be analyzed in this section, because glyphosate is the most important pesticide worldwide and the amount of wetting agent present in glyphosate formulation becomes as high as 150 g/lt.

Glyphosate, N-(phosphonomethyl)glycine, the active ingredient of very well-known herbicide preparations, such as Roundup®, is a systemic and non-selective herbicide utilized for weed control, i.e. in agriculture, forestry, urban areas and even aquaculture (Woodburn 2000; Williams et al. 2000).

The use of this non-selective and broad-spectrum herbicide increased dramatically after the introduction of genetically modified glyphosate-resistant crops in

1997 (Giesy et al. 2000). Although glyphosate is already one of the most used xenobiotics in modern agriculture, we should expect an increasing utilization of glyphosate largely due to the number of transgenic plants developed to be tolerant to this herbicide (May et al. 2002; Nadler-Hassar et al. 2004; Stephenson et al. 2004).

The main formulation of glyphosate is Roundup®, where glyphosate is present as an isopropylamine (IPA) salt and its efficiency is enhanced by addition of the surfactant polyoxyethylene amine (POEA) (Tsui and Chu 2003).

Virtually, every pesticide product contains ingredients other than those identified as the "active" ingredient(s), i.e. the one designed to provide the killing action. These ingredients are misleadingly called "inert". Commercial glyphosate formulations are more acutely toxic than pure glyphosate, since the amount of Roundup® required to kill rats is about 1/3 of the amount of glyphosate alone (Martinez and Brown 1991). Similar results have been obtained in cell division, thus indicating a synergy between glyphosate and Roundup® formulation products (Marc et al. 2002). There are in the literature important studies regarding the possible impact on environment and human health toxicity of glyphosate, particularly since there is a paucity of data regarding chronic exposure to sublethal doses during embryonic developments (Paganelli et al. 2010) but they have not included due to it is out of the scope of this review.

Moreover, regarding the herbicide commercial formulation, the negative impact of polyoxyethylene tallow amine (POEA) surfactant towards examined amphibians was emphasized as correlated with the pH value of water environment. This latter finding is well harmonized either with the literature data (Krogh et al. 2003). POEA represents the group of nonionic alkylamine ethoxylates (ANEOs), which have been employed in agrochemical formulations and as spray adjuvants since the 1970s of twentieth century. For many years these chemicals were treated as non-toxic biodegradable additives; however, over the past two decades the knowledge on their adverse impact, especially on aquatic biota, increased dramatically (Krogh et al. 2003).

The traditional glyphosate surfactant system (called MON 0818) is a blend of components that contains a complex polyethoxylated tallow-amine (POEA) surfactant mixture. The primary component of MON0818 is POEA surfactant (about 75 %) (Wan et al. 1989). The other 25 % of ingredients have been determined to be relatively non-toxic (LC50 > 100 mg l^{-1} for various ingredients to a variety of freshwater species; US Environmental Protection Agency ECOTOX Database), and therefore are unlikely to substantially contribute to the toxicity of MON 0818. The POEA surfactant component consists of a complex mixture of polyethoxylated long-chain aliphatic amines, which can be considered non-ionic in the neat form. In the aqueous environment of the microcosms used in the study, it is at least partially protonated and can be considered cationic.

Wade et al. reported that Mon 0818, a tallow amine surfactant, increased the diffusion of glyphosate into isolated plasma membrane vesicles, suggesting that surfactants could increase the membrane permeability (Wade et al. 1993).

On the other hand, the application of herbicides to control emergent aquatic vegetation or to treat the banks of waterways can produce localized concentrations of glyphosate and surfactant in water that are greater than those from runoff associated with terrestrial uses (e.g., associated with agriculture). Glyphosate has been shown to be slightly toxic or non-toxic to aquatic organisms, whereas the MON 0818 surfactant containing POEA is considered to be moderately toxic (Giesy et al. 2000). Some studies have noted that the concentrations of MON 0818 in shallow water could reach levels that present an elevated risk to aquatic organisms (Wan et al. 1989; Folmar et al. 1979; Servizi et al. 1987). The risk potential of MON 0818 in shallow water is a function of both toxicity and environmental fate. Although numerous studies have been conducted to evaluate the fate and toxicity of glyphosate, limited information is available on the fate and toxicity of the surfactant alone (Giesy et al. 2000; Wan et al. 1989; Folmar et al. 1979; Goldsborough and Brown 1993; Beyers 1995; Gardner and Grue 1996; Franz et al. 1997; Henry et al. 1994; Paveglio et al. 1996). The majority of information available on the environmental fate of non-ionic ethoxylated surfactants pertains to linear alcohol ethoxylates and alkylphenol ethoxylates, which dissipate in the aquatic environment through adsorption to sediment and biodegradation by microbial action (Krogh et al. 2003; Urano et al. 1984; Cano and Dorn 1996).

Since aquatic weed control can occur in relatively shallow water bodies containing various types of sediments, Wang et al. assessed changes in aqueous concentrations of MON 0818 and changes in its toxicity to the cladoceran *Daphnia magna* over time in microcosms with and without natural sediments (Wan et al. 1989). The toxicity of the POEA surfactant, MON 0818, decreased rapidly in water from microcosms containing sediment, and this decrease in toxicity was correlated with the decline of MON 0818 concentrations in the overlying water. These results indicate that toxicity and concentration of the surfactant can be rapidly reduced in shallow water due to interactions with sediments. Although the present study did not directly address the potential for chronic exposure of aquatic organisms to the POEA surfactant, the strong tendency for this surfactant to adsorb rapidly to sediment, together with biodegradation, suggest that risks of chronic toxicity would be low. Furthermore, it also appears that dissipation is increased in the presence of sediment containing higher percent clay and total organic carbon, and higher microbial mass. The information obtained from the chemical analyses and toxicity tests illustrates the utility of applying different approaches to understanding complex surfactant-related issues. The mechanism of dissipation is likely to be related to either sorption or microbial degradation of the POEA surfactant (Giesy et al. 2000).

On the other hand, the organic carbon partition coefficients (K_{oc}) of ^{14}C-labeled POEA surfactant in three different soil types (silt loam, silt clay loam, and sandy loam) ranged from 2,500 to 9,600, suggesting that the surfactant would strongly adsorb to soil (Marvel et al. 1974). In the present study, the POEA surfactant could rapidly adsorb to the surface of the sediment layer and to suspended sediment (a small amount of sediment was suspended into the water column when adding the MON 0818 solution into microcosms and when removing the sheets of aluminum foil from the surface of sediment at the beginning of the study). Microbial degradation can also contribute to the dissipation of POEA from natural water bodies. However, the rate and extent of degradation will depend on the types and amounts of microbes present in the water and sediment. While no study on the microbial degradation of POEA surfactant is available in a water–sediment system, a study using ^{14}C-labeled POEA surfactant in natural water containing suspended sediment reported a half-life of less than 3–4 weeks (Banduhn and Frazier 1978).

The half-lives of POEA surfactant reported by Banduhn and Frazier were obtained in a system of natural water and suspended sediment that likely had a relatively low microbial population (Banduhn and Frazier 1978). Using limited data, Giesy et al. estimated that the aquatic half-life for POEA surfactant would range from 21 to 42 d, which is substantially longer than the half-lives of 13 and 18 h that we estimated for shallow (12 cm) water columns in contact with sediments (Giesy et al. 2000). Sorption and degradation could be expected to occur in the presence of sediment.

Contardo-Jara et al. showed that glyphosate accumulates in L. variegatus, despite the hydrophilic character of the herbicide (Contardo-Jara et al. 2009). The accumulated amounts of glyphosate and the added surfactants in Roundup® Ultra cause an elevation of the biotransformation enzyme sGST at non-toxic concentrations and antioxidant enzyme superoxide dismutase was significantly increased by glyphosate but in particular by Roundup® exposure indicating oxidative stress. The accumulation and the enzymatic response of the worms were clearly higher in the animals exposed to Roundup®, indicating that the formulation Roundup® is of more ecotoxicological relevance than thc glyphosate itself.

The alterations demonstrated by Peixoto[102] in mitochondrial bioenergetics caused by Roundup® cannot be exclusively attributed to the active ingredient, but may as well be the result of other chemicals e.g. POEA, or due the possible synergy between glyphosate and Roundup® formulation products (Peixoto 2005). This synergy between glyphosate and Roundup® inert ingredients has also been reported in cell division, where the delay observed in the cell cycle could be the result of alterations on the mitochondrial bioenergetic reactivity's, which have drastic consequences on cellular function through the perturbation of the bioenergetic charge and balance of the cell (Marc et al. 2002). Therefore, the reduced energetic efficiency of mitochondria may account for some toxic effects resulting from the impairment of the energy requirements of the cell and from the crucial importance of energy metabolism in active tissues, e.g. liver. Marc and coworkers found that the so called "inert ingredients" used in Roundup® are greatly responsible for the observed toxicity at the bioenergetic level. Data obtained in this study clearly demonstrate the ability of Roundup® to impair mitochondrial bioenergetic reactions. Alteration of basic mitochondrial functions was monitored by the detection of changes induced in mitochondrial respiration and membrane energization ($\Delta\psi$). Bearing in mind that mitochondria is provided with a variety of bioenergetic functions mandatory for the regulation of intracellular aerobic energy production and electrolyte homeostasis, these results question the safety of Roundup® on animal health.

7.11 Screening and Efficacy of New Surfactants for Glyphosate

Developing new plant protection formulations is a challenging task. The formulation of the active ingredient must assure a long shelf life stability including very high and low temperatures, a high loading and at the same time, the active ingredient needs to be optimal bioavailable. Greenhouse tests to evaluate the biological performance are regularly done with small plants and spray chambers using nozzles and water amounts which are not in accordance to practical field conditions.

Glyphosate, the most important generic active ingredient produced of the world is the most active patent area by far in order to improve glyphosate performance. Many companies sell glyphosate and the type and amount of surfactant in glyphosate formulations varies greatly. Green and Beestman focused on new formulation and adjuvant technologies available to maximize performance and minimize safety and environmental impact of herbicides (Green and Beestman 2007). Even though governments have not approved many new chemicals for use with agrochemicals during the past decade, there is a rich record of patent applications and product introductions. Formulation technology is the principal mechanism agrochemical companies use to renew products when the initial patents covering the active expire. The adjuvant and formulation industry has done an impressive job finding useful combinations and new utilities of currently approved chemicals.

Current data suggest that surfactant efficacy may be the result of charged surfactants ability to diffuse away from the cuticle into the subtending apoplastic space, where they act directly on the plasma membrane to increase glyphosate uptake. Cationic polyoxyethylene tertiary amine surfactants are more effective than nonionic surfactants at promoting glyphosate phytotoxicity (Wyrill and Burnside 1976; Riechers 1992). Surfactants promoting greatest glyphosate phytotoxicity are required at higher concentrations than necessary for maximum reduction of the spray solution surface tension, indicating that their mode of action is not limited to increasing the spreading characteristics of the spray droplet (Sherrick et al. 1986). This observation suggests that effective surfactants are also involved in increasing the permeability of the cuticle, plasma membrane, or both in increasing foliar uptake of glyphosate and promoting phytotoxicity. Richard and Slife determined that cellular absorption may offer greater resistance to foliar glyphosate uptake than cuticular penetration (Richard and Slife 1979). They suggested that the negative charges associated with the cell wall and the negative membrane potential of the plasma membrane repel the anionic glyphosate molecule. Sherrick et al. hypothesized that the surfactant MON 0818, is able to penetrate the cuticle and act at the plasma membrane (Sherrick et al. 1986). These authors reported that less than 20 % of [^{14}C]MON 0818 applied to field bindweed leaves was recovered in combined surface and chloroform washes, indicating surfactant penetration past the cuticle into underlying tissue.

Riechers et al. suggested that surfactant efficacy involves, at least partially, its ability to diffuse out of the cuticle into the apoplast where it alters membrane permeability to glyphosate (Riechers et al. 1994).

On the other hand, systematic studies conducted by and Nalewaja and co-workers indicate that surfactants of low ethylene oxide (EO) content may be optimal for promoting the uptake of lipophilic pesticides ($\log P > 3$), while those of high EO content were more beneficial for the uptake of hydrophilic compounds ($\log P < 0$) (Stock et al. 1993, 1992; Nalewaja et al. 1996). For compounds of intermediate lipophilicity, uptake is independent on the EO content of the added surfactant. The results were later confirmed by other studies on leaves and also on isolated cuticles (Nalewaja et al. 1996; Sharma et al. 1996; Coret and Chamel 1995).

One of the best way to evaluate the efficacy of glyphosate is measuring the efficiency of EPSPS (5-enolpyruvylshikimate 3-phosphate synthase) inhibition. Starting with a dose that is sprayed from a nozzle, only a fraction is foliarly retained, absorbed, translocated, and ultimately delivered to the plastids where EPSPS is localized. Each of these processes is in turn impacted by many physical, environmental, and physiological variables. For example, foliar retention can be affected by spray interception and droplet rebound which can be affected by plant morphology (e.g., leaf orientation and cuticle structure), environmental conditions (e.g., temperature and humidity), and surfactant concentration and droplet size (De Ruiter et al. 1990; Feng et al. 2003a; Buhler and Burnside 1987; Liu et al. 1996; Casely and Coupland 1985). Absorption is a major barrier for glyphosate efficacy and is affected by both surfactant class and concentration (De Ruiter et al. 1996; Denis and Delrot 1997). The efficacy of the tallow amine based cationic surfactants has been attributed to effective cuticle disruption (Feng et al. 1998; Ryerse 2001; Feng et al. 1999). Absorption is also impacted by the concentration gradient between the droplet and leaf, and efficacy enhancement has been reported at low spray volumes (Cranmer and Linscott 1990; Duncan Yerkes and Weller 1996; Feng et al. 2003a).

Glyphosate is translocated in the phloem from source to sink tissues following the sucrose gradient (McAllister and Haderlie 1985; Gougler and Geiger 1984). The excellent systemicity of glyphosate is derived from its unique combination of three acidic and one basic groups, and loss of any one of its functional groups resulted in decreased translocation (Bromilow and Chamberlain 2000). How glyphosate is loaded into the phloem is not well understood, but may involve phosphate transporters (Morin et al. 1997; Denis and Delrot 1993). Reductions in photoassimilate production and translocation ultimately reduce glyphosate translocation (Geiger et al. 1999). This self-limitation effect reduces the time window in which glyphosate can be efficiently translocated.

In the end, a plant that has been treated with glyphosate will die only if all the tissues have been killed. Any surviving tissue provides the opportunity for the plant to re-grow under favorable environmental conditions. The fate of individual tissues is decided based on whether a sufficient concentration of glyphosate was

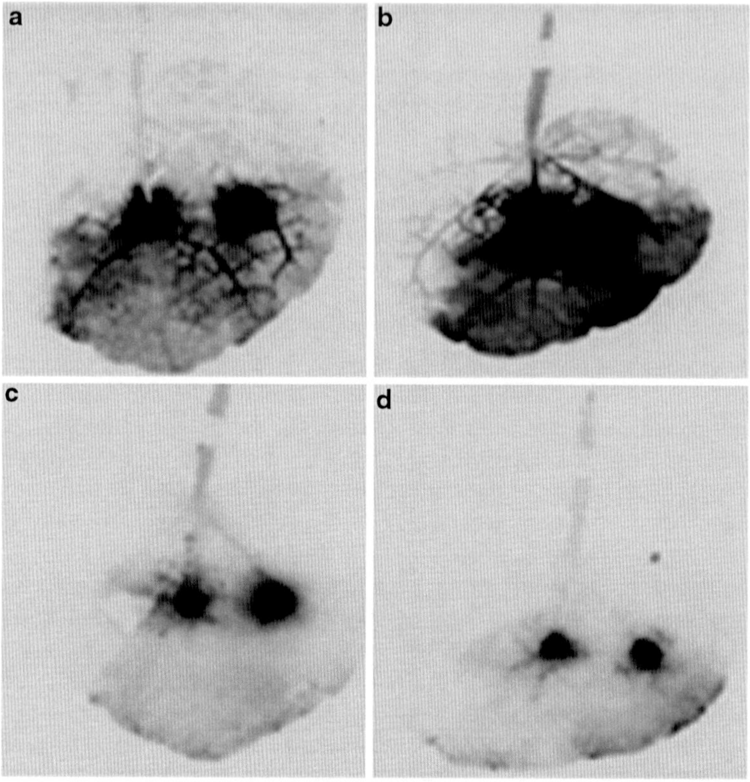

Fig. 7.8 Analysis by autoradiography of leaves with [^{14}C]glyphosate in MON0818 (0.1 %) as a function of harvest time: 2.5 (**a**), 5.0 (**b**), 24 (**c**) and 48 HAT (**d**) (Reprint from Feng et al. (2003b))

attained to inhibit EPSPS. Feng et al. examined the role of glyphosate distribution and tissue sensitivity in overall efficacy (Feng et al. 2003b). The process by which glyphosate is exported out of the treated leaf was examined by autoradiography. A sub-lethal dose of [^{14}C]glyphosate was formulated in 0.1 % MON0818 (tallow amine ethoxylated an nonylohenol ethoxylated mixture) and applied as two 1 μl droplets to the first leaf of 3.5 leaf velvetleaf plants. At various times (2.5, 5.0, 24, and 48 h after treatment, HAT) the treated leaf was excised and washed with water to remove surface residues (Fig. 7.8). Exposure of tissues to X-ray film allowed visualization of glyphosate that was in the leaf and provided a time course of glyphosate leaf export.

The studies used velvetleaf as a model and [^{14}C]glyphosate as the marker, and showed that plant response to glyphosate was impacted by both differential tissue distribution as well as sensitivity. The second conclusion is that the amount of

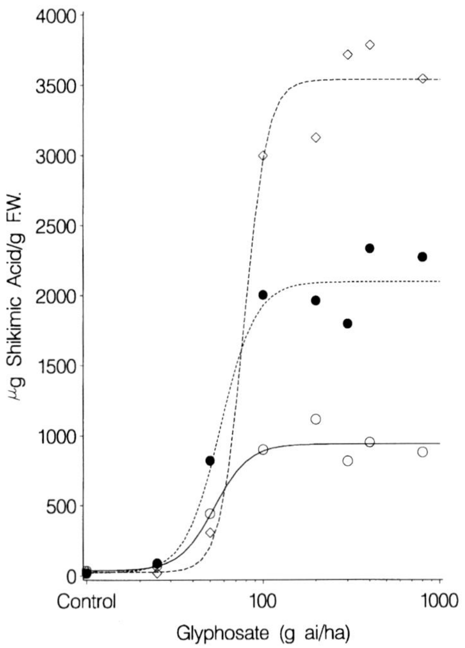

Fig. 7.9 Accumulation on shikimic acid in the third leaf of 3-week-old rape 48 h (◇), 24 h (•) and 6 h (○) after spraying with glyphosate (Reprint from Harring et al. (1998))

glyphosate delivered to tissues is linearly proportional to dose. This means that the dose does not change how glyphosate is distributed only how much is distributed. The current understanding is that whole plant efficacy is dependent on application of a dose of glyphosate that results in sufficient uptake and distribution to attain the lethal threshold in all tissues quickly and before the onset of self-limitation.

On the other hand, accumulation of shikimic acid was related to the dose of glyphosate by a nonlinear logistic dose response models as early as 5 h after spraying (Fig. 7.9). Within the same period shikimic acid accumulation made it possible to distinguish between a formulation of the pure isopropylamine salt of glyphosate and formulations containing different surfactants. ED50 (effective dose 50) estimates based on accumulation of shikimic acid gave a good indication of the relative strength of the evaluated glyphosate formulations. Ranking of ED50 based on accumulation of shikimic acid was the same as achieved by visual assessment of plant death 14 days after spraying.

Wan Kim and Amrhein found glyphosate induced accumulation of shikimic acid within 24 h after treatment of tomato (Wan and Amrhein 1995). Harring et al. studied the glyphosate action on phenylalanine ammonia-lyase activity and shikimic acid accumulation (Harring et al. 1998). Plants sense glyphosate toxicity within hours after application resulting in changes in carbon allocation and photosynthesis (Geiger and Bestman 1990).

7.12 Replacement of Fatty Amines and Nonylphenol Ethoxylated by New Renewable Resources-Based Surfactants

The selection of the type of formulation, active ingredient concentration, selection of the proper adjuvant and amount to maximize performance, maintaining product stability, all while insuring the toxicological properties of the product should meet customer, market and regulatory requirements. There is increasing regulatory and public pressure to decrease the amount of pesticides released into the environment. The challenge is to maintain and safeguard their efficiency with effective adjuvants. The use of petrochemical based surfactants like nonylphenol ethoxylates and fatty amine ethoxylates have declined significantly in the past decade, due to the ecological and toxicological reasons. Hence, there is continuing demand for alternatives based on a renewable source.

Run-off, leaching, adsorption and biodegradation are some of the possible exposure routes and main fate processes. Potential transport routes via rainwater may be either horizontal as run-off ending up in surface waters or vertical leaching to groundwater. Another fate route is adsorption to soil or biological material such as plants, plant roots and plant debris or micro-organisms in the soil. Additionally, the adjuvants may be biodegraded either aerobically or anaerobically.

The effects of nonylphenol ethoxylates and nonylphenol in the environment have been extremely controversial. Concerns first emerged in 1983–1984 when Giger and coworkers from Switzerland established that nonylphenol ethoxylates and products of degradation were more toxic to aquatic life than their precursors (Giger et al. 1984).

The first evidence that alkylphenols could be oestrogenic was published in 1938 by Dodds and Lawson but it was Soto and co-workers who accidentally observed that nonylphenol, which was employed in the manufacture of the test tubes used in their experiments, was capable of initiating proliferation in breast tumour cells as if oestrogens were present (Dodds and Lawson 1938; Soto et al. 1991). Endocrine disrupters can disturb the hormonal system by mimicking the occurrence of natural hormones, blocking their production or by inhibiting or stimulating the endocrine system. Nonylphenol was found to mimic the natural hormone 17β-oestradiol by competing for the binding site of the oestrogen receptor (Fig. 7.10a), due to their structural similarity (Fig. 7.10b) (Lee and Lee 1996; White et al. 1994).

For nonylphenol to exhibit endocrine activity the para-position of the phenolic OH-group and the branched aliphatic side chain appear to be determinant, meaning that not all nonylphenol isomers are capable of inducing oestrogenic activity (Kim et al. 2004; Odum et al. 1997; Tabira et al. 1999). 17β-oestradiol is a natural hormone that influences the development and maintenance of the female sex characteristics, and the maturation and function of accessory sex organs and it is also involved in the neuroendocrine and skeletal systems and is capable of promoting carcinogenicity in target tissues (Lee and Lee 1996; Alberts et al. 1983). Hence, nonylphenol is expected to initiate a variety of responses in organisms. Recently

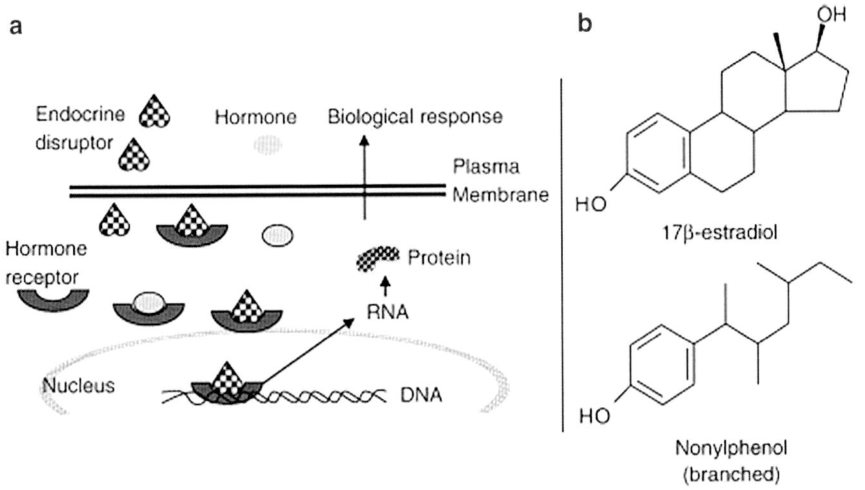

Fig. 7.10 Competition for the oestrogen binding receptor between nonylphenol and 17b-oestradiol in the cell (Reprint from Soares et al. (2005))

it has also been established that nonylphenol has antiandrogenic activity, i.e., is capable of interfering with the proper functioning of androgens that are essential for the normal development of males and their reproductive systems (Lee et al. 2003).

The responses of the endocrine system to nonylphenol have been thoroughly studied in different organisms (White et al. 1994; Laws et al. 2000). For instance, nonylphenol was capable of inducing the production of female proteins in rainbow trout (Oncorhynchus mykiss) at 20.3 mg/l, medaka fish (Oryzias latipes) exposed to 0.1 mg/l nonylphenol, and the platyfish (Xiphophorus maculates) testis morphology and male fertility were negatively affected at nonylphenol concentrations $\geq$0.96 mg/l after 4 weeks of exposure (Jobling et al. 1996; Tabata et al. 2001; Kinnberg et al. 2000).

According to the data above showed, nonylphenol ethoxylates compounds have therefore been added to Annex I to Regulation (EC) No 689/2008 of the European Parliament and of the Council concerning the export and import of dangerous chemicals (OJ L 204, 31.7.2008, p. 1–35). The pesticides agreed as 'priorities for action' are listed here (updated in 2004), as well as a list of substances of 'possible concern' including 98 pesticides (updated in September 2005).

However, many other countries, including China, India and several South American countries use and produce nonylphenolic compounds in large amounts and no action has been taken by any of these countries to reduce or eliminate their usage. Nonylphenol ethoxylates are being replaced by other surfactants in most European, Canadian and Japanese markets, mainly by alcohol ethoxylates (Krogh et al. 1997). These surfactants are less efficient but considered to be environmentally safer as they degrade more rapidly (Campbell 2002). However, few studies have been performed on the metabolites that are produced as a result of degradation of alcohol ethoxylates and it has been recently pointed out that their degradation gives rise to compounds with low water solubility and these adsorb to the sludge solids (Soares et al. 2005).

Meanwhile, tallow amine ethoxylates are the market reference of built-in adjuvants for glyphosate since many years, due to the excellent weed control performance they provide. However due to their toxicity profile there is a growing demand for alternatives with reduced labelling.

Traditionally, tallow amine ethoxylates (TAM-EO) were the built-in adjuvant of choice for standard isopropylamine (IPA) glyphosate 360 a.e./l formulations. These products are mediocre surfactants from a physico-chemical point of view, as indicated by fairly high surface tension and contact angle values. However, they are excellent adjuvants: it is believed that they act by facilitating the penetration of glyphosate, a relatively hydrophilic compound, through the rather hydrophobic cuticle which covers the external surface of higher plants, due to the disruption of cell walls allowing.

TAM-EOs and more especially TAM-15 EO and TAM-20 EO are harmful and hazardous for the environment. Quite unusual for a pesticide formulation is the co-formulant considered to be more toxic than the active ingredient. Glyphosate formulations with reduced labelling but with similar potency are therefore required in today's market.

The main disadvantages of these products is the high risk of damage to the eyes, the high toxicity to aquatic organisms, even in the concentration levels used in the formulation of glyphosate (8–15 %) and its high content of 1.4 dioxane (up to 25,000 ppm) due to the use of ethylene oxide during the process of manufacture of TAM-EO.

Several references were found in the literature about the toxicity of POEA (van Ginkel et al. 1993). TAM-EO has got more toxicity due to their high toxic by-products, without alkyl chain, resulting from the biodegradation process.

Finally in 2000, Bean and Cutler noted the low compatibility of alkoxylated alkylamine surfactants in general with high-strength glyphosate concentrates (Bean and Cutler 2000). This, in turn, has triggered the development of suitable formulants. Another driving force for new glyphosate additive developments is to save costs since the market has polarised into performance and low-cost brands since 2008. Many of the 700 glyphosate patent publications are directed at this issue, but still most of the improved adjuvant systems still contain amine-based surfactants. The current market situation could be called "the era of the small difference" where continually improved formulations are essential for success (Pimentel 1997).

7.13 Trends and Perspectives in Agrochemicals Formulations

The history of agricultural adjuvants dates back to eighteenth and nineteenth centuries when additives such as pitch, resins, flour, molasses, and sugar were used with lime, sulfur, copper, and arsenates to improve "sticking" and biological performance by modifying the physical and chemical characteristics of the applied mixture. Fundamentally, the goal of using adjuvants has stayed the same. Using substances that are inactive when used alone to improve the performance and application of an active ingredient by modifying the physical and chemical characteristics of the spray mixture is a fundamental part of all agrochemical research.

Despite the significance of agrochemicals use for pest control, the environmental problems caused by overuse of agrochemicals have brought scientists and publics much concern in recent years (Dayan et al. 2009). Of the reasons for this are (1) the high toxicity and non-biodegradable of agrochemicals; (2) the lack of scientific formulations. Thus, formulation scientists now are facing the challenge to explore novel green or environmental friendly agrochemical formulations to improve the biologic efficacy and develop techniques that can be employed to reduce pesticide use while maintaining plant protection. The tremendous increase in crop yields associated with the 'green' revolution has been possible in part by the discovery and utilization of chemicals for pest control. However, concerns over the potential impact of pesticides on human health and the environment has led to the introduction of new pesticide registration procedures, such as the Food Quality Protection Act in the United States. These new regulations have reduced the number of synthetic pesticides available in agriculture. Therefore, the current paradigm of relying almost exclusively on chemicals for pest control may need to be reconsidered. New pesticides, including natural product-based pesticides are being discovered and developed to replace the compounds lost due to the new registration requirements. Dayan et al. covered the historical use of natural products in agricultural practices, the impact of natural products on the development of new pesticides, and the future prospects for natural products-based pest management (Dayan et al. 2009).

7.14 New Methodologies in Agrochemicals Formulation

7.14.1 Microemulsions and Nanoemulsions

Microemulsions are considered as thermodynamically stable colloidal dispersions that are optically transparent or translucent with drop size in the range of 100–200 nm (Prince 1977). Thus, microemulsion may be regarded as one phase system (Lindman and Danielson 1981). As indicated above, the cloud point is one of the specific characteristics of the microemulsions. When the temperature of the microemulsion system is increased, the solubility of non-ionic surfactant decreases. The cloud point of microemulsion can be defined as the temperature at which the transparent microemulsion solution becomes cloudy, i.e. from one phase to two phases or three phases (Strey 1996).

It is common knowledge that by hydrating of polyethylene oxide group (PEO) chain, non-ionic surfactant is dissolved in water medium, but it may dissociate with water as a result of dehydration when the temperature exceeds its cloud point. The cloud point increases with increasing the amount of ethylene oxide groups in a chain, i.e. the higher hydrophilicity of non-ionic surfactant, the higher the cloud point. The cloud point is also affected by the concentration of the surfactant solution and the electrolytes in the aqueous solution. In the later case, usually lower the cloud point (Yoshihara et al. 1995; Saito and Shinoda 1967; Minanaperez et al. 1995; Tadros 1994). Chen et al. described, two new-types of pesticide microemulsions used for control *Liriomyza* spp., namely 16 wt % beta-cypermehtrin (inside microemulsion xylene oil droplets) plus monososultap (dissolved in water medium) and 20 % abamectin (inside microemulsion xylene oil droplets) plus monososultap (in the water medium) were prepared (Chen et al. 2000). The effects of agrochemical concentration, various surfactants at various concentrations on the cloud points of microemulsions have been studied. The stability of microemulsions containing 5 wt % abamectin and 1 wt % beta-cypermethrin is also discussed. Similar to the cloud point of surfactant aqueous solution, at constant surfactant concentration the cloud point of the agrochemical microemulsions increases as the hydrophilicity of the surfactant increases. The cloud point of the formulated microemulsions depends on the characteristics of the agrochemical, the kinds and the amounts of the added surfactant and co-surfactant.

The results above described shown that the values of the cloud point of microemulsions were depended on the nature of agrochemical, the surfactants used and the concentration of surfactants. Surfactants with higher HLB value at same concentration produce more stable agrochemical microemulsions, i.e. with higher cloud point. The results showed that the match between surfactant and oil phase or pesticide oil phase was the key to formulate stable microemulsions. However, the water quality was showed almost no effect on the cloud point of the agrochemical microemulsions in the studies.

Meanwhile microemulsions, which contain non-polar agrochemicals, usually have higher cloud points than those of containing polar ones, especially electrolytic agrochemicals. As discussed, higher cloud points can be obtained with an increase in surfactant concentration, but it is more reasonable to use more co-surfactant rather than surfactant, because the former is much cheaper.

On the other hand, Nanoemulsions have uniform and extremely small droplet sizes, typically in the range of 20–100 nm (Forgiarini et al. 2001). In addition, high kinetic stability, low viscosity and optical transparency make them very attractive systems for many industrial applications; for example, in agrochemicals for pesticide delivery (Lee and Tadros 1982). Wang et al. investigated potential applications of the system developed with water insoluble pesticide, β-cypermethrin (β-CP), incorporated into the precursor microemulsion concentrate (Wang et al. 2007). The effect of this active pesticide on stabilities of the concentrate, and the corresponding nanoemulsion, were also investigated. The incorporation of β-CP in the concentrate showed no effect on the phase behavior when present at less than 12 wt %. Compared with the commercial β-CP microemulsion, the excellent stability of sprayed solution diluted from the concentrate makes this system an ideal candidate as a water-insoluble pesticide delivery system. Thus, the application of the new methodology designing by spray formulations of β-CP may enable a reduction in the applied amounts, relative to those formulated as O/W microemulsions. These characteristics make the new methodology promising from both environmental and economical points of view.

7.14.2 *Liposomes*

Liposomes are structures made of lipid bilayers forming one or more concentric spheres which entrap part of the solvent in which they freely float, into their interior. Their unique properties have triggered numerous applications in various fields of science and technology, from basic studies of membrane function to the use as carriers of very different substances. In agriculture, they can be used to improve the efficacy of different biocides and to deliver some essential nutrients (Lasic 1993). Herbicides, fungicides and pesticides are rapidly washed from the leaves of plants and encapsulation in liposomes may prolong the action of these agents on plants and reduce the damage in soil cultures. Pons and Estelrich described the optimization of a cheap and easy method of preparing liposomes (Pons and Estelrich 1996). Among the studied formulations there are some with a high stability and this property makes them very suitable to be used as an agrochemical product.

7.14.3 *Nanomaterials*

Materials with a particle size less than 100 nm in at least one dimension are generally classified as nanomaterials. The development of nanotechnology in conjunction with biotechnology has significantly expanded the application domain of nanomaterials in various fields. However in the field of agriculture, the use of nanomaterials is relatively new and needs further exploration.

Khot et al. summarized the developments and application of novel nanomaterials in agriculture, and described nanopesticides and its applications. Nanopesticides "involve either very small particles of pesticidal active ingredients or other small engineered structures with useful pesticidal properties" and can increase the dispersion and wettability of agricultural formulations (i.e., reduction in organic solvent runoff), and unwanted pesticide movement. Nanomaterials and biocomposites exhibit useful properties such as stiffness, permeability, crystallinity, thermal stability, solubility, and biodegradability needed for formulating nanopesticides. Nanopesticides also offer large specific surface area and hence increased affinity to the target. Nanoemulsions, nanoencapsulates, nanocontainers, and nanocages are some of the nanopesticide delivery techniques that have been discussed recently for plant protection (Khot et al. 2012).

7.15 New Surfactants and Additives Apply to Agrochemicals Formulation

There is increasing regulatory and public pressure to decrease the amount of pesticides released into the environment. The current trends in the development of pesticide formulations are increased enormously to meet the needs of environmental safety, eliminate organic solvents or to improve the activity and persistence of the active ingredient (Knowles 2008).

7.15.1 Alkyl Polyglycosides

Alkyl polyglycosides or APG's are non ionic surfactants with a hydrophilic saccharide instead of an ethylene oxide chain. The alkyl chain has 8–16 carbons. APGs are water soluble with excellent adjuvant properties. APGs are made from renewable raw materials and can be put into very concentrated liquid of dry formulations (Pompeo et al. 2005). They are called "green surfactants" because they are very safe to the environment.

Short chain APGs show decent properties as wetting and penetrating agents and offer a high tolerance to saline solutions though they are non-ionic, and do not exhibit a cloud point typical for alkoxylates (Nickel et al. 1996). The importance of nonionic APGs for glyphosate was first recognized by Syngenta as potentiators for glyphosate (Burval and Chan 1995).

Alkyl polyglycosides (APG) are superior surfactants with outstanding wetting properties and are used as adjuvants for pesticides as they improve the spreading and enhance the uptake of the pesticides (Garst 1997). Nonetheless, the effect of APGs on weed control of glyphosate cannot match industry standards. Advantageously however, as compared to TAM-EO, APGs are generally classified as non-toxic and ready biodegradable. Several options were explored to take advantage of the benefits of APGs as high performance wetting agents whilst optimising the weed control.

A new class of non-ionic surfactants resulting from the direct ethoxylation of alkyl and/or alkenyl polyglycosides was designed as possible alternatives to TAM-EO (Behler and Clasen 2006). Their performance as adjuvants for glyphosate was assessed in greenhouse trials on two model plant species. The new derivatives were compared to TAM-EO and to straight APG. According to the greenhouse data, alkoxylated alkyl polyglycosides showed good weed control almost reaching performance of TAM-EOs and surpassing efficiency of standard APG. Alkoxylated alkyl polyglycosides exhibit also a much better toxicological profile compared to TAM-EOs reducing the risk to end-users.

7.15.2 *Ethoxylated Saccharose Esters*

In 2007, ethoxylated APGs were introduced. Due to their hydrophilic properties, they turned out to be suitable glyphosate potentiators and adjuvants (Abribat et al. 2007). However, their low lipophilicity makes them not compatible in oil based formulations such as emulsifiable concentrate (EC) or concentrated oil-in-water emulsion (EW) and consequently, new lipophilic surfactants based on ethoxylated saccharose esters have been developed.

In a recent application, Mainx and Hofer described alkylene oxide adducts of oligosaccharides like saccharose (Mainx and Hofer 2009). Their esterification leads to a new chemical class of label-free nonionic surfactants. Selected representatives of this class of chemicals showed outstanding results in boosting the performance of crop protection products in greenhouse studies. Indeed, the protective property of azoxystrobin applied with these new ethoxylated saccharose esters was more effective than the industry standards, such as nonylphenol ethoxylated and isotridecyl alcohol 6 EO. In curative tests, the activity of the fungicide epoxiconazole was improved as well, although the effects were less pronounced.

Epoxiconazole was improved as well, although the effects were less pronounced and even the activity of glyphosate was enhanced. Ethoxylated saccharose esters show a good toxicological and ecotoxicological profile. Their behaviour in pesticide formulations and tank mix solutions is excellent, so in total, they can be considered as a novel suitable class of adjuvants and surfactants for pesticide formulations.

The studies above mentioned showed that ethoxylated saccharose esters have an excellent potential as new adjuvants and emulsifiers in agrochemical compositions. Especially with fungicides at low concentrations, they show extraordinary results in improving the activity of the active ingredients suggesting their suitability as both, tank mix adjuvants and inerts for incan formulations. Due to good compatibility with most new solvents, this new chemical class of surfactants can also be considered to become suitable components in all pesticide formulations either alone or in combination with other nonionic or anionic surfactants.

7.15.3 *Dimethylethanolamine Based Esterquats*

Esterquat chemistries are widely used in the commercial laundry market as fabric softeners. They have been used since the 1970s as biodegradable alternatives to dihydrogenated tallow dimethyl ammonium chloride (DHTDMAC) chemistries. However, the use of esterquats in agrochemical formulations has not been thoroughly explored and would represent a biodegradable alternative chemistry (Zoller 2009). In addition, esterquats may offer reduced eye irritation or improved ecotoxicity profiles to an industry that is becoming ever more scrutinized for such formulation improvements.

Malec et al. explored the use of esterquats based on dimethylethanolamine (DMEA) in agricultural formulations (Malec et al. 2009). It was determined that DMEA based esterquats can function as wetters or spreaders, but can also act as emulsifiers. The surface active properties and aquatic toxicity of esterquats as surfactants will be explored as well. The esterquats have lower critical micelle concentrations, lower equilibrium surface tensions, and faster wetting times than traditional agricultural surfactants.

Various formulations were explored including herbicides, insecticides, and fungicides. These formulations included delivery systems such as soluble concentrates (SL) and emulsifiable concentrates (EC). The formulations were evaluated and in many cases provided performance properties that were equal to current standards for commercial products. Greenhouse trials were conducted evaluating the effectiveness of esterquats in glyphosate (*N*-(phosphonomethyl) glycine) formulations. It was found that esterquats were successful glyphosate adjuvants, utilizing lower surfactant use rates while offering equal efficacy to commercial products. Additionally, glyphosate formulations using DMEA esterquats have lower eye irritation compared to other commercial standard glyphosate surfactants. The aquatic toxicity of DMEA esterquats was also studied and the DMEA esterquats were found to be less toxic than TAM-EO.

The studies showed that DMEA esterquats may serve a valuable function in agrochemical formulations. They offer improved biodegradability, lower eye toxicity, improved wetting, spreading, and penetration, and comparable aquatic toxicity to commercial surfactants already in use in agrochemicals.

7.15.4 Chitosan Derivatives

Chitosan based polymeric micelle due to its outstanding biologic properties and functions such as biodegradability, biocompatibility, insecticidal and antibacterial activity, has been widely researched or applied in the fields of agriculture, medicine, pharmaceuticals, functional food in the last decade (Chellat et al. 2000; Risbud and Bhonda 2000; Badawy et al. 2004; Hejazi and Amiji 2003; Kumar et al. 2004; Rabea et al. 2005). However, chitosan has no amphiphilicity, cannot form micelle and load drug directly. In recent years, chitosan-based micelle system has been developed by introducing hydrophobic and/or hydrophilic groups to the chitosan backbone (Zhang et al. 2008). Amphiphilic chitosan derivatives which grafted sulfuryl as hydrophilic moieties and octyl as hydrophobic moieties had been reported (Zhang et al. 2003, 2008).

Lao et al. described novel amphiphilic chitosan derivatives designed and synthesized by grafting octadecanol-1-glycidyl ether to amino groups, sulfate to hydroxyl groups (e.g. N-(octadecanol-1-glycidyl ether)-O-sulfate chitosan (NOSCS)) (Lao et al. 2010). Rotenone as a model drug was chosen and used to assess the potential loading capability of novel chitosan derivatives. The insecticide rotenone was loaded and formed about 116.4–216.0 nm nanoparticles, and its solubility in

NOSCS micelles aqueous solution was increased largely. The highest concentration of rotenone was up to 26.0 mg/mL (NOSCS- 1), which was about 13,000 times that of free rotenone in water (about 0.002 mg/mL).

7.15.5 Ammonium Sulphate and Sulphated Glycerine as Additive

The leaf cuticle and plasma membrane have been identified as barriers limiting glyphosate activity (Riechers et al. 1994; Denis and Delrot 1993). Neither glyphosate nor its different salts are effective in overcoming these barriers easily without appropriate surfactants (Buhler and Burnside 1983b; Jordan 1981). Cationic surfactants have been found to be more effective than nonionic surfactants in increasing efficacy (Wyrill and Burnside 1976; Riechers et al. 1995). The addition of ammonium sulphate (AMS), an inorganic salt, to the glyphosate spray solution improved the efficacy of the herbicide (Blair 1975). Additionally, it has been found that salts dissolved in water used as the carrier for glyphosate may reduce its effectiveness, particularly calcium and magnesium salts. These salts have a positive charge and may associate with the negatively charged glyphosate molecule, replacing the isopropylamine or diammonium salts found in the formulated glyphosate product. AMS reduces the formation of calcium–glyphosate complexes on these leaves and therefore improves performance (Hartzler 2001). The inverse relationships between weed growth parameters and increasing concentration levels of AMS suggest that the ability of AMS to enhance glyphosate activity is to a large extent concentration dependent (Aladesanwa and Oladimeji 2005).

Glycerine and ammonium sulphate have both been popular additives in the long-held practice of adding chemical agents (adjuvants) to improve biological efficacy in crop protection sprays (Gednalske and Herzfeld 1994). Glycerine, a polyol of natural origins (triglycerides), is one of the principle building blocks of plant tissues. This biocompatibility as well as glycerine's humectancy is strong contributors to its adjuvancy with additional benefit as a hydrotrope to homogenize water-based spray solutions (Heldt et al. 2000). Ammonium sulphate is a proven sequestrant of hard water cations as well as a nitrogen source benefiting herbicides of diverse modes of action (Buhler and Burnside 1983a). One difficulty with ammonium sulphate solutions are their ionic strength and the difficulty that presents in creating shelf-stable blends with surfactants (Gednalske and Herzfeld 1994).

Anderson presented the reaction product(s) resulting from sulphating glycerine show strong adjuvant activity with herbicides (Anderson 2010). As with ammonium sulphate, the adjuvant activity of sulphated glycerine includes both, sequestering cations commonly found in hard water as well as increased nitrogen for non-complexing herbicides. The field treatments including sulphated glycerine, demonstrated statistically superior weed control when compared to un-sequestered reference samples. These "hard water" treatments including sulphated glycerine were statistically equivalent to control observed for glyphosate sprays prepared with de-ionized water. The increased efficacy of the herbicide glufosinate was demonstrated when using ammonium sulphate or sulphated glycerine as an adjuvant. Both of these biological observations were confirmed with rates of sulphated glycerine significantly lower than current label rates for ammonium sulphate.

7.16 Conclusion

Uptake of pesticides into plant foliage varies with plants and chemicals, and can be greatly influenced by adjuvants and environmental conditions. Adjuvants usually have multiple functions in relation to pesticide efficacy. Increasing the foliar uptake of active ingredients is of particular importance for herbicides, growth regulators, and defoliants. It is known that the penetration of pesticides into plant leaves is related to the physicochemical properties of the active ingredients, especially molecular size and lipophilicity. However, the uptake rate of a compound cannot be predicted by either of them or even combination of them. For a specific chemical, uptake varies greatly with plant species and there is no simple method at the moment to quickly evaluate the leaf surface permeability of a plant.

Environmental conditions have an important influence on herbicide efficacy. In particular, the effect of humidity on herbicide uptake has been attributed to changes in cuticle hydration and droplet drying. Herbicide uptaking slows or stops when herbicide droplets dry, it was suggested that humectants or wetting agents could enhance herbicide uptake by keeping droplets moist, maintaining the herbicide in solution, and keeping it available for uptake.

Various adjuvants are being used to increase the penetration of pesticides into target plant foliage, but their effects vary with chemicals and plant species. The mechanisms of action of adjuvants in enhancing pesticide uptake remain unclear despite the effort made during the last three decades. Modern analytical and microscopic techniques provide powerful tools to deepen our understanding in this issue. However, a more multidisciplinary approach is urgently needed to elucidate the transcuticular diffusion behaviour of pesticides and the mode of action of adjuvants. A better understanding of the foliar uptake process should lead to a more rational use of pesticides and minimize their negative impact on the environment. Therefore, a systematic study on the complex interaction between active ingredients, surfactants and plants should lead to a more rational and cost-effective use of surfactants in weed control.

On the other hand, commercial glyphosate formulations are more acutely toxic than glyphosate, since the amount of Roundup® required to kill rats is about 1/3 of the amount of glyphosate alone. Similar results have been obtained in cell division, thus indicating a synergy between glyphosate and Roundup® formulation products. Although glyphosate is already one of the most used xenobiotics in modern agriculture, we should expect an increasing utilization of glyphosate largely due to the number of transgenic plants developed to be tolerant to this herbicide. Fatty amine ethoxylates or co-formulant called "inert ingredients" used in Roundup® are greatly responsible for the observed toxicity at the bioenergetic level. Quite unusual for a pesticide formulation is the co-formulant considered to be more toxic than the active ingredient. Glyphosate formulations with reduced labelling but with similar potency are therefore required in today's market.

The fact that the adjuvant could be more toxic than the active ingredient in one of the most active ingredient used pesticides together with the need that more effectiveness is required to reduce the environmental effect of xenobiotics has lead to the development of new adjuvants, e.g. more than 600 patent of glyphosate formulations have been published in the last 12 years.

It is important to point out that there is increasing regulatory and public pressure to decrease the amount of pesticides released into the environment. The challenge is to maintain and safeguard their efficiency with effective adjuvants. The use of petrochemical based surfactants like nonylphenol ethoxylates and fatty amine ethoxylates have declined significantly in the past decade, due to the ecological and toxicological reasons. Hence, there is continuing demand for alternatives based on a renewable source. Suppliers are developing new surfactants and new surfactants formulations based on natural products such as vegetable oils, lecithin, sugars, amino acid and others. Microemulsions, liposomes and nanoemulsions as emerging technologies in agrochemicals formulations have reduced the use of petrochemical solvents, e.g. xylene and improve the efficiency of biocides. The development of nanotechnology in conjunction with biotechnology has significantly expanded the application domain of nanomaterials in various fields. However in the field of agriculture, the use of nanomaterials is relatively new and needs further exploration.

These products will become more popular in the near future as a means of increasing product safety and enabling quicker product regulation.

References

Abribat B, Anderson T, Oester D (2007) Alkoxylated polyglycosides, next generation glyphosate potentiators. In: Gaskin RE (ed) Proceedings of the 8th international symposium on adjuvants for agrochemicals, ISAA, Christchurch

Aladesanwa RD, Oladimeji MO (2005) Optimizing herbicidal efficacy of glyphosate isopropylamine salt through ammonium sulphate as surfactant in oil palm (*Elaeis guineensis*) plantation in a rainforest area of Nigeria. Crop Prot 24:1068–1073

Alberts E, Kalverboer AF, Hopkins B (1983) Mother infant dialog in the first days of life an observational study during breast-feeding. J Child Psychol Psychiatry 24:145–161

Anderson TH (2010) Sulphated glycerine, a new sequestrant and adjuvant for herbicide sprays. In: Baur P, Bonnet M (eds) Proceedings of the 9th international symposium on adjuvants for agrochemicals, Dynevo, Leverkusen, pp 301–306

Annual Book of ASTM Standards (2006) Volume 11.04, E 1519-06a, ASTM International, West Conshohocken. www.astm.org

Badawy MEI, Rabea EI, Rogge TM, Stevens CV, Smagghe G, Steurbaut W, Höfte M (2004) Synthesis and fungicidal activity of new N, O-acyl chitosan derivatives. Biomacromolecules 5:589–595

Baker EA, Hayes AL, Butler RC (1992) Physicochemical properties of agrochemicals: their effects on foliar penetration. Pestic Sci 34:167–182

Banduhn MC, Frazier HW (1978) G 3780A surfactant: biodegradation in nature waters. report no. MSL-0488. Monsanto Company, St. Louis

Bauer H, Schönherr J (1992) Determination of mobilities of organic compounds in plant cuticles and correlation with molar volumes. Pestic Sci 35:1–11

Bean M, Cutler J (2000) Agrochemical composition. EP Patent 1006792

Becker M, Kerstiens G, Schönherr J (1986) Water permeability of plant cuticles: permeance, diffusion and partition coefficients. Trees 1:54–60

Behler A, Clasen F (2006) Method for the alkoxylation of alkyl/alkylene polyglycosides. EP Patent 1716163

Beyers SW (1995) Acute toxicity of Rodeo® herbicide to Rio Grande silvery minnow as estimated by surrogate species: plains minnow and fathead minnow. Arch Environ Contam Toxicol 29:24–26

Blair AM (1975) The addition of ammonium salt or phosphate ester to herbicides to control *Agropyron repens* (L) Beaux. Weed Res 15:101–105

Bromilow RH, Chamberlain K (2000) The herbicide glyphosate and related molecules: physicochemical and structural factors determining their mobility in phloem. Pest Manag Sci 56:368–373

Bromilow RH, Chamberlain K, Evans A (1990) Physicochemical aspects of phloem translocation of herbicides. Weed Sci 38:305–314

Buchholz A, Peter B, Schönherr J (1998) Differences among plant species in cuticular permeabilities and solute mobilities are not caused by differential size selectivities. Planta 206: 322–328

Buhler DD, Burnside OC (1983a) Effect of water quality, carrier volume, and acid on glyphosate phytotoxicity. Weed Sci 31:163–169

Buhler DD, Burnside OC (1983b) Effect of spray components on glyphosate toxicity to annual grasses. Weed Sci 35:124–130

Buhler DD, Burnside OC (1987) Effects of application variables on glyphosate phytotoxicity. Weed Technol 1:14–17

Burghardt M, Schreiber L, Riederer M (1998) Enhancement of the diffusion of active ingredients in barley leaf cuticular wax by monodisperse alcohol ethoxylates. J Agric Food Chem 46:1593–1602

Burval J, Chan JH (1995) Liquid phytoactive compositions. US Patent 5 468 718

Campbell P (2002) Alternatives to nonylphenol ethoxylates, Review of toxicity, biodegradation & technical-economic aspects. In: ToxEcology environmental consulting, Report for Environment Canada, Vancouver

Cano ML, Dorn PB (1996) Sorption of two model alcohol ethoxylate surfactants to sediments. Chemosphere 33:981–994

Casely JC, Coupland D (1985) Environmental and plant factors affecting glyphosate uptake, movement and activity. In: Grossbard E, Atkinson D (eds) The herbicide glyphosate. Butterworth and Co, London, pp 92–123

Chefetz B (2003) Sorption of phenanthrene and atrazine by plant cuticular fractions. Environ Toxicol Chem 22:2492–2498

Chellat F, Tabrizian M, Dumitriu S, Chornet E, Rivard CH, Yahia LH (2000) Study of biodegradation behavior of chitosan-xanthan microspheres in simulated physiological media. J Biomed Mater Res 53:592–599

Chen B, Li Y (2007) Sorption of 1-naphthol by plant cuticular fractions. J Environ Sci 19:1214–1220

Chen F, Wang Y, Zheng F, Wu Y, Liang W (2000) Studies on cloud point of agrochemical microemulsions. Colloids Surf A 175:257–262

Chen B, Li Y, Guo Y, Zhu L, Schnoor JL (2008) Role of the extractable lipids and polymeric lipids in sorption of organic contaminants onto plant cuticles. Environ Sci Technol 42:1517–1523

Contardo-Jara V, Klingelmann E, Wiegand C (2009) Bioaccumulation of glyphosate and its formulation Roundup® Ultra in *Lumbriculus variegatus* and its effects on biotransformation and antioxidant enzymes. Environ Pollut 157:57–63

Cordiner JL (2004) Challenges for the PSE community in formulations. Comput Chem Eng 29:83–92

Coret J, Chamel A (1995) Effects and possible mode of action of some nonionic surfactants on the diffusion of glyphosate and chlorotoluron across isolated plant cuticles. Pestic Sci 43:163–166

Coupland D (1988) Factors affecting the phloem translocation of foliage applied herbicides. In: Aitken RK, Clifford DR (eds) British plant growth regulation 18. British Plant Growth Regulation Group Monograph, pp 85–112

Crafts AS (1960) Evidence for hydrolysis of esters of 2,4-D during absorption by plants. Weeds 8:19–25

Cranmer JR, Linscott DL (1990) Droplet makeup and the effect on phytotoxicity of glyphosate in velvetleaf (*Abutilon theophrasti*). Weed Sci 38:406–410

Dayan FE, Cantrell CL, Duke SO (2009) Natural products in crop protection. Bioorg Med Chem 17:4022–4034

De Ruiter H, Uffing AJM, Meinen E, Prins A (1990) Influence of surfactants and plant species on leaf retention of spray solutions. Weed Sci 38:567–572

De Ruiter H, Straatman K, Meinen E (1993) The influence of a fatty amine surfactant on foliar absorption and translocation of the trolamine salt and *iso*-octyl ester of 2,4-D. Pestic Sci 38:145–154

De Ruiter H, Uffing AJM, Meinen E (1996) Influence of surfactants and ammonium sulfate on glyphosate phytotoxicity to quackgrass (*Elytigia repens*). Weed Technol 10:803–808

Denis MH, Delrot S (1993) Carrier-mediated uptake of glyphosate in broad bean (*Vicia faba*) via a phosphate transporter. Physiol Plantar 87:569–575

Denis MH, Delrot S (1997) Effects of salt and surfactants on foliar uptake and long distance transport of glyphosate. Plant Physiol Biochem 35:291–301

Devine MD, Hall LM (1990) Implications of sucrose transport mechanisms for the translocation of herbicides. Weed Sci 38:299–304

Dodds EC, Lawson W (1938) Molecular structure in relation to estrogenic activity: compounds without a phenanthrene nucleus. Proc R Soc Lond B Biol Sci 125:222–232

Duncan Yerkes CN, Weller SC (1996) Diluent volume influences susceptibility of field bindweed (*Convolvulus arvensis*) biotypes to glyphosate. Weed Technol 10:565–569

Edser C (2007) Multifaceted role for surfactants in agrochemicals. Focus Surfact 3:1–2

Feng PCC, Ryerse JS, Sammons RD (1998) Correlation of leaf damage with uptake and translocation of glyphosate in velvetleaf (*Abutilon theophrasti*). Weed Technol 12:300–307

Feng PCC, Ryerse JS, Jones CR, Sammons RD (1999) Analysis of surfactant leaf damage using microscopy and its relation to glyphosate or deuterium oxide uptake in velvetleaf (*Abutilon theophrasti*). Pest Sci 55:385–386

Feng PCC, Chiu T, Sammons RD, Ryerse JS (2003a) Droplet size affects glyphosate retention, absorption and translocation in corn. Weed Sci 51:443–448

Feng PCC, Chiu T, Sammons RD (2003b) Glyphosate efficacy is contributed by its tissue concentration and sensitivity in velvetleaf (*Abutilon theophrasti*). Pest Biochem Physiol 77:83–91

Fernández Cirelli A, Ojeda C, Castro MJL, Salgot M (2008) Surfactants in sludge- amended agricultural soils. A review. Environ Chem Lett 6:135–148

Folmar LC, Sanders OH, Julin AM (1979) Toxicity of the herbicide glyphosate and several of its formulations to fish and aquatic invertebrates. Arch Environ Contam Toxicol 8:269–278

Forgiarini A, Esquena J, Gonzalez C, Solans C (2001) Formation of nano-emulsions by low-energy emulsification methods at constant temperature. Langmuir 17:2076–2083

Forster WA, Zabkiewicz JA, Riederer M (2004) Mechanisms of cuticular uptake of xenobiotics into living plants: 1. Influence of xenobiotic dose on the uptake of three model compounds, applied in the absence and presence of surfactants into *Chenopodium album*, *Hedera helix* and *Stephanotis floribunda* leaves. Pest Manage Sci 60:1105–1113

Foy CL (1987) Adjuvants: terminology, classification, and mode of action. In: Chow PNP, Grant CA, Hinshalwood AM, Simundson E (eds) Adjuvants and agrochemicals. CRC Press, Boca Raton, pp 1–15

Foy CL (1989) Adjuvants for agrochemicals: introduction, historical overview and future outlook. In: Chow PNP, Grant CA, Hinshalwood AM, Simundsson E (eds) Adjuvants and agrochemicals: II recent development, application, and bibliography of agro adjuvants. CRC Press, Boca Raton, pp 2–9

Franz JE, Mao MK, Sikorski JA (1997) Glyphosate: a unique global herbicide. In: ACS monograph, vol. 189, American Chemical Society, Washington, DC, pp 163–175

Gardner SC, Grue CE (1996) Effects of rodeo® and garlon 3A on nontarget wetland species in central Washington. Environ Toxicol Chem 15:441–451

Garst R (1997) New solutions for agricultural applications. In: Hill K, Rybinski W, Stoll G (eds) Alkyl polyglycosides. VCH, Weinheim, pp 131–137

Gednalske JV, Herzfeld RW (1994) Homogenous herbicidal adjuvant blend comprising glyphosate ammonium sulfate and alkyl polysaccharide. US Patent 5 356 861

Geiger DR, Bestman HD (1990) Self-limitation of herbicide mobility by phytotoxic action. Weed Sci 38:324–329

Geiger DR, Shieh WJ, Fuchs MA (1999) Causes of selflimited translocation of glyphosate in *Beta vulgaris* plants. Pest Biochem Physiol 64:124–133

Giesy JP, Dobson S, Solomon KR (2000) Ecotoxicological risk assessment for Roundup® herbicide. Rev Environ Contam Toxicol 167:35–120

Giger W, Brunner PH, Schaffner C (1984) 4-Nonylphenol in sewage-sludge: accumulation of toxic metabolites from nonionic surfactants. Science 225:623–625

Goldsborough LG, Brown DJ (1993) Dissipation of glyphosate and aminomethylphosphonic acid in water and sediments of boreal forest ponds. Environ Toxicol Chem 12:1139–1147

Gougler JA, Geiger DR (1984) Carbon partitioning and herbicide translocation in glyphosate treated sugar beet (*Beta vulgaris*). Weed Sci 32:546–551

Green JM, Beestman GB (2007) Recently patented and commercialized formulation and adjuvant technology. Crop Prot 26:320–327

Harring T, Streibig JC, Husted S (1998) Accumulation of shikimic acid: a technique for screening glyphosate efficacy. J Agric Food Chem 46:4406–4412

Hartley GS, Brunskill RT (1958) Reflection of water droplets from surfaces. In: Danielli JF, Parkhurst KGA, Riddiford AC (eds) Surface phenomena in chemistry and biology. Pergamon Press, New York, pp 214–223

Hartzler B (2001) Glyphosate- a review. Iowa State University, Ames, Iowa

Heinemann A (2010) Megatrends in agriculture. In: Proceedings of the 8th international symposium on adjuvants for agrochemicals. ISAA Society, Freising, pp 19–20

Hejazi R, Amiji M (2003) Chitosan-based gastrointestinal delivery systems. J Control Release 89:151–165

Heldt N, Zhao J, Friberg S, Zhang Z, Slack G, Li Y (2000) Controlling the size of vesicles prepared from egg lecithin using a hydrotrope. Tetrahedron 56:6985–6990

Henry CJ, Higgins KF, Buhl KJ (1994) Acute toxicity and hazard assessment of Rodeo, X-77 Spreader, and Chem-Trol to aquatic invertebrates. Arch Environ Contam Toxicol 27: 392–399

Hess FD, Foy CL (2000) Interaction of surfactants with plant cuticles. Weed Technol 14:807–813

Hochberg EG (1996) The market for agricultural pesticide inert ingredients and adjuvants. In: Foy CL, Pritchard DW (eds) Pesticide formulation and adjuvant technology. CRC Press, Boca Raton, pp 203–208

Holden WTC (1992) Future formulation trends – the likely impact of regulatory and legislative pressures. In: Brighton crop protection conference, vol. 11, BCPC, Brighton, pp 313–320

Holloway PJ (1993) Structure and chemistry of plant cuticles. Pestic Sci 37:203–232

Jeffree C (1996) Cuticles: an integrated functional approach. In: Kerstiens G (ed) Plant. BIOS scientific, Oxford

Jobling S, Sheahan D, Osborne JA, Matthiessen P, Sumpter JP (1996) Inhibition of testicular growth in rainbow trout (*Oncorhynchus mykiss*) exposed to estrogenic alkylphenolic chemicals. Environ Toxicol Chem 15:194–202

Jordan TN (1981) Effects of diluent volumes and surfactant on the phytotoxicity of glyphosate to bermuda grass (*Cynodon dactylon*). Weed Sci 29:79–83

Khot LR, Sankaran S, Maja JM, Ehsani R, Schuster EW (2012) Applications of nanomaterials in agricultural production and crop protection: a review. Crop Prot 35:64–70

Kim Y-S, Katase T, Sekine S, Inoue T, Makino M, Uchiyama T, Fujimoto Y, Yamashita N (2004) Variation in estrogenic activity among fractions of a commercial nonylphenol by high performance liquid chromatography. Chemosphere 54:1127–1134

Kindall HW, Pimentel D (1994) Constraints on the expansion of the global food supply. Ambio 23:198–205

Kinnberg K, Korsgaard B, Bjerregaard P, Jespersen A (2000) Effects of nonylphenol and 17 beta-estradiol on vitellogenin synthesis and testis morphology in male platyfish Xiphophorus maculatus. J Exp Biol 203:171–181

Kirkwood RC (1999) Recent developments in our understanding of the plant cuticle as a barrier to the foliar uptake of pesticides. Pestic Sci 55:69–77

Knowles A (2008) Recent developments of safer formulations of agrochemicals. Environmentalist 28:35–44

Krogh PH, Holmstrup M, Petersen SO, Jensen J (1997) Ecotoxicological assessment of sewage sludge in agricultural soil. In: Danish ministry for the environment, The Environmental Protection Agency, working report No. 69

Krogh KA, Halling-Sørensen B, Mogensen BB, Vejrup KV (2003) Environmental properties and effects of nonionic surfactant adjuvants in pesticides: a review. Chemosphere 50:871–901

Kumar MNVR, Muzzarelli RA, Muzzarelli C, Sashiwa H, Domb AJ (2004) Chitosan chemistry and pharmaceutical perspectives. Chem Rev 104:6017–6084

Lao SB, Zhang ZX, Xu HH, Jiang GB (2010) Novel amphiphilic chitosan derivatives: synthesis, characterization and micellar solubilization of rotenone. Carbohyd Polym 82:1136–1142

Lasic DD (1993) Liposomes from physics to applications. Elsevier, Amsterdam, pp 507–516

Laws SC, Carey SA, Ferrell JM, Bodman GJ, Cooper RL (2000) Estrogenic activity of octylphenol, nonylphenol, bisphenol A and methoxychlor in rats. Toxicol Sci 54:154–167

Lee PC, Lee W (1996) In vivo estrogenic action of nonylphenol in immature female rats. Bull Environ Contam Toxicol 57:341–348

Lee GWJ, Tadros TF (1982) Formation and stability of emulsions produced by dilution of emulsifiable concentrates. Part I. An investigation of the dispersion on dilution of emulsifiable concentrates containing cationic and non-ionic surfactants. Colloids Surf 5:105–115

Lee JF, Liao PM, Kuo CC, Yang HT, Chiou CT (2000) Influence of a nonionic surfactant (Trion X-100) on contaminant distribution between water and several soil solids. J Colloid Interface Sci 229:445–452

Lee HJ, Chattopadhyay S, Gong EY, Ahn RS, Lee K (2003) Antiandrogenic effects of bisphenol a and nonylphenol on the function of androgen receptor. Toxicol Sci 75:40–46

Lee JF, Hsu MH, Chao HP, Huang HC, Wang SP (2004) The effect of surfactants on the distribution of organic compounds in the soil solid/water system. J Hazard Mater 114:123–130

Li Y, Chen B (2009) Phenanthrene sorption by fruit cuticles and potato periderm with different compositional characteristics. J Agric Food Chem 57:637–644

Lindman B, Danielson I (1981) The definition of microemulsion. Colloid Surf 3:391–392

Liu SH, Campbell RA, Studens JA, Wagner RG (1996) Absorption and translocation of glyphosate in aspen (*Populus tremuloides* Michx.) as influenced by droplet size, droplet number, and herbicide concentration. Weed Sci 44:482–488

Mainx HG, Hofer P (2009) Alkylene oxide adducts of oligosaccharides. Patent WO 2009080215

Malec, AD, Figley, TM, Turpin, KL (2009) Ultra-high loading glyphosate concentrate. US Patent 20090318294

Marc J, Mulner-Lorillon O, Boulben S, Hureau D, Durand G, Belle R (2002) Pesticide Roundup® provokes cell division dysfunction at the level of CDK1/cyclin B activation. Chem Res Toxicol 15:326–331

Martinez TT, Brown K (1991) Oral and pulmonary toxicology of the surfactant used in Roundup® herbicide. Proc West Pharmacol Soc 34:43–46

Marvel JT, Brightwell BB, Suba L (1974) G 3780A surfactant: biodegradation, plant uptake, and ^{14}C-distribution. In: Report 321, Monsanto Company, St. Louis

May OL, Culpepper AS, Cerny RE, Coots CB, Corkern CB, Cothren JT, Croon KA, Ferreira KL, Hart JL, Hayes RM, Huber SA, Martens AB, McCloskey WB, Oppenhuizen ME, Patterson MG, Reynolds DB, Shappley ZW, Subramani J, Witten TK, York AC, Mullinix BG (2002) Transgenic cotton with improved resistance to glyphosate herbicide. Crop Sci 44:234–240

McAllister RS, Haderlie LC (1985) Translocation of ^{14}C-glyphosate and $^{14}C\text{-}CO_2$-labeled photoassimilates in Canada thistle (*Cirsium arvense*). Weed Sci 33:153–159

Mercer GN (2007) A simple diffusion model of the effect of droplet size and spread. Area on foliar uptake of hydrophilic compounds. Pest Biochem Physiol 88:128–133

Minanaperez M, Gracia A, Lachaise J, Salager JL (1995) Solubilization of polar oils with extended surfactants. Colloid Surf A Physicochem Eng Asp 100:217–224

Morin F, Vera V, Nurit F, Tissut M, Marigo G (1997) Glyphosate uptake in *Catharanthus roseus* cells: role of a phosphate transporter. Pest Biochem Physiol 58:13–22

Muzik TJ (1976) Influence of environmental factors on toxicity to plants. In: Audus LJ (ed) Herbicides: physiology, biochemistry, ecology, 2nd edn. Academic, London, pp 203–247

Nadler-Hassar T, Goldshmidt A, Rubin B, Wolf S (2004) Glyphosate inhibits the translocation of green fluorescent protein and sucrose from a transgenic tobacco host to Cuscuta campestris Yunk. Planta 219:790–796

Nalewaja JD, Devilliers B, Matysiak R (1996) Surfactant and salt affect glyphosate retention and absorption. Weed Res 36:241–247

Nickel D, Föster T, von Rybinski W (1996) Physicochemical properties of alkyl polyglycosides. In: Hill K, Rybinski W, Stoll G (eds) Alkyl polyglucosides. VCH, Weinheim, pp 39–69

Odum J, Lefevre PA, Tittensor S, Paton D, Routledge EJ, Beresford NA, Sumpter JP, Ashby J (1997) The rodent uterotrophic assay: critical protocol features, studies with nonyl phenols, and comparison with a yeast estrogenicity assay. Regul Toxicol Pharmacol 25:176–188

Oparka KJ, Roberts AG (2001) Plasmodesmata. A not so open-and-shut case. Plant Physiol 125:123–126

Paganelli A, Gnazzo V, Acosta H, Lopez SL, Carrasco A (2010) Glyphosate-based herbicides produce teretogenic effects on vertebrates by impairing retinoic acid signaling. Chem Res Toxicol 23:1586–1595

Paveglio PL, Kilbride KM, Grue CE, Siemenstad CA, Fresh KL (1996) Use of Rodeo and X-77 spreader to control smooth cordgrass (Spartina alterniflora) in a southwestern Washington estuary: 1 environmental fate. Environ Toxicol Chem 15:961–968

Peixoto F (2005) Comparative effects of the Roundup® and glyphosate on mitochondrial oxidative phosphorylation. Chemosphere 61:1115–1122

Perkins MC, Roberts CJ, Briggs D, Davies MC, Friedmann A, Hart C, Bell G (2005) Macro and microthermal analysis of plant wax/surfactant interactions: plasticizing effects of two alcohol ethoxylated surfactants on an isolated cuticular wax and leaf model. Appl Surf Sci 243:158–165

Petracek PD, Fader RG, Knoche M, Bukovac MJ (1998) Surfactant-enhanced penetration of benzyladenine through isolated tomato fruit cuticular membranes. J Agric Food Chem 46:2346–2352

Pimentel D (1995) Amounts of pesticides reaching the target pests: environmental impacts and ethics. J Agric Environ Ethics 8:17–29

Pimentel D (1997) Techniques for reducing pesticide use. Wiley, New York

Pompeo MP, Mainx HG, Eskuchen R, Rosessing M, Abribat B (2005) APG Granules containing agrochemical active ingredients. US Patent 0,009,707

Pons M, Estelrich J (1996) Liposomes as an agrochemical tool: optimization of their production. Ind Crop Prod 5:203–208

Popp C, Burghardt M, Friedmann A, Riederer M (2005) Characterization of hydrophilic and lipophilic pathways of *Hedera helix* L. cuticular membranes: permeation of water and uncharged organic compounds. J Exp Bot 56:2797–2806
Price CE (1976) Penetration and translocation of herbicides and fungicides in plants. In: McFarlane NR (ed) Herbicides and fungicides-factors affecting their activity, vol 29, The chemical society special publication. Chemical Society, London, pp 42–66
Price CE (1983) The effect of environment on foliage uptake and translocation of herbicides. In: A.o.A. Biologists (ed) Aspects of applied biology 4: influence of environmental factors on herbicide performance and crop and weed biology, vol. 4, The Association of Applied Biologists, Warwick, pp 157–169
Prince LM (1977) Microemulsion. Academic, London
Rabea EI, Badawy MEI, Rogge TM, Stevens CV, Höfte M, Steurbaut W, Smagghe G (2005) Insecticidal and fungicidal activity of new synthesized chitosan derivatives. Pest Manag Sci 61:951–960
Ramsey RJL, Stephenson GR, Hall JC (2005) A review of effects of humidity, humectants, and surfactants composition on the absorption and efficacy of highly water-soluble herbicides. Pest Biochem Physiol 82:162–175
Reynhardt EC, Riederer M (1994) Structures and molecular dynamics of plant waxes. II Cuticular waxes from leaves of *Fagus sylvatica* L. and *Hordeum vulgare* L. Eur Biophys J 23:59–70
Richard EP Jr, Slife FW (1979) In vivo and in vitro characterization of the foliar entry of glyphosate in hemp dogbane *(Apocynum cannabinum)*. Weed Sci 27:426–433
Riechers DE (1992) Surfactant effects on the plasma membrane as a mode of action in promoting glyphosate phytotoxicity. M.Sc. thesis, University of Illinois, Urbana-Champaign
Riechers DE, Wax LM, Liebl RA, Bush DR (1994) Surfactant-increased glyphosate uptake into plasma membrane vesicles isolated from common lambsquarters leaves. Plant Physiol 105:1419–1425
Riechers DE, Wax LM, Liebl RA, Bullock DG (1995) Surfactant effects on glyphosate efficacy. Weed Technol 9:281–285
Riederer M, Markstädter C (1996) Cuticular waxes: a critical assessment of current knowledge. In: Kerstiens G (ed) Plant cuticles – an integrated functional approach. Bios Scientific, Oxford, pp 189–200
Riederer M, Schreiber L (2001) Protecting against water loss: analysis of the barrier properties of plant cuticles. J Exp Bot 52:2023–2032
Risbud MV, Bhonda RR (2000) Polyacrylamide-chitosan hydrogels: in vitro biocompatibility and sustained antibiotic release studies. Drug Deliv 7:69–75
Roberts J (2010) Helena Chemical Company, Presentation to EPA
Ryerse JS, Feng PCC, Sammons RD (2001) Endogenous fluorescence identifies dead cells in plants. Microsc Today 1:22–24
Saito H, Shinoda K (1967) The solubilization of hydrocarbons in aqueous solutions of nonionic surfactants. J Colloid Interface Sci 24:10–15
Santier S, Chamel A (1998) Reassessment of the role of cuticular waxes in the transfer of organic molecules through plant cuticles. Plant Physiol Biochem 36:225–231
Satchivi NM, Stoller EW, Wax LM, Briskin DP (2000a) A nonlinear dynamic simulation model for xenobiotic transport and whole plant allocation following foliar application I. Conceptual foundation for model development. Pestic Biochem Physiol 68:67–84
Satchivi NM, Stoller EW, Wax LM, Briskin DP (2000b) A nonlinear dynamic simulation model for xenobiotic transport and whole plant allocation following foliar application II. Model validation. Pestic Biochem Physiol 68:85–95
Satchivi NM, Stoller EW, Wax LM, Briskin DP (2001) A nonlinear dynamic simulation model for xenobiotic transport and whole plant allocation following foliar application III. Influence of chemical properties, plant characteristics, and environmental parameters on xenobiotic absorption and translocation. Pestic Biochem Physiol 71:77–87

Satchivi NM, Stoller EW, Wax LM, Briskin DP (2006) A nonlinear, dynamic, simulation model for transport, and whole plant allocation of systemic xenobiotics following foliar application IV. Physiochemical properties requirements for optimum absorption and translocation. Pestic Biochem Physiol 84:83–97

Schönherr J (2000) Calcium chloride penetrates plant cuticles via aqueous pores. Planta 212:112–118

Schönherr J (2002) A mechanistic analysis of penetration of glyphosate salts across astomatous cuticular membrane. Pest Manag Sci 58:343–351

Schönherr J, Schreiber L (2004) Size selectivity of aqueous pores in astomatous cuticular membranes isolated from Populus canescens (Aiton) Sm. leaves. Planta 219:405–411

Schönherr J, Baur P, Buchholz A (1999) Modelling foliar penetration: its role in optimising pesticide delivery. In: Brooks GT, Roberst TR (eds) Pesticide chemistry and bioscience. Royal Society of Chemistry, Cambridge, pp 134–151

Schönherr J, Baur P, Uhlig BA (2000) Rates of cuticular penetration of 1-naphthylacetic acid (NAA) as affected by adjuvants, temperature, humidity and water quality. Plant Growth Regul 31:61–74

Schreiber L (1995) A mechanistic approach towards surfactant/wax interactions: effects of octaethyleneglycolmonododecylether on sorption and diffusion of organic chemicals in reconstituted cuticular wax of barley leaves. Pestic Sci 45:1–11

Schreiber L (2006) Review of sorption and diffusion of lipophilic molecules in cuticular waxes and the effects of accelerators on the solute mobilities. J Exp Bot 57:2515–2523

Servizi JA, Gordon RW, Martens DW (1987) Acute toxicity of Garlon 4 and Roundup® herbicides to Salmon, Daphnia, and trout. Bull Environ Contam Toxicol 39:15–22

Sharma SD, Kirkwood RC, Whateley T (1996) Effect of nonionic nonylphenol surfactants on surface physiological properties, uptake and distribution of asulam and diflufenican. Weed Res 36:227–239

Sherrick SL, Holt HA, Hess FD (1986) Absorption and translocation of MON 0818 adjuvant in field bindweed *(Convolvulus amcnsis)*. Weed Sci 34:817–823

Soares A, Vijayram IA, Guieysse B, Murto M, Guieysse B, Mattiasson B (2005) Degradation of non-ionic surfactants under anaerobic conditions. In: Rittmann BE, van Loosdrecht MCM (eds) Third IWA leading-edge conference on water and wastewater treatment technologies. IWA Publishing, Sapporo

Solel Z, Edgington LV (1973) Transcuticular movement of fungicides. Phytopathology 63:505–510

Soto AM, Justicia H, Wray JW, Sonnenschein C (1991) P-Nonylphenol: an estrogenic xenobiotic released from "modified" polystyrene. Environ Health Perspect 92:167–173

Stephenson DO, Patterson MG, Faircloth WH, Lunsford JN (2004) Weed management with fomesafen preemergence in glyphosate-resistant cotton. Weed Technol 18:680–686

Stevens PJG, Bukovac MJ (1987) Studies on octylphenoxy surfactants. Part 1: effects of oxyethylene content on properties of potential relevance to foliar absorption. Pestic Sci 20:9–35

Stevens PJG, Baker EA, Anderson NH (1988) Factors affecting the foliar absorption and redistribution of pesticides. 2. Physicochemical properties of the active ingredient and the role of surfactant. Pestic Sci 24:31–53

Stewart WN (1993) Paleobotany and the evolution of plants, Secondth edn. Cambridge University Press, New York, 405

Stock D, Holloway PJ (1993) Possible mechanisms for surfactant-induced foliar uptake of agrochemicals. Pestic Sci 38:165–177

Stock D, Edgerton BM, Gaskin RE, Holloway PJ (1992) Surfactant enhanced foliar uptake of some organic compounds: interactions with two model polyoxyethylene aliphatic alcohols. Pestic Sci 34:233–242

Stock D, Holloway PJ, Grayson BT, Whitehouse P (1993) Development of a predictive uptake model to rationalise selection of polyoxyethylene surfactant adjuvants for foliage applied agrochemicals. Pestic Sci 37:233–245

Strey R (1996) Water – nonionic surfactant – systems, and the effect of additives. Berich Bunsen Gesell Phys Chem Chem Phys 100:182–189

Tabata A, Kashiwada S, Ohnishi Y, Ishikawa H, Miyamoto N, Itoh M, Magara Y (2001) Estrogenic influences of estradiol-17 beta, p-nonylphenol and bis-phenol-A on Japanese Medaka (*Oryzias latipes*) at detected environmental concentrations. Water Sci Technol 43:109–116

Tabira Y, Nakai M, Asai D, Yakabe Y, Tahara Y, Shinmyozu T, Noguchi M, Takatsuki M, Shimohigashi Y (1999) Structural requirements of para-alkylphenols to bind to estrogen receptor. Eur J Biochem 262:240–245

Tadros TF (1994) Surfactants in agrochemicals. Marcel Dekker, New York

Tadros TF (2005) Applications of surfactants in agrochemicals. In: Applied surfactants: principles and application. Wiley-VCH, Weinheim, pp 503–592

Tamura H, Knoche M, Bukovac MJ (2001) Evidence for surfactant solubilization of plant epicuticular wax. J Agric Food Chem 48:1809–1816

Thomas BA, Spicer RA (1987) The evolution and paleobiology of land plants, vol 2, Ecology, phytogeography and physiology series. Croom Helm, London, pp 309

Tice CM (2001) Selecting the right compounds for screening: does Lipinski's rule of 5 for pharmaceuticals apply to agrochemicals. Pest Manag Sci 57:3–16

Tilman D, Cassman KG, Matson PA, Naylor R, Polasky S (2002) Agricultural sustainability and intensive production practices. Nature 418:671–677

Tominack RL (2000) Herbicide formulations. J Toxicol Clin Toxicol 38:129–135

Trapp S (2004) Plant uptake and transport models for neutral and ionic chemicals. Environ Sci Pollut Res 11:33–39

Tsui MTK, Chu LM (2003) Aquatic toxicity of glyphosate-based formulations: comparison between different organisms and the effects of environmental factors. Chemosphere 52:1189–1197

Tu M, Hurd C, Randall JM (2003) Weed control methods handbook, The Nature Conservancy, Utah State University, Utah

Urano K, Saito M, Murata C (1984) Adsorption of surfactants on sediments. Chemosphere 12:293–300

van Ginkel CG, Stroo CA, Kroon AGM (1993) Biodegradability of ethoxylated fatty amines: detoxification through a central fission of these surfactants. Sci Total Environ 134:689–697

Wade BR, Riechers DE, Libel RA, Wax LM (1993) The plasma membrane as a barrier to herbicide penetration and site for adjuvant action. Pestic Sci 37:195–202

Wan Kim T, Amrhein N (1995) Glyphosate toxicity: long-term analysis of shikimic acid accumulation and chlorophyll degradation in tomato plants. Kor J Weed Sci 15:141–147

Wan MT, Watts RG, Moul DJ (1989) Effects of different dilution water types on the acute toxicity to juvenile Pacific salmon and rainbow trout of glyphosate and its formulated products. Bull Environ Contam Toxicol 43:378–385

Wang CJ, Liu ZQ (2007) Foliar uptake of pesticides – present status and future challenge. Pest Biochem and Physiol 87:1–8

Wang L, Li X, Zhang G, Dong J, Eastoe J (2007) Oil-in-water nanoemulsions for pesticide formulations. J Colloid Interface Sci 314:230–235

White R, Jobling S, Hoare SA, Sumpter JP, Parker MG (1994) Environmentally persistent alkylphenolic compounds are estrogenic. Endocrinology 135:175–182

Williams GM, Kroes R, Munro IC (2000) Safety evaluation and risk assessment of the herbicide Roundup® and its active ingredient, glyphosate, for humans. Regul Toxicol Pharmacol 31: 117–165

Wills GD, McWhorter CG (1988) Absorption and translocation of herbicides effect of environment, adjuvants, and inorganic salts, pesticide formulations. In: Chapter 8, ACS symposium series, Vol 371, American Chemical Society, Washington, pp 90–101

Woodburn A (2000) Glyphosate: production, pricing and use worldwide. Pest Manage Sci 56:309–312

Wyrill JB, Burnside OC (1976) Absorption, translocation, and metabolism of 2,4-D and glyphosate in common milkweed and hemp dogbane. Weed Sci 24:557–566

Yoshihara K, Ohshima H, Momozawa N, Sakai H, Abe M (1995) Binding constants of symmetric or antisymmetric electrolytes and aggregation numbers of oil-in-water type microemulsions with a nonionic surfactant. Langmuir 11:2979–2984

Zabkiewicz JA (2000) Adjuvants and herbicidal efficacy – present status and future prospects. Weed Res 40:139–149

Zabkiewicz JA (2003) Foliar interactions and uptake of agrichemical formulations – present limits and future potential. In: Voss G, Ramos G (eds) Chemistry of crop protection. Wiley VCH, Weinheim, pp 237–251

Zabkiewicz JA (2007) Spray formulation efficacy – holistic and futuristic perspectives. Crop Prot 26:312–319

Zhang C, Ping QN, Zhang H, Shen J (2003) Preparation of N-alkyl-O-sulfate chitosan derivatives and micellar solubilization of taxol. Carbohyd Polym 54:137–141

Zhang C, Qu GW, Sun YJ, Wu XL, Yao Z, Guo QL, Ding Q, Yuan S, Shen Z, Ping Q, Zhou H (2008) Pharmacokinetics, biodistribution, efficacy and safety of N-octyl-O-sulphate chitosan micelles loaded with paclitaxel. Biomaterials 29:1233–1241

Zhu L, Chen B, Tao S (2004) Sorption behavior of polycyclic aromatic hydrocarbons in soil-water system containing nonionic surfactant. Environ Eng Sci 21:263–272

Zhu H, Yu Y, Ozkan HE, Derksen RC, Krause CR (2008) Evaporation and wetted area of single droplets on waxy and hairy leaf surfaces. Commun Agric Appl Biol Sci 73:711–718

Zoller U (2009) Handbook of detergents. Part F: production, vol 142, Surfactant science series. Taylor and Francis Group, New York

Chapter 8
Cheap Materials to Clean Heavy Metal Polluted Waters

Pei-Sin Keng, Siew-Ling Lee, Sie-Tiong Ha, Yung-Tse Hung, and Siew-Teng Ong

Abstract The rapid growth of the human population and industrialization in the world has increased the environmental problems such as water, air and land pollution. Heavy metals can be considered as one of the most problematic pollutants. This chapter reviews developments and technical applicability of treatments methods for the removal of heavy metals. We focus on the use of low-cost materials to remove heavy metals. Isotherm equilibrium models and kinetics are discussed. Designing adsorption columns under continuous flow conditions are evaluated.

Keywords Water pollution • Heavy metal • As • Ni • Pb • Membrane filtration • Ion exchange • Fly ash • Rice husk • Wheat straw • Chitosan • Algae • Bacteria

---

P.-S. Keng, Ph.D.
Department of Pharmaceutical Chemistry, International Medical University, No.126, Jalan Jalil Perkasa 19, Bukit Jalil, 57000 Kuala Lumpur, Malaysia
e-mail: pskeng@yahoo.com

S.-L. Lee, Ph.D., AMIC
Ibnu Sina Institute for Fundamental Science Studies, Universiti Teknologi Malaysia, 81310 Skudai, Johor, Malaysia
e-mail: sllee@ibnusina.utm.my

S.-T. Ha, Ph.D., AMIC • S.-T. Ong, Ph.D., AMIC, FICCE (✉)
Universiti Tunku Abdul Rahman, Jln Universiti, Bandar Barat, 31900 Kampar, Perak, Malaysia
e-mail: hast@utar.edu.my; hast_utar@yahoo.com; ongst@utar.edu.my; ongst_utar@yahoo.com

Y.-T. Hung, Ph.D.
Department of Civil and Environmental Engineering, Cleveland State University, Cleveland, OH, USA
e-mail: yungtsehung@yahoo.com; yungtsehung@gmail.com

E. Lichtfouse et al. (eds.), *Green Materials for Energy, Products and Depollution*, Environmental Chemistry for a Sustainable World 3, DOI 10.1007/978-94-007-6836-9_8,

Contents

8.1 Heavy Metals

In recent years, water pollution by heavy metals poses one of the most serious environmental problems and of the most difficult to solve. Activities such as extensive use of fertilizers, municipal waste to land, electroplating, energy and fuel production, and wastewater from manufacturing industries including textile industry are among the major contributions to the contamination of environment by heavy metals. These metallic agents, once concentrated in the biosphere, generally persist and are not broken down. It is reported that cadmium, lead, cobalt, copper, mercury, chromium, nickel, selenium and zinc are the most important metals related to chain food contamination (McLaughlin et al. 1999). These metals are carcinogenic to human beings if consumed in higher quantity. Therefore, heavy metal pollution poses a great potential threat to the environment and human health.

Heavy metals may be defined as the elements having atomic weights between 63.5 and 200.6 and have a minimum specific gravity of 5 (Srivastava and Majumder 2008). Specific gravity is dimensionless property of matter and its value is obtained by dividing the density of that substance with the density of water. In addition, other experimental conditions, such as temperature, for determining the two densities must be the same. Heavy metals are "trace elements" with individual concentrations that do not exceed 1,000 mg/kg (0.1 %) of earth's crust. They are classified as essential, beneficial, or detrimental. For instances, iron, zinc, copper, chromium, iodine, cobalt, molybdenum, and selenium are recognised as essential for human health, of course with specific recommended daily allowances (RDA) or reference dose (RfD) set by respective authorities. On the other hand, some heavy metals are regarded as purely toxic metals such as cadmium, lead, nickel and mercury, which do not provide any essential or potentially beneficial health effect at any level of exposure (Nordberg et al. 2002). As arsenic, nickel and lead are metals of widespread use and frequently found in industrial wastewater, these metals were discussed in the following section.

8.1.1 Arsenic

Arsenic (As) is a metalloid belonging to group 15 of the periodic table with atomic number of 33 and atomic weight of 74.92. It is found naturally in the environment with estimated crustal abundance of 1.8 mg/kg and estimated oceanic abundance of 3–3.7 mg/L. The common oxidation states for arsenic are 5, 3 and −3, which appear widely in a variety of minerals. Realgar (As_4S_4) and orpiment (As_2S_3) are the two common arsenic sulfides. Arsenic also occurs as arsenide and sulfaresenides of heavy metals such as iron, copper, lead, silver, and gold (Yan Chu 1994). Loellingite ($FeAs_2$), safforlite (CoAs), niccolite (NiAs), arsenopyrite (FeAsS) and enargite (Cu_3AsS_4) are some of the examples of minerals containing arsenide and sulfaresenides. Arsenic can also be found in oxidised form such as in the mineral arsenolite (As_2O_3) (Greenwood 1989)

As a result of its toxicity, arsenic compounds are used mainly in wood preservation and insecticides. It is also employed in the making of special glass and semi-conductors, gallium arsenide (GaAs). GaAs is a semiconductor used in laser diodes and light emitting diodes (LEDs). While arsine gas (AsH_3) has become an important dopant gas in the microchip industry under strict guidelines. The widely used of arsenic in aforementioned activities, mining as well as erosion and weathering of rocks have contributed significantly to the content of this element in the environment.

Ingestion and inhalation of arsenic and its compounds cause various health effects. Arsenic interferes with enzyme action, DNA transcription, and metabolism. It causes chronic weakness, general debility and lassitude, loss of appetite and energy and sometimes a degree of dementia (Hindmarsh and McCurdy 1986; Yamanaka et al. 1996). Uptake of significant amounts of inorganic arsenic is associated with cancer development, especially development of lung, liver and lymphatic cancer. Prolong exposure to arsenic compounds will also cause dermatological effect such as hyperpigmentation and hypopigmentation, progressing to palmar/plantar hyperkeratosis and cause skin cancer eventually (Hindmarsh and McCurdy 1986). In addition, a very high exposure to inorganic arsenic contributes to infertility and miscarriages in women and most importantly to note that the lethal dose of arsenic is reported as 100 mg (Lenntech 2012a).

8.1.2 *Nickel*

Nickel (Ni) is a silvery white, hard, malleable and ferromagnetic metallic element belonging to group 10 of the periodic table with atomic number of 28 and atomic weight of 58.69. Nickel is a fairly good heat and electric conductor. It is a commonly found constituent in most meteorites with iron meteorites (siderites). Besides, nickel is also found sparingly in nature, usually combined with arsenic or iron and sulfur such as niccolite (NiAs) and pentlandite (NiS·2FeS). The estimated crustal abundance of Ni is 8.4×10^1 mg/kg and its estimated oceanic abundance is 5.6×10^{-4} mg/L.

Nickel is corrosion resistant and used primarily for making of stainless steel and many other corrosion resistant alloys. Typical composition of stainless steel is 8 % nickel, 18 % chromium and the remaining is iron. In fact, approximately 65 % of the nickel consumed in the western world is used to make stainless steel (Lenntech 2012b) .Other uses of nickel are in coinage, electroplating and battery production, foundry products as well as catalyst for hydrogenation process.

Nickel is exposed to human beings through inhalation, ingestion and skin contact. Workers in the industries related to nickel applications are the main group to suffer of nickel poisoning via inhalation of nickel-containing aerosols, dusts, fumes, and mists, especially those involved in production, processing, and use of nickel. Smokers are also at high risk to nickel inhalation as cigarettes, on average, contain 1–3 μg nickel (ATSDR 1993; Von Burg 1997) and mainstream smoke from one cigarette contains 0–0.51 μg nickel (Smith et al. 1997). Out of the total inhalation of nickel, about 20–35 % is deposited in the lungs causing lungs inflammation. On the other hand, nickel is also exposed to human population through ingestion of food and drinking water. Drinking water is the main contribution to nickel ingestion with 27 % nickel in drinking water is absorbed in humans compare to food which is approximately 1 %. This is because more soluble nickel compounds have higher absorption efficiencies.

Absorbed nickel is unable to be metabolised by human body. It is eliminated via urine with the elimination half-time of 28 days for the oral route. In fact, serum and urine nickel levels are the most useful monitoring of nickel exposure. According to the Agency for Toxic Substances and Disease Registry (ATSDR) of United States, the reference values for nickel in healthy adults are 0.2 μg/L in serum and 1–3 μg/L in urine (ATSDR 2005). For dermal contact of nickel, it causes dermatitis which is the most adverse health effect associated with nickel exposure. The common dermal contact of nickel is via nickel alloy and nickel-plated products such as stainless steel, coin and jewelry.

8.1.3 *Lead*

Lead (Pb) is a soft, malleable, ductile and dense metal. It has bluish-gray colour and belongs to group 14 with atomic number 82 and atomic weight of 207.2.

Though lead do not occur abundantly in nature but it is readily accessible in the earth's crust as the mineral galena (PbS), anglesite ($PbSO_4$) and cerussite ($PbCO_3$). The estimated crustal abundance is 1.4×10^1 mg/kg and the estimated oceanic abundance is 3×10^{-5} mg/L.

Lead is corrosion resistant, thus historically it was used in plumbing. Nowadays it is used to line tanks that store corrosive materials such as sulfuric acid (H_2SO_4) and as covering for cables. Lead's high density makes it useful as a shield against X-ray and gamma-ray radiation and is used in X-ray machines as well as in the glass of computer and television screens, where it shields the viewer from radiation (Lenntech 2012c). The other major use of lead is in lead-acid batteries manufacturing and formation of lead alloys. For instances, solder, an alloy constituted of half lead and half tin, is a material with a relatively low melting point that is used to join electrical components, pipes and other metallic items.

Like arsenic, lead is toxic in nature. Lead poisoning in human population is due to inhalation, ingestion and dermal contact to lead compounds. The emission of lead particulate in the air is primarily due to anthropogenic activities including automobiles using leaded gasoline. Lead is also found in soil due to adsorption at mineral interfaces, weathering and chipping of lead-based paint from buildings, bridges and other structures. Lead can also leach into drinking water from the leaded pipes in water distribution systems (Madden et al. 2002).

It is reported that children are more vulnerable to the effects of lead than adults. Children have higher gastrointestinal absorption of lead (40–50 %) compare to adults (3–10 %) (ATSDR 2007). A study on the lead content was conducted in Saratov, a town in Russia which its environment is exposed to lead discharge due to automobiles and operations of battery and lead glass factories. Blood lead and environmental samples from 579 children and their homes for soil, water, dust and paint are collected for lead content analysis. The study included the first pediatric blood lead screening among Russian children and the result shows that nearly one-fourth of the children had blood lead levels above the United State level of concern, 10 μg/dL (Rubin et al. 1997). Absorbed and retained of high dose of lead in children will disturb their neurological development. A well-developed literature argues that these neurological impacts reduce cognitive ability and lead to behavioral problems and learning disabilities later in life. Exposure to lead also increases the risk of kidney failure, disruption of central nervous system, brain damage and even death in cases of extremely high exposure (Jones 2012).

8.2 Heavy Metal Removal: Conventional and Emerging Methods

Extensive research and financial resources have been spent on the treatment of heavy metals pollution especially in industrial wastewater which is contaminated by heavy metals. These technologies include conventional physico-chemical methods and biological treatments, where the latter have gradually getting momentum because they are more eco-friendly and cost-effective compare to the conventional techniques. The physical-chemical methods include chemical precipitation, ion-exchange, adsorption, membrane filtration, coagulation-flocculation and electro-chemical methods. Biological treatments involve the use of plants and microbes in any of the above said methods.

8.2.1 *Chemical Precipitation*

Chemical precipitation is the most common and effective method used in the industry for treatment of wastewater because it is relatively easy and inexpensive to operate (Ku and Jung 2001). This technique is employed in almost 75 % of electroplating facilities to treat wastewater (Karthikeyan et al. 1996). Generally, the metals are precipitated as insoluble hydroxides, carbonates, or sulfides and separated from water by sedimentation and filtration (Table 8.1).

Hydroxide Precipitation

Of all the chemical treatments, hydroxide precipitation is the most preferable and being extensively used method in the industry, particularly where complex chemical compounds are not involved and economic recovery is not a consideration. Lime, $Ca(OH)_2$ and caustic soda, NaOH are the preferred choice of bases used to precipitate heavy metals as their respective hydroxides. There are numerous factors govern the efficiency of metal hydroxides precipitation. For instances, the pH of the metals solution, the electrolyte composition of the effluent, the presence of soluble and/or insoluble products following reaction with complexing agents, ionic strength, electrical potential, temperature, and the time of reaction (Lin et al. 2005).

Dean et al. (1972) have found that there is almost complete precipitation and removal of metals such as cobalt, copper, iron, manganese, nickel, and zinc as the hydroxide with no special modification. However addition of soda ash (for lead) or sodium sulfide (for cadmium and mercury) is required to complete the precipitation of cadmium, lead and mercury. Chen et al. (2009) also reported that fly ash is used to enhance lime precipitation of metals. The fly ash-lime-carbonation treatment increased the particle size of the precipitate and significantly improved the efficiency of heavy metal removal. As a result, the removal efficiency of chromium, copper, lead and zinc in effluents had achieved 99 % removal. For strong acidic wastes, limestone ($CaCO_3$) is used as part of the neutralisation together with lime since limestone is less expensive and it reduces the operational cost significantly. However, limestone must be evaluated carefully for each acid waste since its efficiency depends upon particle coating and size, and pH limitation of calcium carbonate (Dean et al. 1972).

Table 8.1 Removal of heavy metals using chemical precipitation

Cation (heavy metal)	Anion (precipitant)	Optimum pH	Contact time (h)	Initial metal concentration (mg/L)	Removal efficiency (%)	Reference
Cu^{2+}, Cr^{4+}	$Ca(OH)_2$, NaOH, Na_2CO_3	12.0, 8.7	2	48.51, 30	98.6, 99.97	Mirbagheri and Hosseini (2005)
Cd^{2+}, Cr^{3+}, Cu^{2+}, Ni^{2+}, Pb^{2+}, Zn^{2+}	$Ca(OH)_2$, NaOH	8.1–8.3	24	0.1–29.7	50–100	Meunier et al. (2006)
Soil, fly ash and sewage leachates (Al^{3+}, Ba^{2+}, Ca^{2+}, Cd^{2+}, Co^{2+}, Cr^{3+}, Cu^{2+}, Fe, Mg^{2+}, Mn, Ni^{2+}, Pb^{2+}, Zn^{2+}, P, S)	NaOH, $Ca(OH)_2$, NaOH/Na_2S, NaOH/Na_2CO_3, NaOH/Na_2HPO_4	5.0 7.5 7.5 8.0	0.3	0.13–719[a] 1.38–11,100[b] 0.23–827[c]	0.0–100[a] 23.9–100[b] 3.5–100[c]	Djedidi et al. (2009)
Ni^{2+}, Co^{2+}, Cu^{2+}	Na_2S	5.81–11.6	0.5–2	50–2,051	66.9–99.9	Lewis and van Hille (2006)
Zn^{2+}	Na_2S	6.5	NA	7,500	99.9 (<0.01 mg/L)	Veeken et al. (2003)
Acid mine drainage (Fe^{2+}, Mn^{2+})	1,3-benzenediamidoethane-thiolate	4.5	20	194 4.65	99.9 (<0.009 ppm), 97.3	Matlock et al. (2002)

NA not available
[a]Refers to the soil leachates
[b]Refers to the fly ash leachates
[c]Refers to the sewage sludge leachates

Another important parameter for precipitation of heavy metals is the adjustment of optimum pH. Heavy metals, like chromium, copper, lead, manganese and zinc do not precipitate at pH below 7.0, thus allow their separation from ferric ion (Fe^{3+}) and aluminium ion (Al^{3+}), which precipitate at pH below 6.5 (Djedidi et al. 2009). The investigation from Mirbagheri and Hosseini (2005) agreed well with Djedidi et al. (2009). In their study of Cu(II) and Cr(IV) removal from wastewater using $Ca(OH)_2$ and NaOH, the Cr(VI) which was converted to Cr(III) using ferrous sulfate showed the maximum precipitation at pH 8.7 with the addition of $Ca(OH)_2$. At this pH value, the concentration of chromate was successfully reduced from 30 to 0.01 mg/L. While the maximum precipitation of copper occurred at pH 12.0 for both $Ca(OH)_2$ and NaOH and the concentration of copper was successfully reduced from 48.51 to 0.694 mg/L.

Although hydroxide precipitation is reported as the most widely used method for heavy metals removal in industry effluent, it also has some limitations. Hydroxide precipitation requires a large amount of chemicals to reduce metals to an acceptable level prior discharge especially if the effluents have high acidity where large amounts of lime are required for neutralisation. In this case, excessive secondary sludge, such as gypsum ($CaSO_4{\cdot}2H_2O$) are produced that require further treatment. These secondary precipitates not only increase the costs of sludge handling but also give long-term environmental impacts (Kurniawan et al. 2006). Secondly, some metal hydroxides are amphoteric, and the mixed metals create a problem using hydroxide precipitation since the ideal pH for one metal may change another metal back into solution. On top of these, some complexing agents present in the wastewater will inhibit the efficacy of metal hydroxide precipitation (Fu and Wang 2011).

Sulfide Precipitation

Sulfide precipitation is another effective method used for removal of heavy metals from wastewater. It has several advantages over hydroxide precipitation. For instances, sulfide precipitates are not amphoteric, hence, the sulfide precipitation process can achieve a high degree of metal removal over a broad pH range compared with hydroxide precipitation. Moreover, the solubilities of the metal sulfide precipitates are intensely lower than hydroxide precipitates. Metal sulfide sludges also exhibit better thickening and dewatering characteristics than the corresponding metal hydroxide sludges, thus reducing the amount and cost of sludges handling (Fu and Wang 2011). The sulfide precipitation process can be employed using various sulfide sources, such as solid FeS, CaS, aqueous Na_2S, NaHS, NH_4S or gaseous H_2S. Some researchers use the degeneration reaction of sodium thiosulfate ($Na_2S_2O_3$) as a source of sulfide for metal precipitation (Lewis 2010).

In fact, metal sulfide precipitation is preferable than hydroxide precipitation in hydrometallurgical processes in recent decades (Bhattacharyya et al. 1979; Lewis and van Hille 2006). The reasons that support its preferential use are based on the high degree of metal removal at relatively low pH values; the low solubility of sulfide precipitates; favourable dewatering characteristics and the stability of the metal sulfides formed. In the batch study of metal recovery conducted by Lewis and van Hille (2006), the maximum metal removal achieved for nickel and cobalt are 99.8 % and 99.9 %, respectively, when the metal: sodium sulfide molar ratio was 1:1. However, when excess sulfide is used, it compromises metal removal efficiency due to the formation of aqueous polysulfide species that will re-dissolve in the solution. The reaction is described as following equation:

$$MS_{(s)} + HS^{-}_{(aq)} \rightarrow MS(HS)^{-}_{(aq)} \quad (8.1)$$

whereby MS: solid metal sulfide

HS^-: aqueous sulfide

$MS(HS^-)$: aqueous polysulphide complex

Hence, sulfide precipitation is not used as widely as it could due to this limitation. Moreover, it is rather difficult to control the dosing of sulfide due to the very low solubility of the metal sulfides and thus the sensitivity of the process to the dose. The colloidal precipitates that formed during metal sulfide precipitation also tend to cause some separation problems in either settling or filtration processes (Veeken et al. 2003).

In order to overcome the disadvantages between hydroxide and sulfide precipitation, nowadays, the combination of these two methods in treatment of heavy metals is being considered (Marchioretto et al. 2005; Djedidi et al. 2009). The set up for this treatment is usually a two-stage processes in which hydroxide precipitation is followed by sulfide precipitation. Djedidi et al. (2009) has reported that the combination of NaOH and Na_2S led to better removal of Pb^{2+}, Zn^{2+}, Cu^{2+} and Mn^{2+} ions in soil, fly ash and sewage sludge leachates than with NaOH/Na_2CO_3 or NaOH/Na_2HPO_4. The metal sulfide precipitations occurred at pH 7.5 with percentage removal of almost 100 %. This combination technique has reduced the reagents consumption.

Chemical Precipitation Combined with Other Methods

There are also reports on the combination of chemical precipitation with other methods in the treatment of heavy metals. González-Muñoz et al. (2006) have studied precipitation of metals using sulfides followed by nanofiltration as a second step in order to optimise the reuse of water and the recovery of heavy metal ions (Zn, Se, As, Cd and Pb). This study showed a promising outcome whereby the heavy metals are being successfully precipitated and the resultant solution after these processes can be directly reused in the plant. Ghosh et al. (2011) investigated a stepwise combination of electro-Fenton process and chemical precipitation to treat rayon industry wastewater contains chemical oxygen demand (COD) and Zn^{2+}. The two-step treatment processes refer to electro-Fenton treatment for COD reduction, followed by removal of zinc by addition of lime, CaO. From the results obtained with industrial wastewater having a COD of 2,400 mg/L and 32 mg/L

Zn^{2+}, approximately 88 % COD was reduced by the electro-Fenton method but this step does not have any appreciable effect on zinc removal. Thus, CaO is added to precipitate zinc as zinc hydroxide, $Zn(OH)_2$ with removal efficiency of 99–99.3 % in the range of pH 9–10. Another study in combination of chemical precipitation and electrocoagulation to remove Cu^{2+} from lithium bromide refrigerant was conducted by Cheng (2006). Cu^{2+} removal from lithium bromide refrigerant was successfully achieved by pretreatment with $Ba(OH)_2$ solution followed by electrocoagulation using carbon steel plates as electrodes. Up to 98.5 % of Cu^{2+} in the contaminated refrigerant could be removed by this combined treatment.

8.2.2 *Ion Exchange*

Ion exchange is the process through which an ion is removed out of an aqueous solution and is replaced by another ionic species. It has been continuously used in the treatment of heavy metals due to several advantages, such as high treatment capacity, high selectivity for certain heavy metal ions and most significantly, fast kinetics and most importantly it reduces the major problem faces in chemical precipitation – handling and disposal of large amount of sediments and sludge (Lacour et al. 2001; Abo-Farha et al. 2009, Al-Enezi et al. 2004). Dąbrowski et al. (2004) concluded that industrial wastewaters containing 0.1 g/L of Cu^{2+}, Cd^{2+} or Hg^{2+} compounds give 10-, 9- and 5-times larger amounts of sediments, respectively, and 6 kg of sediments are obtained from one kilogram of chromates.

Strongly acidic resins with sulfonic acid groups ($-SO_3H$) and weakly acid resins with carboxylic acid groups ($-COOH$) are the most common cation exchangers. Hydrogen ions in the sulfonic group or carboxylic group of the resin can serve as exchangeable ions with metal cations in the wastewater (Lacour et al. 2001; Fu and Wang 2011). The degree of exchange that occurs depends on the size and ionic charge of the metal cations, initial metal concentration, nature (both physical and chemical) of the ion exchange substance, pH and temperature (Lee et al. 2006a; Al-Enezi et al. 2004). Since resins have different swelling characteristics, the total exchange capacity on a volumetric basis ranged from 1.0 to 4.0 meq/mL. Abo-Farha et al. (2009) studied the adsorption behaviour of cation exchange resin purolite C100 with respect to Ce^{4+}, Fe^{3+} and Pb^{2+} from aqueous system. Purolite C100 is a polymer of polystyrene-divinyl benzene, with sulfonic acid groups as the functional groups. They found that the metal ions adsorption sequence can be given as $Ce^{4+} > Fe^{3+} > Pb^{2+}$. Another batch study conducted by Lacour et al. (2001) shows the potential usage of cation-exchange-textiles (CET) – a special class of grafted textiles carrying different functional groups namely carboxylate($R-COO^-,Na^+$), sulfonate ($R{-}SO_3{}^-,Na^+$) or phosphate ($R{-}O{-}PO_3{}^{2-}$, $2Na^+$) in removing Cu^{2+} and Cd^{2+} from the solution. The report reveals the adsorption behaviour of the CET towards Cu^{2+} and Cd^{2+} is influenced by parameters such as the counter-ion and functional group nature, the metal concentration and the presence of different competitors in the solution.

Besides synthetic resins, natural zeolites also show effective removing efficiency towards heavy metal from aqueous solutions under different experimental condition (Inglezakis and Grigoropoulou 2004; Petrus and Warchoł 2005; Barreira et al. 2009). Clinoptilolite is one of the most frequently studied natural zeolites that have received extensive attention due to its selectivity for heavy metals. (Inglezakis and Grigoropoulou 2004) reported that the uptake of Pb^{2+}, Cu^{2+}, Fe^{3+} and Cr^{3+} in aqueous solutions using natural clinoptilolite is influenced by different parameters following this order: metal cations concentration > volumetric flow rate > particle size > modification of the material. In fact, modification of the natural sample by NaCl has led to an increase of removal efficiency by 32–100 %. The study also focused on the fixed bed reactors design rather that batch systems because kinetics in beds is influenced by the whole equilibrium curve and diffusion rates in the material. This indicates the effect of operating conditions carry out in fixed beds is more representative than that obtained in batch systems. Warchoł and Petrus (2006) supported the column study of clinoptilolite by investigating heavy metals removal, ion-exchange equilibrium and column dynamics through mathematical modeling. In fact, the possibility of replacing expensive synthetic ion-exchangers by natural zeolites – low cost sorbents with abundant availability has added value to the application of this method in wastewater treatment.

8.2.3 Membrane Filtration

The ultimate objective of most of the traditional methods in heavy metals treatments is not the recovery of the metal but rather its elimination. However, recovery of heavy metals shows greater benefits. It allows re-use of the metals and thus provides further economic and environmental benefits by reducing the disposal costs and raw material requirements. In this case, membrane processes exert as an alternative for heavy metal recovery, as they can achieve high permeate fluxes and high rejection coefficients with low energy costs and under mild conditions. (Landaburu-Aguirre et al. 2006). Other advantages of this technique including fewer chemicals are used in the process, which helps to minimise the negative impacts of those chemicals on the whole process; produce more reliable performance; simple operation and convenience since it occupies less floor space in comparison to the conventional treatment systems (Paul Chen et al. 2006). Besides, there are many studies show that membrane processes can be easily combined with other separation processes for better removal efficiency (Blöcher et al. 2003; Mimoune and Amrani 2007; Rivas and Palencia 2011).

Membrane separation system consists of a membrane that can separate materials depending on their physical and chemical properties when a driving force, either a gradient of chemical potential (concentration or pressure gradient) or electrical potential, is applied across the membrane. A good membrane must have a narrow range of pore sizes, a high porosity, and a thin layer of material (Paul Chen et al. 2006). The membrane processes used for wastewater treatment containing

Table 8.2 Pore size of the membrane processes

Membrane process	Pore size
Microfiltration	0.1–10 μm (>100 nm)
Ultrafiltration	0.005–0.2 μm (5–200 nm)
Nanofiltration	0.001–0.01 μm (<10 nm)

heavy metals including microfiltration (MF), ultrafiltration (UF), nanofiltration (NF), reverse osmosis (RO) and electrodialysis (ED) depending on the size of the particle that can be retained (Table 8.2). The membrane techniques that are often used for heavy metals removal are UF, NF, RO and ED.

Ultrafiltration

Ultrafiltration (UF) is a membrane technique used to separate heavy metals, organic pollutants or inorganic pollutants from aqueous solution. UF membranes are capable of retaining species in the range of 300–500,000 Da of molecular weight, with pore sizes ranging from 0.005 to 0.2 μm (5–200 nm) (Paul Chen et al. 2006). Since the pore sizes of UF membranes are larger than dissolved metal ions in the form of hydrated ions or as low molecular weight complexes, these ions would pass easily through UF membranes. Hence, two important techniques – micellar enhanced ultrafiltration (MEUF) and polymer enhanced ultrafiltration (PEUF) are proposed for effectively use in removal of metal ions (Table 8.3).

The combination of surfactants and membranes in separation of dilute polluted water has led to the development of MEUF. It is considered an economical alternative available to the conventional membrane separation process, because it reduces the requirement of higher pressure and high membrane costs. In this process, surfactants are added into the aqueous stream at levels equal to or higher than their critical micelle concentrations (CMCs). At this particular surfactant concentration, surfactant monomers will assemble and form aggregates called micelles. The anionic micelles, which are negatively charged, can bind to metal cations, which are positively charged. This micelle solution is then filtered through an ultrafiltration membrane whose pore size is smaller than the micelle size to reject the micelles. The metal cations that adsorbed onto the micelles can thus be removed from the polluted water (Mungray et al. 2012).

MEUF can be used to remove different heavy metals from wastewater containing single metal, mixture of metals or feed mixtures containing metals and organic material (Juang et al. 2003; Häyrynen et al. 2012; Rahmanian et al. 2012). As the binding of metal cations to oppositely charged micelle surface is primarily electrostatic in nature, it is expected that the ions with the same charge are removed with a nearly equal rejection in MEUF. Juang et al. (2003) reported that the rejection of Sr^{2+}, Mn^{2+}, Co^{2+}, Cu^{2+} and Zn^{2+} are 99.1, 78.6, 80.6, 82.1 and 83.7 % respectively. The comparatively high rejection of Sr^{2+} by the GE membrane in this system promotes the application of MEUF for selective separation of Sr^{2+} from low-level waste effluents. Increases of surfactant (sodium dodecyl sulfate) to

Table 8.3 Heavy metal removal by micellar enhanced ultrafiltration (MEUF) and polymer enhanced ultrafiltration (PEUF)

UF type	Membrane material	Surfactant/complexing agent	Heavy metal	Initial concentration	Optimum pH	Removal efficiency (%)	Reference
MEUF	Polyamide, polyethersulfone	Sodium dodecyl sulfate	Cs^{+}, Sr^{2+}, Mn^{2+}, Co^{2+}, Cu^{2+}, Zn^{2+}, Cr^{3+}	50 mg/L	3–12	40.0, 99.1, 78.6, 80.6, 82.1, 83.7, 86.7	Juang et al. (2003)
MEUF	Ceramic	Dodecylbenzenesulfonic acid, dodecylamine	Pb^{2+}, AsO_4^{-}	4.4–7.6 mg/L	7.47	>99.0, 19.0	Ferella et al. (2007)
MEUF	Regenerated cellulose acetate	Sodium dodecyl sulfate	Pb^{2+}	50–150	7.84	99.5	Rahmanian et al. (2012)
MEUF	Polyethersulfone	Sodium dodecyl sulfate	Cd^{2+}, Cu^{2+}	0.37, 0.41	2.7	85, 81	Häyrynen et al. (2012)
PEUF	Ceramic	Chitosan	Cd^{2+}	10 mg/L	7.5	98.5	Llorens et al. (2004)
PEUF	Polysulfone	Chitosan, polyethyleneimine, pectin	Cr^{3+}, Cr^{4+}	10 ppm	7–9	>99, <50	Aroua et al. (2007)
PEUF	Poly(vinyl alcohol)	Poly(ethyleneimine)	Pb^{2+}, Cd^{2+}, Cu^{2+}	100 ppm	5.0	96.0, 99.5, 99.0	Bessbousse et al. (2008)
PEUF	Polyethersulfone	Carboxy methyl cellulose	Cu^{2+}, Cr^{3+}, Ni^{2+}	10 mg/L	7.0	97.6, 99.5, 99.1	Barakat and Schmidt (2010)

metal ratio increase the rejection because the sizes of micelles are relatively larger at higher concentration of surfactant compare to the UF membrane. However the metal rejection shows a decrease of 3–6 % at pH lower than 3 due to competition of H^+ trapped on micelle surface with metal ions. A combination process of micellar enhanced ultrafiltration and activated carbon fibre (ACF) shows nearly complete rejection of copper from the solution. The coupling of MEUF with ACF has the potential to overcome the problems caused by excessive surfactant concentration since activated carbon has proved to be a good adsorbent for surfactants that contained in the permeate (Bade and Lee 2007).

In PEUF, water soluble polymer is used to bind with metal ions to form macromolecular complexes which have a larger molecular weight and size than the membrane. As a result, these macromolecular complexes will be retained and removed while the non-complexed ions will pass through the membrane. So far, polymers with carboxylic or amine groups have been extensively studied for this purpose. PEUF process is more economically competitive because the non-protonated form of the polymers that bind to the metal ions, will release the metal ions when an acid is added. This has led to the regeneration and reuse of the polymer. Hence, pH plays an important role in the binding and removal of metal species from the solution. Llorens et al. (2004) and Aroua et al. (2007) had conducted studies on the removal of cadmium and chromium from aqueous solution using PEUF. For both of the studies, high rejections were obtained at pH 7 or higher and decrease when pH is lowered for all types of the polymers tested namely chitosan, polyethyleneimine and pectin.

Reverse Osmosis

Reverse osmosis (RO) is one of the techniques able to remove a wide range of dissolved species from water. It is not only able to remove suspended solid > 1 μm (conventional filtration), but also dissolved solids, bacteria, viruses, and other germs contained in the water. In the application of RO, when the waste stream flows past the membrane, the solvent (e.g., water) is driven through the membrane (operation pressure ranges from 10 to 70 bars) while the remaining solutes (e.g., organic or inorganic components) do not pass through, and become increasingly concentrated on the feed side of the membrane. The general use of RO is desalination of seawater for its conversion into potable water. It is reported that more than 20 % of the world's desalination process is using this technique (Shahalam et al. 2002). The RO membranes have pore size ranging from approximate 0.5–1.5 nm (5–15 Å). The extremely small size of RO pores allows only the smallest organic molecules and unchanged solutes to pass through the semipermeable membrane along with the solvent (Paul Chen et al. 2006). Nowadays, many studies have been going on to utilise RO process in wastewater treatment.

A study by Ozaki et al. (2002) shows that it is successful to remove Ni^{2+}, Cu^{2+} and Cr^{6+} ions from industrial wastewater using RO membrane separation at lower pressures (350–450 kPa). The rejections 99.03 % of Ni^{2+}, 98.75 % of Cu^{2+} and 99.37 % of Cr^{6+} were achieved at pH 7. Metal rejections are higher at high pH values (7–9) due to the formation of insoluble complexes with OH^- ions, causing the solute changed to a larger molecule and precipitate onto the membrane surface. They also found that presence of co-ions such as calcium and magnesium in the industrial wastewater decreases the rejection. Meanwhile, a pilot-scale combination of membrane bioreactor (MBR) with RO separation is conducted by Dialynas and Diamadopoulos (2009). The combination technique shows high heavy metal removals, whereby Pb and Ni were removed completely, while Cr and Cu were removed by 89 % and 49 %, respectively. However, the performance of RO and its life-time are highly sensitive to wastewater properties. The accumulation of particulate and colloidal matters at the feed side of the membrane surface caused membrane fouling and permeate flux decline (Lee et al. 2006b).

Nanofiltration

Nanofiltration (NF) is the intermediate process between UF and RO and used to separate different fluids or ions. It has larger membrane pore structure, 0.001–0.01 μm (1–10 nm) as compared to the RO membranes, which allows more salt passage through the membrane. The advantages of NF compared to RO processes including operation at lower pressure (7–14 bar) but able to produce high recoveries, thereby conserving total water usage due to a lower concentrate stream flow rate (Paul Chen et al. 2006). Ortega et al. (2008) studied the removal of various metal ions from an acidic leachate solution by two different nanofiltration membranes (Desal5 DK and NF-270). The results show good metal ion rejection (62–100 %) where divalent ions were better rejected than monovalent ions. In recent years, Murthy and Chaudhari (2008) have conducted numerous studies on the removal of heavy metal ions from aqueous solution using NF membrane. They found that Ni^{2+} can be removed by a thin-film composite polyamide NF membrane. Rejection of nickel ions increases with increase in feed pressure and decreases with increase in feed concentration at constant feed flow rate. The maximum rejection of the metal is found to be 98 % and 92 % for an initial feed concentration of 5 and 250 ppm, respectively (Murthy and Chaudhari 2008). At the same year, Bouranene et al. (2008) also proved that Co^{2+} and Pb^{2+} ions can be removed by polyamide NF membrane. The rejection rate of 97 % and 81 % were observed for cobalt and lead, respectively, at constant permeate flux of 122 ± 2 Lh^{-1} m^{-2} and ion concentration of 1,600 mgL^{-1} in mixed-salt solution. In the following year, Murthy and Chaudhari (2009) continue to investigate removal efficiency of cadmium and nickel in a binary system. The maximum observed solute rejection of nickel and cadmium ions are 98.94 % and 82.69 %, respectively, for an initial feed concentration of 5 ppm.

Electrodialysis

Electrodialysis (ED), which requires electrical energy as a driving force, allows transportation of ions through ion permeable membranes from one solution to another due to the potential gradient that created. Since ED membranes have the ability to selectively transport ions having positive or negative charge and reject ions of the opposite charge, the removal or separation of ions/electrolytes can be achieved by this technique. Cifuentes et al. (2009) reported on the effectiveness of ED in separating Cu^{2+} and Fe^{2+} ions and subsequently water recovery in industrial copper electrowinning operations. Approximately 96.6 % of Cu and 99.5 % of Fe were separated from the working solution which constituted of initial concentration of 9 g/L and 5 g/L of Cu^{2+} and Fe^{2+}, respectively. Removal rates for Cu^{2+} and Fe^{2+} from the working solution increased linearly with cell current density and electrolyte agitation. Removal of Cr^{3+} from the tanning process in the leather industry is crucial due to the environmental issue and also to its industry. It is because presence of chromium ion has a negative effect on the quality of the tanned leather. Lambert et al. (2006) had proposed a two-step Cr(III) separation–concentration process. In the first step, a cation-exchange membrane, Nafion® 117, modified by electrodeposition of polyethylenimine was used to separate successfully Cr(III) from NaCl solutions. In the second step, conventional electrodialysis with three different membranes permits to increase the concentration of Cr(III) in the solutions produced in the separation stage.

8.2.4 Adsorption

Adsorption is known as an efficient, convenient and economic method for wastewater treatment. The flexibility in design and operation, high removal efficiency and the possibility for most of the adsorbents to be regenerated and reuse by suitable desorption process have made adsorption process gaining its popularity (Krishnani et al. 2008; Elouear et al. 2008; Afkhami et al. 2010; Salehi et al. 2012). Adsorption is a mass transfer process whereby a substance is transferred from a gas or liquid phase to the surface of a solid and bound by physical and/or chemical and/or electrostatic interactions. Physical adsorption is caused by the van der Waals forces whereas chemical adsorption involved the electronic interactions between specific surface sites and solute molecules. This kind of electrostatic interaction is generally referred to coulombic attractive forces between ions and charged functional groups. The three steps involved in the adsorption process are: (i) the transport of the solute from the bulk solution to the sorbent surface; (ii) adsorption on the particle surface; and (iii) transport within the sorbent particle.

Activated carbon (AC) is being recognised as one of the most popular adsorbents used in the removal of heavy metal contaminants. AC has been used efficiently in both batch and column studies due to its large micropore and mesopore volumes. This kind of structure also resulted in a high surface area. However, the drawback of

AC is it remains as an expensive material because of depleting source of commercial coal-based AC. Higher cost is accounted for higher quality of activated carbon. Another limitation of AC adsorption in heavy metal removal is its high affinity towards organic molecules. As a result, these kinds of high molecular weight organic compounds will block the heavy metal ions from reaching the adsorbents bed.

The problems revealed by AC adsorption have led to the production of alternative adsorbents to replace the costly AC. Attention has been focused on various adsorbents that have metal-binding capacities, especially if it is a source that is readily abundant. Besides this, utilization of inexpensive sources is also being viewed as attractive alternatives. For this, many researchers have shown interest to convert carbonaceous materials into AC for heavy metals remediation. Momčilović et al. (2011) prepared activated carbon from the cones of the European Black pine to remove lead (II) ions from aqueous solutions. The maximum adsorption capacity of 27.53 mg/g is observed and the optimum adsorbent dosage was established to be 2.0 mg/l. Namasivayam et al. (2007) had converted jatropha husk, an agricultural solid waste, generated from bio-diesel industries, to jatropha husk activated carbon. This material is an effective adsorbent for the removal of toxic ions, dyes and organics. Besides AC, other adsorbents that are used for adsorption of heavy metal ions include natural minerals, modified biopolymers as well as low cost materials, which are derived from agricultural waste and industrial by products. Among these compounds, adsorbent from low cost materials have gained considerable attention because of their low cost and local availability, which will be elaborated further in next chapter.

8.2.5 *Biological Treatment*

The method of biological treatment was introduced in the early years of the twentieth century. It involved confining naturally occurring bacteria at very much higher concentrations in tanks. These bacteria, together with some protozoa and other microbes, are collectively referred to as activated sludge. The initial objective of biological treatment is to remove small organic carbon molecules by the bacteria. As a result, the bacteria grow, and the wastewater is cleansed. Lately, this method is applied in the removal of heavy metals from wastewater where biosorption phenomenon itself plays an important role. Biosorption is a metabolism-independent binding of heavy metals to living cells, non-living biomass, or microbial extracellular polymers (Volesky and Holan 1995).

Lipczynska-Kochany and Kochany (2009) have investigated the influence of humic substances (humate, HS) on the biological treatment of wastewater containing heavy metals (Cr, Cu, Fe, Mn, Ni, and Zn). Humic substances (HS) are naturally occurring compounds resulting from microbial and chemical transformation of organic debris. The removal of ammonia and metals was 99 % and over 90 %, respectively. Chipasa (2003) studied the accumulation and removal of heavy metals (Cd, Cu, Pb and Zn) through a biological wastewater treatment system. The results revealed that the removal of heavy metals from wastewater is influenced by their initial contents in the influent. The reduction in heavy metal contents was reported to follow this sequence: Cd < Pb < Cu < Zn. Biological wastewater treatment is a complex process depending not only on many biological and physicochemical conditions but also on process operating conditions and design. Consequently, all these are also influential variables in controlling the removal efficiency of heavy metals.

8.3 Utilization of Low-Cost Adsorbents in Heavy Metals Removal

Heavy metals such as chromium, lead, copper, nickel, zinc and iron are not biodegradable and they are toxic to human health and environment. Due to their tendency to accumulate in organisms and cause various disorders and diseases, these heavy metals are recognized as an emergent class of human carcinogenic compounds. Besides, the presence of high concentration of these metals in the environment may lead to serious problems including water pollution and soil degradation. Obviously, an effective method in heavy metal removal is urgently needed to lessen the pollution loads before the discharge enters the surface water.

Techniques currently used for the removal of heavy metals from wastewaters are chemical precipitation, ion exchange, membrane filtration, electrolytic reduction, coagulation, solvent extraction and catalysis. Owing to the drawbacks found in these techniques: generation of residues, low efficiency, applicability to a wide range pollutants, need of pre-treatments and high cost, they may not be suitable solutions at industrial scale. Since the last decades, adsorption process has been viewed as a highly effective method for removal of various heavy metals due to ease of operation. In particular, high surface area materials like activated carbon, silica aerogel and other mesoporous materials have been proven for their excellent adsorption capability for heavy metals. Unfortunately, production of these adsorbents is usually costly, making them not attractive to be applied by small and medium scale industries. Thus, the use of low cost materials as adsorbent for metal ion removal from the wastewaters has been highlighted. In general, the low-cost adsorbent materials are referred to those available at free cost and exist abundantly in nature. They may also represent the unused resources after a manufacturing process with serious disposal problems. Apparently, utilization of naturally occurring material or locally available agricultural waste materials or industrial by-products as the adsorbents in removing heavy metals from wastewaters is not only cost effective in heavy metal removal, but it also contributes to a zero waste situation in the environment.

A numerous studies have been carried out to investigate as well as to further improve the usage of different low cost adsorbent materials for adsorption of individual or multiple heavy metals in an aqueous solution. These low cost adsorbents range from industrial by-products, agriculture waste to biosorbents, for example wood bark, sawdust, mangoesteen peel, durian husk, oil palm empty fruit bunch, tea leaves and exhausted coffee etc. In this section, some of the intensively studied low cost adsorbents such as fly ash, rice husk, wheat straw, chitin and chitosan composites as well as algae and bacteria were discussed in terms of their efficiency for heavy metals removal. Recent reported adsorption capacities of the selected adsorbents are presented in Tables 8.4, 8.5, 8.6, 8.7, 8.8, and 8.9 to give some idea of adsorbent effectiveness. It is interesting to note that the reported values are strongly dependent on the experimental condition. Therefore it is advisable for the reader to refer to the original work in order to have a better understanding on the specific conditions that have been imposed on the adsorption process.

8.3.1 Fly Ash

Fly ash is a waste material produced during the combustion of various materials, such as coal, plant biomass and municipal solid waste in the electrical generation process. This industrial by-product is usually grey in color, abrasive, mostly alkaline and refractory in nature. It contains mainly silica (SiO_2) of ~47 %, alumina (Al_2O_3) of ~25 %, iron oxide (Fe_2O_3) of ~16 %, and calcium oxide (CaO) of ~5 % with varying amounts of carbon, calcium, magnesium and sulphur. The variations that are commonly found in fly ash, either in terms of their chemical composition or physical properties is most likely duc to differcnt sources of coal as well as diversity in the design of coal-fired boilers. However, an empirical formula for fly ash based on the dominance of certain key elements has been proposed as below (Iyer and Scott 2001): $Si_{1.0}Al_{0.45}Ca_{0.51}Na_{0.047}Fe_{0.039}Mg_{0.020}K_{0.013}Ti_{0.011}$.

Depending on the ratio of calcium to silica and ratio of alumina to iron in the ash, fly ash can be classified into two types: Type C and Type F. The Type C fly ash is normally produced from the burning of low-rank coals, such as lignites or sub-bituminous coals. This fly ash has cementitious properties as it is self-hardening upon reacted with H_2O. Meanwhile, the Type F fly ash is generated from the burning of higher-rank coals like bituminous coals or anthracites. This type of fly ash is pozzolanic in nature where it is hardening when reacted with $Ca(OH)_2$ and H_2O.

In fact, the fly ash has been used effectively as raw material in cement and brick production, soil amendment and as filler in road works. Besides, the fly ash can be converted into zeolite, another versatile material which is widely used as catalyst and adsorbent due to its outstanding high surface area and porosity. However, all the fly ash produced from the power industry cannot be reused by these applications. Consequently, most of the fly ash generated is disposed of as landfill, a practice which is under examination for environmental concerns. Therefore, continuing research efforts have been made to explore other applications prior to discarding in order to lessen the environmental burden. Since fly ash is enriched with SiO_2 and contains a portion of unburned carbon, this material is potentially used as low cost adsorbent to remove various hazardous elements from wastewaters. A lot of

Table 8.4 Adsorption capacities of metals by fly ash[a]

Metals	Adsorbent	Adsorption capacity[b]	Temperature (°C)	References
As(III)	Fly ash coal-char	3.7–89.2	25	Pattanayak et al. (2000)
	Fly ash brown coal-char	25.1–32.1	25	Polowczyk et al. (2011)
As(V)	Fly ash	7.7–27.8	20	Diamadopoulos et al. (1993)
	Fly ash coal-char	0.02–34.5	25	Pattanayak et al. (2000)
Cd(II)	Fly ash	1.6–8.0	–	Ayala et al. (1998)
	Fly ash zeolite	95.6	20	Ayala et al. (1998)
	Fly ash	0.67–0.83	20	Bayat (2002a)
	Afsin-Elbistan fly ash	0.08–0.29	20	Bayat (2002b)
	Seyitomer fly ash	0.0077–0.22	20	Bayat (2002b)
	Fly ash	198.2	25	Apak et al. (1998)
	Fly ash-washed	195.2	25	Apak et al. (1998)
	Fly ash-acid	180.4	25	Apak et al. (1998)
	Bagasse fly ash	1.24–2.0	30–50	Gupta et al. (2003)
	Fly ash	0.05	25	Weng and Huang (1994)
	Coal Fly ash	18.98	25	Papandreou et al. (2007)
	Coal Fly ash pellets	18.92	–	Papandreou et al. (2007)
	Bagasse fly ash	6.19	–	Ho et al. (1989)
	Fly ash zeolite X	97.78	–	Apiratikul and Pavasant (2008b)
	NaOH activated fly ash	30.21	20–25	Visa et al. (2012)
Co(II)	Fly ash zeolite 4A	13.72	–	Hui et al. (2005)
Cr(III)	Fly ash	52.6–106.4	20–40	Cetin and Pehlivan (2007)
	Fly ash porous pellet	22.88	25	Papandreou et al. (2011)
	Bagasse fly ash	4.35	–	Gupta and Ali (2004)
	Fly ash zeolite 4A	41.61	–	Hui et al. (2005)
Cr(VI)	Fly ash + wollastonite	2.92	–	Panday et al. (1984)
	Fly ash + China clay	0.31	–	Panday et al. (1984)
	Fly ash-chitosan composite	33.27	15	Wen et al. (2011)
	Fly ash	1.38	30–60	Banerjee et al. (2004)
	Fe impregnated fly ash	1.82	30–60	Banerjee et al. (2004)
	Al impregnated fly ash	1.67	30–60	Banerjee et al. (2004)
	Afsin-Elbistan fly ash	0.55	20	Bayat (2002b)
	Seyitomer fly ash	0.82	20	Bayat (2002b)
	Bagasse fly ash	4.25–4.35	30–50	Gupta and Ali (2004)
	Fly ash	23.86	–	Bhattacharya et al. (2008)
Cs(I)	Fly ash zeolite	443.9	25	Mimura et al. (2001)
Cu(II)	Fly ash	0.1825	–	Salam et al. (2011)
	Fly ash	1.39	30	Panday et al. (1985)
	Fly ash + wollastonite	1.18	30	Panday et al. (1985)
	Fly ash	1.7–8.1	–	Ayala et al. (1998)
	Afsin-Elbistan fly ash	0.34–1.35	20	Bayat (2002c)
	Seyitomer fly ash	0.09–1.25	20	Bayat (2002c)
	Fly ash	207.3	25	Ricou et al. (1999)
	Fly ash-washed	205.8	25	Ricou et al. (1999)

(continued)

Table 8.4 (continued)

Metals	Adsorbent	Adsorption capacity[b]	Temperature (°C)	References
	Fly ash-acid	198.5	25	Ricou et al. (1999)
	Fly ash	0.63–0.81	25	Lin and Chang (2001)
	Bagasse fly ash	2.26–2.36	30–50	Gupta and Ali (2000)
	Fly ash	0.76	32	Rao et al. (2003)
	Fly ash	7.5	–	Ricou et al. (1999)
	Coal Fly ash pellets	20.92	25	Papandreou et al. (2007)
	Fly ash zeolite 4A	50.45	–	Hui et al. (2005)
	Fly ash	7.0	–	Hossain et al. (2005)
	Coal fly ash (CFA)	178.5–249.1	30–60	Hsu et al. (2008)
	CFA-600	126.4–214.1	30–60	Hsu et al. (2008)
	CFA–NAOH	76.7–137.1	30–60	Hsu et al. (2008)
	Fly ash zeolite X	90.86	–	Apiratikul and Pavasant (2008b)
	Fly ash	7.0	–	Gupta (1998)
	Fly ash mixed willow saw dusk	7.7160 ppm/g	–	Lucaci et al. (2011)
	Fly ash mixed white poplar saw dusk	6.2305 ppm/g	–	Lucaci et al. (2011)
Hg(II)	Fly ash	2.82	30	Sen and De (1987)
	Fly ash	11.0	30–60	Banerjee et al. (2004)
	Fe impregnated fly ash	12.5	30–60	Banerjee et al. (2004)
	Al impregnated fly ash	13.4	30–60	Banerjee et al. (2004)
	Sulfo-calcic fly ash	5.0	30	Ricou et al. (1999)
	Silico-aluminous ashes	3.2	30	Ricou et al. (1999)
	Fly ash-C	0.63–0.73	5–21	Kapoor and Viraraghavan (1992)
Ni(II)	Fly ash	9.0–14.0	30–60	Banerjee et al. (2003)
	Fe impregnated fly ash	9.8–14.93	30–60	Banerjee et al. (2003)
	Al impregnated fly ash	10–15.75	30–60	Banerjee et al. (2003)
	Afsin-Elbistan fly ash	0.40–0.98	20	Bayat (2002c)
	Seyitomer fly ash	0.06–1.16	20	Bayat (2002c)
	Bagasse fly ash	1.12–1.70	30–50	Gupta et al. (2003)
	Fly ash	3.9	–	Ricou et al. (1999)
	Fly ash zeolite 4A	8.96	–	Hui et al. (2005)
	Afsin–Elbistan fly ash	0.98	–	Bayat (2002c)
	Seyitomer fly ash	1.16	–	Bayat (2002c)
	Bagasse fly ash	6.48	–	Ho et al. (1989)
	Fly ash	0.03	–	Rao et al. (2002)
Pb(II)	Fly ash zeolite	70.6	20	Gan (2000)
	Fly ash	444.7	25	Yadava et al. (1987)
	Fly ash-washed	483.4	25	Yadava et al. (1987)
	Fly ash-acid	437.0	25	Yadava et al. (1987)
	Fly ash	753	32	Yadava et al. (1987)
	Bagasse fly ash	285–566	30–50	Goswami and Das (2000)
	Fly ash	18.8	–	Diamadopoulos et al. (1993)

(continued)

Table 8.4 (continued)

Metals	Adsorbent	Adsorption capacity[b]	Temperature (°C)	References
	Fly ash zeolite X	420.61	–	Apiratikul and Pavasant (2008b)
	Fly ash porous pellet	45.54	25	Papandreou et al. (2011)
	NaOH activated fly ash	2,000.0	20–25	Visa et al. (2012)
Zn(II)	Fly ash	6.5–13.3	30–60	Banerjee et al. (2003)
	Fe impregnated fly ash	7.5–15.5	30–60	Banerjee et al. (2003)
	Al impregnated fly ash	7.0–15.4	30–60	Banerjee et al. (2003)
	NaOH activated fly ash	18.87	20–25	Visa et al. (2012)
	Fly ash	0.25–2.8	20	Bayat (2002a)
	Afsin-Elbistan fly ash	0.25–1.19	20	Bayat (2002c)
	Seyitomer fly ash	0.07–1.30	20	Bayat (2002c)
	Bagasse fly ash	2.34–2.54	30–50	Gupta and Ali (2000)
	Bagasse fly ash	13.21	30	Gupta and Sharma (2003)
	Fly ash	4.64	23	Weng and Huang (1990)
	Fly ash	0.27	25	Weng and Huang (1994)
	Fly ash	0.068–0.75	0–55	Weng and Huang (2004)
	Fly ash	3.4	–	Ricou et al. (1999)
	Fly ash porous pellet	17.66	25	Papandreou et al. (2011)
	Fly ash zeolite 4A	30.80	–	Hui et al. (2005)
	Bagasse fly ash	7.03	–	Hui et al. (1989)
	Fly ash	11.11	–	Gupta (1998)
	Rice husk ash	14.30	–	Bhattacharya et al. (2006)
	Fly ash	7.84	–	Gupta (1998)
	Fly ash	0.1806	–	Salam et al. (2011)

[a]These reported adsorption capacities are values obtained under specific conditions. Readers are encouraged to refer to the original articles for information on experimental conditions
[b]In the unit of (mg g^{-1}) unless specified

investigations have been reported on usage of fly ash in adsorption of individual pollutants in an aqueous solution or flue gas. Overall, the results obtained when using these particular fly ashes are encouraging for the removal of heavy metals and organics from industrial wastewater. Some of the researches showed that the unburned carbon in fly ash is partly activated during the combustion process, resulting in a relatively high surface area and good porosity (Wang and Li 2007; Lu et al. 2008). These characteristics of unburned carbon are similar to that of activated carbon, hence fly ash could be a potential adsorbent and substitute of commercial activated carbon. Recently, biomass fly ash has attracted attention to be used as adsorbent in wastewaters treatment (Pengthamkeerati et al. 2008). It could be a better adsorbent as compared to other fly ash due to its higher percentage of unburned carbon.

In real practical, the fly ash is always submitted to heating treatment and/or modification before being applied as adsorbent. Adsorption capacities of metals by different fly ashes are provided in Table 8.4.

Table 8.5 Adsorption capacities of metals by rice husk[a]

Metals	Adsorbent	Adsorption capacity[b]	Temperature (°C)	References
As(III)	Copolymer of iron and aluminum impregnated with active silica derived from rice husk ash	146	–	Abo-El-Enein et al. (2009)
As(V)	Rice husk	615.11	–	Roy et al. (1993)
	Quaternized rice husk	18.98	–	Lee et al. (1999)
Au(I)	Rice husk	64.10	40	Nakbanpote et al. 2002)
	Rice husk	50.50	30	Nakbanpote et al. (2002)
	Rice husk	39.84	20	Nakbanpote et al. (2002)
	Rice husk ash	21.2	–	Nakbanpote et al. (2000)
Cd(II)	Partial alkali digested and autoclaved rice husk	16.7	–	Krishnani et al. (2008)
Cd(II)	Phosphate-treated rice husk	103.09	20	Ajmal et al. (2003)
	Rice husk	73.96	–	Ye et al. (2010)
	Rice husk	21.36	–	Roy et al. (1993)
	Rice husk	4	–	Tarley et al. (2004)
	Rice husk	8.58 ± 0.19	–	Kumar and Bandyopadhyay (2006)
	Rice husk	0.16	–	Munaf and Zein (1997)
	Rice husk	0.32	–	Marshall et al. (1993)
	NaOH activated rice husk	125.94	–	Ye et al. (2010)
	NaOH activated rice husk	7	–	Tarley et al. (2004)
	NaOH activated rice husk	20.24 ± 0.44	–	Kumar and Bandyopadhyay (2006)
	$NaHCO_3$ activated rice husk	16.18 ± 0.35	–	Kumar and Bandyopadhyay (2006)
	Epichlorohydrin treated rice husk	11.12 ± 0.24	–	Kumar and Bandyopadhyay (2006)
	Rice husk ash	3.04	–	Srivastava et al. (2008)
	Polyacrylamide grafted rice husk	0.889	–	Sharma et al. (2009)
	HNO_3, K_2CO_3 treated rice husk	0.044 ± 0.1 mmol g^{-1}	30	Akhtar et al. (2010)
	Partial alkali digested and autoclaved rice husk	9.57	–	Krishnani et al. (2008)
Cr(III)	Rice husk	1.90	–	Marshall et al. (1993)
	Rice husk ash	240.22	–	Wang et al. (2008)
Cr(VI)	Rice husk	164.31	–	Roy et al. (1993)
	Rice husk	4.02	–	Munaf and Zein (1997)
	Rice husk ash	26.31	–	Bhattacharya et al. (2008)

(continued)

Table 8.5 (continued)

Metals	Adsorbent	Adsorption capacity[b]	Temperature (°C)	References
	Rice husk-based activated carbon	14.2–31.5	–	Guo et al. (2002)
	Formaldehyde treated rice husk	10.4	–	Bansal et al. (2009)
	Pre-boiled rice husk	8.5	–	Bansal et al. (2009)
Cu(II)	Rice husk	1.21	–	Marshall et al. (1993)
	Rice husk	0.2	–	Srivastava et al. (2008)
	Rice husk	7.1	–	Nakbanpote et al. (2007)
	Rice husk ash	11.5191	–	Feroze et al. (2011)
	Raw rice hush	4.90	–	Luo et al. (2011)
	Expending rice husk	8.02	–	Luo et al. (2011)
	RH-cellulose	7.7	–	Nakbanpote et al. (2007)
	Rice husk heated to 300 °C (RHA300)	6.5	–	Nakbanpote et al. (2007)
	Rice husk heated to 500 °C (RHA500)	16.1	–	Nakbanpote et al. (2007)
	HNO_3, K_2CO_3 treated rice husk	0.036 ± 0.2 mmol g^{-1}	30	Akhtar et al. (2010)
	Microwave incinerated rice husk ash (800 °C)	3.497	–	Johan et al. (2011)
	Microwave incinerated rice husk ash (500 °C)	3.279	–	Johan et al. (2011)
	Partial alkali digested and autoclaved rice husk	10.9	–	Krishnani et al. (2008)
	Tartaric acid modified rice husk	29	27	Wong et al. (2003a)
	Tartaric acid modified rice husk	22	50	Wong et al. (2003a)
	Tartaric acid modified rice husk	18	70	Wong et al. (2003a)
	Tartaric acid modified rice husk	31.85	–	Wong et al. (2003b)
	Carbonized rice husk	42.1	15	Ye et al. (2012)
	Carbonized rice husk	45.5	25	Ye et al. (2012)
	Carbonized rice husk	55.2	35	Ye et al. (2012)
Fe(II)	Copolymer of iron and aluminum impregnated with active silica derived from rice husk ash	222	–	Abo-El-Enein et al. (2009)
Hg(II)	Rice husk ash	6.72	30	Feng et al. (2004)
	Rice husk ash	9.32	15	Feng et al. (2004)
	Rice husk ash	40.0–66.7	–	Tiwari et al. (1995)
	Polyaniline/rice husk ash nanocomposite	Not determined	–	Ghorbani et al. (2011)
	Partial alkali digested and autoclaved rice husk	36.1	–	Krishnani et al. (2008)
Mn	Copolymer of iron and aluminum impregnated with active silica derived from rice husk ash	158	–	Abo-El-Enein et al. (2009)
	Partial alkali digested and autoclaved rice husk	8.30	–	Krishnani et al. (2008)

(continued)

Table 8.5 (continued)

Metals	Adsorbent	Adsorption capacity[b]	Temperature (°C)	References
Ni(II)	Rice husk	0.23	–	Marshall et al. (1993)
	Rice husk ash	4.71	–	Srivastava et al. (2009)
	Microwave-irradiated rice husk (MIRH)	1.17	30	Pillai et al. (2009)
	Partial alkali digested and autoclaved rice husk	5.52	–	Krishnani et al. (2008)
Pb(II)	Rice husk ash	12.61	30	Feng et al. (2004)
	Rice husk ash	12.35	15	Feng et al. (2004)
	HNO_3, K_2CO_3 treated rice husk	0.058 ± 0.1 mmol g^{-1}	30	Akhtar et al. (2010)
	Rice husk ash	207.50	–	Wang et al. (2008)
	Rice husk ash	91.74	–	Naiya et al. (2009)
	Copolymer of iron and aluminum impregnated with active silica derived from rice husk ash	416	–	Abo-El-Enein et al. (2009)
	Tartaric acid modified rice husk	120.48	–	Wong et al. (2003b)
	Tartaric acid modified rice husk	108	27	Wong et al. (2003a)
	Tartaric acid modified rice husk	105	50	Wong et al. (2003a)
	Tartaric acid modified rice husk	96	70	Wong et al. (2003a)
	Partial alkali digested and autoclaved rice husk	58.1	–	Krishnani et al. (2008)
	Tartaric acid modified rice husk	21.55	–	Tarley et al. (2004)
	Rice husk	6.385	25	Surchi (2011)
	Rice husk	5.69	30	Zulkali et al. (2006)
	Rice husk	45	–	Tarley et al. (2004)
	Rice husk	11.40	–	Roy et al. (1993)
Zn(II)	HNO_3, K_2CO_3 treated rice husk	0.037 ± 0.2 mmol g^{-1}	30	Akhtar et al. (2010)
	Rice husk	30.80	50	Mishra et al. (1997)
	Rice husk	29.69	40	Mishra et al. (1997)
	Rice husk	28.25	30	Mishra et al. (1997)
	Rice husk	26.94	20	Mishra et al. (1997)
	Rice husk ash	14.30	–	Bhattacharya et al. (2006)
	Rice husk ash	7.7221	–	Feroze et al. (2011)
	Rice husk ash	5.88	–	Srivastava et al. (2008)
	Partial alkali digested and autoclaved rice husk	8.14	–	Krishnani et al. (2008)
	Rice husk	0.75	–	Marshall et al. (1993)
	Rice husk	0.173	–	Munaf and Zein (1997)
	Rice husk-polyaniline nanocomposite	24.3	–	Ghorbani et al. (2012)

[a]These reported adsorption capacities are values obtained under specific conditions. Readers are encouraged to refer to the original articles for information on experimental conditions
[b]In the unit of (mg g^{-1}) unless specified

Table 8.6 Adsorption capacities of metals by wheat based materials[a]

Metals	Adsorbent	Adsorption capacity[b]	References
Cd(II)	Wheat straw	39.22	Farooq et al. (2011)
	Wheat straw	14.56	Dang et al. (2009)
	Wheat straw	11.60	Tan and Xiao (2009)
	Wheat straw	40.48	Dhir and Kumar (2010)
	Wheat bran	51.58	Nouri and Hamdaoui (2007)
	Wheat bran	15.71	Nouri et al. (2007)
	Wheat bran	21.0	Farajzadeh and Monji (2004)
	Wheat bran	101	Özer and Pirincci (2006)
Cr(III)	Wheat straw	21.0	Chojnacka (2006)
	Wheat bran	93.0	Farajzadeh and Monji (2004)
Cr(VI)	Wheat straw	295.36	Xu et al. (2011)
	Wheat straw	47.16	Dhir and Kumar (2010)
	Wheat bran	35	Dupont and Guillon (2003)
	Wheat bran	40.8	Wang et al. (2008)
	Wheat bran	310.58	Singh et al. (2009)
	Wheat bran	0.942	Nameni et al. (2008)
Cu(II)	Wheat straw	11.43	Dang et al. (2009)
	Wheat straw-citric acid treated	78.13	Gong et al. (2008)
	Wheat bran	12.7	Dupont et al. (2005)
	Wheat bran	17.42	Aydin et al. (2008)
	Wheat bran	8.34	Basci et al. (2004)
	Wheat bran	6.85	Wang et al. (2009)
	Wheat bran	51.5	Özer et al. (2004)
	Wheat bran	15.0	Farajzadeh and Monji (2004)
Hg(II)	Wheat bran	70.0	Farajzadeh and Monji (2004)
Ni(II)	Wheat straw	41.84	Dhir and Kumar (2010)
	Wheat bran	12.0	Farajzadeh and Monji (2004)
Pb(II)	Wheat bran	87.0	Bulut and Baysal (2006)
	Wheat bran	62.0	Farajzadeh and Monji (2004)
	Wheat bran	79.4	Özer and Pirincci (2006)
Zn(II)	Wheat bran	16.4	Dupont et al. (2005)
U(VI)	Wheat straw	19.2–34.6	Wang et al. (2011)

[a]These reported adsorption capacities are values obtained under specific conditions. Readers are encouraged to refer to the original articles for information on experimental conditions
[b]In the unit of (mg g^{-1})

8.3.2 *Rice Husk*

Rice, a cereal grain, is the most important staple food for a large part of the world's human population, especially in Asia and the West Indies. In fact, the rice is grown on every continent except Antarctica and ranks second only to maize (corn) in terms of worldwide area and production in 2010. In rice milling industries, rice husk which is the hard outer shell of the rice grain is generated as the major by-product or waste,

Table 8.7 Adsorption capacities of metals by chitosan and chitosan composites[a]

Metal	Adsorbent	Adsorption capacity[b]	Temperature (°C)	References
As(III)	Chitosan/ceramic alumina	56.50	25	Veera et al. (2008b)
As(V)	Chitosan/ceramic alumina	96.46	25	Veera et al. (2008b)
Au(III)	Chitosan/cotton fibers (via Schiff base bond)	76.82	25	Qu et al. (2009b)
	Chitosan/cotton fibers (via C–N single bond)	88.64	25	Qu et al. (2009b)
Cd(II)	Chitosan/clay beads	72.31	25	Tirtom et al. (2012)
	Chitosan/cotton fibers (via Schiff base bond)	15.74	25	Zhang et al. (2008)
	Chitosan/glutaraldehyde	32.9	25	Vitali et al. (2008)
	Chitosan/perlite	178.6	25	Shameem et al. (2006)
	Chitosan/PVA	142.9	50	Kumar et al. (2009)
	Chitosan/xanthate	357.14	RT	Sankararamakrishnan et al. (2007)
Co(II)	Chitosan/clinoptilolite	467.90	25	Dinu and Dragan (2010)
Cr(III)	Chitosan/Reactive Blue 2	11.2	25	Vasconcelos et al. (2007)
Cr(VI)	Chitosan coated acid treated oil palm shell charcoal (CCAB)	60.25	–	Nomanbhay and Palanisamy (2005)
	Chitosan coated oil palm shell charcoal (CCB)	52.68	–	Nomanbhay and Palanisamy (2005)
	Chitosan/cellulose	13.05	25	Sun et al. 2009)
	Chitosan/ceramic alumina	153.8	25	Veera et al. (2003)
	Magnetic chitosan	69.40	–	Huang et al. (2009)
	Chitosan/montmorillonite	41.67	25	Fan et al. (2006)
	Chitosan/perlite	153.8	25	Shameem et al. (2003)
Cu(II)	Chitosan/alginate	67.66	–	Wan Ngah and Fatinathan (2008)
	Chitosan/cellulose	26.50	25	Sun et al. (2009)
	Chitosan/ceramic alumina	86.20	25	Veera et al. (2008a)
	Chitosan/clinoptilolite	574.49	–	Dragan et al. (2010)
	Chitosan/clinoptilolite	719.39	25	Dinu and Dragan (2010)
	Chitosan/cotton fibers (via Schiff base bond)	24.78	25	Zhang et al. (2008)
	Chitosan/H2fmbme	113.6	25	Vasconcelos et al. (2008)
	Chitosan/heparin	81.04	25	Coelho et al. (2007)
	Chitosan/perlite	196.07	–	Kalyani et al. (2005)
	Chitosan/perlite	104.0	25	Shameem et al. (2008)
	Chitosan/PVA	47.85	–	Wan Ngah et al. (2004)
	Chitosan/PVC	87.9	–	Srinivasa et al. (2009)
	Chitosan/Reactive Blue 2	57.0	25	Vasconcelos et al. (2007)
	Chitosan/sand	10.87	–	Wan et al. (2007)
	Chitosan/sand	8.18	–	Wan et al. (2010)
Fe(III)	Chitosan/nano-hydroxyapatite	6.75	–	Kousalya et al. (2010)

(continued)

Table 8.7 (continued)

Metal	Adsorbent	Adsorption capacity[b]	Temperature (°C)	References
Hg(II)	Chitosan/cotton fibers (via Schiff base bond)	104.31	35	Qu et al. (2009a)
	Chitosan/cotton fibers (via C–N single bond)	96.28	25	Qu et al. (2009a)
	Chitosan/glutaraldehyde	75.5	RT	Vieira and Beppu (2006)
Ni(II)	Chitosan/clay beads	32.36	25	Tirtom et al. (2012)
	Chitosan/calcium alginate	222.2	–	Vijaya et al. (2008)
	Chitosan/cellulose	13.21	25	Sun et al. (2009)
	Chitosan/ceramic alumina	78.10	25	Veera et al. (2008a)
	Chitosan/clinoptilolite	247.03	25	Dinu and Dragan (2010)
	Chitosan/cotton fibers (via Schiff base bond)	7.63	25	Zhang et al. (2008)
	Chitosan/magnetite	52.55	–	Tran et al. (2010)
	Chitosan/perlite	114.94	–	Kalyani et al. (2005)
	Chitosan/PVC	120.5	–	Srinivasa et al. (2009)
	Chitosan/silica	254.3	–	Vijaya et al. (2008)
Pb(II)	Chitosan/cellulose	26.31	25	Sun et al. (2009)
	Chitosan/cotton fibers (via Schiff base bond)	101.53	25	Zhang et al. (2008)
	Chitosan/magnetite	63.33	–	Tran et al. (2010)
	Chitosan/sand	12.32	–	Wan et al. (2010)
Th(IV)	Poly(methacrylic acid) grafted-chitosan/bentonite	110.5	30	Thayyath et al. (2010)
Zn(II)	Chitosan/cellulose	19.81	25	Sun et al. (2009)

[a]These reported adsorption capacities are values obtained under specific conditions. Readers are encouraged to refer to the original articles for information on experimental conditions
[b]In the unit of (mg g^{-1})

accounting for almost 20 % of the rice production. Throughout the world, the annual generation of rice husk was approximately 120 million tonnes in 2009, leading to serious environmental problems as the rice husk is usually disposed in the end of the production.

In nature, rice husk is tough, woody, insoluble in water, has good chemical stability and high mechanical strength. It is considered as a lignocellulosic agricultural by-product that contains approximately 32 % cellulose, 21 % hemicelluloses, 21 % lignin, 20 % silica and 3 % crude proteins (Nadzi et al. 2007). After a complete combustion process, the resulted rice husk ash consists of about 96 % silica (Rahman and Ismail 1993). The high percentage of silica coupled with a large amount of the structural polymer, lignin has made the rice husk not only resistant to water penetration and fungal decomposition, but also resistant to the best efforts of man to dispose it since husk does not biodegrade easily.

Obviously, the utilization of this abundant scaly residue will yield economic as well as environmental dividends. Therefore, exploration of possible usage of

Table 8.8 Adsorption capacities of metals by untreated and pretreated algae-based materials[a]

Metals	Algae	Adsorption capacity[b]	References
Al(III)	Sargassum fluitans-NaOH treated	0.950–3.740	Lee and Volesky (1999)
Cd(II)	Ascophyllum nodosum (B)	0.338–1.913	Holan et al. (1993)
	Ascophyllum nodosum-$CaCl_2$ treated	0.930	Yu et al. (1999)
	Ascophyllum nodosum-Divinil sulfone treated	1.139	Holan et al. (1993)
	Ascophyllum nodosum-formaldehyde treated	0.750–0.854	Leusch et al. (1995), Chong and Volesky (1995)
	Ascophyllum nodosum-formaldehyde ($3CdSO_4$, H_2O) treated	1.121	Holan et al. (1993)
	Ascophyllum nodosum-formaldehyde + urea treated	1.041	Holan et al. (1993)
	Ascophyllum nodosum-formaldehyde $Cd(CH_3COO)_2$ treated	1.326	Holan et al. (1993)
	Ascophyllum nodosum-glutaraldehyde treated	1.259	Holan et al. (1993)
	Ascophyllum nodosum-glutaraldehyde treated	0.463–0.480	Leusch et al. (1995)
	Caulerpa lentillifera (G)-dried macroalgae	0.026–0.042	Pavasant et al. (2006)
	Caulerpa lentillifera (G)-dried macroalgae	0.0381	Apiratikul and Pavasant (2008a)
	Chaetomorha linum (G)	0.48	Hashim and Chu (2004)
	Chlorella vulgaris (G)	0.30	Klimmek et al. (2001)
	Corallina officinalis (R)	0.2642	Basso et al. (2002)
	Durvillaea potatorum (B) -$CaCl_2$ treated	0.260	Matheickal et al. (1999)
	Durvillaea potatorum-$CaCl_2$ treated	1.100–1.130	Matheickal et al. (1999)
	Codium fragile (G)	0.0827	Basso et al. (2002)
	Ecklonia maxima (B) -$CaCl_2$ treated	1.150	Yu et al. (1999)
	Ecklonia radiate (B) -$CaCl_2$ treated	1.040	Yu et al. (1999)
	Fucus vesiculosus (B)	0.649	Holan et al. (1993)
	Gracilaria edulis (R)	0.24	Hashim and Chu (2004)
	Gracilaria Salicornia (R)	0.16	Hashim and Chu (2004)
	Laminaria hyperbola (B)-treated $CaCl_2$	0.820	Yu et al. (1999)
	Laminaria japonica (B)- treated $CaCl_2$	1.110	Yu et al. (1999)

(continued)

Table 8.8 (continued)

Metals	Algae	Adsorption capacity[b]	References
	Lessonia flavicans (B) -treated $CaCl_2$	1.160	Yu et al. (1999)
	Lessonia nigresense (B)-treated $CaCl_2$	1.110	Yu et al. (1999)
	Padina sp.(B)	0.53	Kaewsarn and Yu (2001)
	Padina sp-$CaCl_2$ treated	0.52	Kaewsarn and Yu (2001)
	Padina tetrastomatica	0.53	Hashim and Chu (2004)
	Porphira columbina (R)	0.4048	Basso et al. (2002)
	Sargassum baccularia (B)	0.74	Hashim and Chu (2004)
	Sargassum fluitans-formaldehyde treated	0.9519	Leusch et al. (1995)
	Sargassum fluitans-glutaraldehyde treated	1.0676	Leusch et al. (1995)
	Sargassum fluitans-protonated biomass	0.710	Davis et al. (2000)
	Sargassum sp. (B)	1.40	Tobin et al. (1984)
	Sargassum natans (B)	1.174	Holan et al. (1993)
	Sargassum siliquosum (M)	0.73	Hashim and Chu (2004)
	Sargassum vulgare-protonated biomass	0.790	Davis et al. (2000)
	Ulva lactuca (G)	0.1.52	Areco et al. (2012)
Cr(IV)	Chlorella vulgaris-artificial cultivation	1.525	Dönmez et al. (1999)
Cr(VI)	Chlorella vulgaris	0.534	Aksu et al. (1997)
	Chlorella vulgaris	1.525	Dönmez et al. (1999)
	Sargassum sp.	1.3257	Cossich et al. (2002)
	Sargassum sp.	1.30	Silva et al. (2003)
	Scenedesmus obliquus	1.131	Dönmez et al. (1999)
	Scenedesmus obliquus-artificial cultivation	1.131	Dönmez et al. (1999)
Cu(II)	Ascophyllum nodosum-$CaCl_2$ treated	1.090	Yu et al. (1999)
	Ascophyllum nodosum-formaldehyde treated	0.990	Chong and Volesky (1995)
	Ascophyllum nodosum-formaldehyde treated	1.306–1.432	Leusch et al. (1995)
	Ascophyllum nodosum-glutaraldehyde treated	0.803–0.850	Leusch et al. (1995)
	Caulerpa lentillifera (G)-dried macroalgae	0.042–0.088	Pavasant et al. (2006)

(continued)

Table 8.8 (continued)

Metals	Algae	Adsorption capacity[b]	References
	Caulerpa lentillifera (G)-dried macroalgae	0.112	Apiratikul and Pavasant (2008b)
	Chlorella miniata (G)	0.366	Lau et al. (1999)
	Chlorella vulgaris	0.295	Lau et al. (1999)
	Chlorella vulgaris	0.254–0.549	Aksu et al. (1997)
	Chlorella vulgaris	0.758	Dönmez et al. (1999)
	Chlorella vulgaris-artificial cultivation	0.759	Dönmez et al. (1999)
	Durvillaea potatorum-$CaCl_2$ treated	0.040–1.310	Matheickal and Yu (1999)
	Ecklonia radiate-$CaCl_2$ treated	0.070–1.110	Matheickal and Yu (1999)
	Ecklonia maxima-$CaCl_2$ treated	1.220	Yu et al. (1999)
	Laminaria hyperbola-treated $CaCl_2$	1.220	Yu et al. (1999)
	Laminaria japonica-treated $CaCl_2$	1.200	Yu et al. (1999)
	Lessonia flavicans-treated $CaCl_2$	1.250	Yu et al. (1999)
	Lessonia nigresense-treated $CaCl_2$	1.260	Yu et al. (1999)
	Padina sp-$CaCl_2$ treated	0.8	Kaewsarn (2002)
	Sargassum sp.	1.08	Silva et al. (2003)
	Sargassum fluitans-glutaraldehyde treated	1.574	Leusch et al. (1995)
	Sargassum fluitans-formaldehyde treated	1.7938	Leusch et al. (1995)
	Sargassum fluitans-NaOH treated	0.650–1.540	Lee and Volesky (1999)
	Sargassum fluitans-protonated biomass	0.800	Davis et al. (2000)
	Sargassum vulgare-protonated biomass	0.930	Davis et al. (2000)
	Scenedesmus obliquus (G)	0.524	Dönmez et al. (1999)
	Scenedesmus obliquus-artificial cultivation	0.524	Dönmez et al. (1999)
	Ulva lactuca (G)	0.34	Areco et al. (2012)
Fe(III)	Chlorella vulgaris	0.439	Aksu et al. (1997)
Ni(II)	Ascophyllum nodosum	1.346–2.316	Holan and Volesky (1994)
	Ascophyllum nodosum-formaldehyde treated	1.431–1.618	Leusch et al. (1995)

(continued)

Table 8.8 (continued)

Metals	Algae	Adsorption capacity[b]	References
	Ascophyllum nodosum-formaldehyde + CH_3COOH treated	0.409	Holan and Volesky (1994)
	Ascophyllum nodosum-formaldehyde + urea treated	0.511	Holan and Volesky (1994)
	Ascophyllum nodosum-glutaraldehyde treated	0.920–1.959	Leusch et al. (1995)
	Chlorella miniata	0.237	Lau et al. (1999)
	Chlorella vulgaris	0.205–1.017	Klimmek et al. (2001)
	Chlorella vulgaris	1.017	Dönmez et al. (1999)
	Chlorella vulgaris	0.205	Lau et al. (1999)
	Chlorella vulgaris-artificial cultivation	1.017	Dönmez et al. (1999)
	Chondrus crispus (R)	0.443	Holan and Volesky (1994)
	Codium taylori (G)	0.099	Holan and Volesky (1994)
	Cystoserta Indica (B)	0.76	Pahlavanzadeh et al. (2010)
	Durvillaea potatorum-$CaCl_2$ treated	0.17–1.13	Yu and Kaewsarn (2000)
	Fucus vesiculosus	0.392	Holan and Volesky (1994)
	Fucus vesiculosus-formaldehyde treated	0.559	Holan and Volesky (1994)
	Galaxaura marginata (R)	0.187	Holan and Volesky (1994)
	Galaxaura marginata-$CaCO_3$ treated	0.187	Holan and Volesky (1994)
	Nizmuddinia Zanardini (B)	0.76	Pahlavanzadeh et al. (2010)
	Padina australis (B)	0.37	Pahlavanzadeh et al. (2010)
	Padina gymnospora (B)	0.170	Holan and Volesky (1994)
	Padina gymnospora-$CaCO_3$ treated	0.238	Holan and Volesky (1994)
	Sargassum fluitans (B)	0.409	Holan and Volesky (1994)
	Sargassum fluitans-epichlorohyridin treated	0.337	Holan and Volesky (1994)
	Sargassum fluitans-formaldehyde treated	1.9932	Leusch et al. (1995)
	Sargassum fluitans-formaldehyde + HCl treated	0.749	Holan and Volesky (1994)
	Sargassum fluitans-glutaraldehyde treated	0.7337	Leusch et al. (1995)

(continued)

Table 8.8 (continued)

Metals	Algae	Adsorption capacity[b]	References
	Sargassum glaucescens (B)	0.76	Pahlavanzadeh et al. (2010)
	Sargassum natans	0.409	Holan and Volesky (1994)
	Sargassum vulgare (M)	0.085	Holan and Volesky (1994)
	Scenedesmus obliquus	0.5145	Dönmez et al. (1999)
	Scenedesmus obliquus-artificial cultivation	0.514	Dönmez et al. (1999)
Pb(II)	Ascophyllum nodosum	1.313–2.307	Holan and Volesky (1994)
	Ascophyllum nodosum-$CaCl_2$ treated	1.150	Yu et al. (1999)
	Ascophyllum nodosum-Bis(ethenil)sulfone treated	1.733	Holan and Volesky (1994)
	Ascophyllum nodosum-formaldehyde treated	1.3755	Leusch et al. (1995)
	Ascophyllum nodosum-formaldehyde + CH_3COOH treated	1.308	Holan and Volesky (1994)
	Ascophyllum nodosum-formaldehyde + urea treated	0.854	Holan and Volesky (1994)
	Ascophyllum nodosum-glutaraldehyde treated	1.318	Holan and Volesky (1994)
	Ascophyllum nodosum-glutaraldehyde treated	0.816–0.898	Leusch et al. (1995)
	Caulerpa lentillifera (G)-dried macroalgae	0.076–0.139	Pavasant et al. (2006)
	Caulerpa lentillifera (G)-dried macroalgae	0.142	Apiratikul and Pavasant (2008a)
	Chlorella vulgaris	0.47	Klimmek et al. (2001)
	Cladophora glomerata (G)	0.355	Jalali et al. (2002)
	Chondrus crispus treated with 1-chloro-2,3-epoxipropane	1.009	Holan and Volesky (1994)
	Chondrus crispus	0.941	Holan and Volesky (1994)
	Codium taylori	1.815	Holan and Volesky (1994)
	Durvillaea potatorum-$CaCl_2$ treated	0.020–1.550	Matheickal and Yu (1999)
	Ecklonia maxima-$CaCl_2$ treated	1.400	Yu et al. (1999)
	Ecklonia radiate-$CaCl_2$ treated	0.050–1.260	Matheickal and Yu (1999)
	Fucus vesiculosus	1.105–2.896	Holan and Volesky (1994)

(continued)

Table 8.8 (continued)

Metals	Algae	Adsorption capacity[b]	References
	Fucus vesiculosus-formaldehyde treated	1.752	Holan and Volesky (1994)
	Fucus vesiculosus-formaldehyde + HCl treated	1.453	Holan and Volesky (1994)
	Galaxaura marginata	0.121	Holan and Volesky (1994)
	Galaxaura marginata-$CaCO_3$ treated	1.530	Holan and Volesky (1994)
	Gracilaria corticata (R)	0.2017–0.2606	Jalali et al. (2002)
	Laminaria hyperbola-treated $CaCl_2$	1.350	Yu et al. (1999)
	Laminaria japonica-treated $CaCl_2$	1.330	Yu et al. (1999)
	Lessonia flavicans-treated $CaCl_2$	1.450	Yu et al. (1999)
	Lessonia nigresense-treated $CaCl_2$	1.460	Yu et al. (1999)
	Padina gymnospora	0.314	Holan and Volesky (1994)
	Padina gymnospora-$CaCO_3$ treated	0.150	Holan and Volesky (1994)
	Padina tetrastomatica (B)	1.049	Jalali et al. (2002)
	Polysiphonia violacea (R)	0.4923	Jalali et al. (2002)
	Sargassum fluitans	1.594	Holan and Volesky (1994)
	Sargassum fluitans-epichlorohyridin treated	0.975	Holan and Volesky (1994)
	Sargassum fluitans-formaldehyde treated	1.8244	Leusch et al. (1995)
	Sargassum fluitans-glutaraldehyde treated	1.6603	Leusch et al. (1995)
	Sargassum hystrix (B)	1.3755	Jalali et al. (2002)
	Sargassum natans	1.221	Holan and Volesky (1994)
	Sargassum natans	1.1487	Jalali et al. (2002)
	Sargassum vulgare	1.100	Holan and Volesky (1994)
	Ulva lactuca (G)	0.61	Jalali et al. (2002)
	Ulva lactuca (G)	0.145	Areco et al. (2012)
	Undaria pinnatifida (B)	1.945	Kim et al. (1999)
Zn(II)	Ascophyllum nodosum-formaldehyde treated	0.680	Chong and Volesky (1995)
	Ascophyllum nodosum-formaldehyde treated	0.719	Leusch et al. (1995)

(continued)

Table 8.8 (continued)

Metals	Algae	Adsorption capacity[b]	References
	Ascophyllum nodosum-formaldehyde treated	0.8718	Leusch et al. (1995)
	Ascophyllum nodosum-glutaraldehyde treated	0.138–0.367	Leusch et al. (1995)
	Caulerpa lentillifera (G)-dried macroalgae	0.021–0.141	Pavasant et al. (2006)
	Chlorella vulgaris	0.37	Klimmek et al. (2001)
	Sargassum fluitans-formaldehyde treated	0.9635	Leusch et al. (1995)
	Sargassum fluitans-glutaraldehyde treated	0.9942	Leusch et al. (1995)
	Ulva lactuca (G)	0.145	Areco et al. (2012)

B brown alga, *G* green alga, *R* red alga

[a]These reported adsorption capacities are values obtained under specific conditions. Readers are encouraged to refer to the original articles for information on experimental conditions

[b]In the unit of (mmol g^{-1})

the rice husk is of great significance in many rice producing countries. It is quite unlikely to use rice husk as animal feed because of its relatively low content of total digestible nutrients (less than 10 %). Besides, the husk has a very low protein and available carbohydrates values, and yet, at the same time, high in crude fiber and crude ash. Due to its abrasive character, poor nutritive value, low bulk density and high ash content which would sometimes cause harmful effect, the rice husk cannot be considered as a safe and suitable animal feed.

Owing to its excellent properties including good heat insulation, ability in emitting smell or gases, and resistant to corrosive, the rice husk can be seen as a potential raw material used to insulate walls, floors and roof cavities. Unfortunately, the cost of building materials manufactured using rice husk as the aggregate is considerably high and not economic as compared to that of using other aggregates, hence its application as building material is not practical in real production. Recently, the rice husk ash has been used as silica source in synthesis of silica aerogel, MCM-41 and other silica- based mesoporous materials which have higher commercial value. Nevertheless, the quality of the resulted materials is hard to be controlled due to variation of silica content and presence of different elements in the rice husk ash depending on the rice species and the combustion process.

Thus, efforts have been made to use this cheap and readily available resource as a low cost adsorbent in the removal of heavy metals from aqueous environment. The high content of cellulose, a polymer contains three reactive hydroxyl groups has made rice husk a potential adsorbent material. However, both lignin and silica that present in rice husk tends to reduce the binding between accessible functional group

Table 8.9 Adsorption capacities of metals by bacterials[a]

Metals	Adsorbent	Adsorption capacity[b]	References
Ag(I)	Geobacillus thermodenitrificans	37.86	Chatterjee et al. (2010)
Cd(II)	Aeromonas caviae	155.3	Loukidou et al. (2004)
	Brevundimonas sp. ZF12 Strain	49.01	Masoudzadeh et al. (2011)
	Enterobacter sp.	46.2	Lu et al. (2006)
	Geobacillus thermodenitrificans	42.9	Chatterjee et al. (2010)
	Ochrobactrum anthropi	–	Ozdemir et al. (2003)
	Pseudomonas aeruginosa	42.4	Chang et al. (1997)
	Pseudomonas putida	8.0	Pardo et al. (2003)
	Pseudomonas putida	500.00	Hussein et al. (2004)
	Pseudomonas sp.	278.0	Ziagova et al. (2007)
	Sphingomonas paucimobilis	–	Tangaromsuk et al. (2002)
	Staphylococcus xylosus	250.0	Ziagova et al. (2007)
	Streptomyces pimprina	30.4	Puranik et al. (1995)
	Streptomyces rimosus	64.9	Selatnia et al. (2004a)
Co(II)	Geobacillus thermodenitrificans	69.76	Chatterjee et al. (2010)
Cr(III)	Geobacillus thermodenitrificans	70.7	Chatterjee et al. (2010)
Cr(VI)	Aeromonas caviae	284.4	Loukidou et al. (2004)
	Bacillus coagulans	39.9	Srinath et al. (2002)
	Bacillus megaterium	30.7	Srinath et al. (2002)
	Bacillus coagulans	39.9	Srinath et al. (2002)
	Bacillus licheniformis	69.4	Zhou et al. (2007)
	Bacillus megaterium	30.7	Srinath et al. (2002)
	Bacillus thuringiensis	83.3	Sahin and Öztürk (2005)
	Pseudomonas sp.	95.0	Ziagova et al. (2007)
	Pseudomonas fluorescens.	111.11	Hussein et al. (2004)
	Staphylococcus xylosus	143.0	Ziagova et al. (2007)
	Zoogloea ramigera	2	Nourbakhsh et al. (1994)
Cu(II)	Bacillus firmus	381	Salehizadeh and Shojaosadati (2003)
	Bacillus sp.	16.3	Tunali et al. (2006)
	Bacillus subtilis	20.8	Nakajima et al. (2001)
	Enterobacter sp.	32.5	Lu et al. (2006)
	Geobacillus thermodenitrificans	50.0	Chatterjee et al. (2010)
	Micrococcus luteus	33.5	Nakajima et al. (2001)
	Pseudomonas aeruginosa	23.1	Chang et al. (1997)
	Pseudomonas cepacia	65.3	Savvaidis et al. (2003)
	Pseudomonas putida	6.6	Pardo et al. (2003)
	Pseudomonas putida	96.9	Uslu and Tanyol (2006)
	Pseudomonas putida	15.8	Chen et al. (2005)
	Pseudomonas putida	163.93	Hussein et al. (2004)
	Pseudomonas stutzeri	22.9	Nakajima et al. (2001)

(continued)

Table 8.9 (continued)

Metals	Adsorbent	Adsorption capacity[b]	References
	Sphaerotilus natans	60	Beolchini et al. (2006)
	Sphaerotilus natans	5.4	Beolchini et al. (2006)
	Streptomyces coelicolor	66.7	Öztürk et al. (2004)
	Thiobacillus ferrooxidans	39.8	Liu et al. (2004)
Fe(III)	Geobacillus thermodenitrificans	79.9	Chatterjee et al. (2010)
	Streptomyces rimosus	122.0	Selatnia et al. (2004b)
Ni(II)	Bacillus thuringiensis	45.9	Öztürk (2007)
	Pseudomonas putida	556	Hussein et al. (2004)
	Streptomyces rimosus	32.6	Selatnia et al. (2004d)
Pb(II)	Bacillus sp.	92.3	Tunali et al. (2006)
	Bacillus firmus	467	Salehizadeh and Shojaosadati (2003)
	Corynebacterium glutamicum	567.7	Choi and Yun (2004)
	Enterobacter sp.	50.9	Lu et al. (2006)
	Geobacillus thermodenitrificans	32.26	Chatterjee et al. (2010)
	Pseudomonas aeruginosa	79.5	Chang et al. (1997)
	Pseudomonas aeruginosa	0.7	Lin and Lai (2006)
	Pseudomonas putida	270.4	Uslu and Tanyol (2006)
	Pseudomonas putida	56.2	Pardo et al. (2003)
	Streptomyces rimosus	135.0	Selatnia et al. (2004c)
Pd(II)	Desulfovibrio desulfuricans	128.2	De Vargas et al. (2004)
	Desulfovibrio fructosivorans	119.8	De Vargas et al. (2004)
	Desulfovibrio vulgaris	106.3	De Vargas et al. (2004)
Pt(IV)	Desulfovibrio desulfuricans	62.5	De Vargas et al. (2004)
	Desulfovibrio fructosivorans	32.3	De Vargas et al. (2004)
	Desulfovibrio vulgaris	40.1	De Vargas et al. (2004)
Th(IV)	Arthrobacter nicotianae	75.9	Nakajima and Tsuruta (2004)
	Bacillus licheniformis	66.1	Nakajima and Tsuruta (2004)
	Bacillus megaterium	74.0	Nakajima and Tsuruta (2004)
	Bacillus subtilis	71.9	Nakajima and Tsuruta (2004)
	Corynebacterium equi	46.9	Nakajima and Tsuruta (2004)
	Corynebacterium glutamicum	36.2	Nakajima and Tsuruta (2004)
	Micrococcus luteus	77.0	Nakajima and Tsuruta (2004)
	Zoogloea ramigera	67.8	Nakajima and Tsuruta (2004)
U(VI)	Arthrobacter nicotianae	68.8	Nakajima and Tsuruta (2004)
	Bacillus licheniformis	45.9	Nakajima and Tsuruta (2004)
	Bacillus megaterium	37.8	Nakajima and Tsuruta (2004)
	Bacillus subtilis	52.4	Nakajima and Tsuruta (2004)
	Corynebacterium equi	21.4	Nakajima and Tsuruta (2004)
	Corynebacterium glutamicum	5.9	Nakajima and Tsuruta (2004)
	Micrococcus luteus	38.8	Nakajima and Tsuruta (2004)
	Nocardia erythropolis	51.2	Nakajima and Tsuruta (2004)
	Zoogloea ramigera	49.7	Nakajima and Tsuruta (2004)

(continued)

Table 8.9 (continued)

Metals	Adsorbent	Adsorption capacity[b]	References
Zn(II)	Aphanothece halophytica	133.0	Incharoensakdi and Kitjaharn (2002)
	Bacillus firmus	418	Salehizadeh and Shojaosadati (2003)
	Geobacillus thermodenitrificans	48.26	Chatterjee et al. (2010)
	Pseudomonas putida	6.9	Pardo et al. (2003)
	Pseudomonas putida	17.7	Chen et al. (2005)
	Streptomyces rimosus	30.0	Mameri et al. (1999)
	Streptomyces rimosus	80.0	Mameri et al. (1999)
	Streptoverticillium cinnamoneum	21.3	Puranik and Paknikar (1997)
	Thiobacillus ferrooxidans	82.6	Celaya et al. (2000)
	Thiobacillus ferrooxidans	172.4	Liu et al. (2004)

[a]These reported adsorption capacities are values obtained under specific conditions. Readers are encouraged to refer to the original articles for information on experimental conditions
[b]In the unit of (mg g^{-1})

on the surface of rice husk and adsorbate ions/molecules. Therefore, considerable researches have been carried out to modify or treat the rice husk via different methods prior to the application in heavy metals removal. Adsorption capacities of metals by untreated and treated rice husk are presented in Table 8.5.

8.3.3 Wheat Straw and Wheat Bran

Wheat (*Triticum aestivum*) is the major food crop of the world. The World Agricultural Supply and Demand Estimates (WASDE) has estimated the total consumption of wheat worldwide was 652 million tons for year 2010. Straw and bran are generated as by-products in the wheat milling industry. Every year, over 200 million tons of wheat straw are produced and every 100 million ton of wheat can create 25 million tons of bran. In addition to their use as feed and fuel, large amounts of redundant wheat straw and bran must be disposed of. Therefore, finding new applications for these abundantly available waste materials is a challenge to the wheat production countries.

The wheat straw is a lignocellulosic agricultural waste consisting of about 34–40 % cellulose, 20–35 % hemicellulose, 8–15 % lignin and sugars as well as other compounds carrying different functional groups like carboxyl, hydroxyl, sulphydryl, amide and amine. Although the substances are almost similar, the percentage composition of different substances may vary in different parts of the world depending on the wheat species and growing conditions. Wheat straw has been used as fodder and in paper industry to produce low quality boards or packing materials. The stems are usually burnt directly in some parts of the world for fuel supply, leading to serious atmospheric pollution and wastage of resources.

The wheat bran is the shell of the wheat seed that contains most nutrients of wheat. Despite its high nutritious value, this bran is always removed in the processing of wheat into flour and discarded as waste. Most recently, wheat bran is found to be a valuable material which is applicable in fermentation industry, pharmaceutics and biomedical research. This by-product can also serve as a potential nutritious and cheap raw material for fermentation industry. In addition, its amazing properties of anti-oxidative and anti-inflammatory can bring about a revolution in the fields of melanoma and cancer research. However, more researches and investigation should be carried out to make its applications realistic in human daily life.

As attempt to fully utilize these industry by-products, both wheat straw and wheat bran have been investigated for their adsorption behaviour towards metal ions (Table 8.6). The high adsorption capability of straw is closely related to high content of cellulose, presence of different functional groups and porosity of the material. The reported metal adsorption capacities of wheat based materials could depend on the structure of wheat bran used in different studies, along with other parameters.

8.3.4 *Chitin, Chitosan and Chitosan Composites*

Chitin is a naturally and abundant occurring mucopolysaccharide from exoskeletons of insects, crustaceans shells, fungi cellular walls, annelids and molluscs. It is the second most abundant polymerin nature after cellulose. Chitin contains 2-acetamido-2-deoxy-β-D-glucose through a β $(1 \rightarrow 4)$ linkage. Through deacetylation process of chitin, a type of natural poly(aminosaccharide), so called chitosan is produced. The chitosan consists mainly of poly $(1 \rightarrow 4)$-2 amino-2-deoxy-D-glucose unit and this derivative biopolymer has been known as a promising polymeric material of great scientific interest due to its special characteristics, for instances, hydrophilicity, biocompatibility, biodegradability, non-toxicity, good adsorption properties and wide range of applications.

The chitosan has drawn particular attention for its feasible application in a variety of forms, from flake-types to gels, beads and fibers. Due to high content of amino and hydroxyl groups in chitosan, this by-product material has been found to be able on adsorbing several heavy metals, including copper, chromium, silver, platinum and lead physically or chemically. In fact, chitosan has binding capacities of more than 1 mmol/g for heavy metals which is far greater than that of activated carbon. The extend of metal adsorption depends strongly on the chitosan source, deacetylation degree, nature of metal ion, variations in crystallinity, amino content and solution pH.

Chitosan can either form gel or dissolve depending on the pH values. Its high sensitivity towards pH has constituted an obstacle to chitosan's performance as a biosorbent in wastewater treatment. To solve this problem, chemical modification of chitosan using cross-linking reagents, for instance glyoxal, formaldehyde, glutaraldehyde, epichlorohydrin, ethylene glycon diglycidyl ether and isocyanates have been used to stabilize chitosan in acidic solutions. Several chemical changes are possible in the chitosan polymeric matrix because of the existence of reactive amino and hydroxyl groups. It has been demonstrated that the chemical modification does not only increase the chemical stability of the adsorbent in acidic media, and especially, decrease its solubility in most mineral and organic acids, but also enhances its mechanical properties. However, the adsorption ability of the modified chitosan may be reduced after the cross-linking procedure due to increment in the rigidity of the structural conformation of the polymer.

A great number of chitosan derivatives have been obtained by introducing selected functional groups into the polymeric matrix of chitosan in order to enhance its interaction with variety of metallic ions, thus increase the adsorption capacity and selectivity for metal ions in solution. The reported chitosan derivatives are those containing nitrogen, phosphorus and sulfur as heteroatoms, and other derivatives such as chitosan crown ethers and chitosan ethylenediaminetetraacetic acid (EDTA)/diethylenetriaminepentaacetic acid (DTPA) complexes. Apart from that, chitosan composites have also been extensively investigated for heavy metals removal. Various types of substance have been used to form composite with chitosan, including activated clay, bentonite, montmorillonite, polyurethane, poly vinyl alcohol, poly vinyl chloride, kaolinite, oil palm ash and perlite. Table 8.7 shows the heavy metal removal capacities through adsorption process using these modified chitosan materials.

8.3.5 Algae

The term "algae" refers to a large and diverse assemblage of organisms that contain chlorophyll and carry out oxygenic photosynthesis. Algae are simple plant-like organisms, ranging from unicellular to multicellular forms, which are easily found in aquatic habitats, freshwater, marine and moist soil. The largest and most complex marine forms are called seaweeds. This abundantly available biomass has been extensively studied for its variety applications as fertilizer, energy sources, pollution control, pigment, stabilizing substances, nutrition and biodegradable plastic production.

Algae can be classified according to several characteristics, including the nature of the chlorophyll(s) present, the carbon reserve polymers produced, the cell wall structure, and the motility type. Although all algae consist of chlorophyll a, there are some, which contain other chlorophylls that differ in minor ways from chlorophylls a. The presence of these additional chlorophylls is typical of particular algal groups. Chrysophyta (golden-brown algae and diatoms), Euglenophyta (euglenoids, is also considered as protozoa), Pyrrophyta (dino-flagellates), Chlorophyta (green algae), Phaeophyta (brown algae) and Rhodophyta (red algae) are some of the major groups of algae.

The algal cell is surrounded by thin, rigid cell walls which are usually made of multilayered microfibrillar framework generally consisting of cellulose and intersperse with amorphous material (Madigan et al. 2000). These walls are essentially impermeable to larger molecules or to macromolecules. However, existence of pores about 3–5 nm wide on the algal cell wall has allowed low molecular-weight constituents such as water, ions, gases, and other nutrients to pass through freely for metabolism and growth.

Due to their high surface area to volume ratio, various algae have been investigated as potential materials for heavy metal removal in aqueous solutions. Actually, the metal biosorption by algae mainly depend on the components on the cell, especially through cell surface and the spatial structure of the cell wall. The cell walls of algae generally contain cellulose, alginic acid and sulphated polysaccharides. These compounds consist of several functional groups, such as carboxyl, hydroxyl, sulfate, phosphate, and amino groups which play a very important role in biosorption process of metal. The mechanism of binding metal ions by algal biomass is always affected by the biomass characteristics, species and ionic charges of the targeted metal ions, and other external environmental factors such as solution pH and temperature.

Among the three groups of algae (red, green and brown algae), brown alga is the most widely studied. This could be associated to the high sorption capability of the brown algae as compared to other two algae species (Romera et al. 2006; Brinza et al. 2007). Very often, the algae undergo treatment in different ways to enhance their sorption capacity (Romera et al. 2006). Adsorption capacities of metals by untreated and treated algae are presented in Table 8.8. In brief, algae have been proven to be both economic and eco-friendly because they are ubiquitous occurrence in nature, can be regenerated after use and have lesser volume of chemical and/or biological sludge to be disposed of.

8.3.6 Bacteria

Bacteria are a large domain of prokaryotic microorganisms. They exist at almost everywhere on the Earth, including in soil, acidic hot springs, radioactive waste, and deep in the Earth's crust, as well as in organic matter and the live bodies of animals and plants. There are about five nonillion (5×10^{30}) bacteria present on the Earth, forming a biomass that exceeds that of all plants and animals. Bacteria have simple morphology and have a wide range of shapes, ranging from spheres or ovoid (coccus) to rods (bacillus, with a cylindrical shape) and spirals (spirillum). They are recognized as microscale organisms whose single cells have neither a membrane-bounded nucleus nor other membrane-bounded organelles like mitochondria and chloroplasts. Due to differences in genetics and ecology, however, bacteria may vary in size as much as in shape. The smallest bacteria are about 0.3 μm, and a few bacteria become fairly large, e.g. some spirochetes occasionally reach 500 μm in length.

In the past decades, there has been an increasing interest to use bacteria in degrading or mineralizing several textile dyes. Bacterial decolorization was attributed to the presence of oxidoreductive enzymes, such as laccase, NADH-DCIP reductase and azoreducases in the bacteria. However, the effectiveness of decolorization using bacteria depends strongly on the adaptability and the activity of the selected bacterium.

Recently, the feasibility of bacteria as biosorbent in heavy metals was explored. It has been demonstrated that composition of cell wall of bacteria is one of the key factors influencing bacterial biosorbing properties. The anionic functional groups exist in the teichoic acids and peptidoglican of gram-positive bacteria, lipopolysaccharides and phospholipids of gram-negative bacteria are the important elements responsible for the metal binding ability of the cell wall. Besides, the metal binding ability of bacteria would be further increased with the involvement of extra-cellular polysaccharides. Usually, the availability of this compound depends on the bacterial species and growth conditions.

Advantages such as availability, small size, ability to grow under controlled conditions and resiliency to a wide range of environmental situations are some of the reasons that made bacteria gaining its popularity in the development of new biosorbents materials. Adsorption capacities of metals by bacterial are provided in Table 8.9.

8.4 Equilibrium, Continuous and Modelling Studies of Heavy Metals Adsorption

8.4.1 Batch Adsorption Experiments

An efficient contact between adsorbent and adsorbate for a period of time is desired in a batch adsorption experiment. This allows the concentration of adsorbate in solution to attain equilibrium with the adsorbate on the surface. In most of the cases, particle size, pH, concentration and agitation are the key factors for the time of equilibrium attainment. Since the last decades, porous adsorbents such as activated carbon and mesoporous silica materials are widely used in batch equilibrium operations for their extremely high surface area, bringing into more effective adsorbent-adsorbate contact and a reducing of diffusional resistance inside the pores. Separation process is always carried out after the adsorption process. The adsorbed adsorbate on the solid phase can be separated from the adsorbate residue in solution via several methods such as settling, filtration, or centrifugation. In industry, the used adsorbent is either regenerated for further use or disposed, depending on the cost involved. Both adsorption isotherm and kinetics modelling are commonly applied in discussion of batch adsorption studies.

8.4.2 Adsorption Isotherm Models in a Batch System

In general, the adsorption isotherms describe how adsorbates interact with adsorbents and therefore they are the basic as well as the most necessity in an adsorption design. The adsorption isotherms are always referred to adsorption properties and equilibrium data obtained in an adsorption system. Apparently, establishing an appropriate and correct correlation for the equilibrium data is important to understand adsorbate-adsorbent interaction, subsequently contributing to improvement of an adsorption system. Besides, the data must fit to a mathematical model for a reliable prediction on adsorption parameters. The compliance of data would also allow a quantitative comparison on adsorption activities of variety adsorption systems under different operating conditions.

During an adsorption process, adsorbate is adsorbed onto the adsorbent. At the same time, desorption process of adsorbate from the adsorbent might also occur with a slower rate. When the rate of both adsorption and desorption processes is the same, the adsorption equilibrium is said to be achieved. Usually, the equilibrium condition is characterized in graph of adsorbate concentration in solid phase versus that in liquid phase. One or more isotherm models may be used to explain the distribution of adsorbate molecules between the adsorbent and the liquid phase, which provides information of equilibrium position in the adsorption process. The suitability, so-called 'favorable' degree of an isotherm model to the adsorption system is often verified from the shape generated in isotherm graph. The isotherm model is 'favored' by the adsorption system if the value of R^2 of graph slope is equal or very close to unity. Besides, the isotherm shape also gives qualitative information on the nature of the solute–surface interaction. In adsorption system, the performance of different adsorbents is always evaluated from their maximum adsorption capacity which can be obtained from the fitted adsorption isotherm. The capacity information also allows selection of the most appropriate adsorbent for a particular adsorption application under certain conditions.

Two- and three- parameter models are widely applied in correlating adsorption equilibria in liquid phase adsorption, even though they are originally used for gas phase adsorption. The thermodynamic assumptions made and different equation parameters used within these models often provide insight into surface properties and affinity of the adsorbent as well as the adsorption mechanism. Obviously, a proper correlation of equilibrium data is crucial for an optimized adsorption condition as well as improved adsorbent's activities.

Two-Parameter Adsorption Isotherms

There are a number of available two-parameter adsorption isotherms for description of adsorption equilibria in liquid phase. However, the underlying assumptions within the model often limit the application of each model for an adsorption system. Among the models, Langmuir, Freundlich, and Brunauer-Emmet-Teller (BET) models appear as the most popular isotherms. Further, other two-parameter models such as Temkin and Dubinin-Radushkevich Jovanovic, Halsey and Hurkins-Jura are also presented in this text.

Langmuir Adsorption Isotherm

The Langmuir model has been intensively applied in quantifying the amount of adsorbate being adsorbed on an adsorbent as a function of concentration at a certain temperature in a homogenous adsorption system (Langmuir 1918). This isotherm assumes the adsorption process happens at specific homogeneous sites on the surface of adsorbate. Therefore, the Langmuir model is valid for homogenous adsorption and it is always referred to monolayer coverage of adsorbate over a homogenous adsorbent surface. For a biosorption process, all the sites on the adsorbent are equivalent and no further adsorption can occur at a site once the site has been occupied by an adsorbate molecule. Based on the assumption made, the adsorption is localized based on lattice model and no further adsorption is allowed at equilibrium where a saturation point is reached. In addition, the model makes assumption that the adsorption of each molecule has equal adsorption activation energy and adsorption enthalpy is independent of surface coverage. The Langmuir equation is written below:

$$q_e = \frac{q_{max}\, K_L\, C_e}{1 + K_L\, C_e} \tag{8.2}$$

where q_e = mass of adsorbate adsorbed/mass of adsorbent at equilibrium
q_{max} = mass of adsorbate adsorbed/mass of adsorbent for a complete monolayer
K_L = Langmuir constant related to the enthalpy of adsorption
C_e = concentration of adsorbate at equilibrium

Apparently, characteristic of the Langmuir isotherm can be represented by a plateau in the plot q_e versus C_e. In addition, the Langmuir isotherm is traditionally used in quantifying the maximum uptake and estimating the adsorption capacity q_{max} of different adsorbents. Since the Langmuir assumes all the surface sites possess equal affinity for the adsorbate, the q_{max} should be temperature independent. In practical, however, the presence of functional groups in the adsorbent has brought to modest changes in adsorption capacity with temperature. Besides, some other factors including the number of active sites on the adsorbent, the chemical state of the sites, the accessibility of the adsorbate to the sites and the affinity between the sites such as binding strength has greatly affected the adsorption capacity of an adsorbent in real situation.

An adsorption process can be classified as physisorption or chemisorption. In physisorption, physical forces are dominant in the bonding between adsorbates and the surface. The forces become weaken with increasing of temperature. Therefore, the exothermal nature of the adsorption process can be proven from the decrease of K_L value with elevating temperature (Ho and Ofomaja 2006; Shaker 2007; Padmavathy 2008; Djeribi and Hamdaoui 2008). On the other hand, the thermal energy is needed for the strong interaction between the adsorbate and the adsorbent in generating new types of electronic bonds in a chemisorption process. Thus, the rise in temperature is more favourable for the chemisorption which involves the binding of adsorbates to active sites, making it an endothermic process (Febrianto et al. 2009). Usually, the van't Hoff plots are used to further verify the exothermal or endothermal nature of the adsorption process by considering the enthalpy of adsorption. An integrated van't Hoff equation relates the Langmuir constant, K_L to the temperature as below.

$$K_L = K_0 \exp\left(-\frac{\Delta H}{RT}\right) \tag{8.3}$$

where K_0 = parameter of the van't Hoff equation

H = enthalpy of adsorption

The determination of the adsorption parameters may not be easy since the Eq. 8.2 is in non-linear form. In order to overcome this problem, the Langmuir equation is often expressed in different linear forms as following:

$$\frac{C_e}{q_e} = \frac{1}{q_{max}} C_e + \frac{1}{K_L q_{max}} \tag{8.4}$$

$$\frac{1}{q_e} = \left(\frac{1}{K_L q_{max}}\right) \frac{1}{C_e} + \frac{1}{q_{max}} \tag{8.5}$$

$$q_e = q_{max} - \left(\frac{1}{K_L}\right) \frac{q_e}{C_e} \tag{8.6}$$

$$\frac{q_e}{C_e} = K_{Lq\,max} - K_{Lqe} \tag{8.7}$$

As compared to the non-linear equation, the four Langmuir linear equations (Eqs. 8.4, 8.5, 8.6, and 8.7) are more favourable among the researchers due to their simplicity and convenience in acquiring isotherm data. Nevertheless, one should realise that the isotherm parameters obtained using the four Langmuir linear equations could be varied in some cases, but they are identical when the non-linear method is applied (Ho 2006).

In fact, the essential characteristics of the Langmuir adsorption isotherm can be expressed in terms of dimensionless constant equilibrium parameter, R_L (Hall et al. 1966). The R_L is also recognized as separation factor and it is defined as:

$$R_L = \frac{1}{1 + K_L C_o} \tag{8.8}$$

where C_o = maximum initial concentration of adsorbate.

The isotherm shape can be interpreted according to the value of calculated R_L (Table 8.10).

Even though the Langmuir adsorption is considered as conventional method in studying an adsorption system, it has suffered from some disadvantages as the model ignores adsorbate-adsorbent interactions. In fact, there are two types of adsorbate-adsorbent interaction, namely direct interaction and indirect interaction

Table 8.10 Langmuir adsorption isotherm constant equilibrium parameter, R_L

R_L value	Type of isotherm
$R_L > 1$	Unfavorable
$R_L = 1$	Linear
$0 < R_L < 1$	Favorable
$R_L = 1$	Irreversible

present in real system. In direct interactions, the adjacent adsorbed molecules tend to make adsorbing near another adsorbate molecule more or less favourable. Meanwhile, indirect interaction influences the adsorption behaviour of the nearly sites by changing surface around the adsorbed site. Since the existence of both adsorbate-adsorbent interactions has been evidenced in heat of adsorption data, they should be taken into consideration in proper estimation of the adsorption parameters. Furthermore, the Langmuir adsorption model considers only the adsorbate of smooth and homogenous surface, making it fails to account for the rough surface of the adsorbate. Since a rough inhomogeneous surface may create different types of sites and change the properties of the sites, such as heat of adsorption, a significant deviation from this model is expected in many cases.

Freundlich Adsorption Isotherm

Freundlich isotherm (Freundlich 1906) is the oldest of the non-linear isotherms. Unlike the Langmuir isotherm which assumes homogenous site energies and limited levels of adsorption, it suggests adsorption onto heterogeneous surface. Since the multilayer adsorption is allowed in this model, concentration of adsorbate on the adsorbent surface rises with increasing of adsorbate concentration in the system. The Freundlich equation relates the concentration of an adsorbate on the surface of an adsorbent to the concentration of the adsorbate in the liquid which it is in contact. The empirical Freundlich equation is mathematically written as:

$$q_e = K_F C_e^{1/n} \tag{8.9}$$

where q_e = mass of adsorbate adsorbed/mass adsorbent
C_e = concentration of adsorbate at equilibrium
K_F = Freundlich constant related to adsorption capacity at particular temperature
n = Freundlich constant related to adsorption intensity at particular temperature, ($n > 1$)

Equation 8.9 can also be presented in a linearized logarithmic form:

$$\log q_e = \log K_F + \frac{1}{n} \log C_e \tag{8.10}$$

K_F is important in providing insight on adsorption capacity which appears as the most significant property of an adsorbent. Usually, adsorption capacity is affected by pH and temperature of the system as well as nature of the adsorbent, such as pore and particle size distribution, specific surface area, cation exchange capacity and surface functional groups. As shown in Eq. 8.10, both Freundlich constants of K_F and n can be calculated from the intercept and slope, respectively in the graph of log q_e versus log C_e. The K_F value depends on the units upon which q_e and C_e are expressed if $1/n \neq 1$. K_F is equivalent to q_e when C_e equals to 1. In the case of $1/n$ equals unity, identical adsorption energy for all sites is observed associated to linear adsorption occurrence (Site 2001). The constant n is the Freundlich equation exponent that characterizes quasi-Gaussian energetic heterogeneity of the adsorption surface (Bansal and Goyal 2005). Usually, larger value of n implies a stronger adsorbent-adsorbate interaction. For a favourable adsorption, Freundlich constant n should have values lying in the range of 1–10.

In fact, Freundlich isotherm is one of the most established isotherm models in biosorption study because it fits nearly all experimental adsorption–desorption data especially in highly heterogeneous adsorbent systems. Although Freundlich isotherm is inappropriate over a wide concentration range, this limitation is always ignored by researchers since moderate concentration is always used in most biosorption studies.

Brunauer–Emmer–Teller (BET) Model

In 1938, Stephen Brunauer, Paul Hugh Emmett, and Edward Teller had introduced the first isotherm for multimolecular layer adsorption (Brunauer et al. 1938). This adsorption theory, so called Brunauer–Emmer–Teller (BET) theory, is extension of Langmuir theory where it proposes multilayer adsorption of adsorbate on the surface of adsorbate. This BET model suggests a random distribution of sites covered by one or more adsorbate molecules by assuming the adsorbent surface consists of fixed individual sites and molecules can be adsorbed more than one layer on the surface of the adsorbent. Besides, Langmuir theory can be applied to each layer and the interaction between each adsorption layer is ignored. In other words, the same kinetics concept proposed by Langmuir is applicable to this multiple layering process, i.e., the rate of adsorption and desorption on any layer is equal.

The BET model is widely used in systems involving heterogeneous material and simple non-polar gases. However, it may not suitable for complex systems dealing with heterogeneous adsorbent such as biosorbents and adsorbates because of the assumption of all sites are energetically identical along with no horizontal interaction between adsorbed molecules. Therefore, this model should not be used in the interpretation of liquid phase adsorption data for complex solids. The simplified form of BET equation is given as following:

$$q_e = q_{\max} \frac{K_B C_e}{(C_e - C_s)\left[1 + (K_B - 1)(C_e/C_s)\right]} \tag{8.11}$$

where q_{max} = mass of adsorbate adsorbed/mass of adsorbent for a complete monolayer

K_B = constant related to energy of adsorption

C_s = concentration of adsorbate at saturation of all layers

After rearrangement, the Eq. 8.11 can be written into a linear form:

$$\frac{C_e}{(C_s - C_e)q} = \frac{1}{K_B q_{\max}} + \left(\frac{K_B - 1}{K_B q_{\max}}\right)\left(\frac{C_e}{C_s}\right) \tag{8.12}$$

Temkin Adsorption Isotherm

After considering the limitation of Langmuir adsorption isotherm, the Temkin model takes into accounts of indirect interactions of adsorbate-adsorbate molecules on adsorption isotherms (Temkin 1941). As the surface of adsorbent is occupied by adsorbate, the Temkin isotherm assumes that the heat of adsorption of all molecules in the layer would decrease linearly with coverage due to the indirect interactions. Thus, the Temkin model is dissimilar with Freundlich model which suggests a logarithmical decrease in adsorption. The Temkin equation is written as below:

$$q_e = \frac{RT}{b} \ln a\, C_e \tag{8.13}$$

where a = Temkin isotherm constant

b = Temkin constant related to heat of adsorption

The Temkin equation proposes the increase in degree of completion of the adsorption centres on an adsorbent may bring to a linear decrease of adsorption energy. The linearized form of Temkin equation is expressed below:

$$q_e = \frac{RT}{b} \ln a + \frac{RT}{b} \ln C_e \tag{8.14}$$

Obviously, both constants a and b can be determined from a plot of q_e versus ln C_e. The Temkin equation is useful to analyse the adsorption data at moderate concentration. Nevertheless, this equation which is originally applied in gas adsorption system, is not well-suited for a complex phenomenon involved in liquid phase adsorption. In liquid phase adsorption, the adsorbed molecules are not organized in a tightly packed structure with identical orientation like what happen in gas phase adsorption. The system becomes more complicated with the presence of solvent molecules and the formation of micelles from adsorbed molecules in liquid phase. Besides, liquid phase adsorption is also greatly influenced by other factors such as temperature, pH, solubility of adsorbate in the solvent and surface chemistry of the adsorbent. Therefore, the Temkin equation is seldom used for experimental data description of complex systems and it receives less attention as compared to three isotherms discussed previously.

Dubinin-Radushkevich Adsorption Isotherm

Dubinin and Radushkevich (D-R) adsorption isotherm has been widely used to express the adsorption isotherms in micropores (Dubinin and Radushkevich 1947). Based on the assumption made of inhomogeneous surface or constant adsorption potential, this model is applicable in estimation of porous structure of the adsorbent and the characteristics of adsorption. Besides, apparent free energy of the adsorption process can also be determined by using this model. The D-R isotherm is given by:

$$q_e = Q_m \exp(-K\varepsilon^2) \tag{8.15}$$

where Q_m = adsorption capacity of adsorbent per unit mass
K = constant related to the affinity coefficient and the adsorption energy
ε = Polanyi's adsorption potential which is correlated to temperature
The D-R equation can be also expressed in a linear form:

$$\ln q_e = \ln Q_m - K\varepsilon^2 \tag{8.16}$$

By plotting ln q_e versus ε^2, constant K and adsorption capacity, Q_m can be determined from the slope and the intercept, respectively. The constant K, defined as $1/\beta E$, is correlated to the affinity coefficient (β) which is closely related to the adsorbate-adsorbent interaction. It is also depends on mean free energy of adsorption (E) per mole of the adsorbate during the transportation process from infinite distance in solution to the surface of the adsorbent. Therefore, E can be calculated from the K value using the following equation:

$$E = \frac{1}{\sqrt{2K}} \tag{8.17}$$

Besides, the mean free energy can be computed using the following relationship:

$$\varepsilon = \mathrm{RT} \ln \left(1 + \frac{1}{C_e}\right) \tag{8.18}$$

Since the D-R isotherm is temperature dependent, a characteristic curve with all the suitable data lying on the same curve is expected by plotting the adsorption data at varied temperatures (ln q_e vs. ε^2). In other words, the applicability of the D-R equation in expressing the adsorption equilibrium data is confirmed with the attainment of the identical curve. Thus, it would be doubtful on the validity of the ascertained parameters when deviation in the characteristic curve generated from analyzed data is observed, even though the fitting procedure gives high correction value. Practically, however, the examination on the characteristic curve of biosorption systems is rarely done as the experiments are usually conducted at particular temperature. The D-R isotherm is applicable for not only single adsorbate but also bi-adsorbate adsorption system (Jaroniec and Derylo 1981). Unfortunately, the D-R isotherm is only appropriate for only an intermediate range of adsorbate concentrations since it may exhibit unrealistic asymptotic behavior.

Jovanovic Adsorption Isotherm

The Jovanovic model keeps the same assumptions contained in the Langmuir model which suggests monolayer adsorption on homogeneous solid surface. However, this model considers also the possibility of mechanical contact between the adsorbing and desorbing molecules (Wrudzinski and Wojciecbowski 1977). A general equation derived by Jovanovic is shown as below:

$$q_e = q_{max}[1 - \exp(-aC\nu)] \tag{8.19}$$

where C = concentration of the adsorbate in the bulk

ν = the heterogeneity parameter (with $0 < \nu < 1$) which emphasizes the allowance for

a = constant correlates to function of the sole temperature

The constant a characterizes the magnitude of the adsorbate–adsorbent interaction energy (Quinones and Guichon 1998). It is defined as (Quinones et al. 1999):

$$\mathrm{a} = \mathrm{K}\ \exp(\mathrm{E_a}/\mathrm{RT}) \tag{8.20}$$

where K = low-pressure equilibrium constant or Henry constant

E_a = characteristic energy of the distribution

The Jovanovic adsorption isotherm emphases the surface binding vibrations of an adsorbed species where the rate of desorption is proportional to the adsorbate surface coverage which can be determined from q_e/q_{max}.

Halsey Adsorption Isotherm

The Halsey model is applicable to multilayer adsorption (Halsey 1948). Its ability to confirm the heteroporous nature of the adsorbent makes it differs from the Freundlich adsorption isotherm which is also used for multilayer adsorption. The Halsey equation is written as below:

$$q_e = \mathrm{Exp}\left(\frac{\ln k_H - \ln C_e}{n}\right) \tag{8.21}$$

where k_H = Halsey isotherm constant

n = Halsey isotherm exponent

Harkin-Jura Adsorption Isotherm

The Harkin-Jura (H-J) adsorption isotherm proposes the presence of a heterogeneous pore distribution in the adsorbent (Harkins and Jura 1944). Similar to Freundlich and Hasley adsorption isotherms, this model is useful in explanation of multilayer adsorption. The H-J isotherm is expressed as below:

$$q_e = \sqrt{\frac{A_H}{B_2 + \log C_e}} \tag{8.22}$$

where A_H = isotherm parameter
B_2 = isotherm constant

Three-Parameter Adsorption Isotherms

In order to have better understanding in adsorbate-adsorbent interaction and adsorption behaviour, models involving more than two parameters are necessary to interpret the data. It is very often for researchers to make comparison studies between a series of two-parameter and three-parameter isotherms in describing the equilibrium data because more than one model might be needed to explain the biosorption mechanism in some cases. Some commonly used three-parameter isotherms for the prediction of biosorption experimental data are presented.

Redlich–Peterson Adsorption Isotherm

The Redlich-Peterson (R-P) isotherm is the combination of the Langmuir and Freundlich equations (Redlich and Peterson 1959). This model proposes a non ideal monolayer adsorption and the adsorption mechanism is a hybrid of the two models. The R-P model is applicable in adsorption equilibrium data description over a wide concentration range. The R-P equation is represented below:

$$q_e = \frac{K_{RP} C_e}{1 + a_{RP} {C_e}^{\beta}} \tag{8.23}$$

where K_{RP}, a_{RP} and β are the Redlich-Peterson parameters.

The exponent β ranges between 0 and 1. When $\beta = 0$, the equation resembles Henry's Law.

$$q_e = \frac{K_{RP} C_e}{1 + a_{RP}} \tag{8.24}$$

If $\beta = 1$, the R-P equation is similar to Langmuir equation.

$$q_e = \frac{K_{RP} C_e}{1 + a_{RP} C_e} \tag{8.25}$$

In most biosorption cases, the β values are close to unity. Thus, the Langmuir adsorption is the better model to correlate the adsorption data.

Equation 8.23 can also be expressed in linear form as:

$$\ln\left(K_R \frac{C_e}{q_e} - 1\right) = \ln a_{RP} + \beta \ln C_e \tag{8.26}$$

Since R-P isotherm incorporates three parameters, it is not possible to obtain the parameters of the R-P isotherms from the linearized equation. To overcome this problem, the parameters of Eq. 8.26 have to be verified by maximizing the correlation coefficients between the experimental data points and those from theoretical model predictions with the solver add-in function of the Microsoft Excel. Nevertheless, the linear equation of R-P isotherm is useful in determination of the parameters of the Langmuir and Freundlich adsorption models.

Sips Adsorption Isotherm

Sips isotherm was proposed after considering the continuing increase in the adsorbed amount with rising concentration (Sips 1948). It has solved the constraint observed in Freundlich model. In fact, the Sips expression (Eq. 8.27) resembles the Freundlich adsorption isotherm, and differs only on the finite limit of adsorbed amount at adequately high concentration of adsorbate.

$$q_e = q_{max} \frac{(K_S C_e)^\gamma}{1 + (K_S C_e)^\gamma} \tag{8.27}$$

where K_S = Sips isotherm constant

On the other hand, the Sips equation, Eq. 8.27 is similar to the Langmuir equation, Eq. 8.2. An additional parameter, γ which describes heterogeneity of the system is present in the Sips equation. The heterogeneity could arise from the adsorbent or the adsorbate, or both of them. When γ equals unity, Eq. 8.27 is same with Eq. 8.5.

Toth Adsorption Isotherm

Both Sips and Freundlich equations have their limitations in describing an adsorption data at certain concentration range. The Sips equation is inappropriate for adsorption equilibria data expression the low concentration end, while the

Freundlich equation is invalid at intense concentration. In order to solve this problem, Toth isotherm which was derived from the potential theory is proposed (Toth 1971). It makes assumption that most sites have adsorption energy less than the mean value due to an asymmetrical quasi-Gaussian energy distribution with its left-hand side widened. This model obeys Henry's at low concentration and reaches an adsorption maximum at high concentration. The Toth equation is written as:

$$q_e = q_{\max} \frac{C_e}{\left[a_t + C_e^t\right]^{1/t}} \tag{8.28}$$

where a_T = adsorptive potential constant

t = heterogeneity coefficient of the adsorbent ($0 < t \leq 1$)

Toth equation is capable to characterize adsorption for heterogeneous systems as it possesses a heterogeneity coefficient, t. When $t = 1$, the surface of adsorbent is homogeneous and the Toth equation is reduced to the Langmuir equation.

8.4.3 Adsorption Kinetics Models in Batch System

The efficiency of the adsorption process is greatly affected by the rate of the adsorbate to be attached on surface of adsorbent. Obviously, an excellent adsorbent should possess not only high adsorption capacity but also fast adsorption rate. Therefore, kinetic studies involving reaction rate investigation are crucial in identifying the ideal adsorbent for a particular adsorbate. In fact, kinetic studies are also carried out to indentify factors affecting the adsorption process, leading to an optimized adsorption system.

In an adsorption process, the kinetic models are always used to examine the rate determining mechanism as well as the role of adsorption surface, chemical reaction involved and/or diffusion mechanisms. In general, the adsorption mechanism is illustrated as the following steps: (i) bulk diffusion which involves transportation of adsorbate from the bulk solution to the surface of the adsorbent; (ii) film diffusion of adsorbate through the boundary layer to the surface of the adsorbent; (iii) pore diffusion or intraparticle diffusion of adsorbate from the surface to within the particle's pores; lastly (iv) adsorption of adsorbate on the active sites that available on the internal surface of the pores. In most of the cases, the bulk diffusion in step (i) is negligible by providing sufficient stirring to avoid particle and solute gradients in the batch system. Thus, the adsorption dynamics can be approximated by three consecutive steps [(ii)–(iv)] only. A rapid uptake always occurs in adsorption process which is the step (iv) of the mechanism. This immeasurable fast process can be considered as an instantaneous process especially in the case of physical adsorption. As a result, the overall rate of adsorption process is determined by either film or intraparticle diffusion in steps (ii) and (iii), respectively, or a combination of both.

By considering the film diffusion is the rate-controlling step in adsorption mechanism, the change of adsorbate concentration with respect to time can be presented as follows:

$$\frac{dC}{dt} = -k_L A(C - C_s) \tag{8.29}$$

where C = bulk liquid phase concentration of adsorbate at any time t

C_s = surface concentration of adsorbate

k_L = external mass transfer coefficient

A = specific surface area for mass transfer

During the initial stage of adsorption, it is assumed that the transport is mainly due to film diffusion mechanism and the intraparticle resistance can be ignored. At $t = 0$, $C = C_0$ and the surface concentration of adsorbate, C_s is negligible. With the assumptions made above, Eq. 8.29 can be simplified as:

$$\left[\frac{d(C/C_0)}{dt}\right] = -k_L A \tag{8.30}$$

Unlike physical reactions, the adsorption rate may be controlled by its own kinetic rates in the case of chemical reactions. As the adsorption kinetics provide valuable insights to the practical application of the process design and operation control, a complete modeling of kinetics should consider the diffusion equations, boundary conditions and adsorption isotherm equation. It has caused a complicated system of equations. This problem is, however, possible to be resolved by separating the diffusion steps. The diffusion mechanisms can be considered independently by assuming initial adsorption rate is controlled by intraparticle diffusion and it is characterized by external diffusion.

The kinetic studies are carried out practically in batch reactions using various adsorbent dosages, particle sizes of adsorbent, initial adsorbate concentrations, agitation rate, pH values and temperatures along with different adsorbent and adsorbate types. A linear regression is used to determine the best-fitting kinetic rate equation. The compliance of experimental data to the kinetic rate equations is often confirmed using the linear least-squares method that applied to the linearly transformed kinetic rate equations. Lagergren model, pseudo-second-order kinetic model, Elovich model and intraparticle diffusion model are the four commonly applied kinetic models among the researchers.

Lagergren Model

Lagergren model, which is also recognized as pseudo-first order kinetic model, appears as simplified kinetic model which has been widely used in adsorption kinetics examination (Lagergren 1898). The well fitting of experimental data to the pseudo-first order kinetic model suggests film diffusion is the rate-limiting step. Based on solid capacity, the Lagergren first-order rate expression is presented as follows:

$$\frac{dq}{dt} = k_1(q_e - q_t) \tag{8.31}$$

where q_e = adsorption capacity at equilibrium state
q_t = adsorption capacity at time t
k_1 = rate constant of pseudo-first-order adsorption
At $t = 0$, $q_t = 0$, and at $t = t$, $q_t = q_t$, Eq. 8.31 can be integrated as:

$$\ln(q_e - q_t) = \ln q_e - k_1 t \tag{8.32}$$

Equation 8.32 can be written in non-linear form as below:

$$q_t = q_e(1 - \exp(-k_1 t)) \tag{8.33}$$

A straight line of $\ln(q_e - q_t)$ versus t is good indicator of the applicability of this kinetic model. The q_e value obtained by this method should be compared with the experimental value to identify the deviation. A time lag resulted from external mass transfer or boundary layer diffusion at the beginning of the adsorption process may cause the difference in q_e values. If a large discrepancy in the q_e values is observed, even though the least square fitting process yields high correlation coefficient, a reaction still cannot be classified as first-order. In this case, q_e and k_l can be estimated via non-linear procedure fitting of Eq. 8.33. On the other hand, both rate constant and equilibrium adsorbate uptake can be determined from the straight-line plots of ln $(q_e - q_t)$ against t of Eq. 8.32 at different initial adsorbate concentrations.

Pseudo-Second-Order Kinetic Model

Ho and McKay introduced the pseudo-second order model on the basis of the adsorption capacity of the solid phase (Ho and McKay 1999). The well fitting of experimental data to the pseudo-second order kinetic model suggests chemisorption is the rate-limiting step. Similar to Lagergren model or pseudo-first order kinetic model, this model takes account of all the steps of adsorption including external film diffusion, intraparticle diffusion and adsorption. In particular, this pseudo-second order model is useful to the adsorption systems using various adsorbent dosages and adsorbate concentrations. This model is written as:

$$\frac{dq}{dt} = k_2(q_e - q_t)^2 \tag{8.34}$$

where k_2 = rate constant of pseudo-second-order adsorption

At $t = 0$, $q = 0$, and at $t = t$, $q_t = q_t$, integration of Eq. 8.34 with the boundary conditions yields

$$\frac{1}{q_e - q} = \frac{1}{q_e} + k_2 t \tag{8.35}$$

The linear form of Eq. 8.35 is shown as below:

$$\frac{t}{q} = \frac{t}{q_e} + \frac{1}{k_2 q_e^2} \tag{8.36}$$

It is noteworthy that pseudo-second order model enables the determination of adsorption capacity, pseudo-second-order rate constant, and initial adsorption rate without prior knowledge of experimental parameters.

Elovich Model

The Elovich model is a kinetic model for chemisorptions that has been widely used to describe the adsorption of gas onto solid systems (Zeldowitsch 1934; Rudzinski and Panczyk 2000). Since last decade, this model has also been applied to discuss the adsorption behaviour of heavy metal removal from aqueous solutions (Cheung et al. 2001). The Elovich equation is generally expressed as:

$$\frac{dq_t}{dt} = \alpha \exp(-\beta q_t) \tag{8.37}$$

Where α = initial adsorption rate
β = the desorption constant

By assuming $\alpha\beta t \gg t$ and by applying the boundary conditions $q_t = 0$ at $t = 0$ and $q_t = q_t$ at $t = t$, Eq. 8.36 can be simplified into (Chien and Clayton 1980):

$$q_t = \frac{1}{\beta} \ln(\alpha\beta) + \frac{1}{\beta} \ln(t) \tag{8.38}$$

If the adsorption data fit the Elovich model, a plot of q_t against $\ln(t)$ should yield a linear relationship. The β and α can be calculated from the slope and intercept, respectively.

Intraparticle Diffusion Model

By assuming intraparticle diffusion is the only rate determining step, the intraparticle diffusion model correlates adsorption capacity to effective diffusivity of adsorbate within the particle (Weber and Morris 1963). The well fitting of experimental data to the intraparticle diffusion model suggests pore diffusion is the rate-limiting step. The model mathematically is expressed in the equation below:

$$q_t = f\left(\frac{Dt}{r_p^2}\right)^{1/2} = K_{WM}t^{1/2} \tag{8.39}$$

where q_t = adsorption capacity at time t

D = effective diffusivity of solutes within the particle

r_p = particle radius

K_{WM} = intraparticle diffusion rate constant

Theoretically, the plot of q versus $t^{1/2}$ should give a straight line passing through the origin. The slope of the straight line provides the intraparticle diffusion rate constant K_{WM}. However, detection of multi linear plots from adsorption data may imply the occurrence of some other mechanisms during the adsorption process. Accordingly, a significant external resistance to mass transfer surrounding the particles in the early stage of adsorption is evidenced by the first shaper portion. Meanwhile, the intraparticle diffusion happens in the second linear portion, which is a steady adsorption stage. Eventually, the final equilibrium stage is achieved when the rate of intraparticle diffusion becomes slow associated to extremely low solute concentration in solution in the third portion. Apparently, a good correlation of rate data in this model is important to rationalize the adsorption mechanism, subsequently to determine K_{WM} values by linearization the curve $q = f(t^{0.5})$.

The rate of diffusion within the particle is much slower compared to that of movement of the adsorbate from solution to the external solid surface. This is because of the restraining chemical attractions between adsorbate and adsorbent as well as the greater mechanical hindrance to movement exist at the surface molecules or surface layers. Usually, adsorbate molecules reach at the adsorbent surface more fast than they can diffuse into the solid adsorbent, resulting in accumulation of the adsorbate at the surface. Therefore, a (pseudo)-equilibrium may be established. Nevertheless, further adsorption of adsorbate can occur at the same rate as the surface concentration is reduced by inward adsorption.

8.4.4 Continuous Packed-Bed System in Biosorption of Heavy Metals

The batch adsorption method is only application for adsorption system involving small volumes of adsorbate. Thus, continuous flow treatments should be applied for large scale application of biosorption process. In this method, adsorbates in solution are flown continuously from either the top or the bottom of a stationary bed of solid adsorbent which is more recognized as column. Since this method has closer simulation of commercial systems in industry, it is commonly used in the assessment of the suitability of an adsorbent for a particular adsorbate. After the adsorption process, the adsorbent can be easily separated from the effluent and the used adsorbent is usually regenerated. Among the different experimental set-ups, the packed-bed column is possibly the most effective device for continuous operations.

In a down-flow packed bed column, the adsorbate solution is in contact with adsorbent when it moves through the column. As flowing down the column, adsorption process occurs where most of the adsorbate is adsorbed gradually from liquid onto the adsorbent. As a result, concentration of adsorbate in the effluent is either very low or even untraceable at the end of the column. Depending on the value of adsorbate concentration desired for its lower boundary, the length of the adsorption zone in the column is somewhat random. When more adsorbate solution is fed into the column, concentration of adsorbate in the effluent increases slowly due to equilibrium and kinetic factors. Usually, the adsorption zone will shift down the column like a slowly moving wave when the upper portion of packing adsorbent is saturated with adsorbate. Ultimately, the lower edge of the adsorption zone reaches the bottom of the column, resulting in a significant rise in adsorbate concentration in the effluent. As little additional adsorption takes place with the entire bed approaching an equilibrium state with the feed at this breakthrough point, the flow is stopped. However, one should realize that the equilibrium between adsorption and desorption process is seldom achieved under this continuous flow condition. In most cases, the breakthrough point increases with increasing of bed height, reducing of adsorbent's particle size and deceasing of flow rate.

A breakthrough curve, which is plot of adsorbate effluent concentration versus time, is applicable for description of continuous packed bed performance. The general position of the breakthrough curve along the time or volume axis is indicative of loading behaviour of the adsorbate to be removed from a solution in a fixed bed. A normalized concentration as a function of time or volume of effluent (V_{eff}) for a given bed height is used to represent the breakthrough position. In real practical, the normalized concentration is referred to the ratio of effluent adsorbate concentration to inlet adsorbate concentration (C/C_0). An adsorption isotherm is favoured with infinite adsorption rate the breakthrough curve approaches a straight vertical line. The breakthrough curve becomes less sharp as decreasing of the mass transfer rate. Since mass transfer is always finite, a breakthrough curve of S-shape would be expected. Both the breakthrough point and breakthrough curve are greatly influenced by nature of the adsorbate and the adsorbent, geometry of the column and operating conditions.

In order to evaluate and predict the dynamic behavior of a continuous packed-bed system, a number of simple mathematical models underlying different assumptions have been established. Some commonly used models in characterization of the fixed bed performance for biosorption process are presented here.

Adams–Bohart Model

The Adams-Bohart model was developed by assuming the adsorption rate is proportional to adsorbent concentration and its residual capacity (Bohart and Adams 1920). This model was originally applied to describe the relationship between C/C_o and t for the adsorption of chlorine on charcoal in fixed bed column. With the assumption of the adsorption rate is limited by external mass transfer, the model equation is expressed below:

$$\ln \frac{C}{C_0} = k_{AB} C_0 t - k_{AB} N \frac{Z}{U_0} \tag{8.40}$$

where $C =$ adsorbate concentration remaining at each contact time

$C_0 =$ initial adsorbate concentration

$k_{AB} =$ Adams–Bohart kinetic constant

$N =$ adsorbate concentration in the bulk liquid

$Z =$ bed depth of column

$U_0 =$ linear velocity calculated by dividing the flow rate by the column's sectional area

Equation 8.40 is generally valid in the initial part of the breakthrough as it is derived based on the assumption of low concentration field where $C < 0.15C_o$. Thus, this model is always utilized in describing the initial part of the breakthrough curve only. The plot of ln C/C_0 against t at a given bed height and flow rate provides values that characterizes operational parameters of the column. As increasing of t, concentration of metal in the bulk liquid tends to approach a saturation condition where maximum capacity of adsorption is achieved.

The Adams-Bohart model is applicable in fixed bed column of different biosorption applications due to its ability to provide solution of the differential equations for mass transfer rate in solid and liquid phases. Consequently, its overall approach is suitable for quantitative description of other systems regardless the phase of adsorbate.

Bed Depth/Service Time Model

The Bed-Depth-Service-Time (BDST) model is often used to examine the capacity of fixed bed at various breakthrough values (Bohart and Adams 1920). By assuming a direct adsorption of adsorbate onto the surface of the adsorbent, this model correlates the service time (t) with the process variables by ignoring intraparticle mass resistance and external film resistance. The BDST model states that the service time of a column is given by:

$$t = \frac{N_0}{C_0 U_0} Z - \frac{1}{K_a C_0} \ln \left(\frac{C_0}{C} - 1 \right) \tag{8.41}$$

where $N_0 =$ adsorption capacity

$K_a =$ rate constant in BDST

At 50 % breakthrough $\left(\frac{C_o}{C}\right) = 2$, and $t = t_{0.5,}$ the Eq. 8.41 can be written as

$$t_{0.5} = \left(\frac{N_o}{C_o U_o}\right) Z \qquad (8.42)$$

or

$$t_{0.5} = \text{constant} \times \text{Z} \qquad (8.43)$$

A straight line passing through the origin is expected in a plot of BDST at 50 % breakthrough against bed depth using Eq. 8.43 if the adsorption data fits the model. Obviously, this model has solved the constraint in Adams and Bohard model which is only appropriate for initial part of the breakthrough.

Yoon–Nelson Model

Yoon-Nelson (Y-N) model suggests the decrease rate in the probability of adsorption for each adsorbate molecule is comparative to the probability of adsorption of adsorbate and the probability of adsorbate breakthrough on the adsorbent (Yoon and Nelson 1984). This model can be applied without knowing the detailed information about adsorbent type, characteristics of both adsorbate and adsorbent, and physical properties of adsorption bed adsorbent. Therefore, it appears as a reasonably simple theoretical model. Expression of the Y-N equation for a single-component system is given as:

$$\ln \frac{C}{C_0 - C} = k_{YN} t - \tau k_{YN} \qquad (8.44)$$

where t = time required for 50 % adsorbate breakthrough
τ = breakthrough (sampling) time
k_{YN} = Yoon and Nelson rate constant

The parameters τ and k_{YN} for the adsorbate can be obtained from the intercept and slope, respective of the ln $C/(C_0 - C)$ versus sampling time (t) plot. These values are important for the calculation of theoretical breakthrough curves for a single-component system.

Thomas Model

The Thomas model is widely used for determination of rate constant for the maximum adsorption capacity of an adsorbate in a column. This model is derived based on assumption of Langmuir kinetics of adsorption-desorption and no axial dispersion (Thomas 1944). The Thomas equation takes into account of the rate driving force obeys second order reversible reaction kinetics, and it is written as:

$$\frac{C}{C_0} = \frac{1}{1 + \exp(k_{Th}/Q(q_0X - C_0V_{eff}))} \tag{8.45}$$

Where k_{Th} = Thomas rate constant
Q = flow rate
q_0 = maximum solid-phase concentration of the solute
X = amount of adsorbent in the column
V_{eff} = effluent volume
The Thomas equation can also be presented in linear form as below:

$$\ln\left(\frac{C_0}{C} - 1\right) = \frac{k_{Th}q_0X}{Q} - \frac{k_{TH}C_0}{Q}V_{eff} \tag{8.46}$$

By plotting ln $[(C_0/C) - 1]$ versus t at a given flow rate, the kinetic coefficient k_{Th} and the adsorption capacity of the bed q_0 can be determined. The Thomas model allows the prediction of the concentration-time profile or breakthrough curve for the effluent, hence contributing to a successful design of a column sorption process.

Clark Model

Clark (Clark 1987) proposed a new simulation for study of a continuous packed-bed system based on the (i) hypothesis of the flow is piston type (Hamdaoui 2006); and (ii) the combination of both mass transfer concept and Freundlich adsorption isotherm (Ayoob and Gupta 2007). By ignoring the phenomenon of dispersion, the equation is expressed as:

$$\frac{C}{C_0} = \left(\frac{1}{1 + Ae^{-rt}}\right)^{1/n-1} \tag{8.47}$$

with

$$\text{A} = \left(\frac{C_0^{n-1}}{C_{break}^{n-1}} - 1\right)e^{rt_{break}} \tag{8.48}$$

and

$$R(n-1) = r \text{ and } R = \frac{K_{Cl}}{U_0}v \tag{8.49}$$

where C_{break} = outlet concentration at breakthrough (or limit effluent concentration)
t_{break} = time at breakthrough
n = Freundlich parameter
k_{Cl} = Clark rate constant
v = migration rate

Both constants of the model A and r of a particular adsorption process on a fixed bed and a selected treatment objective can be obtained from Eq. 8.49 by non-linear regression analysis. These values are critical for the prediction of the breakthrough curve according to the relationship between C/C_0 and t in Eq. 8.47.

Wolborska Model

Wolborska model is a simplified adsorption representation which emphasizes on the general mass transfer for the diffusion mechanism for low concentration range only for breakthrough curve (Wolborska 1989). Due to its limitation in describing the whole adsorption process, this model is seldom applied by the researchers. The equation for the mass transfer in the fixed bed adsorption is as follow:

$$\ln\left(\frac{C}{C_0}\right) = \frac{\beta_a C_0}{N_0}t - \frac{\beta_a Z}{U_0} \tag{8.50}$$

with

$$\beta_a = \frac{U_0^2}{2D}\left[\left(1 + \frac{4\beta_0 D}{U_0}\right)^{1/2} - 1\right] \tag{8.51}$$

where β_a = Kinetics coefficient of the external mass transfer
N_0 = saturation concentration
Z = height of column
U_0 = specific velocity
D = axial dispersion coefficient

Kinetics coefficient of the external mass transfer β_a can be estimated through the breakthrough curve. When the flow rate of solution through the bed is high or in short beds, axial diffusion D is negligible and $\beta_a = \beta_0$. The Wolborska model resembles Adams-Bohart model if the coefficient β_a/N_0 is equal to k_{AB}.

8.4.5 *Response Surface Methodology*

Response surface methodology (RSM) is a multivariate technique that mathematically fits the experimental domain studied in the theoretical design through a response function. It is used for designing experiments, building numerical models, evaluating the effects of variables and searching for the optimum combinations of factors for a system (Myers and Montgomery 2002).

Generally, the targeted response in a designed of experiments is governed by several independent variables. The response refers to the performance measure or quality characteristic of the product or process, whereas independent variables are sometimes called input variables. An experiment is a series of tests, called runs, in which changes are made in the input variables in order to identify the reasons for changes in the output response.

Conventional methods of investigating a process by varying one factor whilst maintaining all other factors involved at constant levels does not describe the combined effect of all the factors involved. This method is also time consuming and of low efficiency as it requires a number of experiments to determine the optimum levels, which are unreliable. These limitations of a classical method can be eliminated by optimising all the affecting parameters collectively by statistical technique such as RSM (Elibol 2002).

Two commonly used designs in RSM are central composite design (CCD) and Box-Behnken design (BBD) (Garg et al. 2008; Fu et al. 2009; Cerino Córdova et al. 2011). BBD is an independent quadratic design that does not contain an embedded factorial or fractional factorial design. In this design, the treatment combinations are at the center and midpoints of edges of the process space. These designs are either rotatable or near rotatable and require three levels of each factor. BBD is considered as an efficient option in RSM and an ideal alternative to CCD because it involved less input variables. Because of fewer treatment combinations, BBD is also a more time and energy saving design as compared to CCD.

Whilst CCD contains an imbedded factorial or fractional factorial design with center points that is augmented with a group of 'star points' that allow estimation of curvature. The star points represent new extreme values (low and high) for each factor in the design. The CCD is rotatable and requires five levels for each factor.

The similarity in both designs is that it allows the estimation of all the regression parameters required and represent in quantitative form as linear (first-order polynomial model), quadratic or cubic function (higher order polynomial models). The equations below show the linear and quadratic models which are most commonly used in RSM. A first-order model with n experimental runs carrying out on k design variables and a single response y can be expressed as follows:

$$y = \beta_0 + \beta_1 x_{i1} + \beta_2 x_{i2} + \ldots\ldots\ldots + \beta_k x_{ik} + \varepsilon_i (i = 1, 2, \ldots\ldots, n) \quad (8.52)$$

where y (predicted response) is a function of the design variables $x_1, x_2, \ldots, x_j$, β_0 is the constant coefficient, $\beta_1, \beta_{2\ldots}, \beta_j$ is the linear coefficients, and ε is the experimental error. This model is appropriate for describing a flat surface with or without tilted surfaces.

When there is a curvature in the response surface, the first-order model is insufficient to describe the response and in this case, a higher order polynomial such as the quadratic model may be used. This is represented by Eq. 8.53.

$$y = \beta_o + \sum_{i=0}^{n} \beta_i X_i + \sum_{i<j} \sum \beta_{ij} X_i X_j \sum_{i=1}^{n} \beta_{ii} X_i^2 + \in \quad (8.53)$$

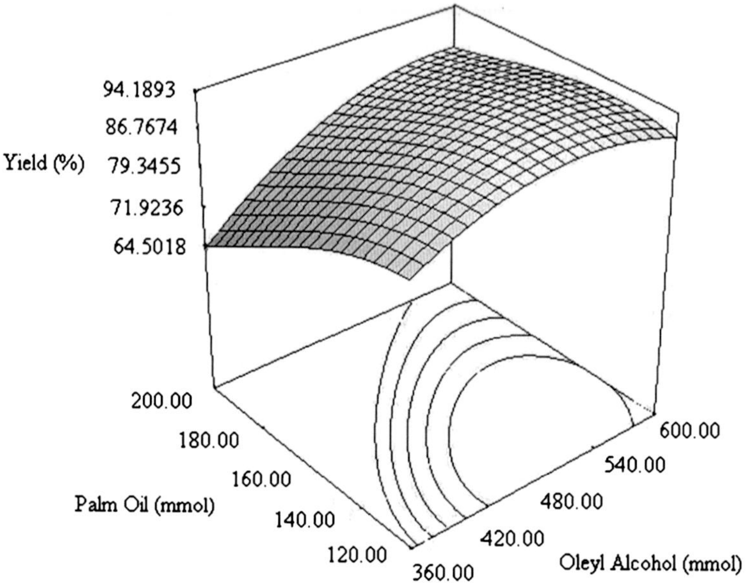

Fig. 8.1 3D response surface plot showing effect of amount of substrates on the percentage yield of palm-based wax ester production (Keng 2008)

where y is the predicted response, β_0 is the constant coefficients, β_I is the linear coefficients of the input factor x_i, β_{ij} is the quadratic coefficients between the input factor x_i and x_j, β_{ii} is the interaction coefficients of input factor x_i and ε is the experimental error. (Box and Hunter 1957). The second-order model is important to provide a good prediction throughout the region of interest. It is rotatable so the variance of the predicted response is the same at all points. Since the objective of RSM is optimisation and as the location of the optimum is unknown prior to running the experiment, it makes sense to use the design that provides equal precisions of estimation in all directions (Tan et al. 2008).

Generally, to perform experimental optimisation with RSM, it involves three major steps: (1) design and perform the experiments; (2) response surface modeling through regression analysis; (3) optimisation and checking the adequacy of the model. In the first step, important independent input variables and desired responses with the design constrains are identified and experimental design are adopted. Such designs are generated using computer software like Statistica, SAS or Design Expert. This is followed by regression analysis and statistical analysis of variance (ANOVA) of the experimental data in order to identify parameters that are significantly affects the desired response and develop a mathematical model having the best fit to the data obtained. Lastly, optimal combination of parameters is identified and confirmation experiments are conducted to verify the optimal parameters and mathematical model generated. For visualisation of the investigated factors, contour and response surface plots in 3D are generated that show the linear, interaction and quadratic effects of two or more parameters (Fig. 8.1).

8.5 Conclusion

Removal of heavy metals from aqueous environment is indeed a challenging problem in the control of environmental pollution. A wide range of treatment technologies such as chemical precipitation, ion exchange, membrane filtration and adsorption can be employed to remove heavy metals but these treatment methods still suffer from some inherent limitations. For instance, chemical precipitation, which has been traditionally and widely used to remove heavy metals from contaminated sources, is often associated with huge sludge production. Ion exchange process is generally applicable to treat or remove heavy metals at low concentration. As for membrane filtration, problems such as membrane fouling, high maintenance cost and low permeate flux are often unavoidable. Taking all these into considerations, the adsorption process thus appeared as one of the most promising methods.

Various low cost adsorbents have been investigated intensively in order to avoid the usage of costly activated carbon. It is evident that in most of the reported works, the study is still being centralised in the maximum adsorption capacities of the materials. While most of the low cost adsorbents have shown its potential in heavy metals removal without any modifications, but it is interesting to note that their removal capabilities can generally be improved through certain chemical modifications. Also, more than one factors such as pH, contact time, initial influent concentration, temperature, adsorbent dosage and etc. can significantly affect the adsorption process.

Numerous equations based on two-parameter or three parameter adsorption isotherms have been employed to describe the adsorption process under equilibrium condition. These include Langmuir, Freundlich, Brunauer-Emmet-Teller (BET), Temkin, Dubinin-Radushkevich, Jovanovic, Hasley, Harkin-Jura, Redlich-Peterson, Sips and Toth models. As for the kinetics studies, the two commonly used rate expressions are the pseudo-first and pseudo-second order kinetic models. In most of the cases, especially those involve chemisorption, pseudo-second order kinetic model is able to provide a better description than the pseudo-first order kinetic model as the former takes the interaction of adsorbent-adsorbate through valency forces into consideration.

To overcome the limitation of batch condition which is applicable only to the treatment under equilibrium condition, it is necessary to carry out flow tests using columns to obtain a design models which would be applicable to commercial systems. Besides this, developing an appropriate mathematical model through statistical approach is also important to obtain the optimum operating condition of the adsorption system.

Several low cost adsorbents can be seen as attractive alternatives for the removal of heavy metals based on their outstanding removal ability. However, in most of the cases, attention seems to be focused on the maximum adsorption capacity of the low cost adsorbents and the model pollutants are mainly synthetic types. In order to fully utilize the application of these materials in wastewater treatment, their removal efficiency should be further tested in effluents from industries. Besides, the characterization of the adsorbents as well as the study of the sorption mechanism is equally important. More effort is required to provide insight into the direction of modeling, the effects of adsorbent recycling and regeneration and the recovery of heavy metals from wastewater. The possibility of immobilized these adsorbents can also be viewed as a feasible approach as suspended system requires a filtration step which is not suited for practical applications. If modification is needed, it is desirable to have a material with a wide range of metal affinities as this will be particularly useful for industrial effluents which carry more than one type of metals. Further research involving the combination of methodologies/system should also be encouraged as this would minimize or overcome the problem associated with a single treatment process.

References

Abo-El-Enein SA, Eissa MA, Diafullah AA, Rizk MA, Mohamed FM (2009) Removal of some heavy metals ions from wastewater by copolymer of iron and aluminum impregnated with active silica derived from rice husk ash. J Hazard Mater 172:574–579

Abo-Farha SA, Abdel-Aal AY, Ashour IA, Garamon SE (2009) Removal of some heavy metal cations by synthetic resin purolite C100. J Hazard Mater 169:190–194

Afkhami A, Saber-Tehrani M, Bagheri H (2010) Simultaneous removal of heavy-metal ions in wastewater samples using nano-alumina modified with 2,4 dinitrophenylhydrazine. J Hazard Mater 181:836–844

Agency for Toxic Substances and Disease Registry (ATSDR) (2005) Toxicological profile for nickel. U.S. Department of Health and Human Services, Public Health Service, Atlanta

Agency for Toxic Substances and Disease Registry (ATSDR) (2007) Toxicological profile for lead. U.S. Department of Health and Human Services, Public Health Services, Atlanta

Agency for Toxic Substances and Disease Registry (ATSDR) (1993) Toxicological profile for nickel (update). US Department of Health and Human Services, Public Health Service, Washington, DC, pp 1–147

Ajmal M, Rao RAK, Anwar S, Ahmad J, Ahmad R (2003) Adsorption studies on rice husk: removal and recovery of Cd(II) from wastewater. Bioresour Technol 86:147–149

Akhtar M, Iqbal S, Kausar A, Bhanger MI, Shaheen MA (2010) An economically viable method for the removal of selected divalent metal ions from aqueous solutions using activated rice husk. Colloids Surf B Biointerfaces 75:149–155

Aksu Z, Açikel Ü, Kutsal T (1997) Application of multicomponent adsorption isotherms to simultaneous biosorption of iron(III) and chromium(VI) on C. vulgaris. J Chem Technol Biotechnol 70:368–378

Al-Enezi G, Hamoda MF, Fawzi N (2004) Ion exchange extraction of heavy metals from wastewater sludges. J Environ Sci Health A Tox Hazard Subst Environ Eng 39(2):455–464

Apak R, Tutem E, Hugul M, Hizal J (1998) Heavy metal cation retention by unconventional sorbents (red muds and fly ashes). Water Res 32:430–440

Apiratikul R, Pavasant P (2008a) Batch and column studies of biosorption of heavy metals by Caulerpa lentillifera. Bioresour Technol 99:2766–2777

Apiratikul R, Pavasant P (2008b) Sorption of Cu^{2+}, Cd^{2+}, and Pb^{2+} using modified zeolite from coal fly ash. Chem Eng J 144:245–258

Areco MM, Hanela S, Duran J, Afonso MS (2012) Biosorption of Cu(II), Zn(II), Cd(II) and Pb(II) by dead biomasses of green alga Ulva lactuca and the development of a sustainable matrix for adsorption implementation. J Hazard Mater 213–214:123–132

Aroua MK, Mohamed Zuki F, Sulaiman NM (2007) Removal of chromium ions from aqueous solutions by polymer-enhanced ultrafiltration. J Hazard Mater 147:752–758
Ayala J, Blanco F, Garcia P, Rodriguez P, Sancho J (1998) Asturian fly ash as a heavy metals removal material. Fuel 77:1147–1154
Aydın H, Bulut Y, Yerlikaya C (2008) Removal of copper (II) from aqueous solution by adsorption onto low-cost adsorbents. J Environ Manage 87:37–45
Ayoob S, Gupta AK (2007) Sorptive response profile of an adsorbent in the defluoridation of drinking water. Chem Eng J 133:273–281
Bade L (2007) Micellar enhanced ultrafiltration and activated carbon fibre hybrid processes for copper removal from wastewater. Korean J Chem Eng 24(2):239–245
Banerjee SS, Jayaram RV, Joshi MV (2003) Removal of nickel and zinc(II) from wastewater using fly ash and impregnated fly ash. Sep Sci Technol 38:1015–1032
Banerjee SS, Joshi MV, Jayaram RV (2004) Removal of Cr(VI) and Hg(II) from aqueous solution using fly ash and impregnated fly ash. Sep Sci Technol 39:1611–1629
Bansal RC, Goyal M (2005) Activated carbon adsorption. CRC Press, Boca Raton
Bansal M, Garg U, Singh D, Garg VK (2009) Removal of Cr(VI) from aqueous solutions using pre-consumer processing agricultural waste: a case study of rice husk. J Hazard Mater 162:312–320
Barakat MA, Schmidt E (2010) Polymer-enhanced ultrafiltration process for heavy metals removal from industrial wastewater. Desalination 256:90–93
Barreira LD, Lito PF, Antunes BM, Otero M, Lin Z, Rocha J, Pereira E, Duarte AC, Silva CM (2009) Effect of pH on cadmium (II) removal from aqueous solution using titanosilicate ETS-4. Chem Eng J 155:728–735
Basci N, Kocadagistan E, Kocadagistan B (2004) Biosorption of copper (II) from aqueous solutions by wheat shell. Desalination 164:135–140
Basso MC, Cerrella EG, Cukierman AI (2002) Empleo de algas marinas para la biosorción de metales pesados de aguas contaminadas. Avances en Energías Renovables y Medio Ambiente 6:69–74
Bayat B (2002a) Combined removal of Zinc (II) and Cadmium (II) from aqueous solutions by adsorption onto high calcium Turkish fly ash. Water Air Soil Pollut 136:6992
Bayat B (2002b) Comparative study of adsorption properties of Turkish fly ashes: II. The case of chromium (VI) and cadmium (II). J Hazard Mater 95:275–290
Bayat B (2002c) Comparative study of adsorption properties of Turkish fly ashes: I. The case of nickel(II), copper(II) and zinc(II). J Hazard Mater 95:251–273
Beolchini F, Pagnanelli F, Toro L, Vegliò F (2006) Ionic strength effect on copper biosorption by Sphaerotilus natans: equilibrium study and dynamic modelling in membrane reactor. Water Res 40:144–152
Bessbousse H, Rhlalou H, Verchére JF, Lebrun L (2008) Removal of heavy metal ions from aqueous solutions by filtration with a novel complexing membrane containing poly(ethyleneimine) in a poly(vinyl alcohol) matrix. J Membr Sci 307:249–259
Bhattacharya AK, Mandal SN, Das SK (2006) Adsorption of Zn(II) from aqueous solution by using different adsorbents. Chem Eng J 123:43–51
Bhattacharya AK, Naiya TK, Mandal SN, Das SK (2008) Adsorption, kinetics and equilibrium studies on removal of Cr(VI) from aqueous solutions using different low-cost adsorbents. Chem Eng J 137:529–541
Bhattacharyya D, Jumawan AB, Grieves RB (1979) Separation of toxic heavy metals by sulphide precipitation. Sep Sci Technol 14:441–452
Blocher C, Dorda J, Mavrov V, Chmiel H, Lazaridis NK, Matis KA (2003) Hybrid flotation – membrane filtration process for the removal of heavy metalions from wastewater. Water Res 37:4018–4026
Bohart G, Adams EQ (1920) Some aspects of the behaviour of charcoal with respect to chlorine. J Am Chem Soc 42:523–544
Bouranene S, Fievet P, Szymczyk A, Samar ME, Alain Vidonne A (2008) Influence of operating conditions on the rejection of cobalt and lead ions in aqueous solutions by a nanofiltration polyamide membrane. J Membr Sci 325:150–157

Box GEP, Hunter JS (1957) Multi factor experimental designs for exploring response surfaces. Ann Math Stat 28:195–241

Brinza L, Dring MJ, Gavrilescu M (2007) Marine micro- and macro-algal species as biosorbents for heavy metals. Environ Eng Manage J 6:237–251

Brunauer S, Emmett PH, Teller E (1938) Adsorption of gases in multimolecular layers. J Am Chem Soc 60:309–319

Bulut Y, Baysal Z (2006) Removal of Pb(II) from wastewater using wheat bran. J Environ Manage 78:107–113

Celaya RJ, Noriega JA, Yeomans JH, Ortega LJ, Ruiz-Manríquez A (2000) Biosorption of Zn(II) by Thiobacillus ferrooxidans. Bioprocess Eng 22:539–542

Cerino Córdova FJ, García León AM, Garcia Reyes RB, Garza González MT, Soto Regalado E, Sánchez González MN, Quezada Lopez I (2011) Response surface methodology for lead biosorption on Aspergillus terreus. Int J Environ Sci Technol 8(4):695–704

Cetin C, Pehlivan E (2007) The use of fly ash as a low cost, environmentally friendly alternative to activated carbon for the removal of heavy metals from aqueous solutions. Colloids Surf A Physicochem Eng Aspects 298:83–87

Chang JS, Law R, Chang CC (1997) Biosorption of lead, copper and cadmium by biomass of Pseudomonas aeruginosa PU21. Water Res 31:1651–1658

Chatterjee SK, Bhattacharjee I, Chandra G (2010) Biosorption of heavy metals from industrial waste water by Geobacillus thermodenitrificans. J Hazard Mater 175:117–125

Chen XC, Wang YP, Lin Q, Shi JY, Wu WX, Chen YX (2005) Biosorption of copper(II) and zinc(II) from aqueous solution by Pseudomonas putida CZ1. Colloids Surf B Biointerfaces 46:101–107

Chen QY, Luo Z, Hills C, Xue G, Tyrer M (2009) Precipitation of heavy metals from wastewater using simulated flue gas: sequent additions of fly ash, lime and carbon dioxide. Water Res 43:2605–2614

Cheng H (2006) Cu(II) removal from lithium bromide refrigerant by chemical precipitation and electrocoagulation. Sep Purif Technol 52:191–195

Cheung CW, Porter JF, McKay G (2001) Sorption kinetic analysis for the removal of cadmium ions from effluents using bone char. Water Res 35(3):605–612

Chien SH, Clayton WR (1980) Application of Elovich equation to the kinetics of phosphate release and sorption on soils. Soil Sci Soc Am J 44:265–268

Chipasa KB (2003) Accumulation and fate of selected heavy metals in a biological wastewater treatment system. Waste Manag 23:135–143

Choi SB, Yun YS (2004) Lead biosorption by waste biomass of Corynebacterium glutamicum generated from lysine fermentation process. Biotechnol Lett 26:331–336

Chojnacka K (2006) Biosorption of Cr(III) ions by wheat straw and grass: a systematic characterization of new biosorbents. Pol J Environ Stud 15:845–852

Chong KH, Volesky B (1995) Description of two-metal biosorption equilibria by Langmuir-type models. Biotechnol Bioeng 47:451–460

Cifuentes L, García I, Arriagada P, Casas JM (2009) The use of electrodialysis for metal separation and water recovery from $CuSO_4$–H_2SO_4–Fe solutions. Sep Purif Technol 68:105–108

Clark RM (1987) Evaluating the cost and performance of field-scale granular activated carbon systems. Environ Sci Technol 21:573–580

Coelho TC, Laus R, Mangrich AS, Fa'vere VT de, Laranjeira MCM (2007) Effect of heparin coating on epichlorohydrin cross-linked chitosan microspheres on the adsorption of copper (II) ions. React Funct Polym 67:468–475

Cossich ES, Tavares CRG, Ravagnani TMK (2002) Biosorption of chromium(III) by Sargassum sp. Biomass. Electron J Biotechnol 5:44–52

Dąbrowski A, Hubicki Z, Podkościelny P, Robens E (2004) Selective removal of the heavy metal ions from waters and industrial wastewaters by ion-exchange method. Chemosphere 56:91–106

Dang VBH, Doan HD, Dang-Vu T, Lohi A (2009) Equilibrium and kinetics of biosorption of cadmium (II) and copper (II) ions by wheat straw. Bioresour Technol 100:211–219

Davis TA, Volesky B, Vieira RHSF (2000) Sargassum seaweed as biosorbent for heavy metals. Water Res 34:4270–7278
De Vargas I, Macaskie LE, Guibal E (2004) Biosorption of palladium and platinum by sulfate-reducing bacteria. J Chem Technol Biotechnol 79:49–56
Dean JG, Bosqui FL, Lanouette KH (1972) Removing heavy metals from wastewaters. Environ Sci Technol 6(6):518–522
Dhir B, Kumar R (2010) Adsorption of heavy metals by Salvinia biomass and agricultural residues. Int J Environ Res 4:427–432
Dialynas E, Diamadopoulos E (2009) Integration of a membrane bioreactor coupled with reverse osmosis for advanced treatment of municipal wastewater. Desalination 238:302–311
Diamadopoulos E, Loannidis S, Sakellaropoulos GP (1993) As(V) removal from aqueous solutions by fly ash. Water Res 27:1773–1777
Dinu MV, Dragan ES (2010) Evaluation of Cu2+, Co2+, and Ni2+ ions removal from aqueous solution using a novel chitosan/clinoptilolite composites: kinetics and isotherms. Chem Eng J 160:157–163
Djedidi Z, Bouda M, Souissi MA, Cheikh RB, Mercier G, Tyagi RD, Blais J (2009) Metals removal from soil, fly ash and sewage sludge leachates by precipitation and dewatering properties of the generated sludge. J Hazard Mater 172:1372–1382
Djeribi R, Hamdaoui O (2008) Sorption of copper(II) from aqueous solutions by cedar sawdust and crushed brick. Desalination 225:95–112
Dönmez GC, Aksu Z, Öztürk A, Kutsal T (1999) A comparative study on heavy metal biosorption characteristics of some algae. Process Biochem 34:885–892
Dragan ES, Dinu MV, Timpu D (2010) Preparation and characterization of novel composites based on chitosan and clinoptilolite with enhanced adsorption properties for Cu2+. Bioresour Technol 101:812–817
Dubinin MM, Radushkevich LV (1947) Equation of the characteristic curve of activated charcoal. Proc Acad Sci Phys Chem USSR 55:331–333
Dupont L, Guillon E (2003) Removal of hexavalent chromium with a lignocellulosic substrate extracted from wheat bran. Environ Sci Technol 37:4235–4241
Dupont L, Bouanda J, Dumonceau J, Aplincourt M (2005) Biosorption of Cu(II) and Zn(II) onto a lignocellulosic substrate extracted from wheat bran. Environ Chem Lett 2:165–168
Elibol M (2002) Response surface methodological approach for inclusion of perfluorocarbon in actinorhodin fermentation medium. Process Biochem 38:667–773
Elouear Z, Bouzid Z, Boujelben J, Feki N, Jamoussi M, Montiel FA (2008) Heavy metal removal from aqueous solutions by activated phosphate rock. J Hazard Mater 156:412–420
Fan DH, Zhu XM, Xu MR, Yan JL (2006) Adsorption properties of chromium (VI) by chitosan coated montmorillonite. J Biol Sci 6:941–945
Farajzadeh MA, Monji AB (2004) Adsorption characteristics of wheat bran towards heavy metal cations. Sep Purif Technol 38:197–207
Farooq U, Khan MA, Atharc M, Kozinskia JA (2011) Effect of modification of environmentally friendly biosorbent wheat (Triticum aestivum) on the biosorptive removal of cadmium(II) ions from aqueous solution. Chem Eng J 171:400–410
Febrianto J, Kosasih AN, Sunarso J, Ju YS, Indraswati N, Ismadi S (2009) Equilibrium and kinetic studies in adsorption of heavy metals using biosorbent: a summary of recent studies. J Hazard Mater 162:616–645
Feng Q, Lin Q, Gong F, Sugita S, Shoya M (2004) Adsorption of lead and mercury by rice husk ash. J Colloid Interface Sci 278:1–8
Ferella F, Prisciandaro M, De Michelis I, Veglio' F (2007) Removal of heavy metals by surfactant-enhanced ultrafiltration from wastewaters. Desalination 207:125–133
Feroze N, Ramzan N, Khan A, Cheema II (2011) Kinetic and equilibrium studies for Zn (II) and Cu (II) metal ions removal using biomass (rice husk) Ash. J Chem Soc Pak 33:139–146
Freundlich H (1906) Adsorption in solution. Phys Chem Soc 40:1361–1368
Fu F, Wang Q (2011) Removal of heavy metal ions from wastewaters: a review. J Environ Manage 92:407–418

Fu JF, Zhao YQ, Xue XD, Li WC, Babatunde AO (2009) Multivariate-parameter optimization of acid blue-7 wastewater treatment by Ti/TiO_2 photoelectrocatalysis via Box-Behnken design. Desalination 243(1–3):42–51

Gan Q (2000) A case study of microwave processing of metal hydroxide sediment sludge from printed circuit board manufacturing wash water. Waste Manage 20:695–701

Garg UK, Kaur MP, Garg VK, Sud D (2008) Removal of nickel (II) from agueous solution by agricultural waste biomass using a response surface methodological approach. Bioresour Technol 99(5):1325–1331

Ghorbani M, Lashkenari MS, Eisazadeh H (2011) Application of polyaniline nanocomposite coated on rice husk ash for removal of Hg(II) from aqueous media. Synth Met 161:1430–1433

Ghorbani M, Eisazadeh H, Ghoreyshi AA (2012) Removal of zinc ions from aqueous solution using polyaniline nanocomposite coated on rice husk. Iran J Energy Environ 3(1):66–71

Ghosh P, Samanta AN, Ray S (2011) Reduction of COD and removal of Zn^{2+} from rayon industry wastewater by combined electro-Fenton treatment and chemical precipitation. Desalination 266:213–217

Gong R, Guan R, Zhao J, Liu X, Ni S (2008) Citric acid functionalizing wheat straw as sorbent for copper removal from aqueous solution. J Health Sci 54:174–178

González-Muñoz MK, Rodrguez MA, Luque S, Álvarez JR (2006) Recovery of heavy metals from metal industry wastewaters by chemical precipitation and nanoflitration. Desalination 200(1–3):742–744

Goswami D, Das AK (2000) Removal of arsenic from drinking water using modified fly-ash bed. Int J Water 1:61–70

Greenwood NN, Earnshaw A (1989) Chemistry of the elements. Pergamon Press, New York

Guo Y, Qi J, Yang S, Yu K, Wang Z, Xu H (2002) Adsorption of Cr(VI) on micro- and mesoporous rice husk-based active carbon. Mater Chem Phys 78:132–137

Gupta VK (1998) Equilibrium uptake, sorption dynamics, process development, and column operations for the removal of copper and nickel from aqueous solution and wastewater using activated slag, a low-cost adsorbent. Ind Eng Chem Res 37:192–202

Gupta VK, Ali I (2000) Utilisation of bagasse fly ash (a sugar industry waste) for the removal of copper and zinc from wastewater. Sep Purif Technol 18:131–140

Gupta VK, Ali I (2004) Removal of lead and chromium from wastewater using bagasse fly ash-a sugar industry waste. J Colloid Interface Sci 271:321–328

Gupta VK, Sharma S (2003) Removal of zinc from aqueous solutions using bagasse fly ash – a low cost adsorbent. Ind Eng Chem Res 42:6619–6624

Gupta VK, Ali I, Jain CK, Sharma M, Saini VK (2003) Removal of cadmium and nickel from wastewater using bagasse fly ash- a sugar industry waste. Water Res 37:4038–4044

Hall KR, Eagleton LC, Acrivos A, Vermeulen T (1966) Pore and solid-diffusion kinetics in fixed-bed adsorption under constant-pattern conditions. Ind Eng Chem Fund 5(2):212–223

Halsey G (1948) Physical adsorption on non-uniform surfaces. J Chem Phys 16:931–937

Hamdaoui O (2006) Dynamic sorption of methylene blue by cedar sawdust and crushed brick in fixed bed columns. J Hazard Mater 138:293–303

Harkins WD, Jura EJ (1944) The decrease of free surface energy as a basis for the development of equations for adsorption isotherms; and the existence of two condensed phases in films on solids. J Chem Phys 12:112–113

Hashim MA, Chu KH (2004) Biosorption of cadmium by brown, green, and red seaweeds. Chem Eng J 97:249–255

Häyrynen P, Landaburu-Aguirre J, Eva Pongrácz E, Keisk RL (2012) Study of permeate flux in micellar-enhanced ultrafiltration on a semi-pilot scale: simultaneous removal of heavy metals from phosphorous rich real wastewaters. Sep Purif Technol 93:59–66

Hindmarsh JT, McCurdy RF (1986) Clinical and environmental aspects of arsenic toxicity. Crit Rev Clin Lab Sci 23:315–347

Ho Y (2006) Isotherms for the sorption of lead onto peat: comparison of linear and non-linear methods. Pol J Environ Stud 15:81–86

Ho YS, McKay G (1999) Pseudo-second order model for sorption processes. Process Biochem 34:451–465

Ho Y, Ofomaja AE (2006) Biosorption thermodynamics of cadmium on coconut copra meal as biosorbent. Biochem Eng J 30:117–123
Ho GE, Mathew K, Newman PWG (1989) Leachate quality from gypsum neutralized red mud applied to sandy soils. Water Air Soil Pollut 47:1–18
Holan ZR, Volesky B (1994) Biosorption of lead and nickel by biomass of marine algae. Biotechnol Bioeng 43:1001–1009
Holan ZR, Volesky B, Prasetyo I (1993) Biosorption of cadmium by biomass of marine algae. Biotechnol Bioeng 41:819–825
Hossain MA, Kumita M, Michigami Y, More S (2005) Optimization of parameters for Cr(VI) adsorption on used black tea leaves. Adsorption 11:561–568
Hsu TC, Yu CC, Yeh CM (2008) Adsorption of Cu^{2+} from water using raw and modified coal fly ashes. Fuel 87:1355–1359
Huang GL, Zhang HY, Jeffrey XS, Tim AGL (2009) Adsorption of chromium(VI) from aqueous solutions using cross-linked magnetic chitosan beads. Ind Eng Chem Res 48:2646–2651
Hui KK, Chao CY, Kot SC (2005) Removal of mixed heavy metal ions in wastewater by zeolite 4A and residual products from recycled coal fly ash. J Hazard Mater 127:89–101
Hussein H, Ibrahim SF, Kandeel K, Moawad H (2004) Biosorption of heavy metals from waste water using Pseudomonas sp. Electron J Biotechnol 17:38–46
Incharoensakdi A, Kitjaharn P (2002) Zinc biosorption from aqueous solution by a halotolerant cyanobacterium Aphanothece halophytica. Curr Microbiol 45:261–264
Inglezakis VJ, Grigoropoulou H (2004) Effects of operating conditions on the removal of heavy metals by zeolite in fixed bed reactors. J Hazard Mater B112:37–43
Iyer RS, Scott JA (2001) Power station fly ash-a review of value-added utilization outside of the construction industry. Resour Conserv Recy 31:217–228
Jalali R, Ghafourian H, Asef Y, Davarpanah SJ, Sepehr S (2002) Removal and recovery of lead using nonliving biomass of marine algae. J Hazard Mater 92:253–262
Jaroniec M, Derylo A (1981) Application of Dubinin-Radushkevich-tyoe equation for describing bisolute adsorption from dilute aqueous solutions on activated carbon. J Colloid Interface 84(1):191–195
Johan NA, Kutty SRM, Isa MH, Muhamad NS, Hashim H (2011) Adsorption of copper by using microwave incinerated rice husk ash (MIRHA). Int J Civil Environ Eng 3:211–215
Jones DJ (2012) Primary prevention and health outcomes: treatment of residential lead-based paint hazards and the prevalence of childhood lead poisoning. J Urban Econ 71:151–164
Juang R, Xu Y, Chen C (2003) Separation and removal of metal ions from dilute solutions using micellar-enhanced ultrafiltration. J Membr Sci 218:257–267
Kaewsarn P (2002) Biosorption of copper(II) from aqueous solutions by pre-treated biomass of marine algae Padina sp. Chemosphere 47:1081–1085
Kaewsarn P, Yu Q (2001) Cadmium(II) removal from aqueous solutions by pre-treated biomass of marine alga Padina sp. Environ Pollut 112:209–213
Kalyani S, Ajitha PJ, Srinivasa RP, Krishnaiah A (2005) Removal of copper and nickel from aqueous solutions using chitosan coated on perlite as biosorbent. Sep Sci Technol 40:1483–1495
Kapoor A, Viraraghavan T (1992) Adsorption of mercury from wastewater by fly ash. Adsorp Sci Technol 9:130–147
Karthikeyan KG, Elliott HA, Cannon FS (1996) Enhanced metal removal from wastewater by coagulant addition. In: Proceedings of 50th Purdue Industrial Waste Conference. 50: 259–267.
Keng PS (2008) Scale-up production of palm-based wax esters using lipozyme rm im and characterisation of the esters. Ph.D. thesis, Universiti Putra Malaysia
Kim YH, Yeon Park J, Yoo YJ, Kwak JW (1999) Removal of lead using xanthated marine brown alga, Undaria pinnatifida. Process Biochem 34:647–652
Klimmek S, Stan HJ, Wilke A, Bunke G, Buchholz R (2001) Comparative analysis of the biosorption of cadmium, lead, nickel, and zinc by algae. Environ Sci Technol 35:4283–4288
Kousalya GN, Muniyappan RG, Sairam SC, Meenakshi S (2010) Synthesis of nano-hydroxyapatite chitin/chitosan hybrid biocomposites for the removal of Fe(III). Carbohydr Polym 82:549–599

Krishnani KK, Meng X, Christodoulatos C, Boddu VM (2008) Biosorption mechanism of nine different heavy metals onto biomatrix from rice husk. J Hazard Mater 153:1222–1234
Ku Y, Jung IL (2001) Photocatalytic reduction of Cr(VI) in aqueous solutions by UV irradiation with the presence of titanium dioxide. Water Res 35:135–142
Kumar U, Bandyopadhyay M (2006) Sorption of cadmium from aqueous solution using pretreated rice husk. Bioresour Technol 97:104–109
Kumar M, Bijay PT, Vinod KS (2009) Crosslinked chitosan/polyvinyl alcohol blend beads for removal and recovery of Cd(II) from wastewater. J Hazard Mater 172:1041–1048
Kurniawan TA, Chan YSG, Lo W, Babel S (2006) Physico–chemical treatment techniques for wastewater laden with heavy metals. Chem Eng J 118:83–98
Lacour S, Bollinger J, Serpaud B, Chantron P, Arcos R (2001) Removal of heavy metals in industrial wastewaters by ion-exchanger grafted textiles. Anal Chim Acta 428:121–132
Lagergren S (1898) Zur theorie der sogenannten adsorption gelöster stoffe, Kungliga Svenska Vetenskapsakademiens. Handlingar 24:1–39
Lambert J, Rakib M, Durand G, Avila-Rodríguez M (2006) Treatment of solutions containing trivalent chromium by electrodialysis. Desalination 191:100–110
Landaburu-Aguirre J, García V, Pongrácz E, Keiski R (2006) Applicability of membrane technologies for the removal of heavy metals. Desalination 200:272–273
Langmuir I (1918) The adsorption of gases on plane surfaces of glass, mica and platinum. J Am Chem Soc 40:1361–1368
Lau PS, Lee HY, Tsang CCK, Tam NFY, Wong YS (1999) Effect of metal interference, pH and temperature on Cu and Ni biosorption by Chlorella vulgaris and Chlorella miniata. Environ Technol 20:953–961
Lee HS, Volesky B (1999) Interference of aluminum in copper biosorption by an algal biosorbent. Water Qual Res J Can 34:519–533
Lee CK, Low KS, Liew SC, Choo CS (1999) Removal of arsenic(V) from aqueous solution by quaternized rice husk. Environ Technol 20:971–978
Lee IH, Kuan Y, Chern J (2006a) Factorial experimental design for recovering heavy metals from sludge with ion-exchange resin. J Hazard Mater B138:549–559
Lee J, Kwon T, Moon I (2006b) Performance of polyamide reverse osmosis membranes for steel wastewater reuse. Desalination 189:309–322
Lenntech. Arsenic – As. Website http://www.lenntech.com/periodic/elements/as.htm#ixzz1zEkpE6Pj. Accessed 30 June 2012
Lenntech. Lead – Pb. Website http://www.lenntech.com/periodic/elements/pb.htm#ixzz1zNv6HhkP. Accessed 6 July 2012
Lenntech. Nickel – Ni. Website http://www.lenntech.com/periodic/elements/ni.htm#ixzz1zIOSNjZx. Accessed 6 July 2012
Leusch A, Holan Z, Volesky B (1995) Biosorption of heavy metals (Cd, Cu, Ni, Pb, Zn) by chemically-reinforced biomass of marine algae. J Chem Technol Biotechnol 62:279–288
Lewis AE (2010) Review of metal sulphide precipitation. Hydrometallurgy 104:222–234
Lewis A, Van Hille R (2006) An exploration into the sulphide precipitation method and its effect on metal sulphide removal. Hydrometallurgy 81:197–204
Lin CJ, Chang JE (2001) Effect of fly ash characteristics on the removal of Cu(II) from aqueous solution. Chemosphere 44:1185–1192
Lin CC, Lai YT (2006) Adsorption and recovery of lead(II) from aqueous solutions by immobilized Pseudomonas Aeruginosa PU21 beads. J Hazard Mater 137:99–105
Lin X, Burns RC, Lawrance GA (2005) Heavy metals in wastewater: the effect of electrolyte composition on the precipitation of cadmium(II) using lime and magnesia. Water Air Soil Pollut 165:131–152
Lipczynska-Kochany E, Kochany J (2009) Effect of humate on biological treatment of wastewater containing heavy metals. Chemosphere 77:279–284
Liu HL, Chen BY, Lan YW, Cheng YC (2004) Biosorption of Zn(II) and Cu(II) by the indigenous Thiobacillus thiooxidans. Chem Eng J 97:195–201

Llorens J, Pujolà M, Sabaté J (2004) Separation of cadmium from aqueous streams by polymer enhanced ultrafiltration: a two-phase model for complexation binding. J Membr Sci 239:173–181

Loukidou MX, Karapantsios TD, Zouboulis AI, Matis KA (2004) Diffusion kinetic study of cadmium(II) biosorption by Aeromonas caviae. J Chem Technol Biotechnol 79:711–719

Lu WB, Shi JJ, Wang CH, Chang JS (2006) Biosorption of lead, copper and cadmium by an indigenous isolate *Enterobacter* sp. J1 possessing high heavy-metal resistance. J Hazard Mater 134:80–86

Lu Z, Maroto-Valer MM, Schobert HH (2008) Role of active sites in the steam activation of high unburned carbon fly ashes. Fuel 87:2598–2605

Lucaci D, Visa M, Duta A (2011) Copper removal on wood-fly ash substrates-thermodynamic study. Revue Roumaine de Chimie 56(10–11):1067–1074

Luo X, Deng Z, Lin X, Zhang C (2011) Fixed-bed column study for Cu^{2+} removal from solution using expanding rice husk. J Hazard Mater 187:182–189

Madden EF, Sexton MJ, Smith DR, Fowler BA (2002) Lead. In: Sarkar B (ed) Heavy metals in the environment. Marcel Dekker, New York

Madigan MT, Martinko JM, Parker J (2000) Brock biology of microorganisms, 9th edn. Pearson Prentice Hall, Upper Saddle River

Mameri N, Boudries N, Addour L, Belhocine D, Lounici H, Grib H, Pauss A (1999) Batch zinc biosorption by a bacterial nonliving Streptomyces rimosus biomass. Water Res 33:1347–1354

Marchioretto MM, Bruning H, Rulkens W (2005) Heavy metals precipitation in sewage sludge. Sep Sci Technol 40(16):3393–3405

Marshall WE, Champagne ET, Evans WJ (1993) Use of rice milling byproducts (hulls and bran) to remove metal ions from aqueous solution. J Environ Sci Health A28:1977–1992

Masoudzadeh N, Zakeria F, Lotfabad T, Sharafi H, Masoomi F, Zahiri HS, Ahmadian G, Noghabi KA (2011) Biosorption of cadmium by *Brevundimonas* sp. ZF12 strain, a novel biosorbent isolated from hot-spring waters in high background radiation areas. J Hazard Mater 197:190–198

Matheickal JT, Yu Q (1999) Biosorption of lead(II) and copper(II) from aqueous solutions by pretreated biomass of Australian marine algae. Bioresour Technol 69:223–229

Matheickal JT, Yu Q, Woodburn GM (1999) Biosorption of cadmium(II) from aqueous solutions by pre-treated biomass of marine alga Durvillaea potatorum. Water Res 33:335–342

Matlock MM, Howerton BS, Atwood DA (2002) Chemical precipitation of heavy metals from acid mine drainage. Water Res 36:4757–4764

McLaughlin MJ, Parker DR, Clarke JM (1999) Metals and micronutrients-food safety issues. Field Crop Res 60:143–163

Meunier N, Drogui P, Montané C, Hausler R, Mercier G, Blais J (2006) Comparison between electrocoagulation and chemical precipitation for metals removal from acidic soil leachate. J Hazard Mater B137:581–590

Mimoune S, Amrani F (2007) Experimental study of metal ions removal from aqueous solutions by complexation–ultrafiltration. J Membr Sci 298:92–98

Mimura H, Yokota K, Akiba K, Onodera Y (2001) Alkali hydrothermal synthesis of zeolites from coal fly ash and their uptake properties of cesium ion. J Nucl Sci Technol 38:766–772

Mirbagheri SA, Hosseini SN (2005) Pilot plant investigation on petrochemical wastewater treatment for the removal of copper and chromium with the objective of reuse. Desalination 171:85–93

Mishra SP, Tiwari D, Dubey RS (1997) The uptake behaviour of rice (Jaya) husk in the removal of Zn(II) ions-A radiotracer study. Appl Radiat Isot 48:877–882

Momčilović M, Purenović M, Bojić A, Zarubica A, Ranđelović M (2011) Removal of lead(II) ions from aqueous solutions by adsorption onto pine cone activated carbon. Desalination 276:53–59

Munaf E, Zein R (1997) The use of rice husk for removal of toxic metals from waste water. Environ Technol 18:359–362

Mungray AA, Kulkarni SV, Mungray AK (2012) Removal of heavy metals from wastewater using micellar enhanced ultrafiltration technique: a review. Cent Eur J Chem 10(1):27–46

Murthy ZVP, Chaudhari LB (2008) Application of nanofiltration for the rejection of nickel ions from aqueous solutions and estimation of membrane transport parameters. J Hazard Mater 160:70–77

Murthy ZVP, Chaudhari LB (2009) Separation of binary heavy metals from aqueous solutions by nanofiltration and characterisation of the membrane using Spiegler–Kedem model. Chem Eng J 150:181–187

Myers RH, Montgomery DC (2002) Response surface methodology: process and product optimization using experiments, 2nd edn. Wiley, New York

Nadzi BS, Nyahumwa C, Tesha J (2007) Chemical and thermal stability of rice husks against alkali treatment. BioResources 3(4):1267–1277

Naiya TK, Bhattacharya AK, Mandal S, Das SK (2009) The sorption of lead(II) ions on rice husk ash. J Hazard Mater 163:1254–1264

Nakajima A, Tsuruta T (2004) Competitive biosorption of thorium and uranium by Micrococcus luteus. J Radioanal Nucl Chem 260:13–18

Nakajima A, Yasuda M, Yokoyama H, Ohya-Nishiguchi H, Kamada H (2001) Copper biosorption by chemically treated Micrococcus luteus cells. World J Microbiol Biotechnol 17:343–347

Nakbanpote W, Thiravavetyan P, Kalambaheti C (2000) Preconcentration of gold by rice husk ash. Miner Eng 13:391–400

Nakbanpote W, Thiravavetyan P, Kalambaheti C (2002) Comparison of gold adsorption by Chlorella vulgaris, rice husk and activated carbon. Miner Eng 15:549–552

Nakbanpote W, Goodman BA, Thiravetyan P (2007) Copper adsorption on rice husk derived materials studied by EPR and FTIR. Colloids Surf A Physicochem Eng Aspects 304:7–13

Namasivayam C, Sangeetha D, Gunasekaran R (2007) Removal of anions, heavy metals, organics and dyes from water by adsorption onto a new activated carbon from Jatrophahusk, an agro-industrial solid waste. Trans IChemE B Process Saf Environ Prot 85(B2):181–184

Nameni M, Moghadam MRA, Aram M (2008) Adsorption of hexavalent chromium from aqueous solutions by wheat bran. Int J Environ Sci Technol 5:161–168

Nomanbhay SM, Palanisamy K (2005) Removal of heavy metal from industrial wastewater using chitosan coated oil palm shell charcoal. Electron J Biotechnol 8:43–53

Nordberg GF, Sandström B, Becking G, Goyer RA (2002) Essentiality and toxicity of metals. In: Sarkar B (ed) Heavy metals in the environment. Marcel Dekker, New York

Nourbakhsh M, Sag Y, Özer D, Aksu Z, Kutsal T, Çaglar A (1994) A comparative study of various biosorbents for removal of chromium(VI) ions from industrial waste waters. Process Biochem 29:1–5

Nouri L, Hamdaoui O (2007) Ultrasonication-assisted sorption of cadmium from aqueous phase by wheat bran. J Phys Chem A 111:8456–8463

Nouri L, Ghodbane I, Hamdaoui O, Chiha M (2007) Batch sorption dynamics and equilibrium for the removal of cadmium ions from aqueous phase using wheat bran. J Hazard Mater 149:115–125

Ortega LM, Lebrun R, Blais J, Hausler R (2008) Removal of metal ions from an acidic leachate solution by nanofiltration membranes. Desalination 227:204–216

Ozaki H, Sharmab K, Saktaywirf W (2002) Performance of an ultra-low-pressure reverse osmosis membrane (ULPROM) for separating heavy metal: effects of interference parameters. Desalination 144:287–294

Ozdemir G, Ozturk T, Ceyhan N, Isler R, Cosar T (2003) Heavy metal biosorption by biomass of Ochrobactrum anthropi producing exopolysaccharide in activated sludge. Bioresour Technol 90:71–74

Özer A, Pirincci HB (2006) The adsorption of Cd(II) ions on sulfuric acid-treated wheat bran. J Hazard Mater B137:849–855

Özer A, Özer D, Özer A (2004) The adsorption of copper (II) ions on to dehydrate wheat bran (DWB): determination of the equilibrium and thermodynamic parameters. Process Biochem 39:2183–2191

Öztürk A (2007) Removal of nickel from aqueous solution by the bacterium Bacillus thuringiensis. J Hazard Mater 147:518–523

Öztürk A, Artan T, Ayar A (2004) Biosorption of nickel(II) and copper(II) ions from aqueous solution by Streptomyces coelicolor A3(2). Colloids Surf B Biointerfaces 34:105–111
Padmavathy V (2008) Biosorption of nickel(II) ions by baker's yeast: kinetic, thermodynamic and desorption studies. Bioresour Technol 99:3100–3109
Pahlavanzadeh H, Keshtkar AR, Safdari J, Abadi Z (2010) Biosorption of nickel(II) from aqueous solution by brown algae: equilibrium, dynamic and thermodynamic studies. J Hazard Mater 175:304–310
Panday KK, Prasad G, Singh VN (1984) Removal of Cr(V1) from aqueous solutions by adsorption on fly ash-wollastonite. J Chem Technol Biotechnol 34A:367–374
Panday KK, Prasad G, Singh VN (1985) Copper(II) removal from aqueous solutions by fly ash. Water Res 19:869–873
Papandreou A, Stournaras CJ, Panias D (2007) Copper and cadmium adsorption on pellets made from fired coal fly ash. J Hazard Mater 148:538–547
Papandreou A, Stournaras CJ, Panias D, Paspaliaris I (2011) Adsorption of Pb(II), Zn(II) and Cr(III) on coal fly ash porous pellets. Miner Eng 24:1495–1501
Pardo R, Herguedas M, Barrado E, Vega M (2003) Biosorption of cadmium, copper, lead and zinc by inactive biomass of Pseudomonas Putida. Anal Bioanal Chem 376:26–32
Pattanayak J, Mondal K, Mathew S, Lalvani SB (2000) A parametric evaluation of the removal of As(V) and As(III) by carbon-based adsorbents. Carbon 38:589–596
Paul Chen J, Mou H, Wang LK, Matsuura T (2006) Membrane filtration. In: Wang LK, Hung Y-T, Shammas NK (eds) Handbook of environmental engineering, volume 4: advanced physicochemical treatment processes. The Humana Press, Totowa, pp 203–259
Pavasant P, Apiratikul R, Sungkhum V, Suthiparinyanont P, Wattanachira S, Marhaba TF (2006) Bioresour Technol 97:2321–2329
Pengthamkeerati P, Satapanajaru T, Singchan O (2008) Sorption of reactive dye from aqueous solution on biomass fly ash. J Hazard Mater 153:1149–1156
Petrus R, Warchoł JK (2005) Heavy metal removal by clinoptilolite. An equilibrium study in multi-component systems. Water Res 39:819–830
Pillai MG, Regupathi I, Kalavathy MH, Murugesan T, RoseMiranda L (2009) Optimization and analysis of nickel adsorption on microwave irradiated rice husk using response surface methodology (RSM). J Chem Technol Biotechnol 84:291–301
Polowczyk I, Bastrzyk A, Sawinski W, Kozlecki T, Rudnicki P, Sadowski Z (2011) Sorption properties of fly ash from brown coal burning towards arsenic removal. Tech Trans Chem 8:135–142
Puranik PR, Paknikar KM (1997) Biosorption of lead and zinc from solutions using Streptoverticillium cinnamoneum waste biomass. J Biotechnol 55:113–124
Puranik PR, Chabukswar NS, Paknikar KM (1995) Cadmium biosorption by Streptomyces pimprina waste biomass. Appl Microbiol Biotechnol 43:1118–1121
Qu RJ, Sun CM, Fang M, Zhang Y, Ji CN, Xu Q (2009a) Removal of recovery of Hg(II) from aqueous solution using chitosan-coated cotton fibers. J Hazard Mater 167:717–727
Qu RJ, Sun CM, Wang MH, Ji CN, Xu Q, Zhang Y (2009b) Adsorption of Au(III) from aqueous solution using cotton fiber/chitosan composite adsorbents. Hydrometallurgy 100:65–71
Quinones I, Guiochon G (1998) Extension of a Jovanovic–Freundlich isotherm model to multi-component adsorption on heterogeneous surfaces. J Chromatogr A 796:15–40
Quinones I, Stanley B, Guiochon G (1999) Estimation of the adsorption energy distributions for the Jovanovic–Freundlich isotherm model with Jovanovic local behaviour. J Chromatogr A 849:45–60
Rahman IA, Ismail J (1993) Preparation and characterization of a spherical gel from a low-cost material. J Mater Chem 3:931–934
Rahmanian B, Pakizeh M, Maskooki A (2012) Optimization of lead removal from aqueous solution by micellar-enhanced ultrafiltration process using Box-Behnken design. Korean J Chem Eng 29(6):804–811

Rao M, Parwate AV, Bhole AG (2002) Removal of Cr^{6+} and Ni^{2+} from aqueous solution using bagasse and fly ash. Waste Manage 22:821–830
Rao M, Parwate AV, Bhole AG, Kadu PA (2003) Performance of low-cost adsorbents for the removal of copper and lead. J Water Supply Res Technol AQUA 52:49–58
Redlich OJ, Peterson DL (1959) A useful adsorption isotherm. J Phys Chem 63:1024
Ricou P, Lecuyer I, Cloirec PL (1999) Removal of Cu^{2+}, Zn^{2+} and Pb^{2+} by adsorption onto fly ash and fly ash/lime mixing. Water Sci Technol 39:239–247
Rivas BL, Palencia M (2011) Removal-concentration of pollutant metal-ions by water-soluble polymers in conjunction with double emulsion systems: a new hybrid method of membrane-based separation. Sep Purif Technol 81:435–443
Romera E, Gonzalez F, Ballester A, Blazquez ML, Munoz JA (2006) Biosorption with algae: a statistical review. Crit Rev Biotechnol 26:223–235
Roy D, Greenlaw PN, Shane BS (1993) Adsorption of heavy metals by green algae and ground rice hulls. J Environ Sci Health A28:37–50
Rubin CH, Esteban E, Jones R, Noonan G, Gurvich E, Utz S, Spirin V, Revich B, Kruchkov G, Jackson RJ (1997) Childhood specific lead poisoning in Russia: a site-specific pediatric blood lead evaluation. Int J Occup Environ Health 3:241–248
Rudzinski W, Panczyk T (2000) Kinetics of isothermal adsorption on energetically heterogeneous solid surfaces: a new theoretical description based on the statistical rate theory of interfacial transport. J Phys Chem 104(39):9149–9162
Şahin Y, Öztürk A (2005) Biosorption of chromium(VI) ions from aqueous solution by the bacterium Bacillus thuringiensis. Process Biochem 40:1895–1901
Salam OEA, Reiad NA, Elshafei MM (2011) A study of the removal characteristics of heavy metals from wastewater by low-cost adsorbents. J Adv Res 2:297–303
Salehi E, Madaeni SS, Heidary F (2012) Dynamic adsorption of Ni(II) and Cd(II) ions from water using 8-hydroxyquinoline ligand immobilized PVDF membrane: isotherms, thermodynamics and kinetics. Sep Purif Technol 94:1–8
Salehizadeh H, Shojaosadati SA (2003) Removal of metal ions from aqueous solution by polysaccharide produced from Bacillus firmus. Water Res 37:4231–4235
Sankararamakrishnan N, Sharma AK, Sanghi R (2007) Novel chitosan derivative for the removal of cadmium in the presence of cyanide from electroplating wastewater. J Hazard Mater 148:353–359
Savvaidis I, Hughes MN, Poole RK (2003) Copper biosorption by Pseudomonas cepacia and other strains. World J Microbiol Biotechnol 19:117–121
Selatnia A, Bakhti MZ, Madani A, Kertous L, Mansouri Y (2004a) Biosorption of Cd2+ from aqueous solution by a NaOH-treated bacterial dead Streptomyces rimosus biomass. Hydrometallurgy 75:11–24
Selatnia A, Boukazoula A, Kechid N, Bakhti MZ, Chergui A (2004b) Biosorption of Fe3+ from aqueous solution by a bacterial dead Streptomyces rimosus biomass. Process Biochem 39:1643–1651
Selatnia A, Boukazoula A, Kechid N, Bakhti MZ, Chergui A, Kerchich Y (2004c) Biosorption of lead (II) from aqueous solution by a bacterial dead Streptomyces rimosus biomass. Biochem Eng J 19:127–135
Selatnia A, Madani A, Bakhti MZ, Kertous L, Mansouri Y, Yous R (2004d) Biosorption of Ni2+ from aqueous solution by a NaOH-treated bacterial dead Streptomyces rimosus biomass. Miner Eng 17:903–911
Sen AK, De AK (1987) Adsorption of mercury(II) by coal fly ash. Water Res 21:885–888
Shahalam AM, Al-Harthy A, Al-Zawhry A (2002) Feed water pretreatment in RO systems in the middle east. Desalination 150:235–245
Shaker MA (2007) Thermodynamic profile of some heavy metal ions adsorption onto biomaterial surfaces. Am J Appl Sci 4:605–612
Shameem H, Abburi K, Tushar KG, Dabir SV, Veera MB, Edgar DS (2003) Adsorption of chromium(VI) on chitosan-coated perlite. Sep Sci Technol 38:3775–3793

Shameem H, Abburi K, Tushar KG, Dabir SV, Veera MB, Edgar DS (2006) Adsorption of divalent cadmium (Cd(II)) from aqueous solutions onto chitosan-coated perlite beads. Ind Eng Chem Res 45:5066–5077

Shameem H, Tushar KG, Dabir SV, Veera MB (2008) Dispersion of chitosan on perlite for enhancement of copper(II) adsorption capacity. J Hazard Mater 152:826–837

Sharma N, Kaur K, Kaur S (2009) Kinetic and equilibrium studies on the removal of Cd^{2+} ions from water using polyacrylamide grafted rice (Oryza sativa) husk and (Tectona grandis) saw dust. J Hazard Mater 163:1338–1344

Silva EA, Cossich ES, Tavares CG, Cardozo Filho L, Guirardello R (2003) Biosorption of binary mixtures of Cr(III) and Cu(II) ions by Sargassum sp. Braz J Chem Eng 20:213–227

Singh KK, Hasan HS, Talat M, Singh VK, Gangwar SK (2009) Removal of Cr(VI) from aqueous solutions using wheat bran. Chem Eng J 151:113–121

Sips RJ (1948) On the structure of a catalyst surface. Chem Phys 16:490–495

Site AD (2001) Factors affecting sorption of organic compounds in natural sorbent/water systems and sorption coefficients for selected pollutants: a review. J Phys Chem Ref Data 30:187–439

Smith CJ, Livingston SD, Doolittle DJ (1997) An international literature survey of "IARC Group I carcinogens" reported in mainstream cigarette smoke. Food Chem Toxicol 35:1107–1130

Srinath T, Verma T, Ramteke PW, Garg SK (2002) Chromium (VI) biosorption and bioaccumulation by chromate resistant bacteria. Chemosphere 48:427–435

Srinivasa RP, Vijaya Y, Veera MB, Krishnaiah A (2009) Adsorptive removal of copper and nickel ions from water using chitosan coated PVC beads. Bioresour Technol 100:194–199

Srivastava NK, Majumder CB (2008) Novel biofiltration methods for the treatment of heavy metals from industrial wastewater. J Hazard Mater 151:1–8

Srivastava VC, Mall ID, Mishra IM (2008) Removal of cadmium(II) and zinc(II) metal ions from binary aqueous solution by rice husk ash. Colloids Surf A Physicochem Eng Aspects 312:172–184

Srivastava VC, Mall ID, Mishra IM (2009) Competitive adsorption of cadmium(II) and nickel(II) metal ions from aqueous solution onto rice husk ash. Chem Eng Process Process Intensification 48:370–379

Sun XQ, Peng B, Jing Y, Chen J, Li DQ (2009) Chitosan(chitin)/cellulose composite biosorbents prepared using ionic liquid for heavy metal ions adsorption. Separations 55:2062–2069

Surchi KMM (2011) Agricultural wastes as low cost adsorbents for Pb removal: kinetics, equilibrium and thermodynamics. Int J Chem 3:103–112

Tan G, Xiao D (2009) Adsorption of cadmium ion from aqueous solution by ground wheat stems. J Hazard Mater 164:1359–1363

Tan IAW, Ahmad AL, Hameed BH (2008) Adsorption of basic dye on high-surface-area activated carbon prepared from coconut husk: equilibrium, kinetic and thermodynamic studies. J Hazard Mater 154:337–346

Tangaromsuk J, Pokethitiyook P, Kruatrachue M, Upatham ES (2002) Cadmium biosorption by Sphingomonas paucimobilis biomass. Bioresour Technol 85:103–105

Tarley CRT, Ferreira SLC, Arruda MAZ (2004) Use of modified rice husks as a natural solid adsorbent of trace metals: characterisation and development of an on-line preconcentration system for cadmium and lead determination by FAAS. Microchem J 77:163–175

Temkin MI (1941) Adsorption equilibrium and the kinetics of processes on nonhomogeneous surfaces and in the interaction between adsorbed molecules. Zh Fiz Chim 15:296–332

Thayyath SA, Sreenivasan R, Abdul Rauf T (2010) Adsorptive removal of thorium(IV) from aqueous solutions using poly(methacrylic acid)-grafted chitosan/bentonite composite matrix: process design and equilibrium studies. Colloids Surf A Physicochem Eng Aspects 368:13–22

Thomas HC (1944) Heterogeneous ion exchange in a flowing system. J Am Chem Soc 66:1664–1666

Tirtom VN, Dincer A, Becerik S, Aydemir T, Celik A (2012) Comparative adsorption of Ni(II) and Cd(II) ions on epichlorohydrin crosslinked chitosan-clay composite beads in aqueous solution. Chem Eng J 197:379–386

Tiwari DP, Singh DK, Saksena DN (1995) Hg(II) adsorption from solutions using rice-husk ash. J Environ Eng 121:479–481
Tobin JM, Cooper DG, Neufeld RJ (1984) Uptake of metal ions by Rhizopus arrhizus biomass. Appl Environ Microbiol 47:821–824
Toth J (1971) State equations of the solid gas interface layer. Acta Chem Acad Hung 69:311–317
Tran HV, Tran LD, Nguyen TN (2010) Preparation of chitosan/magnetite composite beads and their application for removal of Pb(II) and Ni(II) from aqueous solution. Mater Sci Eng 30:304–310
Tunali S, Çabuk A, Akar T (2006) Removal of lead and copper ions from aqueous solutions by bacterial strain isolated from soil. Chem Eng J 115:203–211
Uslu G, Tanyol M (2006) Equilibrium and thermodynamic parameters of single and binary mixture biosorption of lead (II) and copper (II) ions onto Pseudomonas putida: effect of temperature. J Hazard Mater 135:87–93
Vasconcelos HL, Fa'vere VT, Goncalves NS, Laranjeira MCM (2007) Chitosan modified with reactive blue 2 dye on adsorption equilibrium of Cu(II) and Ni(II) ions. React Funct Polym 67:1052–1106
Vasconcelos HL, Camargo TP, Goncalves NS, Laranjeira MCM, Fa'vere VT (2008) Chitosan crosslinked with a metal complexing agent: synthesis, characterization and copper(II) ions adsorption. React Funct Polym 68:572–579
Veeken AHM, Akoyo L, Hulshoff Pol LW, Weijma J (2003) Control of the sulfide (S^{2-}) concentration for optimal zinc removal by sulfide precipitation in a continuously stirred tank reactor. Water Res 37:3709–3717
Veera MB, Krishnaiah A, Jonathan LT, Edgar DS (2003) Removal of hexavalent chromium from wastewater using a new composite chitosan biosorbent. Environ Sci Technol 37:4449–4456
Veera MB, Krishnaiah A, Ann JR, Edgar DS (2008a) Removal of copper(II) and nickel (II) ions from aqueous solutions by a composite chitosan biosorbent. Sep Sci Technol 43:1365–1381
Veera MB, Krishnaiah A, Jonathan LT, Edgar DS, Richard H (2008b) Removal of arsenic (III) and arsenic (V) from aqueous medium using chitosan-coated biosorbent. Water Res 42:633–642
Vieira RS, Beppu MM (2006) Interaction of natural and crosslinked chitosan membranes with Hg(II) ions. Colloids Surf A 279:196–207
Vijaya Y, Srinivasa RP, Veera MB, Krishnaiah A (2008) Modified chitosan and calcium alginate biopolymer sorbents for removal of nickel (II) through adsorption. Carbohydr Polym 72:261–271
Visa M, Isac L, Duta A (2012) Fly ash adsorbents for multi-cation wastewater treatment. Appl Surf Sci 258:6345–6352
Vitali L, Laranjeira MCM, Goncalves NS, F'avere VT (2008) Spray-dried chitosan microspheres containing 8-hydroxyquinoline-5 sulphonic acid as a new adsorbent for Cd(II) and Zn(II) ions. Int J Biol Macromol 42:152–157
Volesky B, Holan ZS (1995) Biosorption of heavy metals. Biotechnol Prog 11:235–250
Von Burg R (1997) Nickel and some nickel compounds. J Appl Toxicol 17:425–431
Wan Ngah WS, Fatinathan S (2008) Adsorption of Cu(II) ions in aqueous solution using chitosan beads, chitosan–GLA beads and chitosan–alginate beads. Chem Eng J 143:62–72
Wan Ngah WS, Kamari A, Koay YJ (2004) Equilibrium kinetics studies of adsorption of copper (II) on chitosan and chitosan/PVA beads. Int J Bio Macromol 34:155–161
Wan MW, Kan CC, Lin CH, Buenda DR, Wu CH (2007) Adsorption of copper (II) by chitosan immobilized on sand. Chia-Nan Annu Bull 33:96–106
Wan MW, Kan CC, Buenda DR, Maria LPD (2010) Adsorption of copper(II) and lead(II) ions from aqueous solution on chitosan-coated sand. Carbohydr Polym 80:891–899
Wang S, Li H (2007) Kinetic modelling and mechanism of dye adsorption on unburned carbon. Dye Pigments 72:308–314
Wang LH, Lin CI (2008a) Adsorption of chromium (III) ion from aqueous solution using rice hull ash. J Chin Inst Chem Eng 39:367–373
Wang LH, Lin CI (2008b) Adsorption of lead(II) ion from aqueous solution using rice hull ash. Ind Eng Chem Res 47:4891–4897

Wang XS, Li ZZ, Sun C (2008) Removal of Cr(VI) from aqueous solutions by low cost biosorbents: marine macroalgae and agricultural by-products. J Hazard Mater 153:1176–1184
Wang XS, Li ZZ, Sun C (2009) A comparative study of removal of Cu(II) from aqueous solutions by locally low-cost materials: marine macroalgae and agricultural by-products. Desalination 235:146–159
Wang X, Xia L, Tan K, Zheng W (2011) Studies on adsorption of uranium(VI) from aqueous solution by wheat straw. Environ Prog Sust Energ 31(4):566–576. doi 10.1002/ep.10582
Warchoł J, Petrus R (2006) Modeling of heavy metal removal dynamics in clinoptilolite packed beds. Micropor Mesopor Mater 93:29–39
Weber WJ, Morris JC (1963) Kinetic of adsorption on carbon from solution. J Sanit Eng Div ASCE 89SA2:31–59
Wen Y, Tang Z, Chen Y, Gu Y (2011) Adsorption of Cr(VI) from aqueous solutions using chitosan-coated fly ash composite as biosorbent. Chem Eng J 175:110–116
Weng CH, Huang CP (1990) Removal of trace heavy metals by adsorption onto fly ash. In: Proceedings of the 1990 ASCE Environmental Engineering Specialty Conference. Division of Environmental Engineering, American Society of Civil Engineers (ASCE). Arlington, pp 923–924
Weng CH, Huang CP (1994) Treatment of metal industrial wastewater by fly ash and cement fixation. J Environ Eng 120:1470–1487
Weng CH, Huang CP (2004) Adsorption characteristics of Zn(II) from dilute aqueous solution by fly ash. Colloids Surf A Physicochem Eng Aspects 247:137–143
Wolborska A (1989) Adsorption on activated carbon of p-nitrophenol from aqueous solution. Water Res 23:85–91
Wong KK, Lee CK, Low KS, Haron MJ (2003a) Removal of Cu and Pb by tartaric acid modified rice husk from aqueous solutions. Chemosphere 50:23–28
Wong KK, Lee CK, Low KS, Haron MJ (2003b) Removal of Cu and Pb from electroplating wastewater using tartaric acid modified rice husk. Process Biochem 39:437–445
Wrudzinski W, Wojciecbowski BW (1977) On the Jovanovic model of adsorption. Colloid Polym Sci 255:859–880
Xu X, Gao BY, Tan X, Yue QY, Zhong QQ, Li Q (2011) Characteristics of amine-crosslinked wheat straw and its adsorption mechanisms for phosphate and chromium (VI) removal from aqueous solution. Carbohydr Polym 84:1054–1060
Yadava KP, Tyagi BS, Panday KK, Singh VN (1987) Fly ash for the treatment of Cd(II) rich effluents. Environ Technol Lett 8:225–234
Yamanaka K, Ohtsubo K, Hasegawa A (1996) Exposure to dimethylarsenic acid, a main metabolite of inorganic arsenic, strongly promotes tumorigenesis initiated by 4-nitroquinolone 1 oxide in the lungs of mice. Carcinogenesis 17:767–770
Yan-Chu H (1994) Arsenic distribution in soils. In: Nriagu JO (ed) Arsenic in the environment, part 1: cycling and characterization. Wiley, New York, pp 17–49
Ye H, Zhu Q, Du D (2010) Adsorptive removal of Cd(II) from aqueous solution using natural and modified rice husk. Bioresour Technol 101:5175–5179
Ye H, Zhang L, Zhang B, Wu G, Du D (2012) Adsorptive removal of Cu(II) from aqueous solution using modified rice husk. Int J Eng Res Appl 2(2):855–863
Yoon YH, Nelson JH (1984) Application of gas adsorption kinetics. I. A. Theoretical model for respirator cartridge service time. Am Ind Hyg Assoc J 45:509–516
Yu Q, Kaewsarn P (2000) Adsorption of Ni^{2+} from aqueous solutions by pretreated biomass of marine macroalga Durvillaea potatorum. Sep Sci Technol 35:689–701
Yu Q, Matheickal JT, Yin P, Kaewsarn P (1999) Heavy metal uptake capacities of common marine macro algal biomass. Water Res 33:1534–1537
Zeldowitsch J (1934) Über den mechanismus der katalytischen oxydation von CO an MnO_2. Acta Physicochem URSS 1:364–449
Zhang GY, Qu RJ, Sun CM, Ji CN, Chen H, Wang CH (2008) Adsorption for metal ions of chitosan coated cotton fiber. J Appl Polymer Sci 110:2321–2327

Zhou M, Liu Y, Zeng G, Li X, Xu W, Fan T (2007) Kinetic and equilibrium studies of Cr(VI) biosorption by dead Bacillus licheniformis biomass. World J Microbiol Biotechnol 23:43–48
Ziagova M, Dimitriadis G, Aslanidou D, Papaioannou X, Litopoulou Tzannetaki E, Liakopoulou-Kyriakides M (2007) Comparative study of Cd(II) and Cr(VI) biosorption on Staphylococcus xylosus and Pseudomonas sp. in single and binary mixtures. Bioresour Technol 98:2859–2865
Zulkali MMD, Ahmad AL, Norulakmal NH, Sharifah NS (2006) Comparative studies of oryza sativa L. Husk and chitosan as lead adsorbent. J Chem Technol Biotechnol 81:1324–1327

Chapter 9
Water Quality Monitoring by Aquatic Bryophytes

Gana Gecheva and Lilyana Yurukova

Abstract Bryophytes are non-vascular plants that are in a close relationship with their immediate environment. They often have large biomass in freshwater ecosystem and high level of production. Moreover, their tissues contain elevated amount of C, N and P, and cell walls have high cation exchange capacity. Aquatic bryophytes can be used to assess freshwater pollution as indicators – presence or absence of species – or as monitors for accumulating elements. Consumption of metals and other substances by aquatic bryophytes is an important exposure pathway for consumers. The use of bryophytes for water quality assessment is well documented, but different techniques and approaches prevent standardization and their applicability on the European scale. Thus we review major findings in 'bryomonitoring'. Data were reviewed from a range of countries, mainly in Europe, illustrating the advantages of low cost methods for monitoring water quality.

Here we introduce the term 'bryomonitoring' as a method to assess alterations of the environment. Biomonitoring can be split into passive – observation and analysis of native bryophytes, and active biomonitoring – based on species transplantation for a fixed exposure period. Two widespread northern hemisphere aquatic mosses, *Fontinalis antipyretica* and *Platyhypnidium riparioides,* are the most commonly used biomonitors for river quality assessment. For passive biomonitoring key issues are background and reference level determination, and proper selection of sampling sites. For active monitoring, upper segments of a same age from a reference region should be applied. The actual analytical techniques give in general similar results, but not completely interchangeable.

G. Gecheva (✉)
University of Plovdiv, 24 Tsar Assen Str, BG-4000 Plovdiv, Bulgaria
e-mail: ggecheva@mail.bg

L. Yurukova
Institute of Biodiversity and Ecosystem Research, Bulgarian Academy of Sciences, Acad. G. Bonchev Str., Bl. 23, BG-1113 Sofia, Bulgaria
e-mail: yur7lild@bio.bas.bg

E. Lichtfouse et al. (eds.), *Green Materials for Energy, Products and Depollution*, Environmental Chemistry for a Sustainable World 3, DOI 10.1007/978-94-007-6836-9_9,

Aquatic bryophytes are used to assess the ecological status. They are a stress-tolerant and various species have a wide trophic range. *Fontinalis antipyretica* and *Platyhypnidium ripariodes* have all criteria for biota monitoring in rivers for heavy metals.

Standardization of sampling procedures and analytical techniques in aquatic bryomonitoring is further needed. The number of samples should be fixed based on sampling area surface. Period of exposure time for active biomonitoring should be specified in general. Background levels and ambient metal concentrations have to be observed in parallel.

Keywords Bryophytes • Pollution monitoring • Water pollution • Cs • U • Ra • Th • Hg • Pb • Heavy metals

Contents

Abbreviations

AAS	Atomic Absorption Spectrophotometry
BAF	Biota Accumulation Factor
BCF	Bioconcentration Factor
BMF	Biomagnification Factor
BQE	Biological Quality Element
C	Element concentration
C_{bckg}	Background concentration
CF	Contamination Factor
C_o	Element concentration in an organism
$C_{w/s}$	Element concentration in water/sediment
DEHP	Di(2-ethylhexyl)-phthalate
DL	Detection Limit
EA	Environmental Alteration
EQR	Ecological Quality Ratio
EQS	Ecological Quality Standard

EU	European Union
FAAS	Furnace Atomic Absorption Spectrophotometry
GEA	Global Environmental Alteration
GFAAS	Graphite Furnace Atomic Absorption Spectrophotometry
HCH	Hexachlorocyclohexanes
HPLC	High Pressure Liquid Chromatography
IBMR	Macrophyte Biological Index for Rivers
ICP-AES (ICP-OES)	Inductively Coupled Plasma Atomic Emission Spectroscopy (Inductively Coupled Plasma Optical Emission Spectrometry)
ICP-MS	Inductively Coupled Plasma Mass Spectrometry
MAC	Macrophyte Assessment and Classification
MACPACS	MACrophyte Prediction And Classification System
MLD	Methodological Limit of Determination
MTR	Mean Trophic Rank
NAA	Neutron Activation Analysis
PAH	Polyaromatic Hydrocarbons
PCB	Polychlorinated Biphenyls
pH	Acidity
RHS	River Habitat Survey
RSD%	Relative Standard Deviation in percents
TIM	Trophic Index of Macrophytes
WFD	Water Framework Directive

9.1 Introduction

Pollution is a major environmental issue affecting freshwater habitats and consequently human health. Lotic and lentic ecosystems constantly react to external and internal changes. Their recovery after human alterations depends on a variety of factors, among them living organisms, including bryophyte communities. Aquatic bryophytes as primary producers and habitat providers are important component of freshwater ecosystems and influence both biodiversity and water chemistry. They affect biodiversity by changing environmental conditions and resource availability and providing suitable habitats for new species and species already present. Nutrient dynamics can also be influenced by bryophytes (Stream Bryophyte Group 1999). In mountain sites (especially spring habitats) bryophytes usually cover large areas; sometimes they are able to modify water flow and can be considered as ecosystem engineers (Jones et al. 1994).

A review on monitoring studies of heavy metals with freshwater plants was published by Whitton (2003), comprising, among all phototrophs, data on bryophyte researches during the period 1969 and 2001. Extensive reviews on both terrestrial and aquatic bryophytes as monitors were also presented by Burton (1990) and Tyler (1990).

A variety of assessment methods based on aquatic bryophytes has been proposed, based both on field and/or laboratory data for broad surveys or the investigation of point-source contamination. Research results and outcomes were reviewed from a range of countries, mainly in Europe. Among them numerous laboratory studies on separate heavy metals bioaccumulation kinetics and intra-, extra- and intercellular distribution in recent years have been published (Samson et al. 1998; Vázquez et al. 1999; Vieira et al. 2009), as well as studies on photosynthetic pigments (Cruz de Carvalho et al. 2011; López and Carballeira 1989; Martinez-Abaigar and Núñez-Olivera 1998; Peñuelas 1984; Spitale 2009), and detoxification mechanisms (Dazy et al. 2008) but they will be reviewed in future paper, focusing on exposure of bryophytes under laboratory conditions.

9.2 Biomonitoring: Methodology and Application

Although there are many interpretations of the term, probably most essential is that biomonitoring is a system for long-term observation, assessment and forecast of possible environmental alterations based on biological objects (Martin and Coughtrey 1982). On the other hand, as a scientific term, bioindication has introduced literature at the end of 1960s of the past century. Bioindication is a time-dependent sensitive response of biological measurable parameters to anthropogenic pressure (Stöcker 1980). Thus bioindicator is an organism (or part of an organism or a group of organisms) that contains information on the quality of the environment (or a part of the environment) (Markert 2008). Haseloff (1982) divided them as visual, chemical and physico-biochemical bioindicators: (1) the first type includes species presence, reduced growth, leaf decoloration and population changes, (2) chemical bioindicators are characterized by accumulation of substances, (3) physico-biochemical bioindicators are characterized by alterations of enzymatic activity and physiological functions. Bioindicators were proposed for long-term observations, as well as for planning and management the effects of human activities (Hertz 1991).

In general, the difference between bioindicators and biomonitors is that the former gives a qualitative, and the last one quantitative assessment of the quality of the environment (Manning and Feder 1980; Martin and Coughtrey 1982; Markert 1991; Markert et al. 2003). A biomonitor is always a bioindicator as well, but a bioindicator does not necessarily meet the requirements for a biomonitor (Markert 2008). Biomonitors can be considered as sensitive and accumulative (Steubing 1976; Stöcker 1980). Sensitive biomonitors applied in aquatic ecosystems provide early warning system (Cairns and van der Schalie 1980). Accumulative biomonitors receive major attention towards heavy metal pollution. Tyler (1972) showed that dead organic matter, lichens and especially mosses as low-level plants, accumulate high heavy metals amounts. The main reason is the high stability of the chemical complexes between heavy metal ions and negative charged organic groups. Burton

(1990) underlined the possibility of bryophytes to produce information for reaction towards ecological factors, as well as element concentrations.

It is useful to distinguish between bioindicator and biomonitors (accumulators), but regardless differences in definitions, bioindication and biomonitoring must supply information on the extent of pollution and degradation of freshwater ecosystems.

Both passive and active biomonitoring received widespread popularity and their advantages and disadvantages are profoundly presented by Martin and Coughtrey (1982). Passive biomonitoring consists in observation and analysis of native bryophytes, while active biomonitoring is based on species transplantation for a fixed exposure period.

Numerous studies have reported that aquatic plants often accumulate heavy metals in concentrations much higher than those reached in their aqueous environment, even when those metals are not essential for metabolism or they are potentially toxic. The metal accumulated by a plant gives a better indication of the metal fraction in the environment likely to affect an aquatic ecosystem than most types of direct chemical analysis (Empain et al. 1980). In aquatic systems, metals exist both as free ions and as complexed forms. For many metals it is the free ionic form which is believed to be responsible for toxicity because the possibility for uptake is increased. Thus plant bioaccumulation is the basis for evaluating indirect exposure to other organisms including humans. Metal accumulation in plants is also pointed as common investigated for the biomonitoring of aquatic pollution in the review of Zhou et al. (2008).

In the current review we apply the term bryomonitoring in the context of biomonitoring and underlying the organisms (i.e. bryophytes) by which environmental quality is determined.

Bryomonitoring is founded on two basic approaches: species distribution and chemical analysis of bryophyte tissues (Burton 1986). Bryophytes can accumulate extremely high levels of heavy metals based on their high cation exchange tissue capacity, lack of cuticle and high surface to volume ratio (Tyler 1990). Metal ions are accumulated mainly through the passive ion exchange. Main tolerance mechanism is cell wall considerable efficiency to immobilize heavy metal ions and thus bryophytes are able to accumulate metals to remarkable concentrations. Large interspecific differences in accumulation levels were found in studies carried out at streams with severe contamination (Burton 1990).

In general, bryophyte high accumulative capacity led to their implementation mainly as monitors in regional and local discharges of heavy metals. The simplest way to use information on metal concentrations for monitoring is to compare values for specimens of a particular species at different sites, such as in a river downstream of an effluent or at a range of sites within a particular catchment or geographical region (Whitton 2003). Furthermore, several studies showed aquatic mosses are also suitable for monitoring radioactive contamination (Hongve et al. 2002) and pollution from organic compounds like oxolinic acid, flumequine or oxytetracycline, normally used as antibacterial agents (Delépée et al. 2004), monitoring of polychlorinated biphenyls and hexachlorocyclohexanes (Mouvet et al. 1985), and of polycyclic aromatic hydrocarbons (Roy et al. 1996).

Whatever is the analyzed substance, knowing its bioavailability is crucial, since bioavailability of its forms is tied to the potential effects on living organisms, man included Bioconcentration factors (BCF) or biota/water accumulation factors (BAF) are typically used to describe ratios of contaminants in tissues versus water for aquatic species and they are used to quantify contaminant uptake efficiency. The concept makes assumptions that the environment and the receptor are in pseudo-steady-state conditions, and the ratio is usually normalized for lipid and total organic carbon content of samples. The bioconcentration factor (BCF) is exceptionally high in bryophytes (Vanderpoorten and Goffinet 2009). The equation has the general form below:

$$BCF = C_o / C_{w/s}$$

where: C_o = contaminant concentration in the organism

$C_{w/s}$ = contaminant concentration in the water/sediment

Mouvet et al. (1986) proposed a method for the evaluation of metal contamination based on a Contamination Factor (CF), defined as the ratio of metal concentration in the indicator to the background level in that species. On the basis of CF obtained, Mouvet et al. (1986) suggested five categories of contamination:

- No contamination, up to 2 times the background level;
- Suspected contamination, between 2 and 6 times the background level;
- Moderate contamination, between 6 and 18 times the background level;
- Severe contamination, between 18 and 54 times, and
- Extreme contamination, more than 54 times.

9.2.1 *Passive Biomonitoring*

In this review the nomenclature accepted in Grolle and Long (2000) for liverworts and Hill et al. (2006) for mosses was presented at first appearance of a species and then taxa are cited following original studies.

High accumulation capacity and vast distribution of some bryophyte species has led to intensive growth of publications dealing with passive biomonitoring. Most of the studies covered only species element content without additional data on the surrounding environment (water and/or sediments).

Among aquatic bryomonitors, *Fontinalis* is the most studied genus (Table 9.1). It has sensitivity to Cu, exhibiting tip chlorosis, while it is insensitive to Cd (Glime 2003). The ecological characteristics of *Fontinalis antipyretica* Hedw. are described in details by Say and Whitton (1983). Species shows considerable potential as monitor of heavy metals. Moreover, it has the ability to colonize variety of substrates and to grow under various flow regimes. Two centimeters tips were suggested to reflect recent events and for long-term surveillance, while whole plants were found more useful for preliminary studies to detect water quality.

Table 9.1 Passive biomonitoring – examples

Species	Sample preparation	Substances	Analysis	Number of studied sites, country	Duration of study	Reference
Rhynchostegium riparioides	Drying at 105 °C, Digestion: 2 M HNO_3	Na, Mg, K, Ca, Cr, Mn, Fe, Co, Ni, Cu, Zn, Cd, Ba, Pb		105, England	6 weeks	Wehr and Whitton (1983a)
Fontinalis antipyretica	Drying at 105 °C, Digestion: 2 M HNO_3	Ca, Mn, Fe, Cu, Zn, Cd, Pb		52, Belgium and England		Say and Whitton (1983)
Rhynchostegium riparioides, Scapania undulata, Hygrohypnum duriusculum, Schistidium agassizii, Philonotis seriata	Dried sample, Digestion: 15 ml HNO_3, 5 ml H_2O_2	P, K, Ca, Mg, S, Fe, Al, Mn, Na, Zn, Cu, Pb, Cd, Co, Ni, As, Se, Cr	ICP-AES	10 rivers, 3 lakes, Bulgaria	3 months	Yurukova et al. (1996)
Platyhypnidium riparioides, Scapania sp., *Fontinalis antipyretica*	Dried at 40 °C, Digestion: HNO_3 and H_2O_2	Al, Ba, Cd, Co, Cr, Cu, Fe, Mn, Ni, Pb, Sr, V, Zn, Ca, Mg, K	ICP-AES/AAS (for Cd, Cu, Pb)	41 sites, Germany		Samecka-Cymerman et al. (2002)
Plagiochila porelloides, Scapania undulata, Atrichum undulatum, Bryum pseudotriquetrum, Rhizomnium punctatum, Fontinalis antipyretica, Amblystegium riparium, Sanionia uncinata, Warnsorfia exannulata, Brachythecium velutinum, B. plumosum, Rhynchostegium riparioides	Dried at 40 °C, Digestion: HNO_3 and H_2O_2	N, P, K, Ca, S, Mg, Na, Fe, Al, Mn, Co, Ni, Cu, Zn, Pb, Cd, As	ICP-AES/Nitrogen after Kjeldahl method	23 sites, Bulgaria	5 years	Yurukova and Gecheva (2004)

(continued)

Table 9.1 (continued)

Species	Sample preparation	Substances	Analysis	Number of studied sites, country	Duration of study	Reference
F. antipyretica, *F. squamosa*, *P. riparioides*	Digestion: 10 ml HNO_3 in a microwave oven	Al, As, Ca, Cd, Co, Cr, Cu, Fe, Hg, K, Mg, Mn, Na, Ni, Pb, Se Zn	FAAS/GFAAS (for As, Cd, Hg, Pb and Se)	Biomonitoring network of 121 sites, Spain	1 year survey	Vázquez et al. (2007)
Sanionia uncinata	Digestion: 300 mg dw of moss samples with HNO_3 and $HClO_4$	Na, Cd, Co, Cr Cu, Fe, Mn, Ni, Pb, V, Zn	FAAS (Fe, Mn and Zn)/GFAAS (Cd, Co, Cr, Cu, Ni, Pb and V)/flame photometer (Na)	29 sites, West Spitsbergen (Svalbard)	Summer season	Samecka-Cymerman et al. (2011)

Fontinalis antipyretica is proved biomonitor of many macro- and microelements in European freshwater ecosystems: Belgium (Empain 1976, 1977; Wehr et al. 1983), Bulgaria (Yurukova et al. 1997), Hungary (Kovács and Podani 1986; Kovács 1992), England (Say and Whitton 1983), Germany (Dietz 1972; Bruns et al. 1995), France (Empain 1976; Mouvet 1984), Poland (Samecka-Cymerman and Kempers 1992, 1993). Reported BCFs are extremely high, for example 3,200 for Pb and 9,400 for Zn (Dietz 1972).

Another intensively studied species is *Platyhypnidium riparioides* (Hedw.) Dixon (=*Eurhynchium riparioides* (Hedw.) P.W. Richards, *E. rusciforme* Milde, *P. rusciforme* (*Schimp.*) M. Fleisch., *Rhynchostegium riparioides* (Hedw.) Cardot, *R. rusciforme* Schimp.). It was confirmed as geographically and ecologically widespread, and as excellent species to monitor heavy metals (Wher and Whitton 1983a). Low pH and calcium values prevent species development. This moss was broadly included in biomonitoring researches in Belgium (Wehr et al. 1983), Bulgaria (Gecheva et al. 2011), England (Jackson et al. 1991), Spain (García-Álvaro et al. 2000).

Among liverworts *Scapania undulata* (L.) Dumort. received major attention for biomonitoring purposes. Species is absent in eutrophic waters and dominates streams which combine very low nutrients with very high heavy metal levels (Wher and Whitton 1983b). It was reported as tolerant to heavy metal contamination and as suitable biomonitor (McLean and Jones 1975; Burton and Peterson 1979; Satake et al. 1989). Thus *Scapania undulata* was implemented in many monitoring programmes (Whitton et al. 1982). Terminal shoots were analyzed from sites influenced by past or present mining activities in England, France, Germany and Ireland. Statistical analyses suggest that elevated pH and/or Ca lead to increased accumulation of Zn and Cd and that probably pH reflect also Pb accumulation. Species indicated the presence of perspective polymetallic deposits in Poland (Sudeten Mts) and had considerable mean values of Cd – 68 mg kg^{-1}, Co – 174 mg kg^{-1}, Ni – 232 mg kg^{-1}, Fe – 130,000 mg kg^{-1}, Mn – 19,900 mg kg^{-1}, as maximum accumulated levels near barite zones in the Sowie Mts were for As 2,190 mg kg^{-1}, B 10,100 mg kg^{-1}, Zn 550 mg kg^{-1}, Ni 150 mg kg^{-1}, Co 1,700 mg kg^{-1}, Ge 4,000 mg kg^{-1}, Pb 3,100 mg kg^{-1}, Sn 200 mg kg^{-1}, V 1,300 mg kg^{-1} (Samecka-Cymerman 1991; Samecka-Cymerman and Kempers 1993).

Studies involving several bryophyte species and/or focusing on interspecific differences are increasing. Metal concentration detected in *Rhynchostegium riparioides*, *Fontinalis antipyretica* and *Cinclidotus danubicus* Schiffn. & Baumgartner reflected Cu and Cr fluctuations (Mouvet et al. 1986). *Fontinalis antipyretica*, *Rhynchostegium riparioides*, *Brachythecium rivulare* Schimp., *Plagiothecium ruthei* Limpr. (=*Plagiothecium denticulatum* (Hedw.) Schimp.), *Pellia fabbroniana* (=*Pellia endiviifolia* (Dicks.) Dumort.), *Scapania undulata* and other species were found useful in biogeochemical prospecting for minerals (Samecka-Cymerman and Kempers 1992, 1993; Pirc 2003). Relationships between Zn, Cd and Pb concentrations in algae, liverwort *Scapania undulata* and three mosses *Amblystegium riparium* (=*Leptodictyum riparium* (Hedw.) Warnst.), *Fontinalis antipyretica* and *Rhynchostegium riparioides* were established in Belgium, France, Germany, Ireland, Italy and Great Britain (Kelly and Whitton 1989). López and Carballeira (1993) studied interspecific differences in metal accumulation among *Fontinalis antipyretica*, *Brachythecium rivulare*, *Rhynchostegium riparioides* and *Scapania undulata* to accumulate metals. *Scapania undulata* and *Rhynchostegium riparioides* showed the highest BCFs. Physico-chemical variables with major influence on metal accumulation were sulphate concentration, pH, nitrite, ammonia and filterable reactive phosphate. Yurukova et al. (1996) applied five aquatic bryophytes (*Rhynchostegium riparioides*, *Scapania undulata*, *Hygrohypnum duriusculum* (De Not.) D.W. Jamieson, *Schistidium agassizii* Sull. & Lesq., *Philonotis seriata* Mitt.) as bioconcentrators of 19 macro- and microelements (N, P, K, Ca, Mg, S, Fe, Al, Mn, Na, Zn, Cu, Pb, Cd, Co, Ni, As, Se, Cr), both at river and lake stations in Rila Mountain.

Positive correlation between copper levels in mosses and in ambient river water ($BCF > 10^3$) was established (Empain 1988). Connection between heavy metal concentrations in bryophytes (*Scapania undulata*, *Fontinalis squamosa* Hedw., *Rhynchostegium riparioides*) and water was studied in Scotland (Caines et al. 1985). Increased hydrogen ion concentrations due to soluble organic compounds and rain waters decrease Al, Mn and Zn bioaccumulation in *Scapania undulata*; moreover, accumulated Mn and Al can be released from bryophyte tissues at low pH level (<5.5). *Fontinalis antipyretica* and *Leptodictyum riparium* were suggested to monitor sites disturbed by multiple pollution sources, as in industrial and urban areas (Gecheva et al. 2011). Physico-chemical water variables represent the major component differentiating bryophyte assemblages at affected sites but the relative importance of environmental factors underlying community compositions differed strongly. Thus, in the assessment of surface water quality, bryophyte species composition was found as representative of river hydromorphology, while the content of elements in bryophyte tissue of water chemistry. The concentrations of elements in water and *F. antipyretica* collected in the same station were poorly correlated (Vázquez et al. 2004). *Fontinalis antipyretica* and *F. squamosa* appeared to avoid sites with low pH and high levels of Ca and Mg (pH, Ca and Mg were highly significantly and positively correlated), while prefer high concentrations of Cl, Na, K and Si – also significantly and positively correlated (Vázquez et al. 2007). Nevertheless, Mg and Ca play protective role for plants and reduce the toxic influence of heavy metals (Samecka-Cymerman and Kempers 1994).

Studies on both moss species and sediments are scarce. Natural background levels for Cd, Cr, Cu, Pb and Zn in four moss species (*Fontinalis antipyretica*, *Rhynchostegium riparioides*, *Amblystegium riparium* and *Fontinalis squamosa*) and sediments were reported from Portugal (Gonçalves et al. 1992). Cadmium and zinc were accumulated 107 and 70 times respectively higher than their background levels. Aquatic mosses appeared to reflect more recent conditions in freshwaters than sediments which incorporate particulate matter and are under unknown influence upstream.

An integrated research incorporated aquatic mosses (*Fontinalis squamosa* and *Rhynchostegium riparioides*), river water and sediments (Say et al. 1981).

Fontinalis antipyretica was successfully used in radiological monitoring of Cs^{137}, Cs^{134}, U^{235}, Ra^{236}, Th^{232}, K^{40} in Bulgarian montane river (Mishev et al. 1996). *Cinclidotus danubicus* was applied as monitor of radionuclides in France (Kirchmann and Lambinon 1973) and *Fontinalis* sp. in the U.K. (M A F F 1967; Hunt 1983).

Selecting sampling sites for passive biomonitoring is an important issue. Proper criteria for extensive biomonitoring network were given by Vázquez et al. (2007). When a single site was located in a drainage basin ($\approx$100 km^2), it was sited in the mid–low stretch of the river to integrate as far as possible the inputs received throughout the basin. When two sites were located in a drainage basin ($\approx$300 km^2), one was located at the head of the river and the other close to the mouth of the river. When more than two sites were located in a drainage basin (500 km^2), they were distributed in the different stretches of the main flow and of the main tributaries, covering different sub-basins. Location of sites in stretches possibly affected by reservoirs or immediately downstream of contamination foci (centres of population, industrial areas, etc.) was avoided. Each sampling site consists of a stretch of river approximately 100 m long and a sample was collected from at least five mats of one of the selected species.

Another key issue is background or reference level determination, as the metal bioavailability depends on metal form and concentration in environment. At the same time, the evaluation of environmental contamination is based on calculated CFs, i.e. on the ratio of metal concentration in the indicator to the background level. The background levels are needed also to calculate the. Background levels of course should be assessed on regional level and per species, since substrate lithology significantly differs geographically. Data are available for background levels of 19 macro- and microelements in *Fontinalis antipyretica* and high mountain river water in Bulgaria (Yurukova et al. 1997). *Scapania undulata* and *Platyhypnidium riparioides* had higher reference concentrations of Cd, Cr, Cu, Co, Ni, Pb, Zn, Fe and Mn in comparison with *Fontinalis antipyretica* and *Fissidens polyphyllus* collected from rivers in Spain (Carballeira and López 1997). Background levels of the former two species showed no significant difference with those in *Brachythecium rivulare*. Background concentrations in liverwort *Chiloscyphus* sp., *Fontinalis antipyretica* and *Platyhypnidium rusciforme* collected in French streams embedded in basaltic rocks were presented by Samecka-Cymerman and Kempers (1999). Background levels in *Platyhypnidium riparioides* for Ni, Cr, Co, Zn, Mn, Pb, Cd, Cu, Ba, Al and V were reported also by Samecka-Cymerman et al. (2002). Background levels in *P. riparioides* for Cd, Co, Cr, Fe, Mn, Ni, Pb and Zn presented by Cesa et al. (2010) for Italian River Bacchiglione basin (calcareous upper and alluvial lowland basin, respectively) were similar to those in dominated with limestone and dolomites in Tatra National Park (Samecka-Cymerman et al. 2007), and vice versa background levels in siliceous substrate are higher. Background levels for 25 elements in *F. antipyretica* and in addition element and site specific CFs were obtained by Vázquez et al. (2004).

It could be summarized that *F. antipyretica* and *P. riparioides* are the most commonly used and proven biomonitors in river quality assessment, especially in chronic exposures or long-term effects of chemical pollutants.

9.2.2 Active Biomonitoring

When no bryophyte records could be found in cases of severe pollution or environmental factors, active monitoring approach is applied. In general, mainly alive specimens collected from unaffected habitats, put in nylon cage are exposed at contamination for a period of several days or months. Major problem with moss transplantation appear to be their survival in habitats where local climate or pollution conditions do not meet species optima (Tyler 1990). Such sensitivity was known for *Fontinalis hypnoides* Hartm. (Gimeno and Puche 1999).

Although a great variety of experimental details exists (Table 9.2), the defined time period of exposition is a major advantage (Siebert et al. 1996). Despite equal metal accumulation in alive and dead material, the last can be associated with decay in plants. The number of moss bags per area, their size and attached or suspended exposition are most likely to result in different outcomes. Exposition period range has to be at least 24 h (Kelly et al. 1987), while 1 month is probably the upper limit at highly contaminated freshwaters. According to López et al. (1994) exposition should be at least for 5 days.

Already in 1970 Benson-Evans and Williams (1976) applied aquatic bryophytes to detect river pollution in Great Britain. *Fontinalis antipyretica* and *Eurhynchium riparioides* were transplanted from their natural environment at six selected stations (Table 9.2). *Fontinalis squamosa* and *Scapania undulata* were studied for heavy metal accumulation from McLean and Jones (1975). Elevated levels of Pb, Cu, Zn and Mn were reached after 6 weeks, and after 18 weeks a decay process was observed. At the same time *Scapania undulata* survived after transplantation at region with lower contamination and no considerable differences in metal content were found. Thus authors recommended selecting the commonest species in the survey area for metal monitoring in rivers. When none is characteristic for monitored river sites, then transplants can be applied under the conditions that same-age specimens of a selected bryomonitor are used and additionally physico-chemical parameters influencing accumulation (pH, light conditions and metal ambient concentrations) are monitored. Growth form is also linked to the process. *Fontinalis* has long branches, while *Scapania* formed solid tufts and thus the former species is more constantly exposed to free metal ions and respectively more rapidly accumulates metals. Bryophyte tolerance is considered to be effective until metals are included in complexes or are isolated in safe areas as cell walls (McLean and Jones 1975). When saturation is reached, any increase in metal concentration lead to damages on cell metabolism, and finally death.

Kelly et al. (1987) applied 2 cm apical stems of *Rhynchostegium riparioides* and *Fontinalis antipyretica* in monitoring of Zn, Cd, Pb in England. Alive specimens of

Table 9.2 Active biomonitoring – examples

Species	Moss-bag [g ww]	Sample preparation	Substances	Analysis	Number of studied sites, country	Exposure time	Reference
Eurhynchium riparioides, Fontinalis antipyretica	–	Perspex cylinder in a nylon mesh cage	–	Mosses were assigned to one of 5 breakdown categories – macroscopically and microscopically for morphology and structure	6, Great Britain	1, 2, 4, 6, 12, and 22 days	Benson-Evans and Williams (1976)
Cinclidotus danubicus		Plastic mesh grids (1 × 1 cm)	PCBs and α-, β, and γ-HCH	gas chromatography	France	13, 24 and 51 days	Mouvet et al. (1985)
Fontinalis antipyretica	500	Containers of plastic mesh with 2.5 × 3.5 mm pore size	Zn, Cu, Pb, Ni, Cr		4 sites, Hungary	3 months	Kovács (1992)
Fontinalis antipyretica	10		Cd, Cu, Cr, Pb, Sn		10 sites	15-24-28 days	Mersch and Pihan (1993)
Hygrohypnum ochraceum	8	Plastic mesh bags	Cd, Mn, Pb, Al	Oven-dried at 40 °C for 24 h/9 ml 4 M HNO_3/AAS	13 (including native stream), France	1 month	Claveri et al. (1995)
Fontinalis antipyretica	20	Glass fiber bags	PAHs		5 sites, Lake Kallavesi, Finland	35 days	Roy et al. (1996)
Fontinalis antipyretica	10	2–4 cm long sections in plastic nets	Cd, Pb, Zn, Cu	Dried at 80 °C/microwave digestion/ICP-MS	18, Germany	12, 14, 16 and 21 days (182 days for a lost sample)	Bruns et al. (1997)

(continued)

Table 9.2 (continued)

Species	Moss-bag [g ww]	Sample preparation	Substances	Analysis	Number of studied sites, country	Exposure time	Reference
R. riparioides, *F. antipyretica*, *Cinclidotus danubicus*	20		Zn, Pb, Cu, Cr, Ni		5 sites	11 and 16 days	Mersch and Reichard (1998)
Fontinalis antipyretica		Cages (15 × 15 × 15 cm) made from plastic coated, stainless steel wire (1.5 × 3 mm holes)	As, Cr, Cu	Samples dried at air temperature for 3–5 days, followed by drying at 40 °C for 3–4 days. Digestion: HNO_3; ICP-AES/AAS (for As)	Norway	24, 48, 72 h, 6 days	Rasmussen and Andersen (1999)
F. antipyretica	10	Nylon net, opening light of 1.3 mm	Cd, Cr, Cu, Pb, Hg	Digestion: 2.5 ml HNO_3; AAS (Cd, Pb, Cu, Cr)/CVAAS (Hg)	9 sites (lake and river), Italy	14 and 28 days	Cenci (2000)
Fontinalis antipyretica	10–12	Plastic 1 × 1 cm mesh bags	Al, Co, Cu, Ni, Zn/K, Mg, Ca	Intracellular, extracellular and particulate fractioning/AAS	4 sites (including control station), Spain	1, 4, 11, 21, 35 days (1, 4, 13, 27, 48 days at control station)	Vázquez et al. (2000)
Fontinalis antipyretica	500	Attached to the buoy	Ca, Mn, Cr, Pb, Zn, Fe, Cu, Cd	Dried at 60 °C	1 site, Korea	2 days	Lee et al. (2002)
Fontinalis antipyretica	5 g dry weight	Nylon nets (10 × 10 cm; 1 mm^2 gaps)	N, P, K, Ca, S, Mg, Na, Fe, Al, Mn, Co, Ni, Cu, Zn, Pb, Cd, As	ICP-AES/Nitrogen after Kjeldahl method	1 site, Bulgaria	31, 30, 31, 61 days (5 months in total)	Yurukova and Gecheva (2003, 2004)

Fontinalis antipyretica		Plastic tube, 150 mm long and 12 mm Diameter. Via lateral cut along the tube, only the basal part of the moss is introduced, allowing the remainder of the plant to float free	Cu, Zn, Pb, Fe, Ni, Ca, K, Mg		10 sites, Portugal		Figueira and Ribeiro (2005)
F. antipyretica	5	Spin-dried, nylon mesh bags (4 mm)	N, P, K, Ca, Mg, S, Fe, Al, Ba, Cd, Co, Cr, Cu, Mn, Ni, Pb, V, Zn		15 sites, Poland	60 days	Samecka-Cymerman et al. (2005)
Rhynchostegium riparioides	25–30	30 × 30 cm nylon bags with 7 mm holes	As, Cd, Cr, Cu, Mn, Ni, Pb, Zn	Digestion: 3.5 ml HNO_3, 0.1 ml HCl and 2 ml H_2O_2; AAS	3 sites, Italy	20, 34, 48 and 62 days	Cesa et al. (2006)
R. riparioides	20–30	Plastic net (4 mm holes)	Al, Cd, Co, Cr, Cu, Fe, Mn, Ni, Pb, Sb, Se, V, Zn	2.5 ml HNO_3 and 0.5 ml H_2O_2; AAS	119 stations placed in 64 sites, Italy	4 weeks	Cesa et al. (2010)
Fontinalis antipyretica	100	Collected from a natural stream	10 metals 16 priority PAHs		4 streams, Lisbon, Portugal	3 months	Augusto et al. (2011)
Fontinalis antipyretica	20	0.5 cm mesh size plastic bag	Zn, Cu, Ni, Pb	Mineralization in a microwave furnace with HNO_3 and $H2O_2$; FAAS	3 sites, Morava River, Czech Republic	28 days	Diviš et al. (2012)

Fontinalis antipyretica, collected from reference region and putted in nylon mesh at depth of 10–20 cm successfully reflected heavy metal contamination (Kovács 1992). Similar results were obtained with the same species during a 28-day exposure by López et al. (1994). Metal uptake kinetic reached an equilibrium phase which is mostly correlated to concentration in water. Retransplantation in an unpolluted site led to metal release occurring with a two-phase, mainly regulated by metal concentration in moss tissue at the beginning of the recovery. *F. antipyretica* was used as active biomonitor in Spain (Vázquez et al. 1999, 2000). High acidity did not appear to influence uptake by intracellular structures. During recovery significant changes were noted in extracellular metal levels. The species was recommended to be applied as active biomonitor. Claveri et al. (1995) applied transplants of *Hygrohypnum ochraceum* (Turner ex Wilson) Loeske for monitoring Al, Mn, Pb, Cd. Good correlation was observed at increasing Al and Pb values in water and moss.

As active biomonitor *F. antipyretica* showed good accumulation properties for all metals (Bruns et al. 1997). Authors recommended use of freely suspended samples in the water and noted that during autumn and winter season highest accumulation were observed. A detailed description of use of *F. antipyretica*, including sampling and conditioning was presented by Cenci (2000), and the method was recommended for detection of high risk situation in European water bodies. Transplants of *Fontinalis antipyretica* were used to assess the contamination of an industrial effluent discharge on a river located in south Portugal for 2 years (Figueira and Ribeiro 2005). An increase of water contamination by Cu, Zn was verified with extra and intracellular moss fractions. An interesting transplantation device was proposed in the study (Table 9.2).

F. antipyretica was applied simultaneously in passive and active biomonitoring of heavy metals (Samecka-Cymerman et al. 2005). Transplants accumulated significantly higher amounts of Al, Cr, Cu, Pb, V and Zn than native mosses, while Co and Mn concentrations were higher in the native specimens.

Relatively obscure are data concerning experiments with dead (initially cleaned and dried) material. Moss-bags with *Fontinalis antipyretica* of about 5 g dry weight were studied for a maximum period of 28 days along river sites in Luxembourg (Mersch and Pihan 1993), and for 5 months at impacted river site in Bulgaria respectively(Yurukova and Gecheva 2003, 2004).

Most common species applied for transplants is *Fontinalis antipyretica* not only for heavy metals but also for polycyclic aromatic hydrocarbons (Roy et al. 1996). Glass fiber bags for 35 days were applied at Lake Kallavesi (Finland) to investigate the cause-effect relationship between bioaccumulation of PAHs and the responses of antioxidant enzymes in aquatic moss. Higher activities of antioxidant enzymes and activated forms of oxygen were observed in moss transplanted near the harbor in comparison with moss-bags located upstream at two reference sites.

Regardless dead or alive are transplants, water acidity has significant influence on metal accumulation. Moss-bags accumulated Al, Pb, Fe and first two elements showed strong dependence on pH level (Mersch and Pihan 1993). Release of Cu and Zn was observed and commented in connection with high acidity. A negative correlation between water acidity and metal accumulation was also reported by Vázquez et al. (2000) during exposure of *Fontinalis antipyretica* at affected stations along Spanish rivers. Release of total Ca and Mg, as well as intracellular K at pH < 5 was established. Carballeira et al. (2001) suggested upper shoots of *Fontinalis antipyretica* could be used for assessment of temporal pH decreasing on the basis of preloaded Cd release.

The most intensive work with moss bags in recent years was done with *Rhynchostegium riparioides* in Italy (Cesa et al. 2006, 2008, 2009a, b, 2010). Transplanted moss confirmed ability to detect spatial patterns of bioaccumulation, to reveal Pb and Cu chronic contamination and Cr, Zn and Ni intermittent contamination, and to localize emission sources. Highest uptake ratios were observed for Al, Cu, Cr, Hg, and Pb under laboratory conditions.

Combined technique incorporating transplants of *Cinclidotus nigricans* for short regular periods and mussels towards long-term investigations was proposed (Mersch and Johansson 1993). Mosses and mussels appeared to be complementary in biomonitring due to specific uptake and depuration kinetics. Transfer technique was evaluated as preferable to assess the recent pollution situation in comparison with the native *Fontinalis antipyretica*.

Active biomonitoring has proven its effectiveness, especially in industrial and urban areas, although a great variety of moss-bag size, exposure periods, used material (dead or alive), and biomonitor species was tested. Thus selection of a particular species, upper segments of a same age, from reference region with known level of anthropogenic impact, is recommended. A control moss-bag at a reference station is advisable.

9.2.3 Analyses of Elements, Priority Substances and Additional Pollutants

Whole plants are recommended to be analyzed in broad studies and in tracing contamination sources, while apical shoots (2–4 cm) are more suitable in regular surveys at particular sites (Wehr and Whitton 1983a; López and Carballeira 1993; García-Álvaro et al. 2000; Cesa et al. 2010).

Due to seasonal fluctuation in metal content in bryophyte tissues and in environmental factors, it is advisable to not rely only up on samples collected during winter. Despite bryophytes grow all around the year and have nutrient uptake less seasonal than vascular plants, most species have reduced biomass during winter and this could lead to an increase in metal concentration within moss tissues. *Fontinalis antipyretica*, *Leptodictyum riparium*, and *Platyhypnidium riparioides* have vital stems throughout the year but seasonal variation of metals between species and habitats was reported (Wehr and Whitton 1983b). Found increased accumulation of Pb in the period autumn-winter was a consequence of the lower aqueous concentrations of reactive phosphate and Mg due to extended period of high flows.

Nevertheless, the results showed that seasonal influences on metal accumulation are negligible and changes which appear seasonal are probably due usually to correlated changes in aqueous chemistry.

Environmental factors influence metal uptake and affect the concentration in the plant at the time of sampling (Whitton 2003). Among measured parameters of ambient water pH should be obligatory. As observed by early works, pH of water influences metal ion accumulation, as in lower pH values the accumulation decreases (Caines et al. 1985; Lingsten 1991). In addition, aqueous Ca was pointed out as the principal non-heavy-metal factor reducing the accumulation of Zn and Cd (Kelly and Whitton 1989), while phosphates reduce Zn and Pb accumulation (Wehr and Whitton 1983a).

Chemical analyses of bryophytes should be preceded by removal of organic particles and additional fragments at field. Washing with tap water can alter element levels through cation exchange and therefore the possible cellular locations of the studied elements, this is to be considered before sampling and extraction method to be applied (Bates 2008). Washing by use of demineralized water is strongly recommended.

The representative samples of aquatic mosses by species should be carefully cleaned from other organic matter and mineral particles before analysis. The laboratories are advised to carry out homogenization of mosses according to their normal routines. The use a sample weight of at least 0.5 g, dried at 40 °C moss material, is credential, and the laboratory staff is recommended to use wet-ashing methods, preferably with concentrated or 1:1 nitric acid (with or without hydrogen peroxide or perchloric acid) or not dissolved before analysis (methodology in European Moss Surveys, Steinnes et al. 1997). Acid-digestion of mosses is performed on a hotplate or in a microwave oven using a range of temperatures. Triplicates of each sample are prepared independently.

The element concentrations could be determined by various analytical techniques, under the broad heading of atomic absorption spectrometry (flame or graphite furnace) (FAAS, GFAAS), inductively coupled plasma optical emission spectrometry (ICP-OES) and inductively coupled plasma mass spectrometry (ICP-MS), and neutron activation analysis (NAA).

In any case reference moss materials are applied in each moss series of analyses. A detection limit (DL) equivalent to three SD of the lowest instrumental measurements of the stock standard solutions, blanks – all reagents and all analytical procedures, but without biological material is recommended. The Methodological Limit of Determination (MLD) is calculated on the basis of not less than three measurements for each solution of the digested sample, dilution of digested solutions and weight of each sample, and corresponding blank. The last concentrations were determined according to all measurements and repetitions, dilution of solutions, weight of samples and blanks, usually in $mg/kg^{-1} = \mu g\ g^{-1} = ppm$. RSD% for all elements analysed is obligatory for each moss sample.

As Markert (1996) pointed out a strict differentiation between the terms 'precision' and 'accuracy' should be established in each analytical research with environmental samples. The precision of the data is connected with repeatedly measuring of the analytical signal purposing to eliminate errors, 1–5 % RSD is sufficiently exact. The accuracy of data involves use of one or more plant reference materials with certified values, and use of independent analytical procedures, i.e. inter-laboratory analyses by a wide ring of laboratories.

No one of above mentioned analytical techniques is suitable for all important essential and trace elements accumulated in moss tissues. Inductively coupled plasma optical emission spectrometry is good to be applied for more than 20 elements, in cases with higher level of Cd, Pb, As in mosses; neutron activation analysis without decomposing of the sample gives the widest idea of inorganic content – moreover 46 elements (Au and Sb included), but without important heavy metals as Cd, Cu, Pb; inductively coupled plasma mass spectrometry, available since 1984, is the combination of multi-element capabilities with extremely wide linear dynamic range over nine orders of magnitude, and could be used for 25–30 elements without P, Al, and Mn, Ca, Fe, due to a lower sensitivity thresholds, especially for toxic Al and essential P.

Analyses carried out simultaneously by ICP-MS and ICP-OES in both water and in *F. antipyretica* provided quite similar results (Vázquez et al. 2004). In moss samples the concentrations of Al, Ba, Fe, Sr and Ti measured by both methods were equivalent ($p < 0.05$), whereas those of Mn were significantly different. For the remaining elements the differences were small, obtaining higher values by ICP-OES than by ICP-MS for Co, Cr, Cu, Ni and Zn, and slightly lower for Ca levels. With respect to water samples, both techniques gave equivalent results for Ba, Ca and Zn. The remaining elements (Fe, Mn, Sr and Al), showed values obtained by ICP-OES lower than those obtained by ICP-MS, finding the greatest difference between techniques in the levels of Al measured. Thus, it can be stated that the results obtained with these two analytical methods were in general similar or equal, but not completely interchangeable.

Procedures for the analysis of organic contaminants in bryophytes include extraction from wet or freeze-dried samples with organic solvents, removal or destruction of lipids, clean-up, fractionation, high pressure liquid chromatography (HPLC) or gas chromatographic separation and different kinds of detection, e.g. fluorimetric, electron capture or MS (European Commission 2010). The total fat weight can be determined and used to normalise analytical results; this procedure should be considered as an alternative to weight normalisation.

9.3 Aquatic Ecosystem Assessment with Bryophytes Under the Water Framework Directive

The Water Framework Directive 2000/60/EC (WFD) (European Union 2000) requires Member States of the European Union (EU) to achieve good ecological status by 2015 in all water bodies. Individual water bodies are graded into one of five quality classes (high, good, moderate, poor or bad) reflecting Ecological Quality Ratio (EQR) which evaluates water quality in a score ranging from 0 (worst status) to 1 (reference status).

The WFD requires comparability of the EQR scales between the different EU countries in order to have a common understanding of the good ecological status of surface waters. This will ensure comparability of the classification results derived by various monitoring systems and reliability of the results produced by each classification tool. Biological quality elements (BQE), among them aquatic flora, are the key parameters on which the assessment is based. High ecological status is thus determined via dominance of reference species in type specific vegetation density. The status observed at the monitored station is compared to the status expected under reference/near natural conditions.

9.3.1 Aquatic Bryophytes as Bioindicators in the Context of Macrophyte Metrics

Local habitat characteristics determine river macrophyte communities, particularly light availability, current velocity, sediment patterns and nutrient supply (Birk and Willby 2010).

Compositional patterns of aquatic bryophytes are sensitive to a number of factors such as water flow velocity and level, eutrophication, pollution, and additional pressures. Substrate type also directly affects macrophyte development. Rocks and hard, immobile substrates are associated with bryophytes (Janauer and Dokulil 2006). Coarse substrate and variable flow regime contribute both to the success of bryophytes and to the exclusion of vascular hydrophytes (Scarlett and O'Hare 2006). Bryophytes are a dominant component in lotic ecosystems, especially in undisturbed conditions and their relations with environmental factors were studied (Glime and Vitt 1987; Suren 1996; Duncan et al. 1999; Suren and Duncan 1999; Suren et al. 2000; Scarlett and O'Hare 2006). Combined influence of underlying geology, water physico-chemistry, current velocity and substrate morphology on 17 bryophyte species was investigated in three minimally impacted high-latitude headwater streams in Scotland (Lang and Murphy 2012).

Anthropogenic influence results in reduction of species richness and fragmentation of populations, at the same time tolerant macrophyte taxa exhibit high growth capacity. Thus presence and abundance of aquatic macrophytes can indicate specific water characteristics such as trophic status, ion content, etc.

Aquatic plants were classified based on trophic status already 30 years ago, and for example *Sphagnum* species were allocated to plants of dystrophic, poor-in-nutrients type (Haslam 1982).

In Eastern Europe and Balkan region Papp and Rajczy (1995) suggested bryophytes as indicators of ecosystem alterations along Danube River. Correlation between the bryophyte assemblages and water quality along the Hungarian Danube section (Papp and Rajczy 1998a) and relation between assemblages of various streams in Hungary to the chemical parameters of water were studied (Papp and Rajczy 1998b). These studies report that species composition of the aquatic-riparian bryophyte vegetation and the abundance-frequency values reflect water quality along rivers. Similarly changes in aquatic bryophyte assemblages were applied for assessment of trophic level (Vanderpoorten and Palm 1998). Phosphates appeared not to influence bryophytes at lowland river sites (Scarlett and O'Hare 2006; Gecheva et al. 2010). It was found also that *Leptodictyum riparium* and *Brachythecium rivulare* are associated with high levels of phosphate. Bryophyte species composition of some Greek streams and their correlation with environmental conditions are described in Papp et al. (1998).

Most macrophyte-based assessment systems in Europe evaluate river ecological status focusing on the potential of species to detect eutrophication as main pressure (Birk et al. 2006). As part of the broad group of aquatic macrophytes, bryophytes are included to several assessment methods: French "Indice Biologique Macrophytique en Riviere" (IBMR), German Reference Index, British Mean Trophic Rank and Dutch Macrophyte Score. *Leptodictyum riparium*, *Brachythecium rivulare* and *Fontinalis antipyretica* are listed in Mean Trophic Rank (MTR) and Macrophyte Biological Index for Rivers (IBMR) with species scores: 1.0 and 2.5; 8 and 7.5; 5 and 5 respectively (Szoszkiewicz et al. 2006). *Platyhypnidium riparioides* has score (5) only in MTR.

In summary, *L. riparium* is accepted as tolerant to organic pollution by both indices. MTR examines *F. antipyretica* and *P. riparioides* as species with intermediate tolerance to nutrient enrichment, and *B. rivulare* with considerably low tolerance. In contrary, IBMR evaluates *F. antipyretica* and *B. rivulare* as tolerant to organic pollution.

Four moss species above were not included in Trophic Index of Macrophytes (TIM) since it was designed to indicate the trophic status of rivers as whole ecosystems and combining water and sediment nutrients.

These broadly used biomonitors were also listed as type specific species according Reference Index (Schaumburg et al. 2004), which classifies rivers by regional approach and reflects different kinds of ecological stresses, as well as river pollution.

Mosses and liverworts were also included in development of predictive models to evaluate the streams ecological quality: MACPACS (MACrophyte Prediction And Classification System), MAC (Macrophyte Assessment and Classification) and AQUAFLORA (Aguiar et al. 2011; Feio et al. 2012).

Seven national macrophyte assessment methods were processed to an international data set that covered three European stream types and common reference conditions were defined (Birk and Willby 2010). Type-specific reference community for siliceous mountain brooks includes liverworts (*Scapania undulata* or *Chiloscyphus polyanthus*, and less frequently, *Marsupella emarginata* or *Jungermannia atrovirens*; thallose *Pellia epiphylla*) and acrocarpous mosses (most notably *Racomitrium aciculare*, plus smaller quantities of marginal species, such as *Philonotis fontana* and *Dicranella palustris*, *Fissidens crassipes* and *F. rufulus*). These taxa occur against a backdrop of extensive growths of a range of pleurocarpous mosses, including *Rhynchostegium riparioides*, *Fontinalis squamosa*, *F. antipyretica*, *Hygrophypnum ochraceum* (and occasionally *H. luridum*), *Brachythecium rivulare*, *B. plumosum*, *Hyocomium armoricum*, *Thamnobryum alopecurum* and

Amblystegium fluviatile. Several of these species persist in the lowest quality sites, most notably *Fontinalis antipyretica* and *Rhynchostegium riparioides*, but most bryophytes are replaced by *Amblystegium riparium*.

F. antipyretica and *P. riparioides* were found to form the "core" of bryophyte communities in more stable stream habitats (Lang and Murphy 2012). Benchmark bryophyte community for upland headwater streams of reference conditions was identified. *Scapania undulata* and *Hygrohypnum ochraceum* were established as indicators for oligotrophic upland streams, and *Chiloscyphus polyanthus* and *Hygrohypnum luridum* for calcareous and mineral-rich streams.

Specification of aquatic plant composition and percentage cover at different stages of eutrophication were given in Hime et al. (2009). *Rhynchostegium riparoides*, *Leptodictyum fluviatile* and *Fontinalis antipyretica* accounted for above 65 % of the total cover in highest quality.

Fontinalis antipyretica is among dominant species in lowland rivers, associated with higher quality, while it is pointed as a negative indicator in mountain streams (Birk and Willby 2010). Scoring bryophytes in the context of macrophyte-based metrics should assume wide trophic range of several moss species, including *F. antipyretica* (Vanderpoorten et al. 1999). Moreover, *F. antipyretica* has different ecological preferences in particular regions probably due to the existence of different ecotypes (Vázquez et al. 2007). As pointed out by Glime (2007) osmotic effect plays a major role in bryophyte nutrient needs and toxicity. Except some thallose taxa, bryophytes lack epidermis and waxy in their cuticles, and are especially susceptible to osmotic shock. Thus the same species can respond quite differently under different concentrations of nutrients and heavy metals, i.e. if a plant has grown from spores at a certain nutrient/ion level, then its osmotic potential is more likely to be adjusted.

Studies on effects of hydro-morphological degradation underlined the importance of bryophytes in stream ecosystems and increased knowledge about the effects of flow regulations. Taxonomic richness was found to decrease at flow regulated sites probably due to increasing substrate stability and dominance of strong competitors (Englund et al. 1997). Additional factor for lower richness was suggested by Downes et al. (2003). Regulated streams have limited ranges of water level and as a consequence, large rocks have only narrow zones that are subject to a variety of wetted conditions, which were assumed as more suitable for bryophyte growth and colonization than constant submergence.

Bryophyte species composition is also suitable for evaluating the impact of hydromorphological alteration. Unstable river sites are characterized by small-sized species or by absence of bryophytes in strongly affected habitats such as those amenable to erosion (Gecheva et al. 2011). Quantification of the impact of small hydroelectric schemes on bryophytes and lichens showed that they are largely under-recorded in small streams where small hydroelectric schemes are likely to be developed (Demars and Britton 2011).

River Habitat Survey (RHS) method was developed in the United Kingdom and carried out in several European countries to assess, in broad terms, the physical character of freshwater streams and rivers (Raven et al. 1998). RHS only records plants growing by the riverside and the results show distribution in riparian habitats, in particular hygrophyte bryophyte species.

The potential for a macrophyte tool indicative of hydro-morphological impact is discussed by O'Hare et al. (2006). The abundance of liverworts and mosses was in negative correlation with homogeneity of water depth, deep water (the dominant depth) and fine particle substrate (the dominant substrate). Presence of *Fontinalis antipyretica* and metrics such as the presence of the group 'liverworts/mosses/lichens' may indicate that a site is unimpacted by hydro-morphological degradation.

Mosses *Fontinalis antipyretica*, *Amblystegium riparium* and *Rhynchostegium riparioides* reach their highest abundance in high stream power and coarse bed sediments in British rivers (Gurnell et al. 2009). Mosses can tolerate and may even prefer the most disturbed physical environment (O'Hare et al. 2011). Thus mosses are considered a stress-tolerant group and so they are not disadvantaged by the low nutrient and carbon conditions associated with high specific stream power and coarse substrate.

9.3.2 Establishing Environmental Quality Standards

Directive 2008/105/EC (Environmental Quality Standards Directive) defines the good chemical status also to be achieved by all Member States in 2015 and gives, together with the WFD, the legal basis for the monitoring of priority substances in sediment and biota (European Union 2008). Good chemical status refers to a list of 41 pollutants: 33 priority substances (PSs) and 8 other pollutants. Member States should have the possibility to establish EQS for sediment and/or biota at national level and to provide long-term trend analysis of concentrations of those substances listed in Part A of Annex I, giving particular consideration to Anthracene, Brominated diphenylether, Cadmium and its compounds, C10-13 Chloroalkanes, Di(2-ethylhexyl)-phthalate (DEHP), Fluoranthene, Hexachloro-benzene, Hexachloro-butadiene, Hexachloro-cyclohexane, Lead and its compounds, Mercury and its compounds, Pentachloro-benzene, Polyaromatic hydrocarbons (PAH) and Tributyltin compounds. The frequency of the monitoring has to provide sufficient data for a reliable long-term trend analysis and should take place every 3 years. In this context, perennial and long-lived shuttle bryophyte species are particularly suitable, since their life-strategy is based on 3–4 years (During 1992).

For organic substances, monitoring in biota should be performed when the biomagnification factor (BMF) is >1 or when the bioconcentration factor (BCF) is >100 (European Commission 2010). Biomagnification referring to absorption of the substances via the epithelia of the intestines, is limited to heterotrophic organisms (Markert et al. 2003). Metals from highly volatile compounds as Hg and As, are taken through the respiratory organs. In the selection of biota species, consideration should be given to the main purposes of the EQS Directive: trend monitoring and compliance with EQS.

Under the review of the existing approaches, active monitoring with transplants was evaluated as more suitable (Besse et al. 2012). Macroinvertebrates were chosen as most appropriate organisms in the active approach, as they enable robust control of biotic factors, by using size-homogenous and sex-homogenous indicator species, that lend themselves well to practical, easy-to-handle, infield caging systems. In the same article bryophytes were pointed out as useful monitors but not reliable for checking compliance with biota EQRs. We suppose that the main reason for that statement is that WFD requires biota EQSs to protect humans, top predators such as birds and mammals, and benthic and pelagic predators. Nevertheless, aquatic bryophytes meet all nine criteria stated in the Guidance on chemical monitoring of sediment and biota under the Water Framework Directive (European Commission 2010):

- "A relationship exists between contaminant concentrations in the species and average concentrations in the surrounding environment;
- The sampled organism is a potential food for predatory organisms or humans;
- The species accumulates the contaminants;
- The species is sedentary (migrating species should be avoided) and thus represents the sampling location, and does not originate e.g. from aquaculture plants;
- The species is widespread and abundant in the study region, to allow comparisons between different areas;
- The species lives long enough so that more than 1 year-class can be sampled, if desired;
- The species is large enough to yield sufficient tissue for analysis;
- The species is easy to collect and hardy enough to survive unfavourable conditions;
- The species is easy to identify."

Aquatic bryophyte species such as *Fontinalis antipyretica* and *Platyhypnidium ripariodes* comply with all the above criteria. The Guidance also pointed out that "candidate species for biota monitoring in rivers include: ... the aquatic bryophytes (e.g. genera *Fontinalis*) for heavy metals."

Plants uptake metals and other substances through tissues or organs as bioconcentrators. Unlike the most common route of metal exposure in plants through the roots, bryophytes provide direct exposure pathway (water to bryophytes) which can be used to evaluate exposure to higher trophic levels to which can cause harm. Moreover, plants give an indication of the soluble metal fraction in the environment which is likely to affect major compartments of the aquatic ecosystem. Therefore, when a heavy metal or toxic element is not detected (concentration is below MLD) in the bryophytes analyzed, indicating a low bioavailability, then water quality could be evaluated as "very good".

In addition, experimental and modelling results suggest that cationic composition of water have significant implication in the interpretation of autochthonous aquatic mosses contamination levels (Ferreira et al. 2009).

It should be emphasized that bryophytes can make an important contribution to stream metabolism and influence the distribution of key groups, such as invertebrates. Invertebrate fauna abundance is strongly correlated with bryophyte biomass, since invertebrates rely on detrital or periphyton biomass associated with the plants (Suren 1993). Most bryophyte monitoring surveys are with heavy metals, but organic contaminants were also monitored. Spatial insecticide contamination was revealed by HCH and PCBs accumulated in *Cinclidotus danubicus* moss bags (Mouvet et al. 1985). Concentration factors of 10,000 for PCBs and 300 for gamma-HCH were reported for *Platyhypnidium riparioides* (Frisque et al. 1983). Data are available that *Fontinalis antipyretica* and *F. squamosa* were affected in their photosynthetic production at 10 $\mu g\ l^{-1}$ of herbicide atrazine after 24 h and 20 days, respectively, *F. hypnoides* exhibited a much greater reduction (90 %) in net photosynthesis within 24-h at an exposure of only 2 $\mu g\ l^{-1}$ (Hofmann and Winkler 1990).

Integrated research on concordance among fish, benthic macroinvertebrates and bryophytes in Finland, showed three groups responded to different environment factors (Paavola et al. 2003). Fish community structure correlated with depth, substrate size and oxygen, macroinvertebrate community with stream size and pH, and bryophytes with water colour, nutrients and habitat variability. The last group is strongly regulated by large substrate availability and by reach-scale factors as flow variability.

According to the guidance document no 25 (European Commission 2010) transplants or so called caged organisms provide a time-integrated assessment of environmental quality over the 4-week transplantation period. In that context an Index (Palladio) for trace element alteration was proposed by Cesa et al. (2010) in the absence of autochthonous bryophytes, offering characterization of 13 elements in five classes. The metric environmental alteration (EA) based on transplanted mosses was presented according to the definition of contamination factor (CF) introduced by Mouvet (1986) for autochthonous bryophytes:

$$EA = C_{(i)}/C_{bckg}$$

EA is the ratio between the element concentration (C) at station i and the background concentration (C_{bckg}). A novel interpretation scale was applied, inspired to Mersch and Claveri (1998) and Mouvet (1986):

1. $EA < 2$ condition of naturality, no evidence of alteration (blue)
2. $2 \leq EA < 4$ suspect of alteration (green)
3. $4 \leq EA < 12$ sure alteration (yellow)
4. $12 \leq EA < 24$ severe alteration (orange)
5. $EA \geq 24$ extreme alteration (red)

Elements producing a suspected, sure, severe or extreme alteration established the condition of global environmental alteration (GEA) for each station, except for Al and Fe since their prevalent terrigenous origin.

Five classes of global environmental alteration are as follows:

– I: no evidence of environmental alteration ($EA < 2$), absolute naturality (blue)

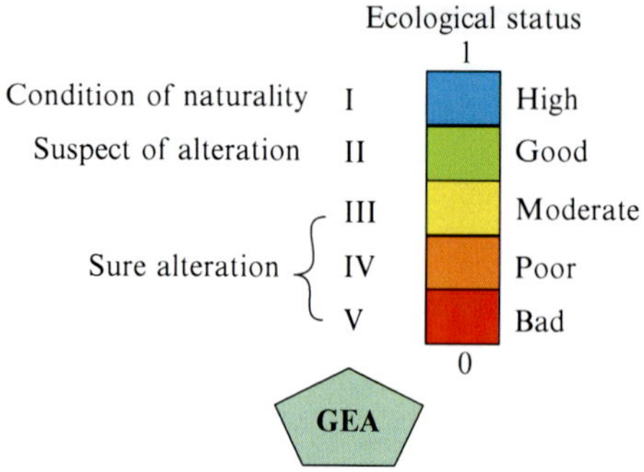

Fig. 9.1 Five classes of global environmental alteration (GEA) and relation to the classification of ecological status according to the water framework directive (WFD)

- II: suspect of alteration ($2 \leq EA < 4$) for some elements (green)
- III: sure alteration ($EA \geq 4$) for one to two elements (yellow)
- IV: sure alteration ($EA \geq 4$) for three to four elements (orange)
- V: sure alteration ($EA \geq 4$) for five or more elements (red)

Among many advantages of the Index Palladio, representation of each class of environmental alteration per station in coloured maps reflecting the WFD five ecological status classes, has to be underlined (Fig. 9.1). The Index can be easily integrated as a tool for environmental monitoring from Public Authorities.

9.4 Bryophytes: Cost-Effective and Rapid Ecological Assessment Tool?

It can be summarized that indigenous and/or transplanted bryophytes as biomonitors have important advantages: suitable sample material can be found throughout the year, low cost and rapid sampling, applicability in all freshwater habitat types, simple process of dissolution and analysis, and possibility to reveal past history (passive monitoring) of contamination or deposition during different time periods (active biomonitoring).

In unstable environment bryophytes often are the only biota representatives. As postulated by Slack and Glime (1985), many aquatic species are to some extent opportunists, since they must withstand currents and abrasion, must survive desiccation and flooding, must spread by vegetative or sexual reproduction.

Additional advantage of bryomonitoring is that most aquatic bryophytes resist freezing (up to – 10 °C) and have low temperature optima for growth (Stream Bryophyte Group 1999). Unlike vascular plants, bryophyte biomass is not grazed by most of the herbivores and thus presents opportunity for representative sample collection throughout the whole year. Moreover, bryophytes are shade-tolerant and can grow under high light intensity. Most species have very low light compensation (the light level at which net photosynthesis is zero) and light saturation points (Glime 2007).

Recent research demonstrated both in field and in laboratory that *F. antipyretica* exhibits partial desiccation tolerance which implies its significant desiccation rate and consequently survival (Cruz de Carvalho et al. 2011).

9.5 Conclusion

Aquatic bryophytes have been studied and used as biological indicators and monitors of water quality for more than 30 years. Freshwater quality can be reflected in individual species abundance as well as by the structure of bryophyte communities. Many species exhibit a high tolerance to contaminants, allowing bioaccumulation. Moreover it is well known that water sample element concentrations are often low and under the detection limits of analytical techniques. Thus bryomonitoring is used to measure bioaccumulation and bioavailability, and to assess the linkages to impacts. In the simplest experiments bioaccumulation can be characterized by measuring concentrations of metals in plant tissues and link exposure and bioaccumulation to potential impacts.

Standardization of sampling procedures in a space- and time-dependent manner and analytical techniques in aquatic bryomonitoring is needed. The number of samples should be fixed based on sampling area surface. Period of exposure time for active biomonitoring should be specified in general. Background levels and ambient metal concentrations have to be observed in parallel.

Two widespread throughout the northern hemisphere, considerably easy to recognize species are recommended in future researches: *Fontinalis antipyretica* and *Platyhypnidium riparioides* both in passive and active biomonitoring. The former is the most easily recognized aquatic species and the last is most tolerant to all kinds of pollution.

Inductively coupled plasma optical emission spectrometry (ICP-OES) could be suggested as more suitable technique in comparison with inductively coupled plasma mass spectrometry (ICP-MS) due to possibility to detect in moss tissues the macroelements (P, K, Ca and S) important for plant physiology.

Acknowledgements This review was completed through the combined efforts of the numerous scientists, who dedicated their researches on bryophytes and monitoring. Those authors have supported our interest in aquatic bryophytes and guided us. To all of them we are exceedingly grateful.

References

Aguiar FC, Feio MJ, Ferreira MT (2011) Choosing the best method for stream bioassessment using macrophyte communities: indices and predictive models. Ecol Indic 11:379–388

Augusto S, Gonzalez C, Vieira R, Máguas C, Branquinho C (2011) Evaluating sources of PAHs in urban streams based on land use and biomonitors. Environ Sci Technol 45:3731–3738

Bates JW (2008) Mineral nutrition and substratum ecology. In: Goffinet B, Shaw AJ (eds) Bryophyte biology, 2nd edn. Cambridge University Press, Cambridge, pp 315–372

Benson-Evans K, Williams PF (1976) Transplanting aquatic bryophytes to assess river pollution. J Bryol 9:81–91

Besse J-P, Geffard O, Coquery M (2012) Relevance and applicability of active biomonitoring in continental waters under the water framework directive. Trends Anal Chem 36:113–127

Birk S, Willby N (2010) Towards harmonization of ecological quality classification: establishing common grounds in European macrophyte assessment for rivers. Hydrobiologia 652:149–163

Birk S, Korte T, Hering D (2006) Intercalibration of assessment methods for macrophytes in lowland streams: direct comparison and analysis of common metrics. Hydrobiologia 566:417–430

Bruns I, Siebert A, Baumbach R, Miersch J, Günther D, Markert B, Krauβ G-J (1995) Analysis of heavy metals and sulphur-rich compounds in the water moss *Fontinalis antipyretica* L. ex Hedw. Fresen J Anal Chem 353:101–104

Bruns I, Friese K, Markert B, Krauss G-J (1997) The use of *Fontinalis antipyretica* L. ex Hedw. as a bioindicator for heavy metals. 2. Heavy metal accumulation and physiological reaction of *Fontinalis antipyretica* L. ex Hedw. in active biomonitoring in the River Elbe. Sci Total Environ 204:161–176

Burton MAS (1986) Biological monitoring. MARC report no. 32. Monitoring and Assessment Research Centre, King's College. University of London, London

Burton MAS (1990) Terrestrial and aquatic bryophytes as monitors of environmental contaminants in urban and industrial habitats. Bot J Lin Soc 104:267–286

Burton MAS, Peterson PJ (1979) Metal accumulation by aquatic bryophytes from polluted mine streams. Environ Pollut 19:39–46

Caines LA, Watt AW, Wells DE (1985) The uptake and release of some trace metals by aquatic bryophytes in acidified waters in Scotland. Environ Pollut 10:1–18

Cairns J Jr, van der Schalie WH (1980) Biological monitoring. Part I: early warning systems. Water Res 14:1179–1196

Carballeira A, López J (1997) Physiological and statistical methods to identify background levels of metals in aquatic bryophytes: dependence on lithology. J Environ Qual 26:980–988

Carballeira A, Vázquez MD, López J (2001) Biomonitoring of sporadic acidification of rivers on the basis of release of preloaded cadmium from the aquatic bryophyte *Fontinalis antipyretica* Hedw. Environ Pollut 111:95–106

Cenci RM (2000) The use of aquatic moss (*Fontinalis antipyretica*) as monitor of contamination in standing and running waters: limits and advantages. J Limnol 60(Suppl 1):53–61

Cesa M, Bizzotto A, Ferraro C, Fumagalli F, Nimis PL (2006) Assessment of intermittent trace element pollution by moss bags. Environ Pollut 144:886–892

Cesa M, Campisi B, Bizzotto A, Ferraro C, Fumagalli F, Nimis PL (2008) A factor influence study of trace element bioaccumulation in moss bags. Arch Environ Contam Toxicol 55(3):386–396

Cesa M, Azzalini G, De Toffol V, Fontanive M, Fumagalli F, Nimis PL, Riva G (2009a) Moss bags as indicators of trace metal contamination in pre-alpine streams. Plant Biosyst 143(1):173–180

Cesa M, Bizzotto A, Ferraro C, Fumagalli F, Nimis PL (2009b) S.T.R.E.A.M., system for trace element assessment with mosses. An equation to estimate mercury concentration in freshwaters. Chemosphere 75:858–865

Cesa M, Bizzotto A, Ferraro C, Fumagalli F, Nimis PL (2010) Palladio, an index of trace element alteration for the river bacchiglione based on *Rhynchostegium riparioides* moss bags. Water Air Soil Pollut 208:59–77

Claveri B, Guérold F, Pihan JC (1995) Use of transplanted mosses and autochthonous liverworts to monitor trace metals in acidic and non-acidic headwater streams (Vosges Mountains, France). Sci Total Environ 175:235–244

Cruz de Carvalho R, Branquinho C, Marques da Silva J (2011) Physiological consequences of desiccation in the aquatic bryophyte *Fontinalis antipyretica*. Planta 234(1):195–205

Dazy M, Béraud E, Cotelle S, Meux E, Masfaraud J-F, Férard J-F (2008) Antioxidantenzyme activities as affected by trivalent and hexavalent chromium species in *Fontinalis antipyretica* Hedw. Chemosphere 73(3):281–290

Delépée R, Pouliquen H, Le Bris H (2004) The bryophyte *Fontinalis antipyretica* Hedw. bioaccumulates oxytetracycline, flumequine and oxolinic acid in the freshwater environment. Sci Total Environ 322(1–3):243–253

Demars BOL, Britton A (2011) Assessing the impacts of small scale hydroelectric schemes on rare bryophytes and lichens. Scottish natural heritage and Macaulay land use institute funded report. Scottish Natural Heritage Commissioned Report no. 421

Dietz J (1972) Die Anreicherung von Schwermetallen in submersen Pflanzen. Geawasser/Abwasser 113:269–273

Diviš P, Machát J, Szkandera R, Dočekalová H (2012) In situ measurement of bioavailable metal concentrations at the downstream on the Morava River using transplanted aquatic mosses and DGT technique. Int J Environ Res 6(1):87–94

Downes BJ, Entwisle TJ, Reich P (2003) Effects on flow regulation on disturbance frequencies and in-channel bryophytes and macroalgae in some upland streams. River Res Appl 19:27–42

Duncan MJ, Suren AM, Brown SLR (1999) Assessment of streambed stability in steep, bouldery streams: development of a new analytical technique. J N Am Benthol Soc 18(4):445–456

During HJ (1992) Ecological classification of bryophytes and lichens. In: Bates JW, Farmer AM (eds) Bryophytes and lichens in a changing environment. Clarendon, Oxford

Empain A (1976) Les bryophytes aquatiques utilisés comme traceurs de la contamination en métaux lourds des eaux douces. In: Symposium Mémoires de la Société Royale de Botanique de Belgique, 20 mars 1976, Bruxelles, 7, pp 141–156

Empain AM (1977) Ecologie des populations bryophytes aquatiques de la Meuse de la Sambre et de la Somme. Doctoral dissertation, University of Liege, Belgium

Empain AM (1988) A posteriori detection of heavy metal pollution of aquatic habitats. In: Glime JM (ed) Methods in bryology. Proceedings of bryol. method. Workshop, Mainz, Hattori Bot Lab, Nichinan, pp 213–220

Empain A, Lambinon J, Mouvet C, Kirchmann R (1980) Utilisation des bryophytes aquatiques et subaquatiques comme indicateurs biologiques de la qualité des eaux courantes. In: Pesson P (ed) La Pollution des Eaux Continentales, 2nd edn. Gauthier-Villars, Paris, pp 195–223

Englund G, Jonsson B-G, Malmqvist B (1997) Effects of flow regulation on bryophytes in North Swedish rivers. Biol Conserv 79:79–86

European Commission (2010) Guidance on chemical monitoring of sediment and biota under the water framework directive. Guidance document no 25. Technical report 2010.3991. Common implementation strategy for the water framework directive. European Commission, Brussels

European Union (2000) Directive 2000/60/EC of the European Parliament and of the council of 23 October 2000 establishing a framework for community action in the field of water policy. Off J Eur Commun L 327:1–72

European Union (2008) Directive 2008/105/EC of the European Parliament and of the council of 16 December 2008 on environmental quality standards in the field of water policy, amending and subsequently repealing Council Directives 82/176/EEC, 83/513/EEC, 84/156/EEC, 84/491/EEC, 86/280/EEC and amending Directive 2000/60/EC of the European Parliament and of the Council. Off J Eur Commun L 348:84–97

Feio MJ, Aguiar FC, Almeida SFP, Ferreira MT (2012) AQUAFLORA: a predictive model based on diatoms and macrophytes for streams water quality assessment. Ecol Indic 18:586–598

Ferreira D, Ciffroy P, Tusseau-Vuillemin M-H, Garnier C, Garnier J-M (2009) Modelling exchange kinetics of copper at the water-aquatic moss (*Fontinalis antipyretica*) interface: influence of water cationic composition (Ca, Mg, Na and pH). Chemosphere 74:1117–1124

Figueira R, Ribeiro T (2005) Transplants of aquatic mosses as biomonitors of metals released by a mine effluent. Environ Pollut 136:293–301

Frisque G, Galoux M, Bernes A (1983) Accumulation de deux micropolluants (les plychlorobiphényles et le gammaHCH) par des bryophytes aquatiques de la Meuse. Meded Fac Landbouwwet Rijksuniv Gent 48:971–983

García-Álvaro MA, Martínez-Abaigar J, Núñez-Olivera E, Beaucourt N (2000) Element concentrations and enrichment ratios in the aquatic moss *Rhynchostegium riparioides* along the River Iregua (La Riojia, Northern Spain). Bryologist 103(3):518–533

Gecheva G, Yurukova L, Cheshmedjiev S, Ganeva A (2010) Distribution and bioindication role of aquatic bryophytes in Bulgarian rivers. Biotechnol Biotechnol Equip 24/2010/SE, 24(2): 164–170

Gecheva G, Yurukova L, Ganeva A (2011) Assessment of pollution with aquatic bryophytes in Maritsa River (Bulgaria). Bull Environ Contam Toxicol 87(4):480–485

Gimeno C, Puche F (1999) Chlorophyll content and morphological changes in cellular structure of *Rhynchostegium riparioides* (Hedw.) Card. (Brachytheciaceae, Musci) and *Fontinalis hypnoides* Hartm. (Fontinalaceae, Musci) in response to water pollution and transplant containers on Palancia river (East Spain). Nova Hedwigia 68:197–216

Glime JM (2003) *Fontinalis*. Available via http://www.bio.umass.edu/biology/conn.river/fontinal.html. Accessed 5 July 2012

Glime JM (2007) Bryophyte ecology. Vol 1. Physiological ecology. Ebook sponsored by Michigan Technological University and the International Association of Bryologists. Available via http://www.bryoecol.mtu.edu. Accessed 13 July 2012

Glime JM, Vitt DH (1987) A comparison of bryophyte species diversity and structure of montane streams and stream banks. Can J Bot 65:1824–1837

Gonçalves EPR, Boaventura RAR, Mouvet C (1992) Sediments and aquatic mosses as pollution indicators for heavy metals in the Ave river basin (Portugal). Sci Total Environ 114:7–24

Grolle R, Long DG (2000) An annotated check-list of the Hepaticae and Anthocerotae of Europe and Macaronesia. J Bryol 22:103–140

Gurnell AM, O'Hare JM, O'Hare MT, Dunbar MJ, Scarlett PM (2009) An exploration of associations between assemblages of aquatic plant morphotypes and channel geomorphological properties within British rivers. Geomorphology 116(1–2):135–144

Haseloff HP (1982) Bioindikatoren und bioindikation. Biol Unserer Zeit 12:20–26

Haslam SM (1982) A proposed method for monitoring river pollution using macrophytes. Environ Technol Lett 3:19–34

Hertz J (1991) Bioindicators for monitoring heavy metals in the environment. In: Merian E (ed) Metals and their compounds in the environment. VCH Verlagsgesellschaft mbH, Weinheim, New York, Basel, Cambridge, pp 221–231

Hill MO, Bell N, Bruggeman-Nannenga MA, Brugués M, Cano MJ, Enroth J, Flatberg KI, Frahm J-P, Gallego MT, Garilleti R, Guerra J, Hedenäs L, Holyoak DT, Hyvönen J, Ignatov MS, Lara F, Mazimpaka V, Muñoz J, Söderström L (2006) An annotated checklist of the mosses of Europe and Macaronesia. J Bryol 28(3):198–267

Hime S, Bateman IJ, Posen P, Hutchins M (2009) A transferable water quality ladder for conveying use and ecological information within public surveys. CSERGE working paper EDM 09–01, University of East Anglia

Hofmann A, Winkler S (1990) Effects of atrazine in environmentally relevant concentrations on submerged macrophytes. Arch Hydrobiol 118:69–80

Hongve D, Brittain JE, Bjørnstad HE (2002) Aquatic mosses as a monitoring tool for ^{137}Cs contamination in streams and rivers – a field study from central southern Norway. J Environ Radioact 60:139–147

Hunt GJ (1983) Radioactivity in surface and coastal waters of the British Isles, 1981. Aquatic environment monitoring report no. 9. Ministry of Agriculture, Fisheries and Food Lowestoft, Suffolk

Jackson PP, Bobinson NJ, Whitton BA (1991) Low molecular weight metal complexes in the freshwater moss *Rhinchostegium riparoides* exposed to elevated concentrations of Zn, Cu, Cd, and Pb in the laboratory and field. Environ Exp Bot 31(3):359–366

Janauer G, Dokulil M (2006) Macrophytes and algae in running waters. In: Ziglio G, Siligardi M, Flaim G (eds) Biological monitoring of rivers. Wiley, Chichester, pp 89–109

Jones CG, Lawton JH, Shachak M (1994) Organisms as ecosystem engineers. Oikos 69:373–386

Kelly MG, Whitton BA (1989) Interspecific differences in Zn, Cd and Pb accumulation by freshwater algae and bryophytes. Hydrobiologia 175:1–11

Kelly MG, Girton C, Whitton BA (1987) Use of moss-bags for monitoring heavy metals in rivers. Water Res 21(11):1429–1435

Kirchmann R, Lambinon J (1973) Plants as bioindicators of the contamination of a river by the effluents of a PWR nuclear power station. Assessment of the radioactive releases of the Sena reactor through aquatic and ripicolous plants of the river Meuse. Bull Soc Roy Bot Belg 106:187–201

Kovács M (1992) Biological indicators of water pollution. In: Kovács M (ed) Biological indicators in environmental protection. Akadémiai Kiadó, Budapest, pp 120–130

Kovács M, Podani J (1986) Bioindication: a short review on the use of plants as indicators of heavy metals. Acta Biol Hung 37:19–29

Lang P, Murphy KJ (2012) Environmental drivers, life strategies and bioindicator capacity of bryophyte communities in high-latitude headwater streams. Hydrobiologia 679:1–17

Lee J, Johnson-Green P, Lee EJ (2002) Chemical analysis of transplanted aquatic mosses and aquatic environment during a fish kill on the Chungnang River, Seoul, Korea. Korean J Biol Sci 6:215–219

Lingsten L (1991) Levels of heavy metals in aquatic mosses in acidified water bodies. Verh Int Ver Limnol 24:2228–2230

López J, Carballeira A (1989) A comparative study of pigment contents and response to stress in five aquatic bryophytes. Lindbergia 15:188–194

López J, Carballeira A (1993) Interspecific differences in metal bioaccumulation and plant-water concentration ratios in five aquatic bryophytes. Hydrobiologia 263:95–107

López J, Vazquez MD, Carballeira A (1994) Stress responses and metal exchange kinetics following transplant of the aquatic moss *Fontinalis antipyretica*. Freshwater Biol 32:185–198

M A F F (Ministry of Agriculture, Fisheries and Food) (1967) Radioactivity of surface and coastal waters of the British Isles. Report FRL I. Ministry of Agriculture, Fisheries and Food, Lowestoft, Suffolk

Manning WJ, Feder WA (1980) Biomonitoring air pollutants with plants. Applied Science Publishers, London, p 142

Markert B (1991) Inorganic chemical investigation in the forest biosphere reserve near Kalinin, USSR. I. Mosses and peat profiles as bioindicators for different chemical elements. Vegetatio 95:127–135

Markert B (1996) Instrumental element and multi-element analyses of plant samples. Wiley, Chichester, 296

Markert B (2008) From biomonitoring to integrated observation of the environment – the multi-markered bioindication concept. Ecol Chem Eng S 15(3):315–333

Markert BA, Breure AM, Zechmeister HG (2003) Definitions strategies and principles for bioindication/biomonitoring of the environment. In: Markert BA, Breure AM, Zechmeister HG (eds) Bioindicators and biomonitors, Elsevier Science Ltd, Oxford, pp 3–39

Martin MH, Coughtrey PJ (1982) Biological monitoring of heavy metal pollution. Applied Science Publishers, London/New York, p 475

Martinez-Abaigar J, Núñez-Olivera E (1998) Ecophysiology of photosynthetic pigments in aquatic bryophytes. In: Bates JW, Ashton NW, Duckett JG (eds) Bryology for the twenty-first century. British Bryological Society, Leeds, pp 277–292

McLean RO, Jones K (1975) Studies of tolerance to heavy metals in the flora of the rivers Ystwyth and Clarach, Wales. Freshwater Biol 5:431–444

Mersch J, Claveri B (1998) Les bryophytes aquatiques comme outil de surveillance de la contamination des eaux courantes par les micropolluants métalliques: modalités d'interprétation des données. Etude Inter-Agences 55, France

Mersch J, Johansson L (1993) Transplanted aquatic mosses and freshwater mussels to investigate the trace metal contamination in the rivers Meurthe and Plaine, France. Environ Technol 14:1027–1036

Mersch J, Pihan JC (1993) Simultaneous assessment of environmental impact on condition and trace metals availability in zebra mussels *Dreissena polymorpha* transplanted into the Wilz River, Luxembourg. Comparison with the aquatic moss *Fontinalis antipyretica*. Arch Environ Contam Toxicol 25:353–364

Mersch J, Reichard M (1998) In situ investigation of trace metal availability in industrial effluents using transplanted aquatic mosses. Arch Environ Contam Toxicol 34:336–341

Mishev P, Damyanova A, Yurukova L. (1996) Mosses as biomonitors of airborne pollution in the northern part of Rila Mountain. Part II. Radioactivity of mosses. In: Carbonnel JP, Stamenov JN (eds) Observatoire de montagne de Moussala – OM2, 4. Sofia, pp 137–141

Mouvet C (1984) Accumulation of chromium and copper by the aquatic moss *Fontinalis antipyretica* L. ex Hedw. transplanted in a metal-contaminated river. Environ Technol Lett 5:541–548

Mouvet C (1986) Métaux lourds et Mousses aquatiques, synthèse méthodologique. Université de Metz – Laboratoire d'Ecologie, rapport de contrat à l'Agence de l'Eau Rhin-Meuse et l'Agence de l'Eau Rhône-Méditerranée-Corse, France

Mouvet C, Galoux M, Bernes A (1985) Monitoring of polychlorinated biphenyls (PCBs) and hexachlorocyclohexanes (HCH) in freshwater using aquatic moss *Cinclidotus danubicus*. Sci Total Environ 44:253–267

Mouvet C, Pattée E, Cordebar P (1986) Utilisation des mousses aquatiques pour l'identification et la localisatiom précise de sources de pollution métallique multiforme. Acta Oecol 7(1):77–91

O'Hare MT, Baattrup-Pedersen A, Nijboer R, Szoszkiewicz K, Ferreira T (2006) Macrophyte communities of European streams with altered physical habitat. Hydrobiologia 566:197–210

O'Hare JM, O'Hare MT, Gurnell AM, Dunbar MJ, Scarlett PM, Laize C (2011) Physical constraints on the distribution of macrophytes linked with flow and sediment dynamics in British rivers. River Res Appl 27:671–683

Paavola R, Muotka T, Virtanen R, Heino J, Kreivi P (2003) Are biological classifications of headwater streams concordant across multiple taxonomic groups? Freshwater Biol 48:1912–1923

Papp B, Rajczy M (1995) Changes of bryophyte vegetation and habitat conditions along a section of the river Danube in Hungary. Cryptog Helv 18:95–105

Papp B, Rajczy M (1998a) The role of bryophytes as bioindicators of water quality in the Danube. Verh Int Ver Theor Limnol 26:1254–1256

Papp B, Rajczy M (1998b) Investigations on the condition of bryophyte vegetation of mountain streams in Hungary. J Hattori Bot Lab 84:81–90

Papp B, Tsakiri E, Babalonas D (1998) Bryophytes and their environmental conditions at Enipeas (Mt Olympos) and Lycorrema (Mt Ossa) streams (Greece). In: Tsekos I, Moustakas M (eds) Progress in botanical research: proceedings of 1st Balkan Botanical Congress, Dordrecht, Boston, pp 129–132

Peñuelas J (1984) Pigments of aquatic mosses of the River Muga, NE Spain, and their response to water pollution. Lindbergia 10:127–132

Pirc S (2003) Aquatic mosses as sampling medium in geochemistry. RMZ-Mater Geoenviron 50(4):735–763

Rasmussen G, Andersen S (1999) Episodic release of arsenic, copper and chromium from awood preservation site monitored by transplanted aquatic moss. Water Air Soil Pollut 109:41–52

Raven PJ, Holmes NTH, Dawson FH, Fox PJA, Everard M, Fozzard I, Rouen KJ (1998) River habitat quality: the physical character of rivers and streams in the UK and the Isle of Man. Environment Agency, Bristol

Roy S, Sen CK, Hanninen O (1996) Monitoring of polycyclic aromatic hydrocarbons using moss bags: bioaccumulation and responses of antioxidant enzymes in *Fontinalis antipyretica* Hedw. Chemosphere 32:2305–2315

Samecka-Cymerman A (1991) Contents of arsenic, vanadium, germanium and other elements in aquatic liverwort *Scapania undulata* (L.) Dum. growing in the Sudety mountains. Pol Arch Hydrobiol 38(1):79–84

Samecka-Cymerman A, Kempers AJ (1992) Anomalous elemental composition of aquatic bryophytes near barite zones in the Sowie Mts. (Poland). J Geochem Explor 43:213–221

Samecka-Cymerman A, Kempers AJ (1993) *Scapania undulata* (L.) Dum. and other aquatic bryophytes as indicators of mineralization in Poland. J Geochem Explor 46:325–334

Samecka-Cymerman A, Kempers AJ (1994) Macro- and microelements in bryophytes from serpentinite and greenstone streams. Pol Arch Hydrobiol 41:431–449

Samecka-Cymerman A, Kempers AJ (1999) Background concentrations of heavy metals in aquatic bryophytes used for biomonitoring in basaltic areas (a case study from central France). Environ Geol 39(2):117–122

Samecka-Cymerman A, Kolon K, Kempers AJ (2002) Heavy metals in aquatic bryophytes from the Ore Mountains (Germany). Ecotoxicol Environ Saf 52:203–210

Samecka-Cymerman A, Kolon K, Kempers AJ (2005) A comparison of native and transplanted *Fontinalis antipyretica* Hedw. as biomonitor of water polluted with heavy metals. Sci Total Environ 341:97–107

Samecka-Cymerman A, Stankiewicz A, Kolon K, Kempers AJ (2007) Self-organizing feature map (neural networks) as a tool in classification of the relations between chemical composition of aquatic bryophytes and types of streambeds in the Tatra national park in Poland. Chemosphere 67:954–960

Samecka-Cymerman A, Wojtun B, Kolon K, Kempers AJ (2011) *Sanionia uncinata* (Hedw.) Loeske as bioindicator of metal pollution in polar regions. Polar Biol 34:381–388

Samson G, Tessier L, Vaillancourt G, Pazdernik L (1998) The use of a first order kinetic model for assessing the cadmium bioaccumulation process by two aquatic mosses. Available via http://www.wmsym.org/archives/1998/html/sess18/18-20/18-20.htm. Assessed 25 June 2012

Satake K, Takamatsu T, Soma M, Shibata K, Nishikawa M, Say PJ, Whitton BA (1989) Lead accumulation and location in the shoots of the aquatic liverwort *Scapania undulata* (L.) Dum. in stream water at Greenside Mine, England. Aquat Bot 33:111–122

Say PJ, Whitton BA (1983) Accumulation of heavy metals by aquatic mosses. I. *Fontinalis antipyretica*. Hydrobiologia 100:245–260

Say PJ, Harding JPC, Whitton BA (1981) Aquatic mosses as monitors of heavy metal contamination in the river Etherow, Great Britain. Environ Pollut 2:295–307

Scarlett P, O'Hare M (2006) Community structure of in-stream bryophytes in English and Welsh rivers. Hydrobiologia 553:143–152

Schaumburg J, Schranz C, Foerster J, Gutowski A, Hofmann G, Meilinger P, Schneider S, Schmedtje U (2004) Ecological classification of macrophytes and phytobenthos for rivers in Germany according to the water framework directive. Limnologica 34:283–301

Siebert A, Bruns I, Krauss G-J, Miersch J, Markert B (1996) The use of the aquatic moss *Fontinalis antipyretica* L. ex Hedw. as a bioindicator for heavy metals. 1. Fundamental investigations into heavy metal accumulation in *Fontinalis antipyretica* L. ex Hedw. Sci Total Environ 177:137–144

Slack NG, Glime J (1985) Niche relationship of mountain stream bryophytes. Bryologist 88:1–18

Spitale D (2009) Spatial distribution of bryophytes along a moisture gradient: an approach using photosynthetic pigments as indicators of stress. Ecol Res 24:1279–1286

Steinnes E, Rühling Å, Lippo H, Mäkinen A (1997) Reference materials for large-scale metal deposition surveys. Accred Qual Assur 2:243–249

Steubing L (1976) Niedere und höhere pflanzen als indikatoren für immissionsbelastungen. Daten Dok Umweltschutz 19:13–27

Stöcker G (1980) Methodishe und theoretishe Grundlangen der Bioindikation (Bioindikation 1). In: Schubert R, Schuh J (eds) Martin Luther Universität, Halle, pp 10–21

Stream Bryophyte Group (1999) Roles of bryophytes in stream ecosystems. J N Am Benthol Soc 18(2):151–184

Suren A (1993) Bryophytes and associated invertebrates in first-order alpine streams of Arthur's Pass, New Zealand. N Z J Mar Freshwater Res 27:479–494

Suren AM (1996) Bryophyte distribution patterns in relation to macro-, meso-, and micro-scale variables in South Island, New Zealand streams. N Z J Mar Freshwater 30:501–523

Suren AM, Duncan MJ (1999) Rolling stones and mosses: effect of substrate stability on bryophyte communities in streams. J N Am Benthol Soc 18(4):457–467

Suren AM, Smart GM, Smith RA, Brown SLR (2000) Drag coefficients of stream bryophytes: experimental determinations and ecological significance. Freshwater Biol 45:309–317

Szoszkiewicz K, Ferreira T, Korte T, Baattrup-Pedersen A, Davy-Bowker J, O'Hare M (2006) European river plant communities: the importance of organic pollution and the usefulness of existing macrophyte metrics. Hydrobiologia 566:211–234

Tyler G (1972) Heavy metals pollute nature may reduce productivity. Ambio 1:57–59

Tyler G (1990) Bryophytes and heavy metals: a literature review. Bot J Linn Soc 104:231–253

Vanderpoorten A, Goffinet B (2009) Introduction to bryophytes. Cambridge University Press, Cambridge

Vanderpoorten A, Palm R (1998) Canonical variables of aquatic bryophyte combinations for predicting water trophic level. Hydrobiologia 386:85–93

Vanderpoorten A, Klein J-P, Stieperaere H, Tremolieres M (1999) Variations of aquatic bryophyte assemblages in the Rhine Rift related to water quality. 1. The Alsatian Rhine floodplain. J Bryol 21:17–23

Vázquez MD, López L, Carballeira A (1999) Uptake of heavy metals to the extracellular and intracellular compartments in three species of aquatic bryophyte. Ecotoxicol Environ Saf 44:12–24

Vázquez MD, Fernández JA, López L, Carballeira A (2000) Effects of water acidity and metal concentration on accumulation and within-plant distribution of metals in the aquatic bryophyte *Fontinalis antipyretica*. Water Air Soil Pollut 120:1–19

Vázquez MD, Wappelhorst O, Markert B (2004) Determination of 28 elements in aquatic moss *Fontinalis antipyretica* Hedw. and water from the upper reaches of the river Nysa (CZ, D), by ICP-MS, ICP-OES and AAS. Water Air Soil Pollut 152:153–172

Vázquez MD, Fernández JÁ, Real C, Villares R, Aboal JR, Carballeira A (2007) Design of an aquatic biomonitoring network for an environmental specimen bank. Sci Total Environ 388(1–3):357–371

Vieira AR, Gonzalez C, Martins-Loução MA, Branquinho C (2009) Intracellular and extracellular ammonium (NH_4^+) uptake and its toxic effects on the aquatic biomonitor *Fontinalis antipyretica*. Ecotoxicology 18:1087–1094

Wehr JD, Whitton BA (1983a) Accumulation of heavy metals by aquatic mosses. 2: *Rhynchostegium riparioides*. Hydrobiologia 100:261–284

Wehr JD, Whitton BA (1983b) Accumulation of heavy metals by aquatic mosses. 3: seasonal changes. Hydrobiologia 100:285–291

Wehr JD, Empain A, Mouvet C, Say PJ, Whitton BA (1983) Methods for processing aquatic mosses used as monitors of heavy metals. Water Res 17(9):985–992

Whitton BA (2003) Use of plants for monitoring heavy metals in freshwaters. In: Ambasht RS, Ambasht NK (eds) Modern trends in applied aquatic ecology. Kluwer Academic/Plenum Publishers, New York, pp 43–63

Whitton BA, Say PJ, Jupp BP (1982) Accumulation of zinc, cadmium and lead by the aquatic liverwort *Scapania*. Environ Pollut 3:299–316

Yurukova L, Gecheva G (2003) Active and passive biomonitoring using *Fontinalis antipyretica* in Maritsa River, Bulgaria. J Balkan Ecol 6(4):390–397

Yurukova L, Gecheva G (2004) Biomonitoring in Maritsa River using aquatic bryophytes. J Environ Prot Ecol 5(4):729–735

Yurukova LD, Ganeva AS, Damyanova AA (1996) Aquatic bryophytes as bioconcentrators of macro- and microelements. In: Carbonnel JP, Stamenov JN (eds) Observatorie de montagne de Moussala, OM2, 4. Bulgarian Academy of Sciences, Sofia, pp 127–136

Yurukova L, Damyanova A, Ivanov K, Janminchev V (1997) Bioaccumulation capacity of *Fontinalis antipyretica* from Musalenska Bistrica River, Rila Mountain. In: Carbonnel JP, Stamenov JN (eds) Observatorie de montagne de Moussala, OM2, 6. Bulgarian Academy of Sciences, Sofia, pp 77–86

Zhou Q, Zhang J, Fu J, Shi J, Jiang G (2008) Biomonitoring: an appealing tool for assessment of metal pollution in the aquatic ecosystem. Anal Chim Acta 606:135–150

Chapter 10
Halogenated PAH Contamination in Urban Soils

Takeshi Ohura, Teru Yamamoto, Kazuo Higashino, and Yuko Sasaki

Abstract Halogenated polycyclic aromatic hydrocarbons (HPAH) such as Cl-PAH and Br-PAH are new ubiquitous environmental aromatic contaminants. We found HPAH in urban soils and river sediments in Japan. We developed the analysis of HPAH by gas chromatography-electron ionization-tandem mass spectrometry. 20 Cl-PAH, 11 Br-PAH, and 16 PAH were analyzed. We found concentrations from 55.6 to 534,000 ng/g for Cl-PAH, 3.14–914 ng/g for Br-PAH and 19,800–1,080,000 ng/g for PAH. Similar compositions of HPAH were observed among some soil samples, suggesting that common emission sources. The highest contaminations of Cl-PAH were observed in old chlor-alkali plant soils. To test HPAH origin from the industry, electrolysis of brine in the presence or absence of pitch were carried out in the laboratory. We found that congener and homolog profiles of HPAH produced by electrolysis are consistent with those of soils from old chlor-alkali plant sites. Consequently, chlor-alkali production facilities including electrolysis of brine in the presence of graphite could be an unintentional source of HPAH in the environment.

Keywords PAH • Cl-PAH • Urban soil pollution • PCB • Br-PAH • GC/MS • Brine electrolysis • Sediments

T. Ohura (✉)
Department of Environmental Bioscience, Meijo University, 1-501 Shiogamaguchi, Nagoya 468-8502, Japan
e-mail: ohura@meijo-u.ac.jp

T. Yamamoto • K. Higashino • Y. Sasaki
The Tokyo Metropolitan Research Institute for Environmental Protection, 1-7-5 Sinsuna, Koto-ku Tokyo 136-0075, Japan

E. Lichtfouse et al. (eds.), *Green Materials for Energy, Products and Depollution*, Environmental Chemistry for a Sustainable World 3, DOI 10.1007/978-94-007-6836-9_10,

Contents

10.1 Introduction

Many organohalogen compounds (OHC) have been used extensively as chemical materials including pesticide and insulator, and are raising our standard of living (Tanabe and Minh 2010). On the other hand, a portion of such compounds emitted in the environment are possible to create various health menaces due to their toxicities (Schecter et al. 2006; Van den Berg et al. 1998). In fact, there are environmental quality standards for organohalogen compounds in the air, water, and soil, in Japan. As one of the hazardous organohalogen compounds, certain polychlorinated dibenzo-*p*-dioxins and dibenzofurans (PCDD/Fs), and polychlorinated biphenyls (PCB) have been adequately recognized by their global distribution, persistence and potent toxicities. (Schecter et al. 2006; Van den Berg et al. 1998). They are therefore concerned to cause problems on adverse effects of not only human health but ecological integrity. The structural characteristics caused by steric hindrance and lipophilicity would be attributed main factors of environmental persistent and bioaccumulative potential.

Polycyclic aromatic hydrocarbons (PAH) have been also known as typical environmental hazardous pollutants because certain species of them possess strong mutagenicity/carcinogenicity and endocrine-disrupting effects, like dioxins and PCBs (Boström et al. 2002). The wide distribution of PAH in the environment could be due to easily productivity during combustion process of organic materials. That is, there are some common features including the sources and toxicities between dioxins and PAH, so that halogenated PAH (HPAH) having structural combination of dioxins and PAH are possibly present in the environment. Based on such hypothesis, author and co-workers have tried to examine the environmental occurrence of HPAH and actually detected many HPAH in the air (Kitazawa et al. 2006; Ohura et al. 2005, 2008, 2009). HPAH in the air showed the behaviors as expected. For example, the concentrations of certain chlorinated PAH (Cl-PAH) associated with particles showed the typical seasonal trend observed in the case of PAH: elevated in winter and decreased in summer (Ohura et al. 2005, 2008). The concentrations of Cl-PAH also showed the significant correlations to those of corresponding parent PAH. These facts suggest that the Cl-PAH in the air are possibly derived from common sources of PAH. Furthermore, much of Cl-PAH showed that the photostability increased with increasing the number of chlorine substituted on the PAH skeleton (Ohura et al. 2004, 2005). It indicates that Cl-PAH may show a higher environmental persistence rather than PAH.

The study on environmental analysis of HPAH except in air and soils is further limited. Recently, the occurrences of HPAH in rice in China were reported and estimated the exposure risks (Ding et al. 2012). Although it suggests that the cancer risk values were below priority risk level (10^{-4}), the overall exposure processes of HPAH remained unclear. There is an incomplete understanding of HPAH's environmental occurrences and emission sources at present, so that lack of knowledge of HPAH would discourage to evaluate the environmental and biological effects. Here, we introduce the occurrences of HPAH contaminations in urban surface soils. We also discuss the possible sources of HPAH.

10.2 Halogenated Polycyclic Aromatic Hydrocarbons Analysis by GC/MS/MS

Although HPAH analysis has been developed using GC/MS in EI mode, those methods were applied to samples collected from air (Ohura et al. 2008, 2009). In general contaminants in air samples including gas and particles contents could be lower than those in soil and sediment samples, so that strict removal of contaminants, i.e. cleanup process, from such environmental samples will need to analyze effectively using GC/MS.

In the case of analysis of dioxins in soils, treatments using sulfuric acid and reduced copper in the cleanup process are frequently performed to reduce pigment compositions from the samples. Although the stabilities of Cl-PAH during the reduced copper treatment are unclear, the sulfuric acid treatment degrades dramatically not only PAH but also Cl-PAH (data not shown). Therefore cleanup media of Cl-PAH should be adopted that used in the case of PAH such as silica gel and alumina (Liu et al. 2007; Ohura et al. 2005, 2008). Indeed, for analysis of Cl-PAH in air, the pretreatment process used silica gel column showed reasonable recovery rates of 92–114 % (Ohura et al. 2005). As other cleanup procedures for PAH analysis in highly contaminated samples, GPC and dialysis used semipermeable membrane devices are also adopted (Czuczwa and Alford-Stevens 1989; Williamson et al. 2002). Although the adjustability for HPAH has been not investigated yet, high selectivity and condensation may be expected to employ those procedures. As instruments in support of the cleanup processes, usage of GC-tandem MS (GC/MS/MS) will be an efficient tool for analysis of samples including matrix such as soils because of possessing the high sensitivity and selectivity. Here, we introduce a new analytical method of HPAH by using GC/MS/MS.

Twenty species of Cl-PAH (Fig. 10.1) and 11 species of Br-PAH (Fig. 10.2) were used as HPAH standard compounds. The HPAH were analyzed on a gas chromatograph (Agilent HP7890A) interfaced with a tandem quadrupole mass

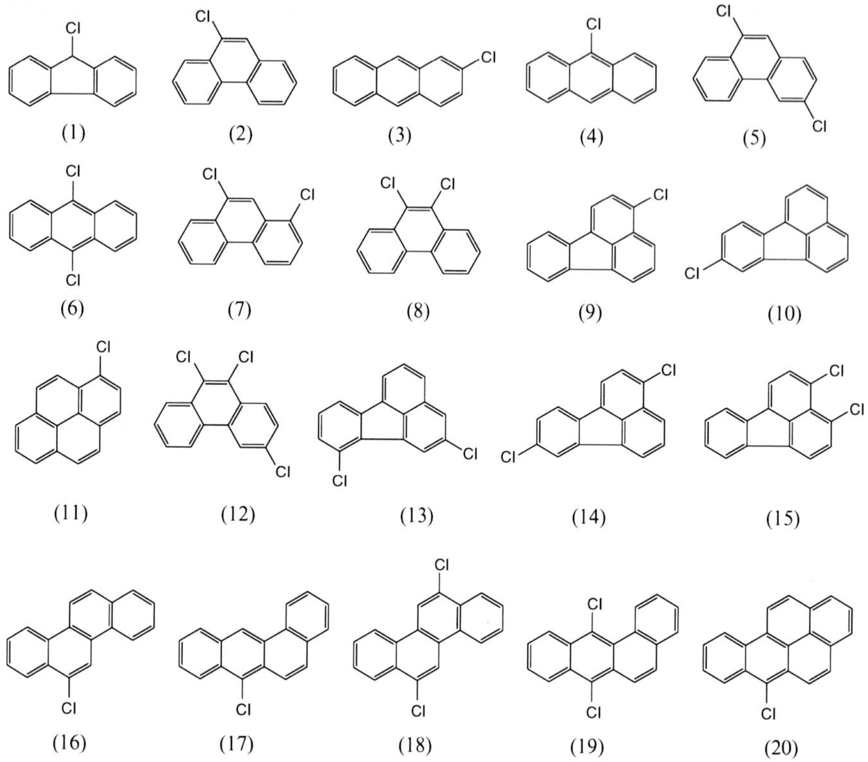

Fig. 10.1 Chemical structures of monitored Cl-PAH. *PAH* polycyclic aromatic hydrocarbons

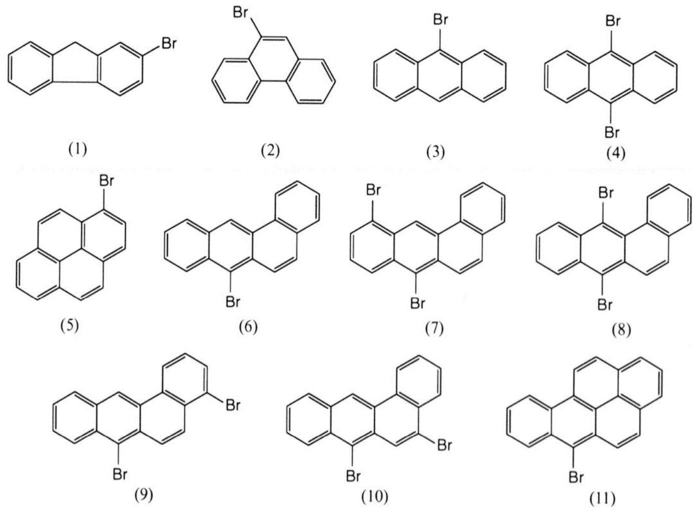

Fig. 10.2 Chemical structures of monitored Br-PAH. *PAH* polycyclic aromatic hydrocarbons

spectrometer (Waters Quattro micro GC), operating in EI mode (70 eV). The analytical conditions were as follows: for both Cl-PAH and Br-PAH, injection temperature was 250 °C and 5 μL of extracts were injected in the system using pulsed splitless mode at pulse time of 2 min and pulsed pressure of 20 psi. An DB-5 ms capillary column (30 m × 0.25 mm i.d. × 0.25 μm film thickness, Agilent) was used. A constant flow of helium at 1.0 mL/min was used as the carrier gas. The GC column temperature was programmed from 100 (initial equilibrium time 1 min) to 200 °C via a ramp of 25 °C/min, 200–300 °C via a ramp of 5 °C/min and maintained at 300 °C for 5 min. GC interface and ionization source temperatures were both 300 °C and the system operated in MS/MS mode using argon as collision gas at a pressure of 2.8×10^{-3} mbar in the collision cell. The developed multiple reaction monitoring (MRM) conditions of targeted HPAH and the detection limits were listed in Table 10.1. The GC/MS/MS chromatograms of each standard mixture of Cl-PAH and Br-PAH were represented in Figs. 10.3 and 10.4, respectively, which achieved good separations.

10.3 Halogenated Polycyclic Aromatic Hydrocarbons in Sediments

There are a few reports that the concentrations of Cl-PAH were investigated in sediment samples (Horii et al. 2009; Ma et al. 2009, Sun et al., 2011). Horii et al. (2009) investigated the residual concentrations and profiles of Cl-PAH in sediments from water bodies near industrialized areas, Tokyo Bay, Japan; the Saginaw River watershed, Michigan, USA; a former chlor-alkali plant, Georgia, USA; and the New Bedford Harbor Superfund site, Massachusetts, USA. Among the soil samples, a high mean concentration of Cl-PAH was observed in sediment collected near a former chlor-alkali plant (8.82 ng/g) in comparison with other sediments affecting multiple sources such as New Bedford Harbor (1.88 ng/g) and Saginaw River watershed (1.14 ng/g). It suggests that chlor-alkali plant could be concerned with formation of Cl-PAH. In addition, Cl-PAH profile observed in the chlor-alkali plant showed that the dominant components were 6-Cl-BaP and 1-Cl-Py. Such profiles have been often observed in those of ambient air samples. This fact suggests that sources of Cl-PAH in the sediment samples are significantly related to those in air samples. Sun et al. (2011) recently reported that the occurrences of Cl-PAH and Br-PAH in surface sediments in Shenzhen, South China. Despite a small number of targeted Cl-PAH (9,10-dichloroanthracene, 9-chlorophenanthrene, and 2-chloroanthracene), the mean concentration of Cl-PAH was 27.6 ng/g, which was >3-times higher than those in the above study. Since no correlation was observed between the concentrations of Cl-PAH and parent PAH, the formation of Cl-PAH may have no relation of that of PAH. Although the authors suggested that the regional urbanization account for increase in HPAH levels in sediment, the direct emission sources remained unclear.

Table 10.1 GC/MS/MS analytical conditions of 20 Cl-PAH and 11 Br-PAH. The running number of compound is corresponded to Fig. 10.1 (Cl-PAH) and 2 (Br-PAH). *GC* gas chromatography, *MS* mass spectrometry, *PAH* polycyclic aromatic hydrocarbons

No	Compound	Abbrev.	RT	MRMI	MRM2	MRM3	MRM4	MRM5	DL (pg)
	ClPAH								
1	9-chlorofluorene	9-ClFlu	8.03	202 > 165 (10)	165 > 164 (16)	165 > 163 (28)	165 > 115 (24)	165 > 139 (24)	16.7
2	9-chlorophenanthrene	9-ClPhe	9.97	212 > 176 (24)	212 > 177 (12)	212 > 151 (28)	212 > 150 (44)	214 > 177 (12)	10.6
3	2-chloroanthracene	2-ClAnt	10.12	212 > 176 (24)	212 > 177 (12)	212 > 151 (28)	212 > 150 (44)	214 > 177 (12)	10.5
4	9-chloroanthracene	9-ClAnt	10.16	212 > 176 (24)	212 > 177 (12)	212 > 151 (28)	212 > 150 (44)	214 > 177 (12)	4.8
5	3,9-dichlorophenanthrene	3,9-ClPhe	12.17	246 > 176 (26)	248 > 176 (26)	176 > 150 (18)	246 > 211 (12)	176 > 175 (16)	23.3
6	9,10-dichloroanthracene	9,10-Cl2Ant	12.43	246 > 176 (26)	248 > 176 (26)	176 > 150 (18)	246 > 211 (12)	176 > 175 (16)	13.2
7	1,9-dichlorophenanthrene	1,9-Cl2Phe	12.43	246 > 176 (26)	248 > 176 (26)	176 > 150 (18)	246 > 211 (12)	176 > 175 (16)	8.5
8	9,10-dichlorophenanthrene	9,10-Cl2Phe	12.64	246 > 176 (26)	248 > 176 (26)	176 > 150 (18)	246 > 211 (12)	176 > 175 (16)	6.2
9	3-chlorofluoranthene	2-ClFluor	13.46	236 > 201 (18)	236 > 200 (32)	238 > 201 (18)	238 > 200 (32)	200 > 199 (16)	20.4
10	8-chlorofluoranthene	8-ClFluor	13.55	236 > 201 (18)	236 > 200 (32)	238 > 201 (18)	238 > 200 (32)	200 > 199 (16)	28.5
11	1-chloropyrene	1-ClPy	14.26	236 > 201 (18)	236 > 200 (32)	238 > 201 (18)	238 > 200 (32)	200 > 199 (16)	13.2
12	3,9,10-trichlorophenanthrene	3,9,10-Cl3Phe	14.95	280 > 210 (24)	282 > 210 (24)	210 > 175 (16)	282 > 175 (42)	282 > 212 (24)	12.4
13	5,7-dichlorofluoranthene	5,7-Cl2Fluor	15.65	270 > 235 (16)	270 > 200 (28)	272 > 200 (28)	200 > 199 (16)	200 > 198 (28)	20.7
14	3,8-dichlorofluoranthene	3,8-Cl2Fluor	16.41	270 > 235 (16)	270 > 200 (28)	272 > 200 (28)	200 > 199 (16)	200 > 198 (28)	7.2
15	3,4-dichlorofluoranthene	3,4-Cl2Fluor	17.28	270 > 235 (16)	270 > 200 (28)	272 > 200 (28)	200 > 199 (16)	200 > 198 (28)	35.7
16	6-chlorochrysene	6-ClChry	18.76	262 > 226 (26)	264 > 226 (22)	226 > 224 (30)	226 > 225 (18)	224 > 223 (18)	25.2
17	7-chlorobenz[a]anthracene	7-ClBaA	18.94	262 > 226 (26)	264 > 226 (22)	226 > 224 (30)	226 > 225 (18)	224 > 223 (18)	23.4
18	6,12-dichlorochrysene	6,12-Cl2Chry	21.37	296 > 226 (26)	298 > 226 (26)	226 > 224 (28)	226 > 225 (18)	224 > 223 (18)	6.1
19	7,12-dichlorobenz[a]anthracene	7,12-Cl2BaA	21.49	296 > 226 (26)	298 > 226 (26)	226 > 224 (28)	226 > 225 (18)	224 > 223 (16)	8.7
20	6-chlorobenzo[a]pyrene	6-ClBaP	24.33	286 > 250 (26)	288 > 250 (32)	250 > 248 (28)	250 > 249 (18)	250 > 224 (22)	11.9

	BrPAH								
1	2-Bromofluorene	2-BrFlu	27.31	244 > 165 (15)	246 > 165 (15)	165 > 164 (15)	165 > 163 (25)	165 > 115 (20)	174.5
2	9-Bromophenanthrene	9-BrPhe	31.12	258 > 176 (25)	256 > 177 (15)	258 > 177 (15)	256 > 176 (25)	258 > 151 (30)	148.6
3	9-Bromoanthracene	9-BrAnt	31.47	258 > 176 (25)	256 > 177 (15)	258 > 177 (15)	256 > 176 (25)	258 > 151 (30)	120.2
4	9,10-Dibromoanthracene	9,10-Br2Ant	37.01	335.9 > 176 (25)	333.9 > 176 (30)	337.9 > 176 (30)	176 > 150 (20)	176 > 175 (15)	145.7
5	1-Bromopyrene	1-BrPy	37.27	280 > 201 (20)	282 > 201 (20)	201 > 200 (15)	280 > 200 (40)	282 > 200 (40)	185.4
6	7-Bromobenz[a]anthracene	7-BrBaA	42.58	306 > 226 (30)	308 > 226 (25)	226 > 224 (25)	226 > 225 (15)	306 > 227 (10)	144.6
7	7,11-Dibromobenz[a]anthracene	7,11-Br2BaA	47.45	383.9 > 226 (30)	387.9 > 226 (30)	226 > 225 (15)	226 > 224 (25)	n/a	75.7
8	7,12-Dibromobenz[a]anthracene	7,12-Br2BaA	47.89	383.9 > 226 (30)	387.9 > 226 (30)	226 > 225 (15)	226 > 224 (25)	n/a	60
9	4,7-Dibromobenz[a]anthracene	4,7-Br2BaA	48.72	383.9 > 226 (30)	387.9 > 226 (30)	226 > 225 (15)	226 > 224 (25)	n/a	59.5
10	5,7-Dibromobenz[a]anthracene	5,7-Br2BaA	48.72	383.9 > 226 (30)	387.9 > 226 (30)	226 > 225 (15)	226 > 224 (25)	n/a	59.7
11	6-Bromobenzo[a]pyrene	6-BrBaP	50.53	330 > 250 (35)	332 > 250 (35)	330 > 251 (15)	332 > 251 (15)	250 > 248 (20)	129.1

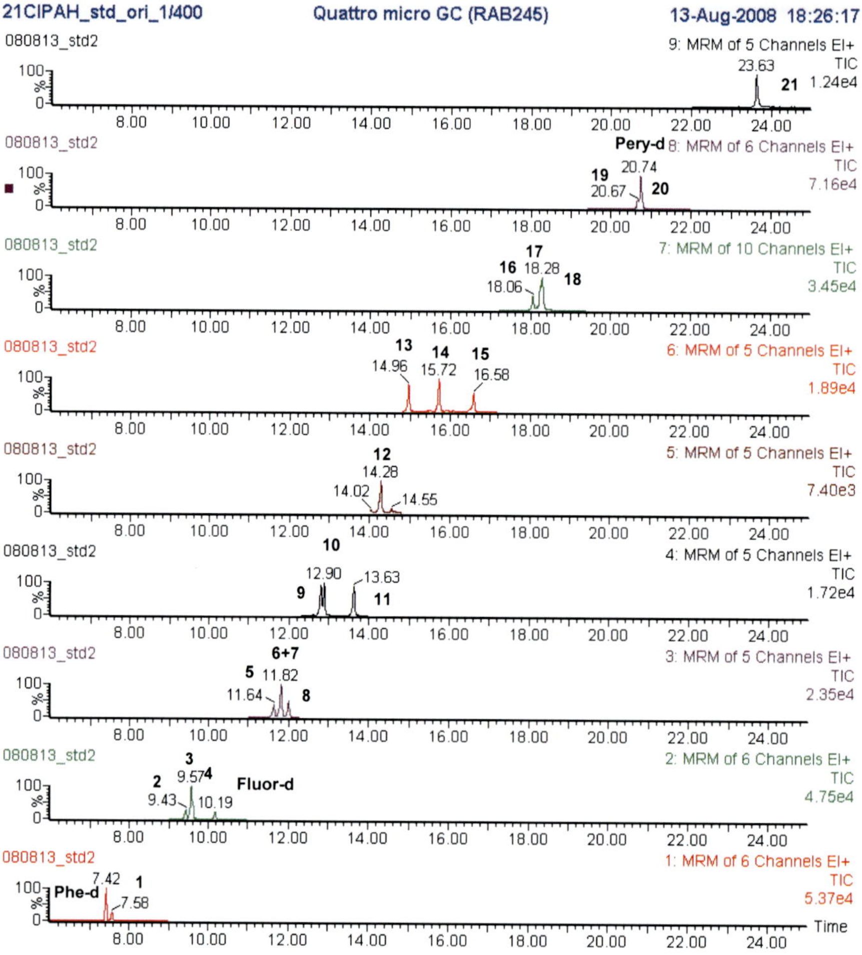

Fig. 10.3 GC/MS/MS chromatogram of 20 Cl-PAH. The number noted in chromatogram is corresponded to Table 10.1 and Fig. 10.3. *PAH* polycyclic aromatic hydrocarbons, *GC* gas chromatography, *MS* mass spectrometry

10.4 Halogenated Polycyclic Aromatic Hydrocarbons in Surface Soils of Old-Factory Sites

The soil samples collected from three former factory sites in a megacity Tokyo, Japan were previously known to be highly polluted (2,900 ~ 2,000,000 pg-TEQ/g) by polychlorinated dibenzo-*p*-dioxins and polychlorinated dibenzofuran (PCDD/F) (Yamamoto et al. 2009). In addition, one river sediment sample was also highly polluted by PCDD/F (5,900 pg-TEQ/g), whereas the possible sources remained unclear. Those highly polluted soil and sediment samples collected previously were

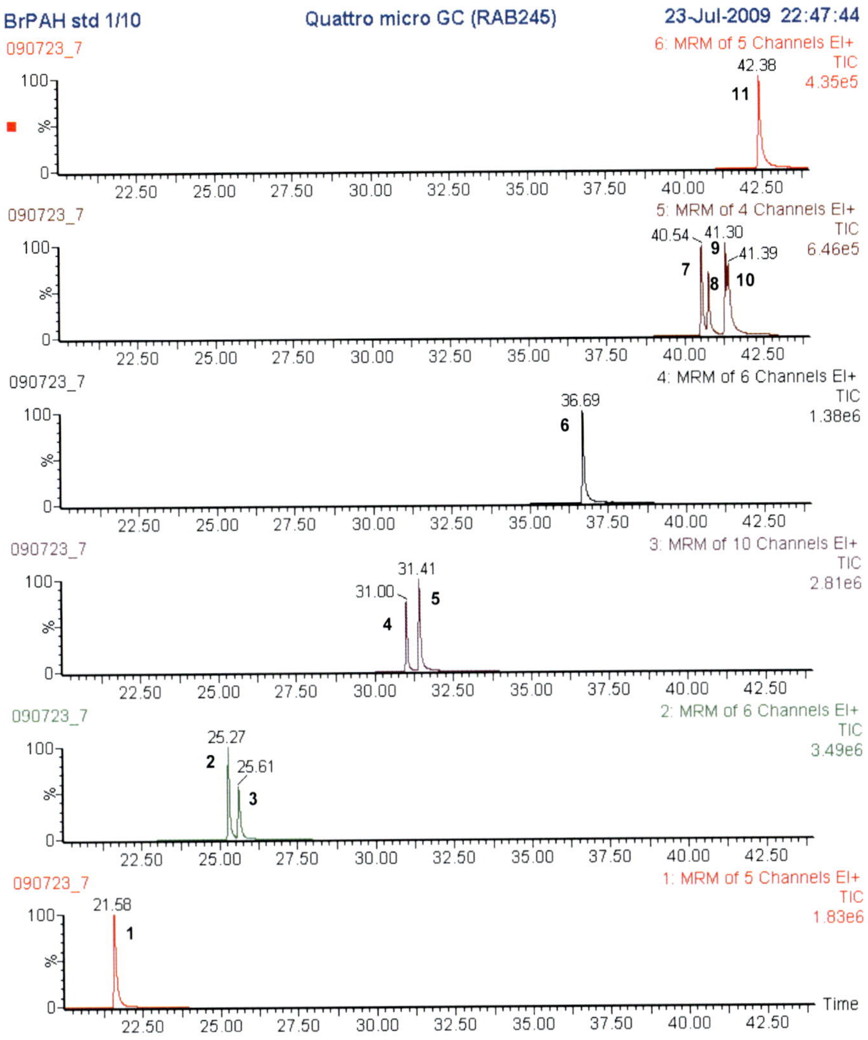

Fig. 10.4 GC/MS/MS chromatogram of 11 Br-PAH. The number noted in chromatogram is corresponded to Table 10.1 and Fig. 10.4. *PAH* polycyclic aromatic hydrocarbons

used in this study, followed by extraction and cleanup. One to ten gram of the dried soil and sediment samples weighted and subsequently soxhlet extraction using toluene for 16 h. Note that the surrogate standards are not added here in the samples because the extracts could be used in other study such as biological assays. The extracts were concentrated at 50 mL, and a part of the solution was subsequently cleaned up using Sep-Pak silica gel (Waters), previously conditioned with 10 mL hexane. The sample extract was eluted with 15 mL of hexane. When the eluate

Table 10.2 Total concentrations and the contributions of 16 species of PAH, 20 species of Cl-PAH, and 11 species of Br-PAH in the samples of surface soils (A-C) and a urban river sediment (D). *PAH* polycyclic aromatic hydrocarbons

Sample code	Sample information	Main factory	Concentration,[a] ng/g (contribution, %)					
			Σ16PAH		Σ20Cl-PAH		Σ11Br-PAH	
A	Soil from landfill	Unknown	117,000	(95.3)	5,770	(4.7)	13.5	(0.01)
B	Soil from former factory site	Metal and rubber processing factory	73,000	(71.2)	29,500	(28.8)	33.4	(0.03)
C	Soil from former factory site	Chlor-alkari plant	1,080,000	(66.9)	534,000	(33.1)	914	(0.06)
D	Sediment from river	Unknown	19,800	(99.7)	55.6	(0.3)	3.14	(0.02)

[a]Concentrations of sample A and C show the average values of two samples

is colored, the decoloration will be successfully accomplished with repeating the cleanup. The eluent was evaporated until ca. 1 mL under nitrogen at 40 °C, followed by addition of internal standards as syringe spike.

Measurements of targeted HPAH and priority PAH were performed in five soils collected from three kinds of surface soil and one sediment samples (Table 10.2). In brief, those soils were from landfill, former metal and rubber processing factory site and former chlor-alkari plant site, and sediment was collected from river nearby previous landfill.

Most of targeted PAH and Cl-PAH rather than Br-PAH were found in the every sample (Table 10.2). The total concentrations of Cl-PAH, Br-PAH, and PAH ranged from 55.6 to 534,000 ng/g, from 3.14 to 914 ng/g, and from 19,800 to 1,080,000 ng/g, respectively. The contributions of PAH were abundant at the range from 66.9 % to 99.7 % in contrast to those of Br-PAH (0.01 ~ 0.06 %). In addition, the total concentrations of each HPAH in the samples showed significant correlations to those of PAH (Fig. 10.5). It indicates that those compounds could be produced at a time even if the emission sources are different among the collected sites.

The relatively high contributions of Cl-PAH were observed in the sample of soil B and C that were collected from former industrial plants. For both Cl-PAH and Br-PAH, the most polluted soil was soil C, followed by soil B, soil A, and sediment D, whereas the extent of PAH as follows: C > A > B > D. Since the soil C was collected from a former chlor-alkari plant site, the industrial activity could be concerned to contamination of PAH and HPAH. Horii et al. (2009) reported that Cl-PAH were detected at relatively high concentration (24.1 ng/g) from marsh sediment collected near a former chlor-alkali plant in Georgia, USA. Comparing Cl-PAH concentrations between two samples corresponding to chlor-alkali plant in Japan and USA, the current data was approximately 20,000 times higher than the sediment sample in USA. Although there was significant difference in the both concentrations, the incontrovertible fact is that chlor-alkali plant could be significant emission sources of not only PCDD/F but also Cl-PAH in the environment.

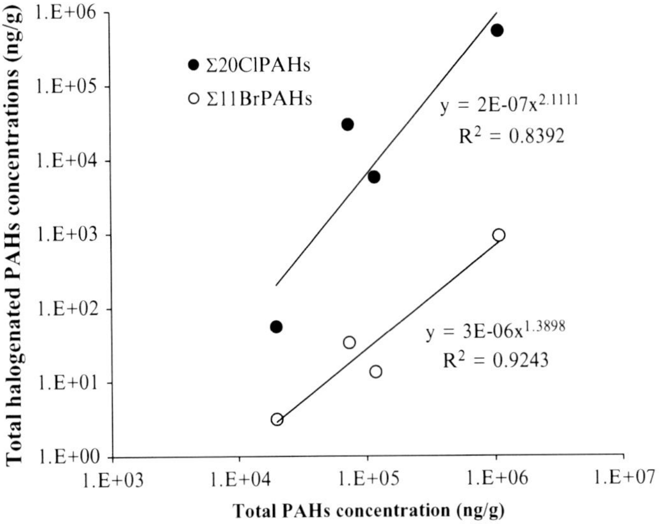

Fig. 10.5 Relationships of total concentrations of Cl-PAH or Br-PAH and PAH in soil and sediment samples. *PAH* polycyclic aromatic hydrocarbons

In our previous studies, the concentrations of Cl-PAH in the air were approximately two orders lower than those of PAH (Ohura et al. 2008). Interestingly, the current case that the content of Cl-PAH in soils is relatively high, suggests that the soils polluted by industrial activities could include relatively high levels of Cl-PAH.

10.5 Halogenated Polycyclic Aromatic Hydrocarbons Production by Brine Electrolysis

As suggested above, we believe that brine electrolysis performed in clor-alkali plant is a crucial emission source of not only Cl-PAH but also Br-PAH. Halogenations of PAH could be occurred during electrochemical treatment of saline aqueous solutions including PAH, which become a serious problem (Muff and Søgaard, 2011). Vizard et al. (2006) reported that the disposal of graphite electrode sludges from brine electrolysis used in old chemical works could provoke high contaminations of PCDD/F in the soils. Indeed, Yamamoto et al. (2009) confirmed the production of PCDD/F during brine electrolysis containing pitch used as the binding agent of graphite electrode. So we investigated whether HPAH are produced by brine electrolysis. For the electrode, artificial graphite was used as anode in the electrolytic cell because it had been typically used in the past industrial chro-alkali process. Furthermore solid pitch used as a material of graphite electrode binder was also added in the brine solution supplied to anode cell. So, the laboratory unit of brine

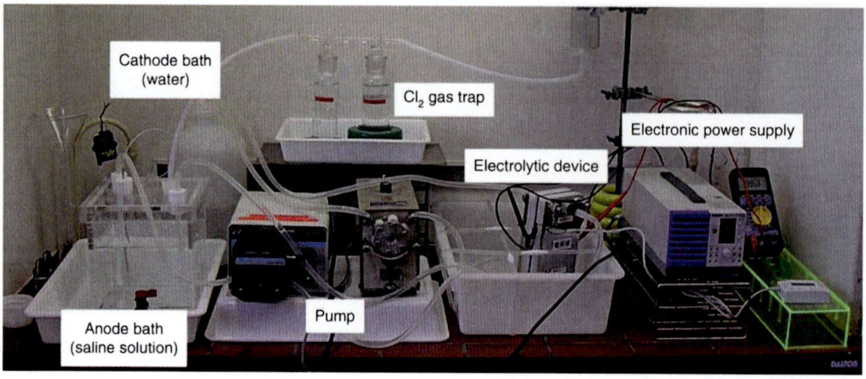

Fig. 10.6 The laboratory unit of brine electrolysis by flow cell system consisted of electrolytic cell, anode/cathode feeder and circulating water system to supply solution

electrolysis by flow cell system was used to investigate the possibility of production of HPAH during the brine electrolysis. The apparatus consisted of the electrolytic cell, anode/cathode feeder and circulating water system to supply solution was installed in an exhaust chamber. Inside the electrolytic cell, an artificial graphite electrode or a titanium electrode used as anode and SUS304 as cathode were separated two compartments by ion-exchange membrane made by perfluorosulfonic acid-based films (Fig. 10.6). Graphite electrode fixed in anode cell was used as in the past typically industrial chlor-alkali processes. Saturated brine solution supplied to anode cell made chlorine ion form chlorine, and hypochlorous acid and hydrochloric acid were also formed depending on pH and temperature condition of analyte. There was also some sodium hydroxide formed by the reaction of sodium ions passing through the separator with hydroxyl ion. Gaseous chlorine formed from anode cell was removed by Cl_2 recovery system. The full scale of experimental apparatus of brine electrolysis is 1.6 m × 0.6 m × 1.1 m (L × W × H) size. Electrolysis was performed in acid during the operation. After dechlorination, the analyte solution included suspended solid was extracted by Empore disk (C18). These were combined and extracted over 16 h with toluene by Soxhlet extraction.

The brine electrolysis was performed at 2.5 kA/m^2 of current density at 48 h. After that, the formations of HPAH were analyzed by using extracts from NaCl solution and added pitch. The toluene extracts from NaCl solution before electrolysis contained PAH with 39,500 ng/L and Cl-PAH with 481 ng/L, but not detection of Br-PAH (Table 10.3). However, after electrolysis, the toluene extracts contained considerably amounts of Cl-PAH and PAH in addition to Br-PAH. The concentrations of Cl-PAH were the highest contribution with 76.2 %, indicated that brine electrolysis induce the formation of Cl-PAH and minimal Br-PAH. A significant amount of PAH included in the pitch dispersed in NaCl solution, so that Cl-PAH could be produced by chlorination of PAH eluted from pitch throughout electrolysis. Similarly, minor amounts of bromine contained in the brine also could induce the bromination of PAH, resulted in production of Br-PAH. Comparing the

Table 10.3 PAH and Cl/Br-PAH concentrations (ng/l) of dichloromethane extracts from NaCl solution (4 L) dispersed pitch (5 g) before (E-1) or after electrolysis at 2.5 kA/m^2, 48 h (E-2), and the concentrations (ng/g) of toluene extracts from the pitch (0.021 g) before electrolysis (E-3). PAH: polycyclic aromatic hydrocarbons

Sample code	Sample information	Σ16PAH	%	Σ20Cl-PAH	%	Σ11Br-PAH	%
E-1	Before electrolysis	39,500	98.8	481	1.2	0	0
E-2	After electrolysis	802,000	23.2	2,640,000	76.2	21,100	0.6
E-3	Pitch extracts	158,000,000	99.9	1,280	<0.1	0	0

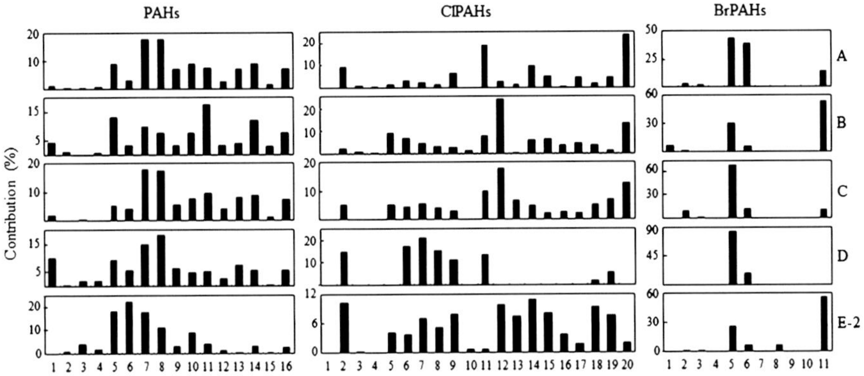

Fig. 10.7 Contributions of individual PAH, Cl-PAH, and Br-PAH in former industrial soils, a urban river sediment, and extract from the NaCl solution after brine electrolysis. The number noted in Cl/Br-PAH is corresponded to Table 10.1. The number noted in PAH is as follows: 1: naphthalene, 2: acenaphthene, 3: acenaphthylene, 4: fluorine, 5: phenanthrene, 6: anthracene, 7: fluoranthene, 8: pyrene, 9: benz[*a*]anthracene, 10: chrysene, 11: benzo[*b*]fluoranthene, 12: benzo[*k*]fluoranthene, 13: benzo[*a*]pyrene, 14: indeno[1,2,3-*cd*]pyrene, 15: dibenz[*a,h*]anthracene, 16: benzo[*ghi*]perylene. *PAH* polycyclic aromatic hydrocarbons

compositions of PAH and HPAH in sample (E-2) obtained from the electrolysis, specific distributions were observed (Fig. 10.7). The compositions of PAH and Cl-PAH detected from the pitch after electrolysis (E-2) were showed slight different patterns to those of environmental samples (A ~ D), but similar pattern for Br-PAH. As further emission source analysis, we here performed cluster analysis using the PAH and HPAH concentrations obtained from environmental samples and electrolytic solution.

10.6 Profiles of Halogenated Polycyclic Aromatic Hydrocarbons in Soils

The homologue profiles of certain compounds in the environment give us useful information such as the emission sources, environmental behaviors and fate. Here the profiles of PAH, Cl-PAH, and Br-PAH group were compared among the samples, because there were significant differences of the concentration levels among the compound group (Fig. 10.7). The profiles of PAH showed similar pattern among the samples of A, C and D, and the main contributors were fluoranthene (#7 in Fig. 10.7) and pyrene (#8). The profiles of sample B showed relative high contributions of high molecular weights PAH such as benzo[*b*]fluoranthene (#11) and indeno[1,2,3-*cd*]pyrene (#14) in compared to the other samples. The relative high compositions of such PAH have been observed in other urban surface soils (Zhang and Wang 2011). PAH pollution at urban sites could be mainly caused by various anthropogenic sources such as traffic and industrial activities. In addition, the occurrence in soils are significantly affected by not only the directly exposures and wet/dry deposition from air but also decay processes such as biodegradation, photodegradation, and effluent. Therefore, it could be difficult to evaluate the sources from profile of PAH in surface soils.

Of Cl-PAH, the profiles showed characteristic pattern rather than those of PAH (Fig. 10.7). The profiles of sample B and C were similar, which illustrating the relative high contributions of 3,9,10-trichlorophenanthrene (#12) and 6-chlorobenzo[*a*]pyrene (#20). It also suggests that those soils may be polluted by the common sources and/or production mechanisms. The profile of Cl-PAH on sample A was highly contributed by 1-chloropyrene (#11) and 6-chlorobenzo[*a*]pyrene (#20). A similar profile was also observed in the surveys of Cl-PAH in particulate matter in urban air, Japan. It suggests that Cl-PAH in soil A and urban air may be emitted from common sources. The profile of sediment sample E was obviously different from those of above soil samples, and high contributions of high molecular weights Cl-PAH such as 3,9,10-trichlorophenanthrene (#12), 3,8-dichlorofluoranthene (#14) and 6,12-dichlorochrysene (#18) were observed. The profile analyzed in sediment of Saginaw river (USA) also showed relatively high contribution of 3,8-dichlorofluoranthene and 6,12- dichlorochrysene. Horii et al. reported that the sediment of Saginaw river receive discharges from multiple industrial sources. Therefore, the pollution of Cl-PAH in sediment E could also be affected by multiple sources rather than a specific source.

For Br-PAH, small amounts were detected in all soils collected with relatively high contributions of 1-bromopyrene (#5), 7-bromobenz[*a*]anthracene (#6), and 6-bromobenzo[*a*]pyrene (#11). The similar trend has been observed in the past study that was surveyed in Tokyo Bay sediment core. It suggests that those selected Br-PAH may be relatively stable, resulted abundant in soils.

In order to clarify the relationship among the soil samples, cluster analysis was performed by using the concentrations of Cl-PAH. The dendrogram indicated that the profiles between sample A and D, and B and C were similar to each other (Fig. 10.8). It suggests that those soils are possible to be polluted by the common sources. Furthermore, the profiles of sample B and C were also similar to the profile of pitch solution after electrolysis (E-2). This fact indicates that sample B and C could be affected by brine electrolysis. Indeed, sample C was collected from a former clor-alkali plant (Table 10.2), elucidating Cl-PAH as useful indicator

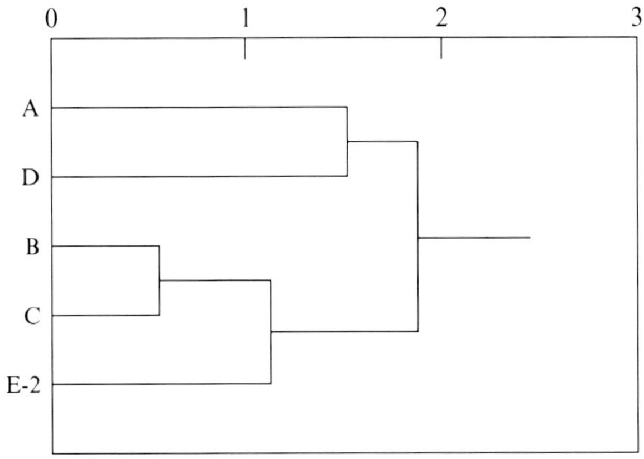

Fig. 10.8 Hierarchical dendrogram of samples contaminated by Cl-PAH. *PAH* polycyclic aromatic hydrocarbons

for these emission sources. The differences in compositions observed among environmental and electrolytic solution samples may be due to the environmental degradation, volatilization, runoff, and so on. Consequently, brine electrolysis is possible to produce HPAH in the process, so that chlor-alkari plant might be one of significant emission source of HPAH in the environment. Since the environmental impact and behavior of HPAH is not adequately evaluated, further investigations on the environmental fate are urgently required.

The concentration ratios of certain PAH have been used as a tool exploring the emission sources and environmental behaviors. If the analysis is performed by other persistent pollutants, relative stable and/or source specific species should be used to avoid environmental losses. Here we introduce the suitability of Cl-PAH as chemical marker. In a previous work we indicated that 3-chlorofluoranthene (3-ClFluor) is ubiquitously present in the environment, and is relative stable to photo-irradiation among ClPAH species. It was also indicated that 6-chlorobenzo[*a*]pyrene (6-ClBaP) in the air exhibit characteristic behaviors compared to other Cl-PAH; its concentration is elevated in summer, and decreased in winter. So we used those Cl-PAH as the indicator to define the profiles among soil samples. Table 10.4 shows the ratios of frequently observed ClPAH to 3-ClFluor. There were no significant differences of each concentration ratio normalized to 3-ClFluor among the soil samples, whereas the ratios normalized to 6-ClBaP and 3,9,10-trichlorophenanthrene (3,9,10-Cl3Phe) showed great variability among the samples. Considering those profile analysis and the source information, the concentration ratio of 3,9,10-Cl3Phe/3,9,10-Cl3Phe + 1-chloropyrene (1-ClPy) could be an appropriate indicator to evaluate the possible sources. It also means that those ClPAH may be useful species to distinguish the characteristics such as sources, origins, and fates in environmental samples.

Table 10.4 Concentration ratios of selected Cl-PAH normalized to 3-Cl-Fluor, 6-Cl-BaP, and 3,9,10-Cl-3Phe and 1-Cl-Py in former industrial soils and urban river sediment, soil samples from an e-waste recycling facility and a chemical industrial complex. *PAH* polycyclic aromatic hydrocarbons

Samples	9-ClPhe/ 3-Fluor	3,9,10-Cl3Phe/ 3-ClFluor	6-ClBaP/ 3-Fluor	9-ClPhe/ 6-ClBaP	3,9,10-Cl3Phe/ 6-ClBaP	1-ClPy/6-ClBaP	3,9,10-Cl3Phe/ 3,9,10-Cl3Phe + 1-ClPy	Reference
A	2.38	0.58	0.84	0.44	0.26	0.90	0.23	This study
B	0.90	0.90	0.84	0.17	1.76	0.57	0.76	This study
C	1.75	0.85	0.39	2.76	8.64	1.50	0.85	This study
D	1.37	nd	nd	nd	*nd*	nd	nd	This study
E-2	1.28	0.55	0.20	5.18	4.97	0.39	0.93	This study
e-waste soil[a]	1.00	3.82	32.7	0.03	0.12	0.25	0.32	Ma
Chemical soils[b]	nd	0.27	110	nd	<0.01	0.06	0.04	Ma

nd not determined
[a]Soils from an e-waste recycling facility
[b]Soils from a chemical industrial complex

10.7 Conclusion

Halogenated PAH are grouped in relatively new contaminants in the environment, so that the environmental behaviors remain unclear. In particular, there is only a few reports that HPAH were surveyed in soils and sediments. In the current chapter, we introduced the analytical method suitable for HPAH analysis in soils and sediment, i.e. GC/MS/MS. The method was applied to the surveys of HPAH in former industrial soils and river sediment in an urban city, Japan. The extends of HPAH contaminations varied in kinds of collected soils, suggesting that the significant sources could be present in the sites collected soil samples. The compositions of HPAH also showed specific patterns in the each sample. Since the soils collected from a former chlor-alkari plant site was included in the sample, the contribution to HPAH contaminants was also investigated. Using lab-scale electrolysis apparatus, brine electrolysis containing pitch induced the production of HPAH in the NaCl solutions. Indeed the profiles of Cl-PAH obtained from a former chlor-alkari plant site showed affinity of those obtained from the electrolysis. These findings suggest that Cl-PAH could become indicator to recognize the sources in various environmental contaminants.

References

Boström CE, Gerde P, Hanberg A, Jernström B, Johansson C, Kyrklund T, Rannug A, Törnqvist M, Victorin K, Westerholm R (2002) Cancer risk assessment, indicators, and guidelines for polycyclic aromatic hydrocarbons in the ambient air. Environ Health Perspect 110(Suppl 3):451–488

Czuczwa JM, Alford-Stevens A (1989) Optimized gel permeation chromatographic cleanup for soil, sediment, wastes, and oily waste extracts for determination of semivolatile organic pollutants and PCBs. J Assoc Off Anal Chem 72:752–759

Ding C, Ni HG, Zeng H (2012) Parent and halogenated polycyclic aromatic hydrocarbons in rice and implications for human health in China. Environ Pollut 168:80–86

Horii Y, Ohura T, Yamashita N, Kannan K (2009) Chlorinated polycyclic aromatic hydrocarbons in sediments from industrial areas in Japan and the United States. Arch Environ Contam Toxicol 57:651–660

Kitazawa A, Amagai T, Ohura T (2006) Temporal trends and relationships of particulate chlorinated polycyclic aromatic hydrocarbons and their parent compounds in urban air. Environ Sci Technol 40:4592–4598

Liu LB, Liu Y, Lin JM, Tang N, Hayakawa K, Maeda T (2007) Development of analytical methods for polycyclic aromatic hydrocarbons (PAH) in airborne particulates: a review. J Environ Sci (China) 19:1–11

Ma J, Horii Y, Cheng J, Wang W, Wu Q, Ohura T, Kannan K (2009) Chlorinated and parent polycyclic aromatic hydrocarbons in environmental samples from an electronic waste recycling facility and a chemical industrial complex in China. Environ Sci Technol 43:643–649

Muff J, Søgaard EG (2011) Identification and fate of halogenated PAH formed during electrochemical treatment of saline aqueous solutions. J Hazard Mater 186:1993–2000

Ohura T, Kitazawa A, Amagai T (2004) Seasonal variability of 1-chloropyrene on atmospheric particles and photostability in toluene. Chemosphere 57:831–837

Ohura T, Kitazawa A, Amagai T, Makino M (2005) Occurrence, profiles, and photostabilities of chlorinated polycyclic aromatic hydrocarbons associated with particulates in urban air. Environ Sci Technol 39:85–91

Ohura T, Fujima S, Amagai T, Shinomiya M (2008) Chlorinated polycyclic aromatic hydrocarbons in the atmosphere: seasonal levels, gas-particle partitioning, and origin. Environ Sci Technol 42:3296–3302

Ohura T, Sawada K, Amagai T, Shinomiya M (2009) Discovery of novel halogenated polycyclic aromatic hydrocarbons in urban particulate matters: occurrence, photostability, and AhR activity. Environ Sci Technol 43:2269–2275

Schecter A, Birnbaum L, Ryan JJ, Constable JD (2006) Dioxins: an overview. Environ Res 101:419–428

Sun JL, Ni HG, Zeng H (2011) Occurrence of chlorinated and brominated polycyclic aromatic hydrocarbons in surface sediments in Shenzhen, South China and its relationship to urbanization. J Environ Monit 13:2775–2781

Tanabe S, Minh TB (2010) Dioxins and organohalogen contaminants in the Asia-Pacific region. Ecotoxicology 19:463–478

Van den Berg M, Birnbaum L, Bosveld AT, Brunström B, Cook P, Feeley M, Giesy JP, Hanberg A, Hasegawa R, Kennedy SW, Kubiak T, Larsen JC, van Leeuwen FX, Liem AK, Nolt C, Peterson RE, Poellinger L, Safe S, Schrenk D, Tillitt D, Tysklind M, Younes M, Waern F, Zacharewski T (1998) Toxic equivalency factors (TEFs) for PCBs, PCDDs PCDFs for humans and wildlife. Environ Health Perspect 106:775–792

Vizard CG, Rimmer DL, Pless-Mulloli T, Singleton I, Air VS (2006) Identifying contemporary and historic sources of soil polychlorinated dibenzo-p-dioxins and polychlorinated dibenzofurans in an industrial urban setting. Sci Total Environ 370:61–69

Williamson KS, Petty JD, Huckins JN, Lebo JA, Kaiser EM (2002) Sequestration of priority pollutant PAH from sediment pore water employing semipermeable membrane devices. Chemosphere 49:717–729

Yamamoto T, Higashino K, Ohura T, Amagai T, Takemori H, Takasuga T, Sasaki Y (2009) Laboratory investigation of PCDD/F and dixin-like compounds formation during chlor-alkari process. Organohalogen Compd 71:863–868

Zhang Y, Wang J (2011) Distribution and source of polycyclic aromatic hydrocarbons (PAHs) in the surface soil along main transportation routes in Jiaxing City, China. Environ Monit Assess 182:535–543

Index

E. Lichtfouse et al. (eds.), *Green Materials for Energy, Products and Depollution*, Environmental Chemistry for a Sustainable World 3, DOI 10.1007/978-94-007-6836-9,

T

U